CALCULUS
ONE VARIABLE

BICENTENNIAL
1807
⊛WILEY
2007
BICENTENNIAL

THE WILEY BICENTENNIAL—KNOWLEDGE FOR GENERATIONS

*E*ach generation has its unique needs and aspirations. When Charles Wiley first opened his small printing shop in lower Manhattan in 1807, it was a generation of boundless potential searching for an identity. And we were there, helping to define a new American literary tradition. Over half a century later, in the midst of the Second Industrial Revolution, it was a generation focused on building the future. Once again, we were there, supplying the critical scientific, technical, and engineering knowledge that helped frame the world. Throughout the 20th Century, and into the new millennium, nations began to reach out beyond their own borders and a new international community was born. Wiley was there, expanding its operations around the world to enable a global exchange of ideas, opinions, and know-how.

For 200 years, Wiley has been an integral part of each generation's journey, enabling the flow of information and understanding necessary to meet their needs and fulfill their aspirations. Today, bold new technologies are changing the way we live and learn. Wiley will be there, providing you the must-have knowledge you need to imagine new worlds, new possibilities, and new opportunities.

Generations come and go, but you can always count on Wiley to provide you the knowledge you need, when and where you need it!

WILLIAM J. PESCE
PRESIDENT AND CHIEF EXECUTIVE OFFICER

PETER BOOTH WILEY
CHAIRMAN OF THE BOARD

STUDENT SOLUTIONS MANUAL

Garret Etgen
University of Houston

to accompany

CALCULUS
ONE VARIABLE

10th Edition

Saturnino Salas
Einar Hille
Garret Etgen
University of Houston

John Wiley & Sons, Inc.

Cover Photo: Steven Puetzer/Masterfile
Bicentennial Logo Design: Richard J. Pacifico

CONTENTS

CHAPTER 1

SECTION 1.2

1. rational   **3.** rational   **5.** rational

7. rational   **9.** rational   **11.** $\dfrac{3}{4} = 0.75$

13. $\sqrt{2} > 1.414$ **15.** $-\dfrac{2}{7} < -0.285714$ **17.** $|6| = 6$

19. $|-3 - 7| = 10$ **21.** $|-5| + |-8| = 13$ **23.** $|5 - \sqrt{5}| = 5 - \sqrt{5}$

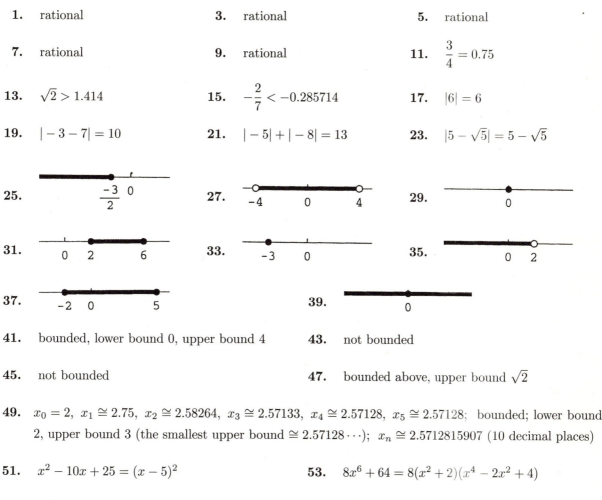

25.   **27.**   **29.**

31.   **33.**   **35.**

37.   **39.**

41. bounded, lower bound 0, upper bound 4   **43.** not bounded

45. not bounded   **47.** bounded above, upper bound $\sqrt{2}$

49. $x_0 = 2$, $x_1 \cong 2.75$, $x_2 \cong 2.58264$, $x_3 \cong 2.57133$, $x_4 \cong 2.57128$, $x_5 \cong 2.57128$; bounded; lower bound 2, upper bound 3 (the smallest upper bound $\cong 2.57128\cdots$); $x_n \cong 2.5712815907$ (10 decimal places)

51. $x^2 - 10x + 25 = (x - 5)^2$ **53.** $8x^6 + 64 = 8(x^2 + 2)(x^4 - 2x^2 + 4)$

55. $4x^2 + 12x + 9 = (2x + 3)^2$ **57.** $x^2 - x - 2 = (x - 2)(x + 1) = 0$; $x = 2, -1$

59. $x^2 - 6x + 9 = (x - 3)^2$; $x = 3$ **61.** $x^2 - 2x + 2 = 0$;   no real zeros

63. no real zeros   **65.** $5! = 120$

67. $\dfrac{8!}{3!5!} = \dfrac{8 \cdot 7 \cdot 6}{3 \cdot 2 \cdot 1} = 56$ **69.** $\dfrac{7!}{0!7!} = \dfrac{7!}{1 \cdot 7!} = 1$

71. Let r be a rational number and s an irrational number. Suppose $r + s$ is rational. Then $(r + s) - r = s$ is rational, a contradiction.

73. The product of a rational and an irrational number may either be rational or irrational; $0 \cdot \sqrt{2} = 0$ is rational, $1 \cdot \sqrt{2} = \sqrt{2}$ is irrational.

75. Suppose that $\sqrt{2} = p/q$ where p and q are integers and $q \neq 0$. Assume that p and q have no common factors (other than ± 1). Then $p^2 = 2q^2$ and p^2 is even. This implies that $p = 2r$ is even. Therefore $2q^2 = 4r^2$ which implies that q^2 is even, and hence q is even. It now follows that p and q are both even, contradicting the assumption that p and q have no common factors.

77. Let x be the length of a rectangle that has perimeter P. Then the width y of the rectangle is given by $y = (1/2)P - x$ and the area is

$$A = x\left(\frac{1}{2}P - x\right) = \left(\frac{P}{4}\right)^2 - \left(x - \frac{P}{4}\right)^2.$$

It follows that the area is a maximum when $x = P/4$. Since $y = P/4$ when $x = P/4$, the rectangle of perimeter P having the largest area is a square.

SECTION 1.3

1. $2 + 3x < 5$

$3x < 3$

$x < 1$

Ans: $(-\infty, 1)$

3. $16x + 64 \leq 16$

$16x \leq -48$

$x \leq -3$

Ans: $(-\infty, -3]$

5. $\frac{1}{2}(1 + x) < \frac{1}{3}(1 - x)$

$3(1 + x) < 2(1 - x)$

$3 + 3x < 2 - 2x$

$5x < -1$

$x < -\frac{1}{5}$

Ans: $(-\infty, -\frac{1}{5})$

7. $x^2 - 1 < 0$

$(x + 1)(x - 1) < 0$

Ans: $(-1, 1)$

9. $x^2 - x - 6 \geq 0$

$(x - 3)(x + 2) \geq 0$

Ans: $(\infty, -2] \cup [3, \infty)$

11. $2x^2 + x - 1 \leq 0$

$(2x - 1)(x + 1) \leq 0$

Ans: $[-1, 1/2]$

13. $x(x - 1)(x - 2) > 0$

Ans: $(0, 1) \cup (2, \infty)$

15. $x^3 - 2x^2 + x \geq 0$

$x(x - 1)^2 \geq 0$

Ans: $[0, \infty)$

17. $x^3(x - 2)(x + 3)^2 < 0$

Ans: $(0, 2)$

19. $x^2(x - 2)(x + 6) > 0$

Ans: $(-\infty, -6) \cup (2, \infty)$

21. $(-2, 2)$

23. $(-\infty, -3) \cup (3, \infty)$

25. $(\frac{3}{2}, \frac{5}{2})$

27. $(-1, 0) \cup (0, 1)$

29. $(\frac{3}{2}, 2) \cup (2, \frac{5}{2})$

31. $(-5, 3) \cup (3, 11)$

33. $(-\frac{5}{8}, -\frac{3}{8})$

35. $(-\infty, -4) \cup (-1, \infty)$

37. $|x - 0| < 3$ or $|x| < 3$

39. $|x - 2| < 5$

41. $|x - (-2)| < 5$ or $|x + 2| < 5$

43. $|x - 2| < 1 \implies |2x - 4| = 2|x - 2| < 2$, so $|2x - 4| < A$ true for $A \geq 2$.

45. $|x + 1| < A \implies |3x + 3| = 3|x + 1| < 3A \implies |3x + 3| < 4$

                 provided that $0 < A \le \frac{4}{3}$

47. (a) $\dfrac{1}{x} < \dfrac{1}{\sqrt{x}} < 1 < \sqrt{x} < x$ (b) $x < \sqrt{x} < 1 < \dfrac{1}{\sqrt{x}} < \dfrac{1}{x}$

49. If a and b have the same sign, then $ab > 0$. Suppose that $a < b$. Then $a - b < 0$ and

$$\frac{1}{b} - \frac{1}{a} = \frac{a - b}{ab} < 0.$$

Thus, $(1/b) < (1/a)$.

51. With $a \ge 0$ and $b \ge 0$

$$b \ge a \implies b - a = (\sqrt{b} + \sqrt{a})(\sqrt{b} - \sqrt{a}) \ge 0 \implies \sqrt{b} - \sqrt{a} \ge 0 \implies \sqrt{b} \ge \sqrt{a}.$$

53. By the hint

$$\big|\, |a| - |b| \,\big|^2 = (|a| - |b|)^2 = |a|^2 - 2|a|\,|b| + |b|^2 = a^2 - 2|ab| + b^2$$

$$\le a^2 - 2ab + b^2 = (a - b)^2.$$

$$(ab \le |ab|)$$

Taking the square root of the extremes, we have

$$\big|\, |a| - |b| \,\big| \le \sqrt{(a - b)^2} = |a - b|.$$

55. With $0 \le a \le b$

$$a(1 + b) = a + ab \le b + ab = b(1 + a).$$

Division by $(1 + a)(1 + b)$ gives

$$\frac{a}{1 + a} \le \frac{b}{1 + b}.$$

57. Suppose that $a < b$. Then

$$a = \frac{a + a}{2} \le \frac{a + b}{2} \le \frac{b + b}{2} = b.$$

$\dfrac{a + b}{2}$ is the midpoint of the line segment $\overline{ab}$.

SECTION 1.4

1. $d(P_0, P_1) = \sqrt{(6 - 0)^2 + (-3 - 5)^2} = \sqrt{36 + 64} = \sqrt{100} = 10$

3. $d(P_0, P_1) = \sqrt{[5 - (-3)]^2 + (-2 - 2)^2} = \sqrt{64 + 16} = 4\sqrt{5}$

5. $\left(\dfrac{2+6}{2}, \dfrac{4+8}{2}\right) = (4,6)$ **7.** $\left(\dfrac{2+7}{2}, \dfrac{-3-3}{2}\right) = \left(\tfrac{9}{2}, -3\right)$

9. $m = \dfrac{5-1}{(-2)-4} = \dfrac{4}{-6} = -\dfrac{2}{3}$ **11.** $m = \dfrac{b-a}{a-b} = -1$

13. $m = \dfrac{0-y_0}{x_0-0} = -\dfrac{y_0}{x_0}$

15. Equation is in the form $y = mx + b$.   Slope is 2; y-intercept is -4.

17. Write equation as $y = \tfrac{1}{3}x + 2$.   Slope is $\tfrac{1}{3}$; y-intercept is 2.

19. Write equation as $y = \tfrac{7}{3}x + \tfrac{4}{3}$.   Slope is $\tfrac{7}{3}$; y-intercept is $\tfrac{4}{3}$.

21. $y = 5x + 2$ **23.** $y = -5x + 2$ **25.** $y = 3$ **27.** $x = -3$

29. Every line parallel to the x-axis has an equation of the form $y = a$ constant. In this case $y = 7$.

31. The line $3y - 2x + 6 = 0$ has slope $\tfrac{2}{3}$. Every line parallel to it has that same slope. The line through $P(2, 7)$ with slope $\tfrac{2}{3}$ has equation $y - 7 = \tfrac{2}{3}(x - 2)$, which reduces to $3y - 2x - 17 = 0$.

33. The line $3y - 2x + 6 = 0$ has slope $\tfrac{2}{3}$. Every line perpendicular to it has slope $-\tfrac{3}{2}$.
The line through $P(2, 7)$ with slope $-\tfrac{3}{2}$ has equation $y - 7 = -\tfrac{3}{2}(x - 2)$, which reduces to $2y + 3x - 20 = 0$.

35. $\left(\tfrac{1}{2}\sqrt{2}, \tfrac{1}{2}\sqrt{2}\right), \left(-\tfrac{1}{2}\sqrt{2}, -\tfrac{1}{2}\sqrt{2}\right)$ [Substitute $y = x$ into $x^2 + y^2 = 1$.]

37. $(3, 4)$ [Write $4x + 3y = 24$ as $y = \tfrac{4}{3}(6 - x)$ and substitute into $x^2 + y^2 = 25$.]

39. $(1, 1)$ **41.** $\left(-\tfrac{2}{23}, \tfrac{38}{23}\right)$

43. We select the side joining $A(1, -2)$ and $B(-1, 3)$ as the base of the triangle.
length of side AB: $\sqrt{29}$;   equation of line through A and B: $5x + 2y - 1 = 0$
equation of line through $C(2, 4) \perp 5x + 2y - 1 = 0$: $y - 4 = \tfrac{2}{5}(x - 2)$
point of intersection of the two lines: $\left(\dfrac{-27}{29}, \dfrac{82}{29}\right)$
altitude of the triangle: $\sqrt{\left(2 + \tfrac{27}{29}\right)^2 + \left(4 - \tfrac{82}{29}\right)^2} = \dfrac{17}{\sqrt{29}}$
area of triangle: $\dfrac{1}{2}\left(\sqrt{29}\right)\left(\dfrac{17}{\sqrt{29}}\right) = \dfrac{17}{2}$

45. Substitute $y = m(x - 5) + 12$ into $x^2 + y^2 = 169$ and you get a quadratic in x that involves m. That quadratic has a unique solution iff $m = -\tfrac{5}{12}$. (A quadratic $ax^2 + bx + c = 0$ has a unique solution iff $b^2 - 4ac = 0$).

47. The slope of the line through the center of the circle and the point P is -2. Therefore the slope of the tangent line is $\frac{1}{2}$. The equation for the tangent line to the circle at P is

$$(y+1) = \frac{1}{2}(x-1), \text{ or } x - 2y - 3 = 0.$$

49. $(2.36, -0.21)$

51. $(0.61, 2.94)$, $(2.64, 1.42)$

53. Midpoint of line segment $\overline{PQ}$: $\left(\frac{5}{2}, \frac{5}{2}\right)$

Slope of line segment $\overline{PQ}$: $\frac{13}{3}$

Equation of the perpendicular bisector: $y - \frac{5}{2} = -\left(\frac{3}{13}\right)\left(x - \frac{5}{2}\right)$ or $3x + 13y - 40 = 0$

55. $d(P_0, P_1) = \sqrt{(-2-1)^2 + (5-3)^2} = \sqrt{13}$, $d(P_0, P_2) = \sqrt{[-2-(-1)]^2 + (5-0)^2} = \sqrt{26}$, $d(P_1, P_2) = \sqrt{[1-(-1)]^2 + (3-0)^2} = \sqrt{13}$.

Since $d(P_0, P_1) = d(P_1, P_2)$, the triangle is isosceles.

Since $[d(P_0, P_1)]^2 + [d(P_1, P_2)]^2 = [d(P_0, P_2)]^2$, the triangle is a right triangle.

57. The line l_2 through the origin perpendicular to $l_1 : Ax + By + C = 0$ has equation $y = \dfrac{B}{A}x$. The lines l_1 and l_2 intersect at the point $P\left(\dfrac{-AC}{A^2+B^2}, \dfrac{-BC}{A^2+B^2}\right)$. The distance from P to the origin is $\dfrac{|C|}{\sqrt{A^2+b^2}}$.

59. The coordinates of M are $\left(\dfrac{a}{2}, \dfrac{b}{2}\right)$; and

$$d(M, (0, b)) = d(M, (0, a)) = d(M, (0, 0)) = \tfrac{1}{2}\sqrt{a^2 + b^2}.$$

61. Denote the points $(1, 0)$, $(3, 4)$ and $(-1, 6)$ by A, B and C, respectively. The midpoints of the line segments $\overline{AB}$, $\overline{AC}$, and $\overline{BC}$ are $P(2, 2)$, $Q(0, 3)$ and $R(1, 5)$.

An equation for the line through A and R is: $x = 1$.

An equation for the line through B and Q is: $y = \frac{1}{3}x + 3$.

An equation for the line through C and P is: $y - 2 = -\frac{4}{3}(x - 2)$.

These lines intersect at the point $(1, \frac{10}{3})$.

63. Let $A(0, 0)$ and $B(a, 0)$, $a > 0$, be adjacent vertices of a parallelogram. If $C(b, c)$ is the vertex opposite B, then the vertex D opposite A has coordinates $(a + b, c)$. [See the figure.]

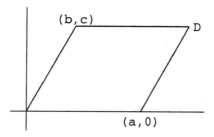

The line through A and D has equation: $y = \dfrac{c}{a+b}\, x$.

The line through B and C has equation: $y = -\dfrac{c}{a-b}\, (x-a)$.

These lines intersect at the point $\left(\dfrac{a+b}{2}, \dfrac{c}{2}\right)$ which is the midpoint of each of the line segments $\overline{AD}$ and $\overline{BC}$.

65.   Since the relation between F and C is linear, $F = mC + b$ for some constants m and C. Setting $C = 0$ and $F = 32$ gives $b = 32$. Thus $F = mC + 32$. Setting $C = 100$ and $F = 212$ gives $m = (212 - 32)/100 = 9/5$. Therefore

$$F = \frac{9}{5} C + 32$$

The Fahrenheit and Centigrade temperatures are equal when

$$C = F = \frac{9}{5} C + 32$$

which implies $C = F = -40°$.

SECTION 1.5

1.   (a) $f(0) = 2(0)^2 - 3(0) + 2 = 2$ (b) $f(1) = 2(1)^2 - 3(1) + 2 = 1$

     (c) $f(-2) = 2(-2)^2 - 3(-2) + 2 = 16$ (d) $f(\tfrac{3}{2}) = 2(3/2)^2 - 3(3/2) + 2 = 2$

3.   (a) $f(0) = \sqrt{0^2 + 2 \cdot 0} = 0$ (b) $f(1) = \sqrt{1^2 + 2 \cdot 1} = \sqrt{3}$

     (c) $f(-2) = \sqrt{(-2)^2 + 2(-2)} = 0$ (d) $f(\tfrac{3}{2}) = \sqrt{(3/2)^2 + 2(3/2)} = \tfrac{1}{2}\sqrt{21}$

5.   (a) $f(0) = \dfrac{2 \cdot 0}{|0 + 2| + 0^2} = 0$ (b) $f(1) = \dfrac{2 \cdot 1}{|1 + 2| + 1^2} = \dfrac{1}{2}$

     (c) $f(-2) = \dfrac{2 \cdot (-2)}{|-2 + 2| + (-2)^2} = -1$ (d) $f(\tfrac{3}{2}) = \dfrac{2 \cdot (3/2)}{|(3/2) + 2| + (3/2)^2} = \dfrac{12}{23}$

7.   (a) $f(-x) = (-x)^2 - 2(-x) = x^2 + 2x$ (b) $f(1/x) = (1/x)^2 - 2(1/x) = \dfrac{1 - 2x}{x^2}$

     (c) $f(a + b) = (a + b)^2 - 2(a + b) = a^2 + 2ab + b^2 - 2a - 2b$

9.   (a) $f(-x) = \sqrt{1 + (-x)^2} = \sqrt{1 + x^2}$ (b) $f(1/x) = \sqrt{1 + (1/x)^2} = \sqrt{1 + x^2}/|x|$

     (c) $f(a + b) = \sqrt{1 + (a + b)^2} = \sqrt{a^2 + 2ab + b^2 + 1}$

11.   (a) $f(a + h) = 2(a + h)^2 - 3(a + h) = 2a^2 + 4ah + 2h^2 - 3a - 3h$

     (b) $\dfrac{f(a + h) - f(a)}{h} = \dfrac{[2(a + h)^2 - 3(a + h)] - [2a^2 - 3a]}{h} = \dfrac{4ah + 2h^2 - 3h}{h} = 4a + 2h - 3$

13. $x = 1, 3$ **15.** $x = -2$ **17.** $x = -3, 3$

19. $\operatorname{dom}(f) = (-\infty, \infty)$; $\operatorname{range}(f) = [0, \infty)$ **21.** $\operatorname{dom}(f) = (-\infty, \infty)$; $\operatorname{range}(f) = (-\infty, \infty)$

23. $\operatorname{dom}(f) = (-\infty, 0) \cup (0, \infty)$; $\operatorname{range}(f) = (0, \infty)$

25. $\operatorname{dom}(f) = (-\infty, 1]$; $\operatorname{range}(f) = [0, \infty)$

27. $\operatorname{dom}(f) = (-\infty, 7]$; $\operatorname{range}(f) = [-1, \infty)$ **29.** $\operatorname{dom}(f) = (-\infty, 2)$; $\operatorname{range}(f) = (0, \infty)$

31. horizontal line one unit above the x-axis.   **33.** line through the origin with slope 2.

35. line through $(0, 2)$ with slope $\frac{1}{2}$.   **37.** upper semicircle of radius 2 centered at the origin.

39. $\operatorname{dom}(f) = (-\infty, \infty)$ **41.** $\operatorname{dom}(f) = (-\infty, 0) \cup (0, \infty)$; $\operatorname{range}(f) = \{-1, 1\}$.

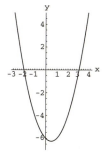   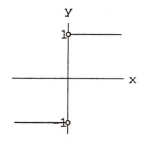

43. $\operatorname{dom}(f) = [0, \infty)$; $\operatorname{range}(f) = [1, \infty)$.

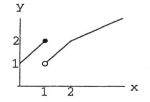

45. The curve is the graph of a function: domain $[-2, 2]$,   range $[-2, 2]$.

47. The curve is not the graph of a function; it fails the *vertical line test*.

49. odd: $f(-x) = (-x)^3 = -x^3 = -f(x)$

51. neither even nor odd: $g(-x) = -x(-x - 1) = x(x + 1)$; $g(-x) \neq g(x)$ and $g(-x) \neq -g(x)$

53. even.   **55.** odd

57. (a)

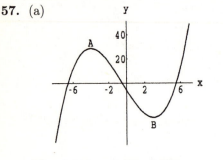

(b) -6.566, -0.493, 5.559

(c) $A(-4, 28.667)$, $\quad B(3, -28.500)$

59. $-5 \le x \le 8$, $\quad 0 \le y \le 100$

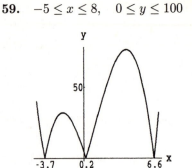

61. Range: $[-9, \infty)$.

(a) $y = x^2 - 4x + 5 = x^2 - 4x + 4 - 9 = (x - 2)^2 - 9$. Therefore $y \ge -9$.

(b) $x = \dfrac{4 \pm \sqrt{36 + 4y}}{2}$ which implies $y \ge -9$.

63. $A = \dfrac{C^2}{4\pi}$, where C is the circumference; $\text{dom}\,(A) = [0, \infty)$

65. $V = s^{3/2}$, where s is the area of a face; $\text{dom}\,V = [0, \infty)$

67. $S = 3d^2$, where d is the diagonal of a face; $\text{dom}\,(S) = [0, \infty)$

69. $A = \dfrac{\sqrt{3}}{4}\,x^2$, where x is the length of a side; $\text{dom}\,(A) = [0, \infty)$

71. Let y be the length of the rectangle. Then

$$x + 2y + \frac{\pi x}{2} = 15 \quad \text{and} \quad y = \frac{15}{2} - \frac{2 + \pi}{4}\,x, \qquad 0 \le x \le \frac{30}{2 + \pi}$$

Area: $A = xy + \frac{1}{2}\pi\,(x/2)^2 = \left(\dfrac{15}{2} - \dfrac{2 + \pi}{4}\,x \right) x + \frac{1}{8}\pi x^2 = \dfrac{15}{2}\,x - \dfrac{x^2}{2} - \dfrac{\pi}{8}\,x^2, \quad 0 < x < \dfrac{30}{2 + \pi}.$

73. The coordinates x and y are related by the equation $\quad y = -\dfrac{b}{a}(x - a)$, $\ 0 \le x \le a$.

The area A of the rectangle is given by $\ A = xy = x\left[-\dfrac{b}{a}(x - a) \right] = bx - \dfrac{b}{a}\,x^2$, $\ 0 \le x \le a$.

75. Let P be the perimeter of the square. Then the edge length of the square is $P/4$ and the area of the square is $A_s = (P/4)^2 = P^2/16$. The circumference of the circle is $28 - P$ which implies that the radius is $\dfrac{1}{2\pi}(28 - \pi)$. Thus, the area of the circle is $A_c = \pi\left[\dfrac{1}{2\pi}(28 - P) \right]^2 = \dfrac{1}{4\pi}(28 - P)^2$ and the total area is $A_s + A_c = \dfrac{P^2}{16} + \dfrac{1}{4\pi}(28 - P)^2$, $\ 0 \le P \le 28$.

77. Set length plus girth equal to 108. Then $\ l = 108 - 2\pi r$, and $\ V = (108 - 2\pi r)\pi r^2$.

SECTION 1.6

1. polynomial, degree 0   **3.** rational function   **5.** neither

7. neither   **9.** neither

11. $\text{dom}\,(f) = (-\infty, \infty)$ **13.** $\text{dom}\,(f) = (-\infty, \infty)$ **15.** $\text{dom}\,(f) = \{x : x \neq \pm 2\}$

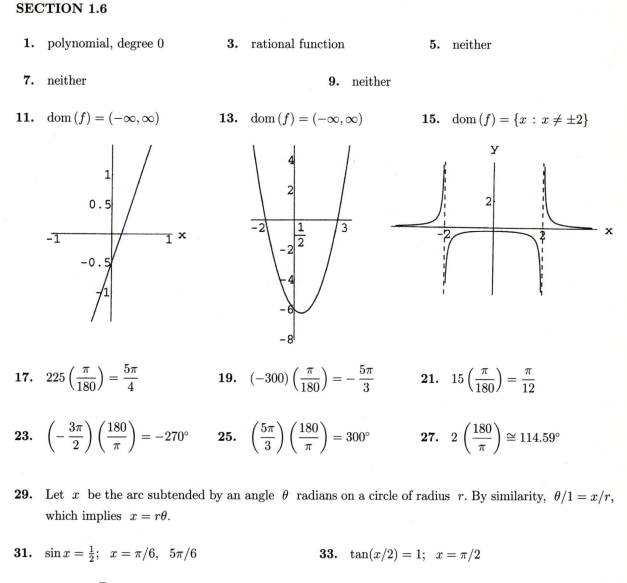

17. $225\left(\dfrac{\pi}{180}\right) = \dfrac{5\pi}{4}$ **19.** $(-300)\left(\dfrac{\pi}{180}\right) = -\dfrac{5\pi}{3}$ **21.** $15\left(\dfrac{\pi}{180}\right) = \dfrac{\pi}{12}$

23. $\left(-\dfrac{3\pi}{2}\right)\left(\dfrac{180}{\pi}\right) = -270°$ **25.** $\left(\dfrac{5\pi}{3}\right)\left(\dfrac{180}{\pi}\right) = 300°$ **27.** $2\left(\dfrac{180}{\pi}\right) \cong 114.59°$

29. Let x be the arc subtended by an angle θ radians on a circle of radius r. By similarity, $\theta/1 = x/r$, which implies $x = r\theta$.

31. $\sin x = \frac{1}{2};\ \ x = \pi/6,\ \ 5\pi/6$ **33.** $\tan(x/2) = 1;\ \ x = \pi/2$

35. $\cos x = \sqrt{2}/2;\ \ x = \pi/4,\ \ 7\pi/4$ **37.** $\cos 2x = 0;\ \ x = \pi/4,\ \ 3\pi/4,\ \ 5\pi/4,\ \ 7\pi/4$

39. $\sin 51° \cong 0.7771$ **41.** $\sin(2.352) \cong 0.7101$

43. $\tan 72.4° \cong 3.1524$ **45.** $\sin x = 0.5231;\ \ x = 0.5505,\ \ \pi - 0.5505$

47. $\tan x = 6.7192;\ \ x = 1.4231,\ \ \pi + 1.4231$ **49.** $\sec x = -4.4073;\ \ x = 1.7997,\ \ \pi + 1.7997$

51. The x coordinates of the points of intersection are: $x \cong 1.31,\ 1.83,\ 3.40,\ 3.93,\ 5.50,\ 6.02$

53. $\text{dom}\,(f) = (-\infty, \infty)$; range $(f) = [0, 1]$ **55.** $\text{dom}\,(f) = (-\infty, \infty)$; range $(f) = [-2, 2]$

57. $\text{dom}(f) = \left(k\pi - \dfrac{\pi}{2}, k\pi + \dfrac{\pi}{2}\right),\ \ k = 0,\ \pm 1,\ \pm 2,\ \ldots;\ $ range $(f) = [1, \infty)$

59. period: $\dfrac{2\pi}{\pi} = 2$

61. period: $\dfrac{2\pi}{1/3} = 6\pi$

63.

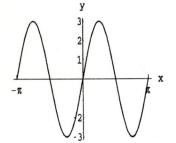

65.

67.

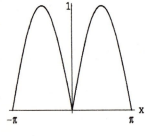

69. odd

71. even

73. odd

75. Assume that $\theta_2 > \theta_1$. Let $m_1 = \tan\theta_1$, $m_2 = \tan\theta_2$. The angle α between l_1 and l_2 is the smaller of $\theta_2 - \theta_1$ and $180° - [\theta_2 - \theta_1]$. In the first case

$$\tan\alpha = \tan[\theta_2 - \theta_1] = \frac{\tan\theta_2 - \tan\theta_1}{1 + \tan\theta_2 \, \tan\theta_1} = \frac{m_2 - m_1}{1 + m_2 m_1} > 0$$

In the second case, $\tan\alpha = \tan[180° - (\theta_2 - \theta_1)] = -\tan(\theta_2 - \theta_1) = -\dfrac{m_2 - m_1}{1 + m_2 m_1} < 0$

Thus $\tan\alpha = \left| \dfrac{m_2 - m_1}{1 + m_2 m_1} \right|$

77. $\left(\frac{23}{37}, \frac{116}{37}\right);$ $\alpha \cong 73°$ $[m_1 = -3 = \tan\theta_1, \; \theta_1 \cong 108°; \quad m_2 = \frac{7}{10} = \tan\theta_2, \; \theta_2 \cong 35°]$

79. $\left(-\frac{17}{13}, -\frac{2}{13}\right);$ $\alpha \cong 82°$ $[m_1 = \frac{5}{6} = \tan\theta_1, \; \theta_1 \cong 40°; \quad m_2 = -\frac{8}{5} = \tan\theta_2, \; \theta_2 \cong 122°]$

81. By similar triangles, $\sin\theta = \dfrac{\sin\theta}{1} = \dfrac{\text{opp}}{\text{hyp}}$, and $\cos\theta = \dfrac{\cos\theta}{1} = \dfrac{\text{adj}}{\text{hyp}}$.

83. From the figure, $h = b\sin C = c\sin B$.

Therefore $\dfrac{\sin B}{b} = \dfrac{\sin C}{c}$

Similarly, you can show that $\dfrac{\sin A}{a} = \dfrac{\sin B}{b}$

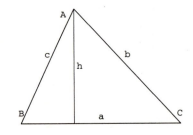

85. By the law of cosines

$$1^2 + 1^2 - 2(1)(1)\cos(\alpha - \beta) = (\cos\beta - \cos\alpha)^2 + (\sin\beta - \sin\alpha)^2$$

$$= \cos^2\beta - 2\cos\beta\cos\alpha + \cos^2\alpha + \sin^2\beta - 2\sin\beta\sin\alpha + \sin^2\alpha$$

$$= 2 - 2\cos\beta\cos\alpha - 2\sin\beta\sin\alpha.$$

The result follows.

87. From the identities $\sin\left(\frac{1}{2}\pi + \theta\right) = \cos\theta$ and $\cos\left(\frac{1}{2}\pi + \theta\right) = -\sin\theta$ we get

$$\sin\left(\frac{1}{2}\pi - \theta\right) = \sin\left[\frac{1}{2}\pi + (-\theta)\right] = \cos(-\theta) = \cos\theta$$

and

$$\cos\left(\frac{1}{2}\pi - \theta\right) = \cos\left[\frac{1}{2}\pi + (-\theta)\right] = -\sin(-\theta) = \sin\theta.$$

89. Replace β by $-\beta$ in the identity of Exercise 88.

91. (a)

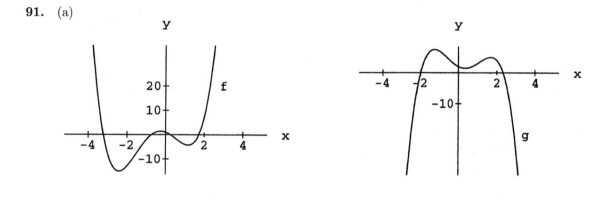

(c)

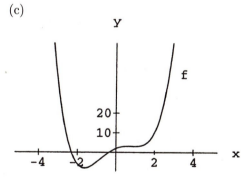
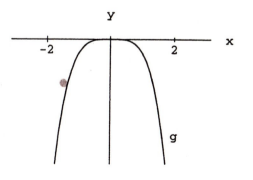

93. (a)

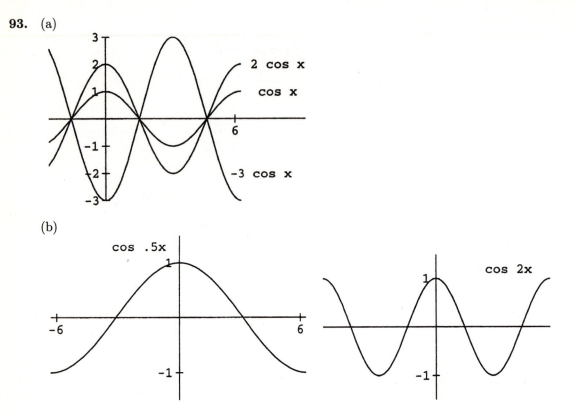

(b)

(c) *A* changes the amplitude; *B* stretches or compresses horizontally

SECTION 1.7

1. $(f+g)(2) = f(2) + g(2) = 3 + \frac{9}{2} = \frac{15}{2}$ **3.** $(f \cdot g)(-2) = f(-2)g(-2) = 15 \cdot \frac{7}{2} = \frac{105}{2}$

5. $(2f - 3g)(\frac{1}{2}) = 2f(\frac{1}{2}) - 3g(\frac{1}{2}) = 2 \cdot 0 - 3 \cdot \frac{9}{4} = -\frac{27}{4}$

7. $(f \circ g)(1) = f[g(1)] = f(2) = 3$

9. $(f+g)(x) = f(x) + g(x) = x - 1;$ $\mathrm{dom}\,(f+g) = (-\infty, \infty)$
   $(f-g)(x) = f(x) - g(x) = 3x - 5;$ $\mathrm{dom}\,(f-g) = (-\infty, \infty)$
   $(f \cdot g)(x) = f(x)g(x) = -2x^2 + 7x - 6;$ $\mathrm{dom}\,(f \cdot g) = (-\infty, \infty)$
   $(f/g)(x) = \dfrac{2x-3}{2-x};$ $\mathrm{dom}\,(f/g) = \{x :\ x \neq 2\}$

11. $(f+g)(x) = x + \sqrt{x-1} - \sqrt{x+1};$ $\mathrm{dom}\,(f+g) = [1, \infty)$
   $(f-g)(x) = \sqrt{x-1} + \sqrt{x+1} - x;$ $\mathrm{dom}\,(f-g) = [1, \infty)$
   $(f \cdot g)(x) = \sqrt{x-1}\,(x - \sqrt{x+1}) = x\sqrt{x-1} - \sqrt{x^2 - 1};$ $\mathrm{dom}\,(f \cdot g) = [1, \infty)$

   $(f/g)(x) = \dfrac{\sqrt{x-1}}{x - \sqrt{x+1}};$ $\mathrm{dom}\,(f/g) = \{x : x \geq 1 \ \text{and} \ x \neq \frac{1}{2}(1 + \sqrt{5})\}$

13.   (a) $(6f + 3g)(x) = 6(x + 1/\sqrt{x}) + 3(\sqrt{x} - 2/\sqrt{x}) = 6x + 3\sqrt{x}; \quad x > 0$

     (b) $(f - g)(x) = x + 1/\sqrt{x} - (\sqrt{x} - 2/\sqrt{x}) = x + 3/\sqrt{x} - \sqrt{x}; \quad x > 0$

     (c) $(f/g)(x) = \dfrac{x\sqrt{x} + 1}{x - 2}; \quad x > 0, \; x \neq 2$

15.

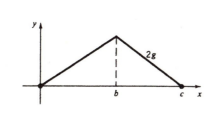

17.

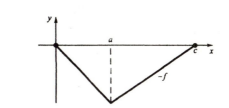

19.

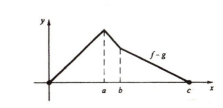

21.

23. $(f \circ g)(x) = 2x^2 + 5; \; \text{dom}\,(f \circ g) = (-\infty, \infty)$ **25.** $(f \circ g)(x) = \sqrt{x^2 + 5}; \; \text{dom}\,(f \circ g) = (-\infty, \infty)$

27. $(f \circ g)(x) = \dfrac{x}{x - 2}; \; \text{dom}\,(f \circ g) = \{x : \; x \neq 0, 2\}$

29. $(f \circ g)(x) = \sqrt{1 - \cos^2 2x} = |\sin 2x|; \; \text{dom}\,(f \circ g) = (-\infty, \infty)$

31. $(f \circ g \circ h) = 4\,[g(h(x))] = 4\,[h(x) - 1] = 4(x^2 - 1); \quad \text{dom}\,(f \circ g \circ h) = (-\infty, \infty)$

33. $(f \circ g \circ h)\dfrac{1}{g(h(x))} = \dfrac{1}{1/[2h(x) + 1]} = 2h(x) + 1 = 2x^2 + 1; \quad \text{dom}\,(f \circ g \circ h) = (-\infty, \infty)$

35.   Take $f(x) = \dfrac{1}{x}$ since $\dfrac{1 + x^4}{1 + x^2} = F(x) = f(g(x)) = f\left(\dfrac{1 + x^2}{1 + x^4}\right).$

37.   Take $f(x) = 2\sin x$ since $2\sin 3x = F(x) = f(g(x)) = f(3x).$

39.   Take $g(x) = \left(1 - \dfrac{1}{x^4}\right)^{2/3}$ since $\left(1 - \dfrac{1}{x^4}\right)^2 = F(x) = f(g(x)) = [g(x)]^3.$

41.   Take $g(x) = 2x^3 - 1$ (or $-(2x^3 - 1)$)   since $(2x^3 - 1)^2 + 1 = F(x) = f(g(x)) = [g(x)]^2 + 1.$

43. $(f \circ g)(x) = f(g(x)) = \sqrt{g(x)} = \sqrt{x^2} = |x|;$

     $(g \circ f)(x) = g(f(x)) = [f(x)]^2 = [\sqrt{x}]^2 = x, \quad x \geq 0$

45. $(f \circ g)(x) = f(g(x)) = 1 - \sin^2 x = \cos^2 x; \quad (g \circ f)(x) = g(f(x)) = \sin f(x) = \sin(1 - x^2).$

47. $(f + g)(x) = f(x) + g(x) = f(x) + c; \; quad g(x) = c.$

49. $(fg)(x) = f(x)g(x) = c\,f(x)$ implies $g(x) = c$.

51. (a) The graph of g is the graph of f shifted 3 units to the right. $\text{dom}\,(g) = [3, a+3]$, $\text{range}\,(g) = [0, b]$.

(b) The graph of g is the graph of f shifted 4 units to the left and scaled vertically by a factor of 3. $\text{dom}\,(g) = [-4, a-4]$, $\text{range}\,(g) = [0, 3b]$.

(c) The graph of g is the graph of f scaled horizontally by a factor of 2. $\text{dom}\,(g) = [0, a/2]$, $\text{range}\,(g) = [0, b]$.

(d) The graph of g is the graph of f scaled horizontally by a factor of $\frac{1}{2}$. $\text{dom}\,(g) = [0, 2a]$, $\text{range}\,(g) = [0, b]$.

53. fg is even since $(fg)(-x) = f(-x)g(-x) = f(x)g(x) = (fg)(x)$.

55. (a) If f is even, then

$$f(x) = \begin{cases} -x, & -1 \le x < 0 \\ 1, & x < -1. \end{cases}$$

(b) If f is odd, then

$$f(x) = \begin{cases} x, & -1 \le x < 0 \\ -1, & x < -1. \end{cases}$$

57. $g(-x) = f(-x) + f[-(-x)] = f(-x) + f(x) = g(x)$

59. $f(x) = \dfrac{1}{2}\underbrace{[f(x) + f(-x)]}_{\text{even}} + \dfrac{1}{2}\underbrace{[f(x) - f(-x)]}_{\text{odd}}$

61. (a) $(f \circ g)(x) = \dfrac{5x^2 + 16x - 16}{(2-x)^2}$ (b) $(g \circ k)(x) = x$ (c) $(f \circ k \circ g)(x) = x^2 - 4$

63. (a) For fixed a, varying b varies the y-coordinate of the vertex of the parabola.

(b) For fixed b, varying a varies the x-coordinate of the vertex of the parabola

(c) The graph of $-F$ is the reflection of the graph of F in the x-axis.

65. (a) For $c > 0$, the graph of cf is the graph of f scaled vertically by the factor c; for $c < 0$, the graph of cf is the graph of f scaled vertically by the factor $|c|$ and then reflected in the x-axis.

(b) For $c > 1$, the graph of $f(cx)$ is the graph of f compressed horizontally; for $0 < c < 1$, the graph of $f(cx)$ is the graph of f stretched horizontally; for $-1 < c < 0$, the graph of $f(cx)$ is the graph of f stretched horizontally and reflected in the y-axis; for $c < -1$, the graph of $f(cx)$ is the graph of f compressed horizontally and reflected in the y-axis.

SECTION 1.8

1. Let S be the set of integers for which the statement is true. Since $2(1) \le 2^1$, S contains 1. Assume now that $k \in S$. This tells us that $2k \le 2^k$, and thus

$$2(k+1) = 2k + 2 \le 2^k + 2 \le 2^k + 2^k = 2(2^k) = 2^{k+1}.$$

$$\underset{(k \ge 1)}{\uparrow}$$

This places $k + 1$ in S.

We have shown that

$$1 \in S \quad \text{and that} \quad k \in S \quad \text{implies} \quad k + 1 \in S.$$

It follows that S contains all the positive integers.

3. Let S be the set of integers for which the statement is true. Since $(1)(2) = 2$ is divisible by $2, 1 \in S$. Assume now that $k \in S$. This tells us that $k(k+1)$ is divisible by 2 and therefore

$$(k+1)(k+2) = k(k+1) + 2(k+1)$$

is also divisible by 2. This places $k + 1 \in S$.

We have shown that

$$1 \in S \text{ and that } k \in S \text{ implies } k + 1 \in S.$$

It follows that S contains all the positive integers.

5. Use
$$1^2 + 2^2 + \cdots + k^2 + (k+1)^2 = \tfrac{1}{6}k(k+1)(2k+1) + (k+1)^2$$
$$= \tfrac{1}{6}(k+1)[k(2k+1) + 6(k+1)]$$
$$= \tfrac{1}{6}(k+1)(2k^2 + 7k + 6)$$
$$= \tfrac{1}{6}(k+1)(k+2)(2k+3)$$
$$= \tfrac{1}{6}(k+1)[(k+1)+1][2(k+1)+1].$$

7. By Exercise 6 and Example 1

$$1^3 + 2^3 + \cdots + (n-1)^3 = [\tfrac{1}{2}(n-1)n]^2 = \tfrac{1}{4}(n-1)^2 n^2 < \tfrac{1}{4}n^4$$

and

$$1^3 + 2^3 + \cdots + n^3 = [\tfrac{1}{2}n(n+1)]^2 = \tfrac{1}{4}n^2(n+1)^2 > \tfrac{1}{4}n^4.$$

9. Use

$$\frac{1}{\sqrt{1}} + \frac{1}{\sqrt{2}} + \frac{1}{\sqrt{3}} + \cdots + \frac{1}{\sqrt{n}} + \frac{1}{\sqrt{n+1}}$$

$$> \sqrt{n} + \frac{1}{\sqrt{n+1}+\sqrt{n}}\left(\frac{\sqrt{n+1}-\sqrt{n}}{\sqrt{n+1}-\sqrt{n}}\right) = \sqrt{n+1}.$$

11. Let S be the set of integers for which the statement is true. Since

$$3^{2(1)+1} + 2^{1+2} = 27 + 8 = 35$$

is divisible by 7, we see that $1 \in S$.

Assume now that $k \in S$. This tells us that

$$3^{2k+1} + 2^{k+2} \text{ is divisible by 7.}$$

It follows that

$$3^{2(k+1)+1} + 2^{(k+1)+2} = 3^2 \cdot 3^{2k+1} + 2 \cdot 2^{k+2}$$

$$= 9 \cdot 3^{2k+1} + 2 \cdot 2^{k+2}$$

$$= 7 \cdot 3^{2k+1} + 2(3^{2k+1} + 2^{k+2})$$

is also divisible by 7. This places $k+1 \in S$.

We have shown that

$$1 \in S \qquad \text{and that} \qquad k \in S \quad \text{implies} \quad k+1 \in S.$$

It follows that S contains all the positive integers.

13. For all positive integers $n \geq 2$,

$$\left(1 - \frac{1}{2}\right)\left(1 - \frac{1}{3}\right) \cdots \left(1 - \frac{1}{n}\right) = \frac{1}{n}.$$

To see this, let S be the set of integers n for which the formula holds. Since $1 - \frac{1}{2} = \frac{1}{2}$, $\quad 2 \in S$. Suppose now that $k \in S$. This tells us that

$$\left(1 - \frac{1}{2}\right)\left(1 - \frac{1}{3}\right) \cdots \left(1 - \frac{1}{k}\right) = \frac{1}{k}$$

and therefore that

$$\left(1 - \frac{1}{2}\right)\left(1 - \frac{1}{3}\right) \cdots \left(1 - \frac{1}{k}\right)\left(1 - \frac{1}{k+1}\right) = \frac{1}{k}\left(1 - \frac{1}{k+1}\right) = \frac{1}{k}\left(\frac{k}{k+1}\right) = \frac{1}{k+1}.$$

This places $k+1 \in S$ and verifies the formula for $n \geq 2$.

15. From the figure, observe that adding a vertex V_{N+1} to an N-sided polygon increases the number of diagonals by $(N-2) + 1 = N - 1$. Then use the identity $\frac{1}{2}N(N-3) + (N-1) = \frac{1}{2}(N+1)(N+1-3)$.

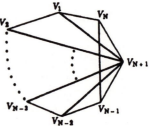

17. To go from k to $k+1$, take $A = \{a_1, \ldots, a_{k+1}\}$ and $B = \{a_1, \ldots, a_k\}$. Assume that B has 2^k subsets: $B_1, B_2, \ldots B_{2^k}$. The subsets of A are then $B_1, B_2, \ldots, B_{2^k}$ together with

$$B_1 \cup \{a_{k+1}\}, \; B_2 \cup \{a_{k+1}\}, \ldots, B_{2^k} \cup \{a_{k+1}\}.$$

This gives $2(2^k) = 2^{k+1}$ subsets for A.

19. $n = 41$

CHAPTER 1. REVIEW EXERCISES

1. rational

3. irrational

5. bounded below by 1

7. bounded; lower bound -5, upper bound 1

9. $2x^2 + x - 1 = (2x - 1)(x + 1); \; x = \frac{1}{2}, -1$

11. $x^2 - 10x + 25 = (x - 5)^2; \; x = 5$

13. $5x - 2 < 0$

$5x < 2$

$x < \frac{2}{5}$

Ans: $\left(-\infty, \frac{2}{5}\right)$

15. $x^2 - x - 6 \geq 0$

$(x - 3)(x + 2) \geq 0$

Ans: $(-\infty, -2] \cup [3, \infty)$

17. $\dfrac{x + 1}{(x + 2)(x - 2)} > 0$

Ans: $(-2, -1) \cup (2, \infty)$

19. $|x - 2| < 1$

$-1 < x - 2 < 1$

Ans: $(1, 3)$

21. $\left| \dfrac{2}{x + 4} \right| > 2$

$\dfrac{2}{x + 4} > 2$ or $\dfrac{2}{x + 4} < -2$

If $\dfrac{2}{x + 4} > 2$

$x + 4 > 0$ and $2 > 2x + 8$

$-4 < x < -3$

If $\dfrac{2}{x + 4} < -2$

$x + 4 < 0$ and $2 > -2x - 8$

$-5 < x < -4$

Ans: $(-5, -4) \cup (-4, -3)$

23. $d(P, Q) = \sqrt{(1 - 2)^2 + (4 - (-3))^2} = 5\sqrt{2}$;   midpoint: $\left(\dfrac{2 + 1}{2}, \dfrac{4 - 3}{2} \right) = \left(\dfrac{3}{2}, \dfrac{1}{2} \right)$

25. $x = 2$

27. The line $l: 2x - 3y = 6$ has slope $m = 2/3$. Therefore, an equation for the line through $(2, -3)$ perpendicular to l is: $y + 3 = -\frac{3}{2}(x - 2)$ or $3x + 2y = 0$

29.

$$\begin{aligned} x - 2y &= -4 \\ 3x + 4y &= 3 \end{aligned} \quad \Rightarrow \; 5x = -5 \quad \Rightarrow \; x = -1; \quad (-1, \; \tfrac{3}{2}).$$

31. Solve the equations simultaneously:

$$2x^2 = 8x - 6$$

$$2x^2 - 8x + 6 = 0$$

$$2(x - 1)(x - 3) = 0$$

the line and the parabola intersect at $(1, 2)$ and $(3, 18)$.

33. domain: $(-\infty, \infty)$; range: $(-\infty, 4]$

35. domain: $[4, \infty)$; range: $[0, \infty)$

37. domain: $(-\infty, \infty)$; range: $[1, \infty)$

39. domain: $(-\infty, \infty)$; range: $[0, \infty)$

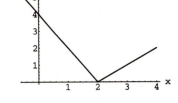

41. $x = \frac{7}{6}\pi, \; \frac{11}{6}\pi$

43. $x = \frac{3\pi}{2}$

45.

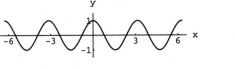

47.

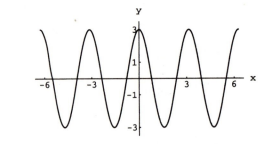

49. $(f + g)(x) = (3x + 2) + (x^2 - 1) = x^2 + 3x + 1, \quad \text{dom}\,(f + g) = (-\infty, \infty).$
$(f - g)(x) = (3x + 2) - (x^2 - 1) = 3 + 3x - x^2, \quad \text{dom}\,(f - g) = (-\infty, \infty).$
$(f \cdot g)(x) = (3x + 2)(x^2 - 1) = 3x^3 + 2x^2 - 3x - 2, \quad \text{dom}\,(f \cdot g) = (-\infty, \infty).$
$\left(\dfrac{f}{g}\right)(x) = \dfrac{3x + 2}{x^2 - 1}, \quad \text{dom}\,(f/g) = (-\infty, -1) \cup (-1, 1) \cup (1, \infty).$

51. $(f+g)(x) = \cos^2 x + \sin 2x, \quad \text{dom } (f+g) = [0, 2\pi].$

$(f-g)(x) = \cos^2 x - \sin 2x, \quad \text{dom } (f-g) = [0, 2\pi].$

$(f \cdot g)(x) = \cos^2 x(\sin 2x) = 2 \cos^3 x \sin x, \quad \text{dom } (f \cdot g) = [0, 2\pi].$

$\left(\dfrac{f}{g}\right)(x) = \dfrac{\cos^2 x}{\sin 2x} = \tfrac{1}{2} \cot x, \quad \text{dom}(f/g): \ x \in (0, 2\pi), x \neq \tfrac{1}{2}\pi, \pi, \tfrac{3}{2}\pi.$

53. $(f \circ g)(x) = \sqrt{(x^2 - 5) + 1} = \sqrt{x^2 - 4}, \quad \text{dom } (f \circ g) = (-\infty, -2] \cup [2, \infty).$

$(g \circ f)(x) = \left(\sqrt{x+1}\right)^2 - 5 = x - 4, \quad \text{dom } (g \circ f) = [-1, \infty).$

55. (a) $y = kx.$

(b) If $b = ka,$ then $\alpha b = \alpha k a = k(\alpha b).$ Hence, Q is a point on $l.$

(c) If $\alpha > 0,$ P, Q are on the same side of the origin; if $\alpha < 0,$ P, Q are on opposite sides of the origin.

57. Since $|a| = |a - b + b| \leq |a - b| + |b|$ by the given inequality, we have $|a| - |b| \leq |a - b|.$

CHAPTER 2

SECTION 2.1

1. (a) 2   (b) −1   (c) does not exist   (d) −3

3. (a) does not exist   (b) −3   (c) does not exist   (d) −3

5. (a) does not exist   (b) does not exist   (c) does not exist   (d) 1

7. (a) 2   (b) 2   (c) 2   (d) −1

9. (a) 0   (b) 0   (c) 0   (d) 0

11. $c = 0, 6$ **12.** −1   **13.** 12   **14.** 1

15. $\frac{3}{2}$ **16.** does not exist   **17.** $\lim\limits_{x \to 3} \dfrac{2x - 6}{x - 3} = \lim\limits_{x \to 3} 2 = 2$

25. $\lim\limits_{x \to 3} \dfrac{x - 3}{x^2 - 6x + 9} = \lim\limits_{x \to 3} \dfrac{x - 3}{(x - 3)^2} = \lim\limits_{x \to 3} \dfrac{1}{x - 3};$ does not exist

27. $\lim\limits_{x \to 2} \dfrac{x - 2}{x^2 - 3x + 2} = \lim\limits_{x \to 2} \dfrac{x - 2}{(x - 1)(x - 2)} = \lim\limits_{x \to 2} \dfrac{1}{x - 1} = 1$

29. does not exist   **30.** $\lim\limits_{x \to 0} \dfrac{2x - 5x^2}{x} = \lim\limits_{x \to 0}(2 - 5x) = 2$

33. $\lim\limits_{x \to 1} \dfrac{x^2 - 1}{x - 1} = \lim\limits_{x \to 1} \dfrac{(x - 1)(x + 1)}{x - 1} = \lim\limits_{x \to 1}(x + 1) = 2$

35. 0   **36.** 1   **37.** 16

38. does not exist   **39.** 4   **40.** does not exist

47.
$$\lim_{x \to 1} \frac{\sqrt{x^2 + 1} - \sqrt{2}}{x - 1} = \lim_{x \to 1} \frac{(\sqrt{x^2 + 1} - \sqrt{2})(\sqrt{x^2 + 1} + \sqrt{2})}{(x - 1)(\sqrt{x^2 + 1} + \sqrt{2})}$$

$$= \lim_{x \to 1} \frac{x^2 - 1}{(x - 1)(\sqrt{x^2 + 1} + \sqrt{2})} = \lim_{x \to 1} \frac{x + 1}{\sqrt{x^2 + 1} + \sqrt{2}} = \frac{2}{2\sqrt{2}} = \frac{1}{\sqrt{2}}$$

49.
$$\lim_{x \to 1} \frac{x^2 - 1}{\sqrt{2x + 2} - 2} = \lim_{x \to 1} \frac{x^2 - 1}{\sqrt{2x + 2} - 2} \frac{\sqrt{2x + 2} + 2}{\sqrt{2x + 2} + 2}$$

$$= \lim_{x \to 1} \frac{(x - 1)(x + 1)\left(\sqrt{2x + 2} + 2\right)}{2x + 2 - 4} = \lim_{x \to 1} \frac{(x - 1)(x + 1)\left(\sqrt{2x + 2} + 2\right)}{2(x - 1)}$$

$$= \lim_{x \to 1} \frac{(x + 1)\left(\sqrt{2x + 2} + 2\right)}{2} = \frac{2 \cdot 4}{2} = 4$$

51.   2   **52.** $\frac{3}{2}$ **53.** (i) 5 (ii) does not exist

54. $(i) -\frac{5}{4}$ (ii) 0   **55.** $c = -1$ **56.** $c = -2$

63. $\lim\limits_{x \to 0} f(x) = 1$;

      $\lim\limits_{x \to 0} g(x) = 0$

SECTION 2.2

1. $\dfrac{1}{2}$ **2.** $\lim\limits_{x \to 0} \dfrac{x(1+x)}{2x^2} = \lim\limits_{x \to 0} \dfrac{1+x}{2x}$;   does not exist

3. $\lim\limits_{x \to 1} \dfrac{x^4 - 1}{x - 1} = \lim\limits_{x \to 1} (x^3 + x^2 + x + 1) = 4$ **4.**   does not exist

5. -1 **6.**   does not exist   **7.**   0

8. $\lim\limits_{x \to 2^+} f(x) = \lim\limits_{x \to 2^+} (x^2 - x) = 2$ **17.**   1   **19.**   1

21. δ_1 **22.** $\delta = \frac{1}{2}\epsilon = .05$ **23.** $2\epsilon = .02$ **24.** $\delta = 1.75$

29.   for $\epsilon = 0.5$ take $\delta = 0.24$;   for $\epsilon = 0.25$ take $\delta = 0.1$

31.   for $\epsilon = 0.5$ take $\delta = 0.75$;   for $\epsilon = 0.25$ take $\delta = 0.43$

33.   for $\epsilon = 0.25$ take $\delta = 0.23$;   for $\epsilon = 0.1$ take $\delta = 0.14$

35.   Since

$$|(2x - 5) - 3| = |2x - 8| = 2|x - 4|,$$

we can take $\delta = \frac{1}{2}\epsilon$:

$$\text{if} \quad 0 < |x - 4| < \tfrac{1}{2}\epsilon \quad \text{then,} \quad |(2x - 5) - 3| = 2|x - 4| < \epsilon.$$

37.   Since

$$|(6x - 7) - 11| = |6x - 18| = 6|x - 3|,$$

we can take $\delta = \frac{1}{6}\epsilon$:

$$\text{if} \quad 0 < |x - 3| < \tfrac{1}{6}\epsilon \quad \text{then} \quad |(6x - 7) - 11| = 6|x - 3| < \epsilon.$$

39.   Since

$$\big||1 - 3x| - 5\big| = \big||3x - 1| - 5\big| \le |3x - 6| = 3|x - 2|,$$

we can take $\delta = \frac{1}{3}\epsilon$:

$$\text{if} \quad 0 < |x - 2| < \tfrac{1}{3}\epsilon \quad \text{then} \quad \big||1 - 3x| - 5\big| \le 3|x - 2| < \epsilon.$$

41. Statements (b), (e), (g), and (i) are necessarily true.

43.   (i) $\lim\limits_{x \to 3} \dfrac{1}{x-1} = \dfrac{1}{2}$ (ii) $\lim\limits_{h \to 0} \dfrac{1}{(3+h)-1} = \dfrac{1}{2}$

      (iii) $\lim\limits_{x \to 3} \left(\dfrac{1}{x-1} - \dfrac{1}{2} \right) = 0$ (iv) $\lim\limits_{x \to 3} \left| \dfrac{1}{x-1} - \dfrac{1}{2} \right| = 0$

45. By (2.2.6) parts (i) and (iv) with $L = 0$

47. Let $\epsilon > 0$. If

$$\lim_{x \to c} f(x) = L,$$

then there must exist $\delta > 0$ such that

(∗)   if $0 < |x - c| < \delta$ then $|f(x) - L| < \epsilon.$

Suppose now that

$$0 < |h| < \delta.$$

Then

$$0 < |(c + h) - c| < \delta$$

and thus by (∗)

$$|f(c + h) - L| < \epsilon.$$

This proves that

$$\text{if} \quad \lim_{x \to c} f(x) = L \quad \text{then} \quad \lim_{h \to 0} f(c + h) = L.$$

If, on the other hand,

$$\lim_{h \to 0} f(c + h) = L,$$

then there must exist $\delta > 0$ such that

(∗∗)   if $0 < |h| < \delta$ then $|f(c + h) - L| < \epsilon.$

Suppose now that

$$0 < |x - c| < \delta.$$

Then by (∗∗)

$$|f(c + (x - c)) - L| < \epsilon.$$

More simply stated,

$$|f(x) - L| < \epsilon.$$

This proves that

$$\text{if} \quad \lim_{h \to 0} f(c + h) = L \quad \text{then} \quad \lim_{x \to c} f(x) = L.$$

49.   (a)   Set $\delta = \epsilon\sqrt{c}$. By the hint,

$$\text{if} \quad 0 < |x - c| < \epsilon\sqrt{c} \quad \text{then} \quad |\sqrt{x} - \sqrt{c}| < \frac{1}{\sqrt{c}}|x - c| < \epsilon.$$

      (b)   Set $\delta = \epsilon^2$. If $0 < x < \epsilon^2$,   then $|\sqrt{x} - 0| = \sqrt{x} < \epsilon$.

51.   Take $\delta = $ minimum of 1 and $\epsilon/7$. If $0 < |x - 1| < \delta$,   then $0 < x < 2$ and $|x - 1| < \epsilon/7$. Therefore

$$|x^3 - 1| = |x^2 + x + 1|\,|x - 1| < 7|x - 1| < 7(\epsilon/7) = \epsilon.$$

53.   Set $\delta = \epsilon^2$.   If $3 - \epsilon^2 < x < 3$,   then $-\epsilon^2 < x - 3$, $0 < 3 - x < \epsilon^2$ and therefore $|\sqrt{3 - x} - 0| < \epsilon$.

55.   Suppose, on the contrary, that $\lim\limits_{x \to c} f(x) = L$ for some particular c. Taking $\epsilon = \frac{1}{2}$, there must exist $\delta > 0$ such that

$$\text{if} \quad 0 < |x - c| < \delta, \quad \text{then} \quad |f(x) - L| < \tfrac{1}{2}.$$

Let x_1 be a rational number satisfying $0 < |x_1 - c| < \delta$ and x_2 an irrational number satisfying $0 < |x_2 - c| < \delta$. (That such numbers exist follows from the fact that every interval contains both rational and irrational numbers.) Now $f(x_1) = L$ and $f(x_2) = 0$. Thus we must have both

$$|1 - L| < \tfrac{1}{2} \quad \text{and} \quad |0 - L| < \tfrac{1}{2}.$$

From the first inequality we conclude that $L > \frac{1}{2}$. From the second, we conclude that $L < \frac{1}{2}$. Clearly no such number L exists.

57.   We begin by assuming that $\lim\limits_{x \to c^+} f(x) = L$ and showing that

$$\lim\limits_{h \to 0} f(c + |h|) = L.$$

Let $\epsilon > 0$. Since $\lim\limits_{x \to c^+} f(x) = L$, there exists $\delta > 0$ such that

(*)   if $c < x < c + \delta$ then $|f(x) - L| < \epsilon$.

Suppose now that $0 < |h| < \delta$. Then $c < c + |h| < c + \delta$ and, by (*),

$$|f(c + |h|) - L| < \epsilon.$$

Thus $\lim\limits_{h \to 0} f(c + |h|) = L$.

Conversely we now assume that $\lim\limits_{h \to 0} f(c + |h|) = L$. Then for $\epsilon > 0$ there exists $\delta > 0$ such that

(**)   if $0 < |h| < \delta$ then $|f(c + |h|) - L| < \epsilon$.

Suppose now that $c < x < c + \delta$. Then $0 < x - c < \delta$ so that, by (**),

$$|f(c + (x - c)) - L| = |f(x) - L| < \epsilon.$$

Thus $\lim\limits_{x \to c^+} f(x) = L$.

59. (a) Let $\epsilon = L$. Since $\lim\limits_{x \to c} f(x) = L$, there exists $\delta > 0$ such that if $0 < |x - c| < \delta$ then

$$L - f(x) \le |L - f(x)| = |f(x) - L| < L$$

Therefore, $f(x) > L - L = 0$ for all $x \in (c - \delta, c + \delta)$; take $\gamma = \delta$.

(b) Let $\epsilon = -L$ and repeat the argument in part (a).

61. (a) Let $\lim\limits_{x \to c} f(x) = L$ and $\lim\limits_{x \to c} g(x) = M$, and let $\epsilon > 0$. There exist positive numbers δ_1 and δ_2 such that

$$|f(x) - L| < \epsilon/2 \quad \text{if} \quad 0 < |x - c| < \delta_1$$

and

$$|g(x) - M| < \epsilon/2 \quad \text{if} \quad 0 < |x - c| < \delta_2$$

Let $\delta = \min(\delta_1, \delta_2)$. Then

$$M - L = M - g(x) + g(x) - f(x) + f(x) - L \ge [M - g(x)] + [f(x) - L]$$

$$\ge -\epsilon/2 - \epsilon/2 = -\epsilon$$

for all x such that $0 < |x - c| < \delta$. Since ϵ is arbitrary, it follows that $M \ge L$.

(b) No. For example, if $f(x) = x^2$ and $g(x) = |x|$ on $(-1, 1)$, then $f(x) < g(x)$ on $(-1, 1)$ except at $x = 0$, but $\lim\limits_{x \to 0} x^2 = \lim\limits_{x \to 0} |x| = 0$.

SECTION 2.3

1. (a) 3 (b) 4 (c) −2 (d) 0 (e) does not exist (f) $\frac{1}{3}$

3. $\lim\limits_{x \to 4} \left(\dfrac{1}{x} - \dfrac{1}{4} \right) \left(\dfrac{1}{x - 4} \right) = \lim\limits_{x \to 4} \left(\dfrac{4 - x}{4x} \right) \left(\dfrac{1}{x - 4} \right) = \lim\limits_{x \to 4} \dfrac{-1}{4x} = -\dfrac{1}{16};$ Theorem 2.3.2 does not apply

since $\lim\limits_{x \to 4} \dfrac{1}{x - 4}$ does not exist.

5. 3 **6.** −3 **7.** 5

8. does not exist **9.** −1 **10.** does not exist

11. $\lim\limits_{h \to 0} h \left(1 + \dfrac{1}{h} \right) = \lim\limits_{h \to 0} (h + 1) = 1$ **12.** $\lim\limits_{x \to 2} \dfrac{x^2 - 4}{x - 2} = \lim\limits_{x \to 2} \dfrac{x + 2}{1} = 4$

21. $\lim\limits_{x \to 4} \dfrac{\sqrt{x} - 2}{x - 4} = \lim\limits_{x \to 4} \dfrac{\sqrt{x} - 2}{x - 4} \cdot \dfrac{\sqrt{x} + 2}{\sqrt{x} + 2} = \lim\limits_{x \to 4} \dfrac{x - 4}{(x - 4)(\sqrt{x} + 2)} = \dfrac{1}{4}$

23. $\lim\limits_{x \to 1} \dfrac{x^2 - x - 6}{(x + 2)^2} = \lim\limits_{x \to 1} \dfrac{(x + 2)(x - 3)}{(x + 2)^2} = \lim\limits_{x \to 1} \dfrac{x - 3}{x + 2} = -\dfrac{2}{3}$

25. $\lim\limits_{h \to 0} \dfrac{1 - 1/h^2}{1 - 1/h} = \lim\limits_{h \to 0} \dfrac{h^2 - 1}{h^2 - h} = \lim\limits_{h \to 0} \dfrac{(h + 1)(h - 1)}{h(h - 1)} = \lim\limits_{h \to 0} \dfrac{h + 1}{h};$ does not exist

27. $\displaystyle\lim_{h\to 0}\frac{1-1/h}{1+1/h}=\lim_{h\to 0}\frac{h-1}{h+1}=-1$

29. $\displaystyle\lim_{t\to -1}\frac{t^2+6t+5}{t^2+3t+2}=\lim_{t\to -1}\frac{(t+1)(t+5)}{(t+1)(t+2)}=\lim_{t\to -1}\frac{t+5}{t+2}=4$

31. $\displaystyle\lim_{t\to 0}\frac{t+a/t}{t+b/t}=\lim_{t\to 0}\frac{t^2+a}{t^2+b}=\frac{a}{b}$

33. $\displaystyle\lim_{x\to 1}\frac{x^5-1}{x^4-1}=\lim_{x\to 1}\frac{(x-1)(x^4+x^3+x^2+x+1)}{(x-1)(x^3+x^2+x+1)}=\lim_{x\to 1}\frac{x^4+x^3+x^2+x+1}{x^3+x^2+x+1}=\frac{5}{4}$

35. $\displaystyle\lim_{h\to 0}h\left(1+\frac{1}{h^2}\right)=\lim_{h\to 0}\frac{h^2+1}{h};$ does not exist

37. $\displaystyle\lim_{x\to -4}\left(\frac{2x}{x+4}+\frac{8}{x+4}\right)=\lim_{x\to -4}\frac{2x+8}{x+4}=\lim_{x\to -4}2=2$

39. (a) $\displaystyle\lim_{x\to 4}\left(\frac{1}{x}-\frac{1}{4}\right)=\lim_{x\to 4}\frac{4-x}{4x}=0$

    (b) $\displaystyle\lim_{x\to 4}\left[\left(\frac{1}{x}-\frac{1}{4}\right)\left(\frac{1}{x-4}\right)\right]=\lim_{x\to 4}\left[\left(\frac{4-x}{4x}\right)\left(\frac{1}{x-4}\right)\right]=\lim_{x\to 4}\left(-\frac{1}{4x}\right)=-\frac{1}{16}$

    (c) $\displaystyle\lim_{x\to 4}\left[\left(\frac{1}{x}-\frac{1}{4}\right)(x-2)\right]=\lim_{x\to 4}\frac{(4-x)(x-2)}{4x}=0$

    (d) $\displaystyle\lim_{x\to 4}\left[\left(\frac{1}{x}-\frac{1}{4}\right)\left(\frac{1}{x-4}\right)^2\right]=\lim_{x\to 4}\frac{4-x}{4x(x-4)^2}=\lim_{x\to 4}\frac{1}{4x(4-x)};$ does not exist

41. (a) $\displaystyle\lim_{x\to 4}\frac{f(x)-f(4)}{x-4}=\lim_{x\to 4}\frac{(x^2-4x)-(0)}{x-4}=\lim_{x\to 4}x=4$

    (b) $\displaystyle\lim_{x\to 1}\frac{f(x)-f(1)}{x-1}=\lim_{x\to 1}\frac{x^2-4x+3}{x-1}=\lim_{x\to 1}\frac{(x-1)(x-3)}{x-1}=\lim_{x\to 1}(x-3)=-2$

    (c) $\displaystyle\lim_{x\to 3}\frac{f(x)-f(1)}{x-3}=\lim_{x\to 3}\frac{x^2-4x+3}{x-3}=\lim_{x\to 3}\frac{(x-1)(x-3)}{x-3}=\lim_{x\to 3}(x-1)=2$

    (d) $\displaystyle\lim_{x\to 3}\frac{f(x)-f(2)}{x-3}=\lim_{x\to 3}\frac{x^2-4x+4}{x-3};$ does not exist

43. $f(x)=1/x,\quad g(x)=-1/x\quad$ with $c=0$

45. True. Let $\displaystyle\lim_{x\to c}[f(x)+g(x)]=L$. If $\displaystyle\lim_{x\to c}g(x)=M$ exists, then $\displaystyle\lim_{x\to c}f(x)=\lim_{x\to c}[f(x)+g(x)-g(x)]=$ $L-M$ also exists. This contradicts the fact that $\displaystyle\lim_{x\to c}f(x)$ does not exist.

47. True. If $\displaystyle\lim_{x\to c}\sqrt{f(x)}=L$ exists, then $\displaystyle\lim_{x\to c}\sqrt{f(x)}\sqrt{f(x)}=L^2$ also exists.

49. False;   for example set $f(x)=x$ and $c=0$

51. False;   for example, set $f(x)=1-x^2,\ g(x)=1+x^2,$ and $c=0$.

53. If $\lim\limits_{x \to c} f(x) = L$ and $\lim\limits_{x \to c} g(x) = L,$ then

$$\lim_{x \to c} h(x) = \lim_{x \to c} \tfrac{1}{2}\{[f(x) + g(x)] - |f(x) - g(x)|\}$$

$$= \lim_{x \to c} \tfrac{1}{2}[f(x) + g(x)] - \lim_{x \to c} \tfrac{1}{2}|f(x) - g(x)|$$

$$= \tfrac{1}{2}(L + L) - \tfrac{1}{2}(L - L) = L.$$

A similar argument works for H.

55. (a) Suppose on the contrary that $\lim\limits_{x \to c} g(x)$ does exist. Let $L = \lim\limits_{x \to c} g(x)$. Then

$$\lim_{x \to c} f(x)g(x) = \lim_{x \to c} f(x) \cdot \lim_{x \to c} g(x) = 0 \cdot L = 0.$$

This contradicts the fact that $\lim\limits_{x \to c} f(x)g(x) = 1$

(b) $\lim\limits_{x \to c} g(x)$ exists since $\lim\limits_{x \to c} g(x) = \lim\limits_{x \to c} \dfrac{f(x)g(x)}{f(x)} = \dfrac{1}{L}.$

57. $f(x) = 2x^2 - 3x, \quad a = 2, \quad f(2) = 2$

$$\lim_{x \to a} \frac{f(x) - f(a)}{x - a} = \lim_{x \to 2} \frac{2x^2 - 3x - 2}{x - 2}$$

$$= \lim_{x \to 2} \frac{(x - 2)(2x + 1)}{x - 2} = \lim_{x \to 2} (2x + 1) = 5$$

59. $f(x) = \sqrt{x}, \quad a = 4, \quad f(4) = 2$

$$\lim_{x \to a} \frac{f(x) - f(a)}{x - a} = \lim_{x \to 4} \frac{\sqrt{x} - 2}{x - 4} = \lim_{x \to 4} \frac{\sqrt{x} - 2}{x - 2} \cdot \frac{\sqrt{x} + 2}{\sqrt{x} + 2}$$

$$= \lim_{x \to 4} \frac{x - 4}{(x - 4)(\sqrt{x} + 2)} = \lim_{x \to 4} \frac{1}{\sqrt{x} + 2} = \frac{1}{4}$$

61. (a) $\lim\limits_{h \to 0} \dfrac{f(x + h) - f(x)}{h} = \lim\limits_{h \to 0} \dfrac{x + h - x}{h} = 1$

(b) $\lim\limits_{h \to 0} \dfrac{f(x + h) - f(x)}{h} = \lim\limits_{h \to 0} \dfrac{(x + h)^2 - x^2}{h} = \lim\limits_{h \to 0} \dfrac{x^2 + 2xh + h^2 - x^2}{h}$

$$= \lim_{h \to 0} (2x + h) = 2x$$

(c) $\lim\limits_{h \to 0} \dfrac{f(x + h) - f(x)}{h} = \lim\limits_{h \to 0} \dfrac{(x + h)^3 - x^3}{h} = \lim\limits_{h \to 0} \dfrac{x^3 + 3x^2h + 3xh^2 + h^3 - x^3}{h}$

$$= \lim_{h \to 0} (3x^2 + 3xh + h^2) = 3x^2$$

(d) $\lim\limits_{h \to 0} \dfrac{f(x + h) - f(x)}{h} = \lim\limits_{h \to 0} \dfrac{(x + h)^4 - x^4}{h} = \lim\limits_{h \to 0} \dfrac{x^4 + 4x^3h + 6x^2h^2 + 4xh^3 + h^4 - x^4}{h}$

$$= \lim_{h \to 0} (4x^3 + 6x^2h + 4xh^2 + h^3) = 4x^3$$

(e) $\lim\limits_{h \to 0} \dfrac{f(x + h) - f(x)}{h} = \lim\limits_{h \to 0} \dfrac{(x + h)^n - x^n}{h} = nx^{n-1}$ for any positive integer n.

SECTION 2.4

1. (a) f is discontinuous at $x = -3,\ 0,\ 2,\ 6$

    (b) at -3, neither; f is continuous from the right at 0; at 2 and 6, neither

2. continuous

3. continuous

4. continuous

5. removable discontinuity

6. jump discontinuity

7. continuous

15. infinite discontinuity

17.

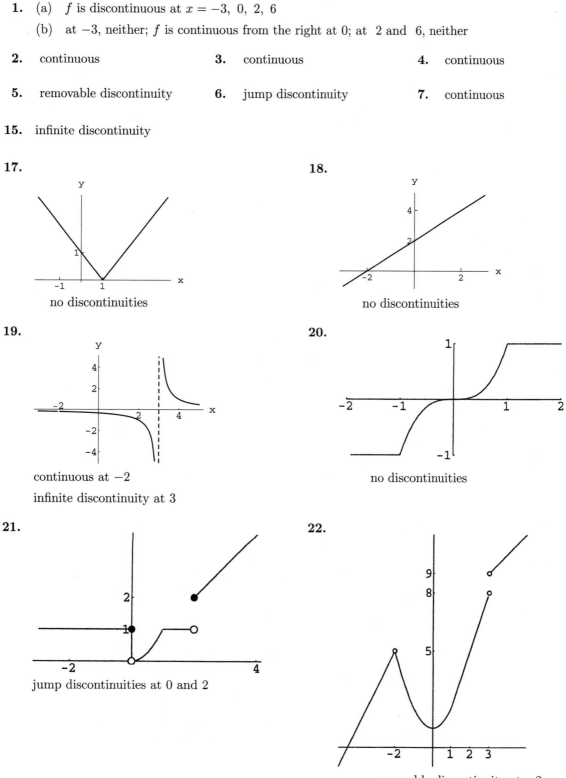

no discontinuities

18.

y

no discontinuities

19.

continuous at -2

infinite discontinuity at 3

20.

no discontinuities

21.

jump discontinuities at 0 and 2

22.

removable discontinuity at -2;

jump discontinuity at 3

29.

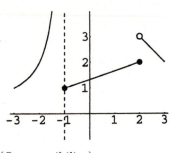

(One possibility)

31. $f(1) = 2$

32. impossible; $\lim_{x \to 1^-} f(x) = -1$; $\lim_{x \to 1^+} f(x) = 1$

35. Since $\lim_{x \to 1^-} f(x) = 1$ and $\lim_{x \to 1^+} f(x) = A - 3 = f(1)$, take $A = 4$.

37. The function f is continuous at $x = 1$ iff

$$f(1) = \lim_{x \to 1^-} f(x) = A - B \quad \text{and} \quad \lim_{x \to 1^+} f(x) = 3$$

are equal;   that is, $A - B = 3$. The function f is discontinuous at $x = 2$ iff

$$\lim_{x \to 2^-} f(x) = 6 \quad \text{and} \quad \lim_{x \to 2^+} f(x) = f(2) = 4B - A$$

are unequal;   that is, iff $4B - A \neq 6$. More simply we have $A - B = 3$ with $B \neq 3$:

$$A - B = 3, \ 4B - A \neq 6 \quad \Longrightarrow \quad A - B = 3, \ 3B - 3 \neq 6 \quad \Longrightarrow \quad A - B = 3, \ B \neq 3.$$

39. $c = -3$

40. $f(5) = \frac{1}{6}$

41. $f(5) = \frac{1}{3}$

45. nowhere;   see Figure 2.1.8

47. $x = 0$, $x = 2$,   and all non-integral values of x

49. Refer to (2.2.6). Use the equivalence of (i) and (ii) setting $L = f(c)$.

51. Suppose that g does not have a non-removable discontinuity at c. Then either g is continuous at c or it has a removable discontinuity at c. In either case, $\lim g(x)$ as $x \to c$ exists. Since $g(x) = f(x)$ except at a finite set of points $x_1, x_2, \ldots, x_n$, $\lim f(x)$ exists as $x \to c$ by Exercise 54, Section 2.3.

53. By implication, f is defined on $(c - p, c + p)$. The given inequality implies that $B \geq 0$. If $B = 0$, then $f \equiv f(c)$ is a constant function and hence is continuous. Now assume that $B > 0$. Let $\epsilon > 0$ and let $\delta = \min\{\epsilon/B, \, p\}$. If $|x - c| < \delta$ then $x \in (c - p, \, c + p)$ and

$$|f(x) - f(c)| \leq B|x - c| < B \cdot \delta \leq B \cdot \frac{\epsilon}{B} = \epsilon$$

Thus, f is continuous at c.

55. $\lim\limits_{h\to 0}[f(c+h)-f(c)] = \lim\limits_{h\to 0}\left[\dfrac{f(c+h)-f(c)}{h}\cdot h\right] = \lim\limits_{h\to 0}\left[\dfrac{f(c+h)-f(c)}{h}\right]\cdot\lim\limits_{h\to 0} h = L\cdot 0 = 0.$
Therefore f is continuous at c by Exercise 47.

57. $\frac{5}{2}$

59. $\lim\limits_{x\to 0^-} f(x) = -1;\quad \lim\limits_{x\to 0^+} f(x) = 1;\quad f$ is discontinuous for all k.

SECTION 2.5

1. $\lim\limits_{x\to 0}\dfrac{\sin 3x}{x} = \lim\limits_{x\to 0} 3\left(\dfrac{\sin 3x}{3x}\right) = 3(1) = 3$
2. $\lim\limits_{x\to 0}\dfrac{3x}{\sin 5x} = \lim\limits_{x\to 0}\dfrac{3}{5}\left(\dfrac{5x}{\sin 5x}\right) = \dfrac{3}{5}(1) = \dfrac{3}{5}$

3. $\lim\limits_{x\to 0}\dfrac{\sin 4x}{\sin 2x} = \lim\limits_{x\to 0}\dfrac{4x}{2x}\cdot\dfrac{\sin 4x}{4x}\cdot\dfrac{2x}{\sin 2x} = 2(1)(1) = 2$

4. $\lim\limits_{x\to 0}\dfrac{\sin x^2}{x} = \lim\limits_{x\to 0} x\left(\dfrac{\sin x^2}{x^2}\right) = \lim\limits_{x\to 0} x\cdot\lim\limits_{x\to 0}\dfrac{\sin x^2}{x^2} = 0(1) = 0$

5. $\lim\limits_{x\to 0}\dfrac{\sin x}{x^2} = \lim\limits_{x\to 0}\dfrac{(\sin x)/x}{x};\quad$ does not exist

11. $\lim\limits_{x\to 0}\dfrac{\sin^2 3x}{5x^2} = \lim\limits_{x\to 0}\dfrac{9}{5}\left(\dfrac{\sin 3x}{3x}\right)^2 = \dfrac{9}{5}(1) = \dfrac{9}{5}$

12. $\lim\limits_{x\to 0}\dfrac{\tan^2 3x}{4x^2} = \lim\limits_{x\to 0}\dfrac{9}{4}\cdot\dfrac{1}{\cos^2 3x}\cdot\dfrac{\sin^2 3x}{(3x)^2} = \dfrac{9}{4}(1)(1) = \dfrac{9}{4}$

13. $\lim\limits_{x\to 0}\dfrac{2x}{\tan 3x} = \lim\limits_{x\to 0}\dfrac{2x\cos 3x}{\sin 3x} = \lim\limits_{x\to 0}\dfrac{2}{3}\left(\dfrac{3x}{\sin 3x}\right)\cos 3x = \dfrac{2}{3}(1)(1) = \dfrac{2}{3}$

15. $\lim\limits_{x\to 0} x\csc x = \lim\limits_{x\to 0}\dfrac{x}{\sin x} = 1$

17. $\lim\limits_{x\to 0}\dfrac{x^2}{1-\cos 2x} = \lim\limits_{x\to 0}\dfrac{x^2}{1-\cos 2x}\cdot\left(\dfrac{1+\cos 2x}{1+\cos 2x}\right) = \lim\limits_{x\to 0}\dfrac{x^2(1+\cos 2x)}{\sin^2 2x}$

$= \lim\limits_{x\to 0}\dfrac{1}{4}\left(\dfrac{2x}{\sin 2x}\right)^2 (1+\cos 2x) = \dfrac{1}{4}(1)(2) = \dfrac{1}{2}$

19. $\lim\limits_{x\to 0}\dfrac{1-\sec^2 2x}{x^2} = \lim\limits_{x\to 0}\dfrac{-\tan^2 2x}{x^2} = \lim\limits_{x\to 0}\dfrac{-\sin^2 2x}{x^2\cos^2 2x} = \lim\limits_{x\to 0}\left[-4\left(\dfrac{\sin 2x}{2x}\right)^2\dfrac{1}{\cos^2 2x}\right] = -4$

21. $\lim\limits_{x\to 0}\dfrac{2x^2+x}{\sin x} = \lim\limits_{x\to 0}(2x+1)\dfrac{x}{\sin x} = 1$

23. $\lim\limits_{x\to 0}\dfrac{\tan 3x}{2x^2+5x} = \lim\limits_{x\to 0}\dfrac{1}{x(2x+5)}\dfrac{\sin 3x}{\cos 3x} = \lim\limits_{x\to 0}\dfrac{3}{2x+5}\left(\dfrac{\sin 3x}{3x}\right)\dfrac{1}{\cos 3x} = \dfrac{3}{5}(1)(1) = \dfrac{3}{5}$

25. $\displaystyle\lim_{x\to 0} \frac{\sec x - 1}{x \sec x} = \lim_{x\to 0} \frac{\dfrac{1}{\cos x} - 1}{x\left(\dfrac{1}{\cos x}\right)} = \lim_{x\to 0} \frac{1 - \cos x}{x} = 0$ **27.** $\dfrac{2\sqrt{2}}{\pi}$

29. $\displaystyle\lim_{x\to \pi/2} \frac{\cos x}{x - \pi/2} = \lim_{h\to 0} \frac{\cos(h + \pi/2)}{h} = \lim_{h\to 0} \frac{-\sin h}{h} = -1$

$\qquad\qquad h = x - \pi/2 \qquad\qquad \cos(h + \pi/2) = \cos h \cos \pi/2 - \sin h \sin \pi/2$

31. $\displaystyle\lim_{x\to \pi/4} \frac{\sin(x + \pi/4) - 1}{x - \pi/4} = \lim_{h\to 0} \frac{\sin(h + \pi/2) - 1}{h} = \lim_{h\to 0} \frac{\cos h - 1}{h} = 0$

$\qquad\qquad\qquad \text{↑} \qquad h = x - \pi/4$

33. Equivalently we will show that $\displaystyle\lim_{h\to 0} \cos(c + h) = \cos c$. The identity

$$\cos(c + h) = \cos c \cos h - \sin c \sin h$$

gives

$$\lim_{h\to 0} \cos(c + h) = \cos c \left(\lim_{h\to 0} \cos h\right) - \sin c \left(\lim_{h\to 0} \sin h\right) = (\cos c)(1) - (\sin c)(0) = \cos c.$$

35. 0   **37.**   1

39. $f(x) = \sin x; \quad a = \pi/4$

$\displaystyle\lim_{h\to 0} \frac{f(a + h) - f(a)}{h} = \lim_{h\to 0} \frac{\sin\left(\frac{\pi}{4} + h\right) - \sin\left(\frac{\pi}{4}\right)}{h} = \lim_{h\to 0} \frac{\sin(\pi/4)\cos h + \cos(\pi/4)\sin h - \sin(\pi/4)}{h}$

$\displaystyle\qquad = \lim_{h\to 0} \frac{-\sin(\pi/4)(1 - \cos h) + \cos(\pi/4)\sin h}{h}$

$\displaystyle\qquad = -\sin(\pi/4) \lim_{h\to 0} \frac{1 - \cos h}{h} + \cos(\pi/4) \lim_{h\to 0} \frac{\sin h}{h} = \cos(\pi/4) = \frac{\sqrt{2}}{2}$

tangent line: $y - \dfrac{\sqrt{2}}{2} = \dfrac{\sqrt{2}}{2}\left(x - \dfrac{\pi}{4}\right)$

41. $f(x) = \cos 2x; \quad a = \pi/6$

$\displaystyle\lim_{h\to 0} \frac{f(a + h) - f(a)}{h} = \lim_{h\to 0} \frac{\cos 2\left(\frac{\pi}{6} + h\right) - \cos\left(2\frac{\pi}{6}\right)}{h} = \lim_{h\to 0} \frac{\cos(2h + \pi/3) - \cos(\pi/3)}{h}$

$\displaystyle\qquad = \lim_{h\to 0} \frac{\cos(\pi/3)\cos 2h - \sin(\pi/3)\sin 2h - \cos(\pi/3)}{h}$

$\displaystyle\qquad = -\cos(\pi/3) \lim_{h\to 0} 2\frac{1 - \cos 2h}{h} - \sin(\pi/3) \lim_{h\to 0} 2\frac{\sin 2h}{h}$

$\displaystyle\qquad = -\cos(\pi/3) \cdot 2 \cdot 0 - \sin(\pi/3) \cdot 2 \cdot 1 = -\sqrt{3}$

tangent line: $y - \dfrac{1}{2} = -\sqrt{3}\left(x - \dfrac{\pi}{6}\right)$

43. For $x \neq 0$, $|x \sin(1/x)| = |x|\,|\sin(1/x)| \leq |x|$. Thus,

$$-|x| \leq |x \sin(1/x)| \leq |x|$$

Since $\displaystyle\lim_{x\to 0}(-|x|) = \lim_{x\to 0} |x| = 0$,   the result follows by the pinching theorem.

45. For x close to 1(radian), $0 < \sin x \le 1$. Thus,

$$0 < |x - 1| \sin x \le |x - 1|$$

and the result follows by the pinching theorem.

47. Suppose that there is a number B such that $|f(x)| \le B$ for all $x \ne 0$. Then $|x\, f(x)| \le B\, |x|$ and

$$-B\, |x| \le x\, f(x) \le B\, |x|$$

The result follows by the pinching theorem.

49. Suppose that there is a number B such that $\left| \dfrac{f(x) - L}{x - c} \right| \le B$ for $x \ne c$. Then

$$0 \le |f(x) - L| = \left| (x - c)\, \frac{f(x) - L}{x - c} \right| \le B|x - c|$$

By the pinching theorem, $\displaystyle \lim_{x \to c} |f(x) - L| = 0$ which implies $\displaystyle \lim_{x \to c} f(x) = L$.

51. $\displaystyle \lim_{x \to 0} \frac{20x - 15x^2}{\sin 2x} = 10$ **52.** $\frac{1}{3}$

SECTION 2.6

1. Set $f(x) = 2x^3 - 4x^2 + 5x - 4$. Then f is continuous on $[1, 2]$ and $f(1) = -1 < 0$. $f(2) = 6 > 0$. By the intermediate-value theorem there is a c in $[1, 2]$ such that $f(c) = 0$.

3. Set $f(x) = \sin x + 2 \cos x - x^2$ Then f is continuous on $[0, \frac{\pi}{2}]$ and $f(0) = 2 > 0$, $f(\frac{\pi}{2}) = 1 - \frac{\pi^2}{4} < 0$. By the intermediate-value theorem there is a c in $[0, \frac{\pi}{2}]$ such that $f(c) = 0$.

5. Set $f(x) = x^2 - 2 + \dfrac{1}{2x}$. Then f is continuous on $[\frac{1}{4}, 1]$ and $f(\frac{1}{4}) = \frac{1}{16} > 0$, $f(1) = -\frac{1}{2} < 0$. By the intermediate-value theorem there is a c in $[\frac{1}{4}, 1]$ such that $f(c) = 0$.

7. Set $f(x) = x^3 - \sqrt{x + 2}$. Then f is continuous on $[1, 2]$ and $f(1) = 1 - \sqrt{3} < 0$, $f(2) = 6 > 0$. By the intermediate-value theorem there is a c in $[1, 2]$ such that $f(c) = 0$.
i.e. $c^3 = \sqrt{c + 2}$.

9. $f(x)$ is continuous on $[0, 1]$; $f(0) = 0 < 1$ and $f(1) = 4 > 1$.
By the intermediate value theorem there is a c in $(0, 1)$ such that $f(c) = 1$.

11. Set $f(x) = x^3 - 4x + 2$. Then $f(x)$ is continuous on $[-3, 3]$.
Checking the integer values on this interval,

$$f(-3) = -13 < 0, \quad f(-2) = 2 > 0, \quad f(0) = 2 > 0, \quad f(1) = -1 < 0, \text{ and } f(2) = 2 > 0.$$

By the intermediate value theorem there are roots in $(-3, -2), (0, 1)$ and $(1, 2)$.

13.   **14.**   **15.**   **16.**

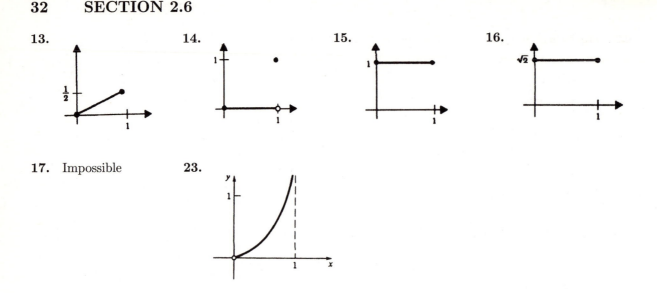

17.   Impossible   **23.**

25.   If $f(0) = 0$ or if $f(1) = 1$, we have a fixed point. If $f(0) \neq 0$ and $f(1) \neq 1$, then set $g(x) = x - f(x)$. g is continuous on $[0, 1]$ and $g(0) < 0 < g(1)$. Therefore there exists a number $c \in (0, 1)$ such that $g(c) = c - f(c) = 0$.

27.   (a) If $f(0) = 1$ or if $f(1) = 0$, we're done. Suppose $f(0) \neq 1$ and $f(1) \neq 0$. The diagonal from $(0, 1)$ to $(1, 0)$ has equation $y = 1 - x$, $0 \leq x \leq 1$. Set $g(x) = (1 - x) - f(x)$. g is continuous on $[0, 1]$ and $g(0) = 1 - f(0) > 0$; $g(1) = -f(1) < 0$. Therefore there exists a number $c \in (0, 1)$ such that $g(c) = (c - 1) - f(c) = 0$.

     (b) Suppose $g(0) = 0$ and $g(1) = 1$. Set $h(x) = g(x) - f(x)$; h is continuous on $[0, 1]$. If $h(0) = 0$ or if $h(1) = 0$, we're done. If not, then $h(0) < 0$ and $h(1) > 0$. Therefore there exists a number $c \in (0, 1)$ such that $h(c) = 0$. The same argument works for the other case.

29.   The cubic polynomial $P(x) = x^3 + ax^2 + bx + c$ is continuous on $(-\infty, \infty)$.. Writing P as

$$P(x) = x^3 \left(1 + \frac{a}{x} + \frac{b}{x^2} + \frac{c}{x^3} \right) \quad x \neq 0$$

it follows that $P(x) < 0$ for large negative values of x and $P(x) > 0$ for large positive values of x. Thus there exists a negative number N such that $P(x) < 0$ for $x < N$, and a positive number M such that $P(x) > 0$ for $x > M$. By the intermediate-value theorem, P has a zero in $[N, M]$.

31.   The function T is continuous on $[4000, 4500]$; and $T(4000) \cong 98.0995$, $T(4500) \cong 97.9478$. Thus, by the intermediate-value theorem, there is an elevation h between 4000 and 4500 meters such that $T(h) = 98$.

33.   Let $A(r)$ denote the area of a circle with radius r, $r \in [0, 10]$. Then $A(r) = \pi r^2$ is continuous on $[0, 10]$, and $A(0) = 0$ and $A(10) = 100\pi \cong 314$. Since $0 < 250 < 314$ it follows from the intermediate value theorem that there exists a number $c \in (0, 10)$ such that $A(c) = 250$.

35. Inscribe a rectangle in a circle of radius R. Introduce a coordinate system with the origin at the center of the circle and the sides of the rectangle parallel to the coordinate axes. Then the area of the rectangle is given by

$$A(x) = 4x\sqrt{R^2 - x^2}, \quad x \in [0, R].$$

Since A is continuous on $[0, R]$, A has a maximum value.

37. $f(-3) = -9$, $f(-2) = 5$; $f(0) = 3$, $f(1) = -1$; $f(1) = -1$, $f(2) = 1$ Thus, f has a zero in $(-3, -2,)$ in $(0, 1)$ and in $(1, 2)$.
$r_1 = -2.4909$, $r_2 = 0.6566$, and $r_3 = 1.8343$

39. $f(-2) = -5.6814$, $f(-1) = 1.1829$; $f(0) = 0.5$, $f(1) = -0.1829$; $f(1) = -0.1829$, $f(2) = 6.681$ Thus, f has a zero in $(-2, -1)$, in $(0, 1)$ and in $(1, 2)$.
$r_1 = -1.3482$, $r_2 = 0.2620$, and $r_3 = 1.0816$

41. f is not continuous at $x = 1$. Therefore f does not satisfy the hypothesis of the intermediate-value theorem. However,

$$\frac{f(-3) + f(2)}{2} \neq \frac{1}{2} = f(c) \text{ for } c \cong 0.163$$

43. f satisfies the hypothesis of the intermediate-value theorem:

$$\frac{f(\pi/2) + f(2\pi)}{2} = \frac{1}{2} = f(c) \text{ for } c \cong 2.38, \; 4.16, \; 5.25$$

45. f is bounded.
$\max(f) = 1 \quad [f(1) = 1]$
$\min(f) = -1 \quad [f(-1) = -1]$

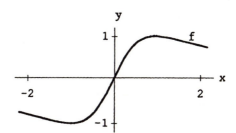

47. f is bounded.
no max f is not defined at $x = 0$
$\min(f) \cong 0.3540$

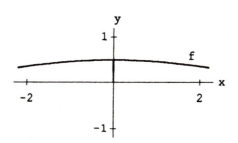

PROJECT 2.6

1. (a) $\dfrac{2-1}{2^n} < 0.001 \implies 2^n > 1000 \implies n > 9.$

The minimum number of iterations required is $n = 10$.

$$\sqrt{2} \simeq \frac{1449}{1024} \simeq 1.415034.$$

(b) $\dfrac{2-1}{2^n} < 0.0001 \implies 2^n > 10{,}000 \implies n > 13.$

The minimum number of iterations required is $n = 14$.

$$\frac{2-1}{2^n} < 0.00001 \implies 2^n > 100{,}000 \implies n > 16.$$

The minimum number of iterations required is $n = 17$.

3. $f(x) = \sin x + x + 3$; $f(-3) \approx -0.142$ and $f(-2) \approx 0.091$. Therefore, $f(c) = 0$ for some $c \in (-3, -2)$.

(a) A minimum of 7 iterations are required; $c \simeq \dfrac{-279}{128} \simeq -2.17969.$

Accurate to 6 decimal places, the root is $c \simeq -2.179727.$

(b) $\dfrac{2-1}{2^n} < 0.00001 \implies 2^n > 100{,}000 \implies n > 16.$

The minimum number of iterations required is $n = 17$.

$$\frac{2-1}{2^n} < 0.000001 \implies 2^n > 1{,}000{,}000 \implies n > 19.$$

The minimum number of iterations required is $n = 20$.

CHAPTER 2. REVIEW EXERCISES

1. $\displaystyle\lim_{x \to 3} \frac{x^2 - 3}{x + 3} = \frac{3^2 - 3}{3 + 3} = 1.$

3. $\displaystyle\lim_{x \to 3} \frac{(x - 3)^2}{x + 3} = \frac{0}{6} = 0.$

5. $\displaystyle\lim_{x \to 2^+} \frac{x - 2}{|x - 2|} = \lim_{x \to 2^+} \frac{x - 2}{x - 2} = 1.$

7. $\displaystyle\lim_{x \to 0} \left(\frac{1}{x} - \frac{1 - x}{x}\right) = \lim_{x \to 0} \frac{x}{x} = 1.$

9. $\displaystyle\lim_{x \to 1^+} \frac{|x - 1|}{x} = \lim_{x \to 1^+} \frac{x - 1}{x} = 0$ and $\lim_{x \to 1^-} \dfrac{|x - 1|}{x} = \lim_{x \to 1^-} \dfrac{-x + 1}{x} = 0.$ $\lim_{x \to 1} \dfrac{|x - 1|}{x} = 0.$

11. $\displaystyle\lim_{x \to 1} \frac{\sqrt{x} - 1}{x - 1} = \lim_{x \to 1} \frac{\sqrt{x} - 1}{(\sqrt{x} - 1)(\sqrt{x} + 1)} = \lim_{x \to 1} \frac{1}{\sqrt{x} + 1} = \frac{1}{2}.$

13. $\displaystyle\lim_{x \to 3^+} \frac{\sqrt{x^2 - 2x - 3}}{x - 3} = \lim_{x \to 3^+} \frac{\sqrt{x + 1}}{\sqrt{x - 3}}$, does not exist.

15. $\lim\limits_{x\to 2} \dfrac{x^3 - 8}{x^4 - 3x^2 - 4} = \lim\limits_{x\to 2} \dfrac{(x-2)(x^2 + 2x + 4)}{(x-2)(x+2)(x^2 + 1)} = \lim\limits_{x\to 2} \dfrac{x^2 + 2x + 4}{(x+2)(x^2 + 1)} = \dfrac{3}{5}.$

17. $\lim\limits_{x\to 0} \dfrac{\tan^2 2x}{3x^2} = \lim\limits_{x\to 0} \dfrac{4}{3\cos^2 2x}\left(\dfrac{\sin 2x}{2x}\right)^2 = \dfrac{4}{3}.$

19. $\lim\limits_{x\to 0} \dfrac{x^2 - 3x}{\tan x} = \lim\limits_{x\to 0} \dfrac{x(x-3)\cos x}{\sin x} = -3$

21. $\lim\limits_{x\to 0} \dfrac{\sin 3x}{5x^2 - 4x} = \lim\limits_{x\to 0} \dfrac{\sin 3x}{3x\left(\dfrac{5}{3}x - \dfrac{4}{3}\right)} = -\dfrac{3}{4}.$

23. $\lim\limits_{x\to -\pi} \dfrac{x + \pi}{\sin x} = \lim\limits_{x\to -\pi} \dfrac{x + \pi}{-\sin(x + \pi)} = -1.$

25. $\lim\limits_{x\to 2^-} \dfrac{x - 2}{|x^2 - 4|} = \lim\limits_{x\to 2^-} -\dfrac{x - 2}{(x-2)(x+2)} = \lim\limits_{x\to 2^-} -\dfrac{1}{x + 2} = -\dfrac{1}{4}.$

27. $\lim\limits_{x\to 1^+} \dfrac{x^2 - 3x + 2}{\sqrt{x - 1}} = \lim\limits_{x\to 1^+} \dfrac{(x-2)(x-1)}{\sqrt{x - 1}} = \lim\limits_{x\to 1^+} (x - 2)\sqrt{x - 1} = 0.$

29. $\lim\limits_{x\to 2} f(x) = 3(2) - 2^2 = 2.$

31. (a) False (b) False (c) True (d) False (e) True

33. (a)

(b) (i) 1 (ii) 0 (iii) does not exist (iv) -6 (v) 4 (vi) does not exist

(c) (i) f is continuous from the left at -1; f is not continuous from the right at -1.

 (ii) f is not continuous from the left at 2; f is continuous from the right at 2.

35. For f to be continuous at 2, we must have

$$\lim\limits_{x\to 2^-} (2x^2 - 1) = 7 = A = \lim\limits_{x\to 2^+} (x^3 - 2Bx) = 8 - 4B.$$

These equations imply that $A = 7$ and $B = \dfrac{1}{4}.$

37. $\lim\limits_{x\to -3} \dfrac{x^2 - 2x - 15}{x + 3} = \lim\limits_{x\to -3} \dfrac{(x - 5)(x + 3)}{x + 3} = -8;$ set $f(-3) = -8.$

39. $\lim\limits_{x\to 0} \dfrac{\sin \pi x}{x} = \lim\limits_{x\to 0} \dfrac{\pi \sin \pi x}{\pi x} = \pi$; set $f(0) = \pi$.

41. (a) False; need f continuous

(b) True; intermediate-value theorem

(c) False; need f continuous on $[a, b]$

(d) False; consider $f(x) = x^2$ on $[-1.1]$.

43. Set $f(x) = 2\cos x - x + 1$. Then $f(1) = 2\cos 1 > 0$ and $f(2) = 2\cos 2 - 1 < 0$. By the intermediate-value theorem, $f(x)$ as zero in $[1, 2]$.

45. If $0 < |x - (-4)| = |x + 4| < 1$, then $2x + 5 < 0$ and $|2x + 5| = -2x - 5$.
Let $\epsilon > 0$.

$$\big| |2x + 5| - 3 \big| = |-2x - 5 - 3| = |-2x - 8| = 2|x + 4|.$$

If $\delta = \min\{1, \epsilon/2\}$, then $\big| |2x + 5| - 3 \big| < \epsilon$. Therefore, $\lim\limits_{x\to -4} |2x + 5| = 3$.

47. Suppose that $\lim\limits_{x\to 0} \dfrac{f(x)}{x} = L$ exists. Then

$$\lim_{x\to 0} f(x) = \lim_{x\to 0} \frac{f(x)\, x}{x} = \lim_{x\to 0} \frac{f(x)}{x}\, x = L \lim_{x\to 0} x = 0.$$

49. (a) Yes. For example, set $f(x) = \begin{cases} \dfrac{\sin \pi x}{x}, & x > 0 \\ \pi, & x = 0 \end{cases}$

(b) No. By the continuity of f (from the right) at 0, $f(1/n) = 0$ for each postive integer implies that $f(0) = 0$.

CHAPTER 3

SECTION 3.1

1. $f'(x) = \lim_{h \to 0} \dfrac{f(x+h) - f(x)}{h} = \lim_{h \to 0} \dfrac{[2 - 3(x+h)] - [2 - 3x]}{h} = \lim_{h \to 0} \dfrac{-3h}{h} = \lim_{h \to 0} -3 = -3$

3. $f'(x) = \lim_{h \to 0} \dfrac{f(x+h) - f(x)}{h} = \lim_{h \to 0} \dfrac{[5(x+h) - (x+h)^2] - (5x - x^2)}{h}$

$= \lim_{h \to 0} \dfrac{5h - 2xh - h^2}{h} = \lim_{h \to 0}(5 - 2x - h) = 5 - 2x$

5. $f'(x) = \lim_{h \to 0} \dfrac{f(x+h) - f(x)}{h} = \lim_{h \to 0} \dfrac{(x+h)^4 - x^4}{h}$

$= \lim_{h \to 0} \dfrac{(x^4 + 4x^3 h + 6x^2 h^2 + 4xh^3 + h^4) - x^4}{h}$

$= \lim_{h \to 0}(4x^3 + 6x^2 h + 4xh^2 + h^3) = 4x^3$

7. $f'(x) = \lim_{h \to 0} \dfrac{f(x+h) - f(x)}{h} = \lim_{h \to 0} \dfrac{\sqrt{x+h-1} - \sqrt{x-1}}{h}$

$= \lim_{h \to 0} \dfrac{(x+h-1) - (x-1)}{h(\sqrt{x+h-1} + \sqrt{x-1})} = \lim_{h \to 0} \dfrac{1}{\sqrt{x+h-1} + \sqrt{x-1}} = \dfrac{1}{2\sqrt{x-1}}$

9. $f'(x) = \lim_{h \to 0} \dfrac{f(x+h) - f(x)}{h} = \lim_{h \to 0} \dfrac{\dfrac{1}{(x+h)^2} - \dfrac{1}{x^2}}{h}$

$= \lim_{h \to 0} \dfrac{x^2 - (x^2 + 2hx + h^2)}{hx^2(x+h)^2} = \lim_{h \to 0} \dfrac{-2x - h}{x^2(x+h)^2} = -\dfrac{2}{x^3}$

11. $f(x) = x^2 - 4x; \quad c = 3$:

difference quotient:

$$\dfrac{f(3+h) - f(3)}{h} = \dfrac{(3+h)^2 - 4(3+h) - (-3)}{h}$$

$$= \dfrac{9 + 6h + h^2 - 12 - 4h + 3}{h} = \dfrac{2h + h^2}{h} = 2 + h$$

Therefore, $\quad f'(3) = \lim_{h \to 0} \dfrac{f(3+h) - f(3)}{h} = \lim_{h \to 0}(2+h) = 2$

13. $f(x) = 2x^3 + 1; \quad c = 1$:

difference quotient:

$$\dfrac{f(-1+h) - f(-1)}{h} = \dfrac{2(-1+h)^3 + 1 - (-1)}{h}$$

$$= \dfrac{2\left[-1 + 3h - 3h^2 + h^3\right] + 2}{h} = \dfrac{6h - 6h^2 + 2h^3}{h} = 6 - 6h + 2^2$$

Therefore, $\quad f'(-1) = \lim_{h \to 0} \dfrac{f(-1+h) - f(-1)}{h} = \lim_{h \to 0}(6 - 6h + 2h^2) = 6$

15. $f(x) = \dfrac{8}{x+4}; \quad c = -2$:
   difference quotient:

$$\frac{f(-2+h) - f(-2)}{h} = \frac{\dfrac{8}{(-2+h)+4} - 4}{h} = \frac{\dfrac{8}{h+2} - 4}{h}$$

$$= \frac{8 - 4h - 8}{h(h+2)} = \frac{-4}{h+2}$$

   Therefore, $f'(-2) = \lim\limits_{h\to 0} \dfrac{f(-2+h) - f(-2)}{h} = \lim\limits_{h\to 0} \dfrac{-4}{h+2} = -2$

17. $f(4) = 4$. Slope of tangent at $(4,4)$ is $f'(4) = -3$.   Tangent $y - 4 = -3(x - 4)$.

19. $f(-2) = \dfrac{1}{4}$. Slope of tangent at $(-2, \dfrac{1}{4})$ is $\dfrac{1}{4}$.   Tangent: $y - \dfrac{1}{4} = \dfrac{1}{4}(x + 2)$.

21.   (a) f is not continuous at $c = -1$ and $c = 1$; f has a removable discontinuity at $c = -1$ and a jump
            discontinuity at $c = 1$.

   (b) f is continuous but not differentiable at $c = 0$ and $c = 3$.

23.   at $x = -1$ **25.**   at $x = 0$ **27.**   at $x = 1$

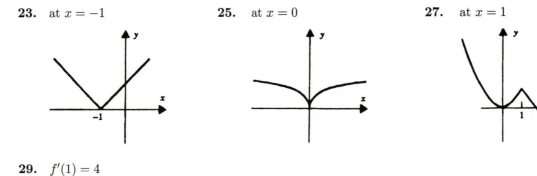

29. $f'(1) = 4$

$$\lim_{h\to 0^-} \frac{f(1+h) - f(1)}{h} = \lim_{h\to 0^-} \frac{4(1+h) - 4}{h} = 4$$

$$\lim_{h\to 0^+} \frac{f(1+h) - f(1)}{h} = \lim_{h\to 0^+} \frac{2(1+h)^2 + 2 - 4}{h} = 4$$

31. $f'(-1)$ does not exist

$$\lim_{h\to 0^-} \frac{f(-1+h) - f(-1)}{h} = \lim_{h\to 0^-} \frac{h - 0}{h} = 1$$

$$\lim_{h\to 0^+} \frac{f(-1+h) - f(-1)}{h} = \lim_{h\to 0^+} \frac{h^2 - 0}{h} = 0$$

33.   **35.**   **37.**

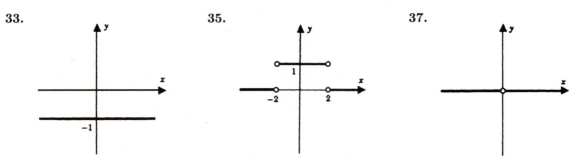

39. Since $f(1) = 1$ and $\lim\limits_{x \to 1^+} f(x) = 2$, f is not continuous at 1. Therefore, by (3.1.3), f is not differentiable at 1.

41. Continuity at $x = 1$: $\lim\limits_{x \to 1^-} f(x) = 1 = \lim\limits_{x \to 1^+} f(x) = A + B$. Thus $A + B = 1$.
Differentiability at $x = 1$:

$$\lim_{h \to 0^-} \frac{f(1+h) - f(1)}{h} = \lim_{h \to 0^-} \frac{(1+h)^3 - 1}{h} = 3 = \lim_{h \to 0^+} \frac{f(1+h) - f(1)}{h} = A$$

Therefore, $A = 3$, $\implies$ $B = -2$.

43. $f(x) = c$, c any constant.

45. $f(x) = |x + 1|$; or $f(x) = \begin{cases} 0, & x \neq -1 \\ 1, & x = -1. \end{cases}$

47. $f(x) = 2x + 5$

49. (a) $\lim\limits_{x \to 2^+} f(x) = \lim\limits_{x \to 2^-} f(x) = f(2) = 2$ Thus, f is continuous at $x = 2$.

(b) $\lim\limits_{h \to 0^-} \dfrac{f(2+h) - f(2)}{h} = \lim\limits_{h \to 0^-} \dfrac{(2+h)^2 - (2+h) - 2}{h} = 3$

$\lim\limits_{h \to 0^+} \dfrac{f(2+h) - f(2)}{h} = \lim\limits_{h \to 0^+} \dfrac{2(2+h) - 2 - 2}{h} = 2$ f is not differentiable at 2.

51. (a) f is not continuous at 0 since $\lim\limits_{x \to 0^-} f(x) = 1$ and $\lim\limits_{x \to 0^+} f(x) = 0$. Therefore f is not differentiable at 0.

(b)

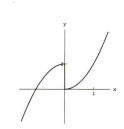

53. (a) If the tangent line is horizontal, then the normal line is vertical: $x = c$.

(b) If the tangent line has slope $f'(c) \neq 0$, then the normal line has slope $-\dfrac{1}{f'(c)}$:

$$y - f(c) = -\frac{1}{f'(c)}(x - c).$$

(c) If the tangent line is vertical, then the normal line is horizontal: $y = f(c)$.

55. The normal line has slope -4: $y - 2 = -4(x - 4)$.

57. $f(x) = x^2$, $f'(x) = 2x$
The tangent line at $(3, 9)$ has equation $y - 9 = 6(x - 3)$ and x-intercept $x = \dfrac{3}{2}$.
The normal line at $(3, 9)$ has equation $y - 9 = -\frac{1}{6}(x - 3)$ and x-intercept $x = 57$.
$s = 57 - \dfrac{3}{2} = \dfrac{111}{2}$.

59. (a) Since $\quad |\sin(1/x)| \le 1 \quad$ it follows that

$$-x \le f(x) \le x \qquad \text{and} \qquad -x^2 \le g(x) \le x^2$$

Thus $\quad \lim_{x \to 0} f(x) = f(0) = 0 \quad$ and $\quad \lim_{x \to 0} g(x) = g(0) = 0, \quad$ which implies that f and g are continuous at 0.

(b) $\lim_{h \to 0} \dfrac{h \sin(1/h) - 0}{h} = \lim_{h \to 0} \sin(1/h) \quad$ does not exist.

(c) $\lim_{h \to 0} \dfrac{h^2 \sin(1/h) - 0}{h} = \lim_{h \to 0} h \sin(1/h) = 0.$ Thus g is differentiable at 0 and $g'(0) = 0.$

61. $f'(1) = \lim_{h \to 0} \dfrac{f(1+h) - f(1)}{h} = \lim_{h \to 0} \dfrac{[(1+h)^2 - 3(1+h)] - (-2)]}{h} = \lim_{h \to 0} \dfrac{-h + h^2}{h} = -1$

$f'(1) = \lim_{x \to 1} \dfrac{f(x) - f(1)}{x - 1} = \lim_{x \to 1} \dfrac{(x^2 - 3x) - (-2)}{x - 1} = \lim_{x \to 1} \dfrac{(x-2)(x-1)}{x-1} = \lim_{x \to 1} (x - 2) = -1$

63. $f'(-1) = \lim_{h \to 0} \dfrac{f(-1+h) - f(-1)}{h} = \lim_{h \to 0} \dfrac{(-1+h)^{1/3} + 1}{h}$

$= \lim_{h \to 0} \dfrac{(-1+h)^{1/3} + 1}{h} \cdot \dfrac{(-1+h)^{2/3} - (-1+h)^{1/3} + 1}{(-1+h)^{2/3} - (-1+h)^{1/3} + 1}$

$= \lim_{h \to 0} \dfrac{h}{h\left[-1+h)^{2/3} - (-1+h)^{1/3} + 1\right]} = \dfrac{1}{3}$

$f'(-1) = \lim_{x \to -1} \dfrac{f(x) - f(-1)}{x - (-1)} = \lim_{x \to -1} \dfrac{x^{1/3} + 1}{x + 1} = \lim_{x \to -1} \dfrac{x^{1/3} + 1}{x + 1} \cdot \dfrac{x^{2/3} - x^{1/3} + 1}{x^{2/3} - x^{1/3} + 1}$

$= \lim_{x \to -1} \dfrac{x + 1}{(x + 1)(x^{2/3} - x^{1/3} + 1)} = \dfrac{1}{3}$

65. (a) $D = \dfrac{(2+h)^{5/2} - 2^{5/2}}{h} \qquad -1 \le h \le 1$

(b) $f'(2) \cong 7.071$

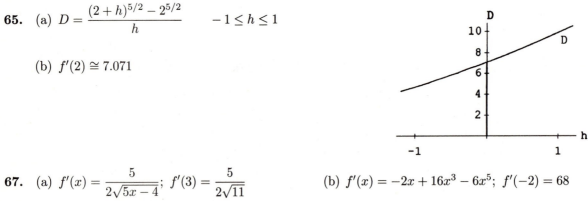

67. (a) $f'(x) = \dfrac{5}{2\sqrt{5x - 4}};\ f'(3) = \dfrac{5}{2\sqrt{11}}$ (b) $f'(x) = -2x + 16x^3 - 6x^5;\ f'(-2) = 68$

(c) $f'(x) = -\dfrac{3(3 - 2x)}{(2 + 3x)^2} - \dfrac{2}{2 + 3x};\ f'(-1) = 13$

69. (c) $f'(x) = 10x - 21x^2$ (d) $f'(x) = 0$ at $x = 0, \frac{10}{21}$

71. (a) $f'(\frac{3}{2}) = -\frac{11}{4};\quad$ tangent line: $y - \frac{21}{8} = -\frac{11}{4}\left(x - \frac{3}{2}\right),\quad$ normal line: $y - \frac{21}{8} = \frac{4}{11}\left(x - \frac{3}{2}\right)$

(c) $(1.453, 1.547)$

SECTION 3.2

1. $F'(x) = -1$ $\qquad$ **3.** $F'(x) = 55x^4 - 18x^2$ $\qquad$ **5.** $F'(x) = 2ax + b$

7. $F'(x) = 2x^{-3}$ $\qquad$ **9.** $G'(x) = (x^2 - 1)(1) + (x - 3)(2x) = 3x^2 - 6x - 1$

11. $G'(x) = \dfrac{(1-x)(3x^2) - x^3(-1)}{(1-x)^2} = \dfrac{3x^2 - 2x^3}{(1-x)^2}$

13. $G'(x) = \dfrac{(2x+3)(2x) - (x^2 - 1)(2)}{(2x+3)^2} = \dfrac{2(x^2 + 3x + 1)}{(2x+3)^2}$

15. $G'(x) = (x^3 - 2x)(2) + (2x + 5)(3x^2 - 2) = 8x^3 + 15x^2 - 8x - 10$

17. $G'(x) = \dfrac{(x-2)(1/x^2) - (6 - 1/x)(1)}{(x-2)^2} = \dfrac{-2(3x^2 - x + 1)}{x^2(x-2)^2}$

19. $G'(x) = (9x^8 - 8x^9)\left(1 - \dfrac{1}{x^2}\right) + \left(x + \dfrac{1}{x}\right)(72x^7 - 72x^8) = -80x^9 + 81x^8 - 64x^7 + 63x^6$

21. $f'(x) = -(x-2)^{-2}; \quad f'(0) = -\frac{1}{4}, \quad f'(1) = -1$

23. $f'(x) = \dfrac{(1+x^2)(-2x) - (1-x^2)(2x)}{(1+x^2)^2} = \dfrac{-4x}{(1+x^2)^2}; \quad f'(0) = 0, \quad f'(1) = -1$

25. $f'(x) = \dfrac{(cx+d)a - (ax+b)c}{(cx+d)^2} = \dfrac{ad - bc}{(cx+d)^2}, \quad f'(0) = \dfrac{ad-bc}{d^2}, \quad f'(1) = \dfrac{ad-bc}{(c+d)^2}$

27. $f'(x) = xh'(x) + h(x); \quad f'(0) = 0h'(0) + h(0) = 0(2) + 3 = 3$

29. $f'(x) = h'(x) + \dfrac{h'(x)}{[h(x)]^2}, \quad f'(0) = h'(0) + \dfrac{h'(0)}{[h(0)]^2} = 2 + \dfrac{2}{3^2} = \dfrac{20}{9}$

31. $f'(x) = \dfrac{(x+2)(1) - x(1)}{(x+2)^2} = \dfrac{2}{(x+2)^2}$,

slope of tangent at $(-4, 2): f'(-4) = \frac{1}{2}$; $\qquad$ equation for tangent: $y - 2 = \frac{1}{2}(x + 4)$

33. $f'(x) = (x^2 - 3)(5 - 3x^2) + (5x - x^3)(2x)$;

slope of tangent at $(1, -8): f'(1) = (-2)(2) + (4)(2) = 4$; $\qquad$ equation for tangent: $y + 8 = 4(x - 1)$

35. $f'(x) = (x - 2)(2x - 1) + (x^2 - x - 11)(1) = 3(x - 3)(x + 1)$;

$f'(x) = 0$ at $x = -1, 3$; $(-1, 27), (3, -5)$

37. $f'(x) = \dfrac{(x^2 + 1)(5) - 5x(2x)}{(x^2 + 1)^2} = \dfrac{5(1 - x^2)}{(x^2 + 1)^2}, \quad f'(x) = 0$ at $x = \pm 1; (-1, -5/2), (1, 5/2)$

39. $f(x) = x^4 - 8x^2 + 3; \quad f'(x) = 4x^3 - 16x = 4x(x^2 - 4) = 4x(x - 2)(x + 2)$

(a) $f'(x) = 0$ at $x = 0, \pm 2$

(b) $f'(x) > 0$ on $(-2, 0) \cup (2, \infty)$

(c) $f'(x) < 0$ on $(-\infty, -2) \cup (0, 2)$

41. $f(x) = x + \dfrac{4}{x^2}; \quad f'(x) = 1 - \dfrac{8}{x^3} = \dfrac{x^3 - 8}{x^3} = \dfrac{(x-2)(x^2 + 2x + 4)}{x^3}$

(a) $f'(x) = 0$ at $x = 2$.

(b) $f'(x) > 0$ on $(-\infty, 0) \cup (2, \infty)$

(c) $f'(x) < 0$ on $(0, 2)$

43. slope of line 4; slope of tangent $f'(x) = -2x$; $-2x = 4$ at $x = -2$; $(-2, -10)$

45. $f(x) = x^3 + x^2 + x + C$ \qquad\qquad **47.** $f(x) = \dfrac{2x^3}{3} - \dfrac{3x^2}{2} + \dfrac{1}{x} + C$

49. We want f to be continuous at $x = 2$. That is, we want

$$\lim_{x \to 2^-} f(x) = f(2) = \lim_{x \to 2^+} f(x).$$

This gives

(1) \qquad\qquad $8A + 2B + 2 = 4B - A.$

We also want

$$\lim_{x \to 2^-} f'(x) = \lim_{x \to 2^+} f'(x).$$

This gives

(2) \qquad\qquad $12A + B = 4B.$

Equations (1) and (2) together imply that $A = -2$ and $B = -8$.

51.

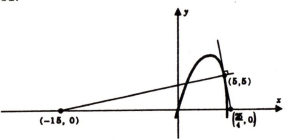

slope of tangent at $(5, 5)$ is $f'(5) = -4$

tangent $y - 5 = -4(x - 5)$ intersects

$\quad$ x-axis at $\left(\frac{25}{4}, 0\right)$

normal $y - 5 = \frac{1}{4}(x - 5)$ intersects

$\quad$ x-axis at $(-15, 0)$

area of triangle is

$$\tfrac{1}{2}(5)\left(15 + \tfrac{25}{4}\right) = \tfrac{425}{8}$$

$(-15, 0)$ \qquad (5,5) \qquad $\left(\frac{25}{4}, 0\right)$

53. If the point $(1, 3)$ lies on the graph, we have $f(1) = 3$ and thus

(∗) \qquad\qquad $A + B + C = 3.$

If the line $4x + y = 8$ (slope -4) is tangent to the graph at $(2, 0)$, then $f(2) = 0$ and $f'(2) = -4$. Thus,

(∗∗) \qquad\qquad $4A + 2B + C = 0 \quad$ and $\quad 4A + B = -4.$

Solving the equations in (∗) and (∗∗), we find that $A = -1, \quad B = 0, \quad C = 4.$

55. Let $f(x) = ax^2 + bx + c$. Then $f'(x) = 2ax + b$ and $f'(x) = 0$ at $x = -b/2a$.

57. Let $f(x) = x^3 - x$. The secant line through $(-1, f(-1)) = (-1, 0)$ and $(2, f(2)) = (2, 6)$ has slope

$m = \dfrac{6 - 0}{2 - (-1)} = 2.$ Now, $f'(x) = 3x^2 - 1$ and $3c^2 - 1 = 2$ implies $c = -1, 1.$

59. Let $f(x) = 1/x$, $x > 0$. Then $f'(x) = -1/x^2$. An equation for the tangent line to the graph of f at the point $(a, f(a))$, $a > 0$, is $y = (-1/a^2)x + 2/a$. The y-intercept is $2/a$ and the x-intercept is $2a$. The area of the triangle formed by this line and the coordinate axes is: $A = \frac{1}{2}(2/a)(2a) = 2$ square units.

61. Let (x, y) be the point on the graph that the tangent line passes through. $f'(x) = 3x^2 - 1$, so $x^3 - x - 2 = (3x^2 - 1)(x + 2)$. Thus $x = 0$ or $x = -3$. The lines are $y = -x$ and $y + 24 = 26(x + 3)$.

63. Since f and $f + g$ are differentiable, $g = (f + g) - f$ is differentiable. The functions $f(x) = |x|$ and $g(x) = -|x|$ are not differentiable at $x = 0$ yet their sum $f(x) + g(x) \equiv 0$ is differentiable for all x.

65. Since

$$\left(\frac{f}{g}\right)(x) = \frac{f(x)}{g(x)} = f(x) \cdot \frac{1}{g(x)},$$

it follows from the product and reciprocal rules that

$$\left(\frac{f}{g}\right)'(x) = \left(f \cdot \frac{1}{g}\right)'(x) = f(x)\left(-\frac{g'(x)}{[g(x)]^2}\right) + f'(x) \cdot \frac{1}{g(x)} = \frac{g(x)f'(x) - f(x)g'(x)}{[g(x)]^2}.$$

67. $F'(x) = 2x\left(1 + \dfrac{1}{x}\right)(2x^3 - x + 1) + (x^2 + 1)\left(\dfrac{-1}{x^2}\right)(2x^3 - x + 1) + (x^2 + 1)\left(1 + \dfrac{1}{x}\right)(6x^2 - 1)$

69. $g(x) = [f(x)]^2 = f(x) \cdot f(x)$
$g'(x) = f(x)f'(x) + f(x)f'(x) = 2f(x)f'(x)$

71. (a) $f'(x) = 0$ at $x = 0, -2$ (b) $f'(x) > 0$ on $(-\infty, -2) \cup (0, \infty)$
(c) $f'(x) < 0$ on $(-2, -1) \cup (-1, 0)$

73. (a) $f'(x) \neq 0$ for all $x \neq 0$ (b) $f'(x) > 0$ on $(0, \infty)$
(c) $f'(x) < 0$ on $(-\infty, 0)$

75. (a) $\dfrac{\sin(0 + 0.001) - \sin 0}{0.001} \cong 0.99999$ $\dfrac{\sin(0 - 0.001) - \sin 0}{-0.001} \cong 0.99999$

$\dfrac{\sin[(\pi/6) + 0.001] - \sin(\pi/6)}{0.001} \cong 0.86578$ $\dfrac{\sin[(\pi/6) - 0.001] - \sin(\pi/6)}{-0.001} \cong 0.86628$

$\dfrac{\sin[(\pi/4) + 0.001] - \sin(\pi/4)}{0.001} \cong 0.70675$ $\dfrac{\sin[(\pi/4) - 0.001] - \sin(\pi/4)}{-0.001} \cong 0.70746$

$\dfrac{\sin[(\pi/3) + 0.001] - \sin(\pi/3)}{0.001} \cong 0.49957$ $\dfrac{\sin[(\pi/3) - 0.001] - \sin(\pi/3)}{-0.001} \cong 0.50043$

$\dfrac{\sin[(\pi/2) + 0.001] - \sin(\pi/2)}{0.001} \cong -0.0005$ $\dfrac{\sin[(\pi/2) - 0.001] - \sin(\pi/2)}{-0.001} \cong 0.0005$

(b) $\cos 0 = 1$, $\cos(\pi/6) \cong 0.866025$, $\cos(\pi/4) \cong 0.707107$, $\cos(\pi/3) = 0.5$, $\cos(\pi/2) = 0$

(c) If $f(x) = \sin x$ then $f'(x) = \cos x$.

SECTION 3.3

1. $\dfrac{dy}{dx} = 12x^3 - 2x$

3. $\dfrac{dy}{dx} = 1 + \dfrac{1}{x^2}$

5. $\dfrac{dy}{dx} = \dfrac{(1+x^2)(1) - x(2x)}{(1+x^2)^2} = \dfrac{1-x^2}{(1+x^2)^2}$

7. $\dfrac{dy}{dx} = \dfrac{(1-x)2x - x^2(-1)}{(1-x)^2} = \dfrac{x(2-x)}{(1-x)^2}$

9. $\dfrac{dy}{dx} = \dfrac{(x^3-1)3x^2 - (x^3+1)3x^2}{(x^3-1)^2} = \dfrac{-6x^2}{(x^3-1)^2}$

11. $\dfrac{d}{dx}(2x-5) = 2$

13. $\dfrac{d}{dx}[(3x^2 - x^{-1})(2x+5)] = (3x^2 - x^{-1})2 + (2x+5)(6x + x^{-2}) = 18x^2 + 30x + 5x^{-2}$

15. $\dfrac{d}{dt}\left(\dfrac{t^4}{2t^3 - 1}\right) = \dfrac{(2t^3-1)4t^3 - t^4(6t^2)}{(2t^3-1)^2} = \dfrac{2t^3(t^3-2)}{(2t^3-1)^2}$

17. $\dfrac{d}{du}\left(\dfrac{2u}{1-2u}\right) = \dfrac{(1-2u)2 - 2u(-2)}{(1-2u)^2} = \dfrac{2}{(1-2u)^2}$

19. $\dfrac{d}{du}\left(\dfrac{u}{u-1} - \dfrac{u}{u+1}\right) = \dfrac{(u-1)(1) - u}{(u-1)^2} - \dfrac{(u+1)(1) - u}{(u+1)^2}$

$$= -\dfrac{1}{(u-1)^2} - \dfrac{1}{(u+1)^2} = -\dfrac{2(1+u^2)}{(u^2-1)^2}$$

21. $\dfrac{d}{dx}\left(\dfrac{x^3+x^2+x+1}{x^3-x^2+x-1}\right) = \dfrac{(x^3-x^2+x-1)(3x^2+2x+1) - (x^3+x^2+x+1)(3x^2-2x+1)}{(x^3-x^2+x-1)^2}$

$$= \dfrac{-2(x^4+2x^2+1)}{(x^2+1)^2(x-1)^2} = \dfrac{-2}{(x-1)^2}$$

23. $\dfrac{dy}{dx} = (x+1)\dfrac{d}{dx}[(x+2)(x+3)] + (x+2)(x+3)\dfrac{d}{dx}(x+1) = (x+1)(2x+5) + (x+2)(x+3)$

At $x = 2$, $\dfrac{dy}{dx} = (3)(9) + (4)(5) = 47$.

25. $\dfrac{dy}{dx} = \dfrac{(x+2)\dfrac{d}{dx}[(x-1)(x-2)] - (x-1)(x-2)(1)}{(x+2)^2}$

$$= \dfrac{(x+2)(2x-3) - (x-1)(x-2)}{(x+2)^2}$$

At $x = 2$, $\dfrac{dy}{dx} = \dfrac{4(1) - 1(0)}{16} = \dfrac{1}{4}$.

27. $f'(x) = 21x^2 - 30x^4,\ f''(x) = 42x - 120x^3$

29. $f'(x) = 1 + 3x^{-2},\ f''(x) = -6x^{-3}$

31. $f(x) = 2x^2 - 2x^{-2} - 3,\quad f'(x) = 4x + 4x^{-3},\quad f''(x) = 4 - 12x^{-4}$

33. $\dfrac{dy}{dx} = x^2 + x + 1$

$\dfrac{d^2y}{dx^2} = 2x + 1$

$\dfrac{d^3y}{dx^3} = 2$

35. $\dfrac{dy}{dx} = 8x - 20$

$\dfrac{d^2y}{dx^2} = 8$

$\dfrac{d^3y}{dx^3} = 0$

37. $\dfrac{dy}{dx} = 3x^2 + 3x^{-4}$

$\dfrac{d^2y}{dx^2} = 6x - 12x^{-5}$

$\dfrac{d^3y}{dx^3} = 6 + 60x^{-6}$

39. $\dfrac{d}{dx}\left[x\dfrac{d}{dx}\left(x - x^2\right)\right] = \dfrac{d}{dx}\left[x(1 - 2x)\right] = \dfrac{d}{dx}\left[x - 2x^2\right] = 1 - 4x$

41. $\dfrac{d^4}{dx^4}\left[3x - x^4\right] = \dfrac{d^3}{dx^3}\left[3 - 4x^3\right] = \dfrac{d^2}{dx^2}\left[-12x^2\right] = \dfrac{d}{dx}\left[-24x\right] = -24$

43. $\dfrac{d^2}{dx^2}\left[(1 + 2x)\dfrac{d^2}{dx^2}\left(5 - x^3\right)\right] = \dfrac{d^2}{dx^2}\left[(1 + 2x)(-6x)\right] = \dfrac{d^2}{dx^2}\left[-6x - 12x^2\right] = -24$

45. $y = x^4 - \dfrac{x^3}{3} + 2x^2 + C$

47. $y = x^5 - \dfrac{1}{x^4} + C$

49. Let $p(x) = ax^2 + bx + c$. Then $p'(x) = 2ax + b$ and $p''(x) = 2a$. Now

$$p''(1) = 2a = 4 \Longrightarrow a = 2; \quad p'(1) = 2(2)(1) + b = -2 \Longrightarrow b = -6;$$

$$p(1) = 2(1)^2 - 6(1) + c = 3 \Longrightarrow c = 7$$

Thus $p(x) = 2x^2 - 6x + 7$.

51. (a) If $k = n$, $f^{(n)}(x) = n!$

(b) If $k > n$, $f^{(k)}(x) = 0$.

(c) If $k < n$, $f^{(k)}(x) = n(n - 1)(n - 2)\cdots(n - k + 1)x^{n-k}$.

53. Let $f(x) = \begin{cases} x^2 & x \geq 0 \\ 0 & x \leq 0 \end{cases}$

(a) $f'_+(0) = \lim\limits_{h \to 0^+} \dfrac{f(0 + h) - f(0)}{h} = \lim\limits_{h \to 0^+} \dfrac{h^2 - 0}{h} = 0,$

$f'_-(0) = \lim\limits_{h \to 0^-} \dfrac{f(0 + h) - f(0)}{h} = \lim\limits_{h \to 0^-} \dfrac{0}{h} = 0$

Therefore, f is differentiable at 0 and $f'(0) = 0$.

(b) $f'(x) = \begin{cases} 2x & x \geq 0 \\ 0 & x \leq 0 \end{cases}$

(d)

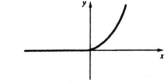

(c) $f''_+(0) = \lim\limits_{h \to 0^+} \dfrac{f'(0 + h) - f'(0)}{h} = \lim\limits_{h \to 0^+} \dfrac{2h - 0}{h} = 2,$

$f''_-(0) = \lim\limits_{h \to 0^-} \dfrac{f'(0 + h) - f'(0)}{h} = \lim\limits_{h \to 0^-} \dfrac{0}{h} = 0$

Since $f''_+(0) \neq f''_-(0),$ $f''(0)$ does not exist.

55. It suffices to give a single counterexample. For instance, if

$f(x) = g(x) = x$, then $(fg)(x) = x^2$ so that $(fg)''(x) = 2$ but

$f(x)g''(x) + f''(x)g(x) = x \cdot 0 + 0 \cdot x = 0$.

57. $f''(x) = 6x$;   (a) $x = 0$ (b) $x > 0$ (c) $x < 0$

59. $f''(x) = 12x^2 + 12x - 24$;   (a) $x = -2, 1$ (b) $x < -2$, $x > 1$ (c) $-2 < x < 1$

61. The result is true for $n = 1$:

$$\frac{d^1 y}{dx^1} = \frac{dy}{dx} = -x^{-2} = (-1)^1 1! \, x^{-1-1}.$$

If the result is true for $n = k$:

$$\frac{d^k y}{dx^k} = (-1)^k k! \, x^{-k-1}$$

then the result is true for $n = k + 1$:

$$\frac{d^{k+1} y}{dx^{k+1}} = \frac{d}{dx}\left[\frac{d^k y}{dx^k}\right] = \frac{d}{dx}\left[(-1)^k k! \, x^{-(k+1)}\right] = (-1)^{(k+1)}(k+1)! \, x^{-(k+1)-1}.$$

63. $\dfrac{d}{dx}(uvw) = uv\dfrac{dw}{dx} + uw\dfrac{dv}{dx} + vw\dfrac{du}{dx}$

65. By the reciprocal rule, $\dfrac{d}{dx}\left[\dfrac{1}{1-x}\right] = \dfrac{-(-1)}{(1-x)^2} = \dfrac{1}{(1-x)^2}$.

$$\frac{d^2}{dx^2}\left[\frac{1}{1-x}\right] = \frac{d}{dx}\left[\frac{1}{(1-x)^2}\right]$$

$$= \frac{d}{dx}\left[\frac{1}{1-x} \cdot \frac{1}{1-x}\right] = \frac{1}{1-x} \cdot \frac{1}{(1-x)^2} + \frac{1}{1-x} \cdot \frac{1}{(1-x)^2}$$

$$= \frac{2}{(1-x)^3}$$

$$\frac{d^3}{dx^3}\left[\frac{1}{1-x}\right] = \frac{d}{dx}\left[\frac{2}{(1-x)^3}\right]$$

$$= 2\frac{d}{dx}\left[\frac{1}{1-x} \cdot \frac{1}{(1-x)^2}\right] = 2\frac{1}{1-x} \cdot \frac{2}{(1-x)^3} + 2\frac{1}{(1-x)^2} \cdot \frac{1}{(1-x)^2}$$

$$= \frac{6}{(1-x)^4} = \frac{3!}{(1-x)^4}$$

and so on.

In general, $\dfrac{d^n}{dx^n}\left[\dfrac{1}{1-x}\right] = \dfrac{n!}{(1-x)^{n+1}}$. You can use induction to prove this result.

67. (b) The lines tangent to the graph of f are parallel to the line $x - 2y + 12 = 0$ at the points $\left(\dfrac{-1}{\sqrt{2}}, \dfrac{1}{2\sqrt{2}}\right)$ and $\left(\dfrac{1}{\sqrt{2}}, \dfrac{-1}{2\sqrt{2}}\right)$.

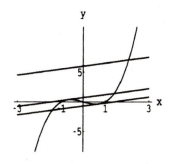

69. (a) $f(x) = x^3 + x^2 - 4x + 1$; $f'(x) = 3x^2 + 2x - 4$.

(b)

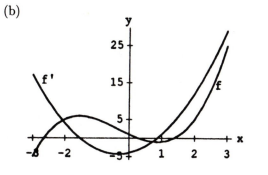

(c) The graph is "falling" when $f'(x) < 0$; the graph is "rising" when $f'(x) > 0$.

71. (a) $f(x) = \frac{1}{2}x^3 - 3x^2 + 3x + 3$; $f'(x) = \frac{3}{2}x^2 - 6x + 3$

(b)

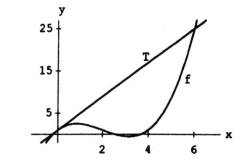

(c) The line tangent to the graph is horizontal; the graph turns from rising to falling, or from falling to rising.

(d) $x_1 \cong 0.586, \;\; x_2 \cong 3.414$

SECTION 3.4

1. $A = \pi r^2$, $\dfrac{dA}{dr} = 2\pi r$. When $r = 2$, $\dfrac{dA}{dr} = 4\pi$.   **3.** $A = \dfrac{1}{2}z^2$, $\dfrac{dA}{dz} = z$. When $z = 4$, $\dfrac{dA}{dz} = 4$.

5. $y = \dfrac{1}{x(1+x)}$, $\dfrac{dy}{dx} = \dfrac{-(2x+1)}{x^2(1+x)^2}$.   At $x = 2$, $\dfrac{dy}{dx} = -\dfrac{5}{36}$.

7. $V = \frac{4}{3}\pi r^3$, $\frac{dV}{dr} = 4\pi r^2 =$ the surface area of the ball.

9. $y = 2x^2 + x - 1$, $\frac{dy}{dx} = 4x + 1$; $\frac{dy}{dx} = 4$ at $x = \frac{3}{4}$.   Therefore $x_0 = \frac{3}{4}$.

11. (a) $w = s\sqrt{2}$, $V = s^3 = \left(\frac{w}{\sqrt{2}}\right)^3 = \frac{\sqrt{2}}{4}w^3$, $\frac{dV}{dw} = \frac{3\sqrt{2}}{4}w^2$.

   (b) $z^2 = s^2 + w^2 = 3s^2$, $z = s\sqrt{3}$. $V = s^3 = \left(\frac{z}{\sqrt{3}}\right)^3 = \frac{\sqrt{3}}{9}z^3$, $\frac{dV}{dz} = \frac{\sqrt{3}}{3}z^2$.

13. (a) $\frac{dA}{d\theta} = \frac{1}{2}r^2$ (b) $\frac{dA}{dr} = r\theta$

   (c) $\theta = \frac{2A}{r^2}$ so $\frac{d\theta}{dr} = \frac{-4A}{r^3} = \frac{-4}{r^3}\left(\frac{1}{2}r^2\theta\right) = \frac{-2\theta}{r}$

15. $y = ax^2 + bx + c$, $z = bx^2 + ax + c$.

$\frac{dy}{dx} = 2ax + b$, $\frac{dz}{dx} = 2bx + a$.

$\frac{dy}{dx} = \frac{dz}{dx}$ iff $2ax + b = 2bx + a$. With $a \neq b$, this occurs only at $x = \frac{1}{2}$.

SECTION 3.5

1. $y = x^4 + 2x^2 + 1$, $y' = 4x^3 + 4x = 4x(x^2 + 1)$
   $y = (x^2 + 1)^2$, $y' = 2(x^2 + 1)(2x) = 4x(x^2 + 1)$

3. $y = 8x^3 + 12x^2 + 6x + 1$, $y' = 24x^2 + 24x + 6 = 6(2x + 1)^2$
   $y = (2x + 1)^3$, $y' = 3(2x + 1)^2(2) = 6(2x + 1)^2$

5. $y = x^2 + 2 + x^{-2}$, $y' = 2x - 2x^{-3} = 2x(1 - x^{-4})$
   $y = (x + x^{-1})^2$, $y' = 2(x + x^{-1})(1 - x^{-2}) = 2x(1 + x^{-2})(1 - x^{-2}) = 2x(1 - x^{-4})$

7. $f'(x) = -1(1 - 2x)^{-2}\frac{d}{dx}(1 - 2x) = 2(1 - 2x)^{-2}$

9. $f'(x) = 20(x^5 - x^{10})^{19}\frac{d}{dx}(x^5 - x^{10}) = 20(x^5 - x^{10})^{19}(5x^4 - 10x^9)$

11. $f'(x) = 4\left(x - \frac{1}{x}\right)^3\frac{d}{dx}\left(x - \frac{1}{x}\right) = 4\left(x - \frac{1}{x}\right)^3\left(1 + \frac{1}{x^2}\right)$

13. $f'(x) = 4(x - x^3 + x^5)^3\frac{d}{dx}(x - x^3 + x^5) = 4(x - x^3 + x^5)^3(1 - 3x^2 + 5x^4)$

15. $f'(t) = 4(t^{-1} + t^{-2})^3\frac{d}{dt}(t^{-1} + t^{-2}) = 4(t^{-1} + t^{-2})^3(-t^{-2} - 2t^{-3})$

17. $f'(x) = 4\left(\frac{3x}{x^2 + 1}\right)^3\frac{d}{dx}\left(\frac{3x}{x^2 + 1}\right) = 4\left(\frac{3x}{x^2 + 1}\right)^3\left[\frac{(x^2 + 1)3 - 3x(2x)}{(x^2 + 1)^2}\right] = \frac{324x^3(1 - x^2)}{(x^2 + 1)^5}$

19. $f'(x) = -\left(\dfrac{x^3}{3} + \dfrac{x^2}{2} + \dfrac{x}{1}\right)^{-2} \dfrac{d}{dx}\left(\dfrac{x^3}{3} + \dfrac{x^2}{2} + \dfrac{x}{1}\right) = -\left(\dfrac{x^3}{3} + \dfrac{x^2}{2} + x\right)^{-2}(x^2 + x + 1)$

21. $\dfrac{dy}{dx} = \dfrac{dy}{du}\dfrac{du}{dx} = \dfrac{-2u}{(1+u^2)^2} \cdot (2)$

At $x = 0$, we have $u = 1$ and thus $\dfrac{dy}{dx} = \dfrac{-4}{4} = -1.$

23. $\dfrac{dy}{dx} = \dfrac{dy}{du}\dfrac{du}{dx} = \dfrac{(1-4u)2 - 2u(-4)}{(1-4u)^2} \cdot 4(5x^2+1)^3(10x) = \dfrac{2}{(1-4u)^2} \cdot 40x(5x^2+1)^3$

At $x = 0$, we have $u = 1$ and thus $\dfrac{dy}{dx} = \dfrac{2}{9}(0) = 0.$

25. $\dfrac{dy}{dt} = \dfrac{dy}{du}\dfrac{du}{dx}\dfrac{dx}{dt} = \dfrac{(1+u^2)(-7) - (1-7u)(2u)}{(1+u^2)^2}(2x)(2)$

$= \dfrac{7u^2 - 2u - 7}{(1+u^2)^2}(4x) = \dfrac{4x(7x^4 + 12x^2 - 2)}{(x^4 + 2x^2 + 2)^2} = \dfrac{4(2t-5)[7(2t-5)^4 + 12(2t-5)^2 - 2]}{[(2t-5)^4 + 2(2t-5)^2 + 2]^2}$

27. $\dfrac{dy}{dx} = \dfrac{dy}{ds}\dfrac{ds}{dt}\dfrac{dt}{dx} = 2(s+3) \cdot \dfrac{1}{2\sqrt{t-3}} \cdot (2x)$

At $x = 2$, we have $t = 4$ so that $s = 1$ and thus $\dfrac{dy}{dx} = 2(4)\dfrac{1}{2\cdot 1}(4) = 16.$

29. $(f \circ g)'(0) = f'(g(0))g'(0) = f'(2)g'(0) = (1)(1) = 1$

31. $(f \circ g)'(2) = f'(g(2))g'(2) = f'(2)g'(2) = (1)(1) = 1$

33. $(g \circ f)'(1) = g'(f(1))f'(1) = g'(0)f'(1) = (1)(1) = 1$

35. $(f \circ h)'(0) = f'(h(0))h'(0) = f'(1)h'(0) = (1)(2) = 2$

37. $(g \circ f \circ h)'(2) = g'(f(h(2)))\ f'(h(2))h'(2) = g'(1)f'(0)h'(2) = (0)(2)(2) = 0$

39. $f'(x) = 4(x^3 + x)^3(3x^2 + 1)$

$f''(x) = 3(4)(x^3 + x)^2(3x^2 + 1)^2 + 4(x^3 + x)^3(6x) = 12(x^3 + x)^2[(3x^2 + 1)^2 + 2x(x^3 + x)]$

41. $f'(x) = 3\left(\dfrac{x}{1-x}\right)^2 \cdot \dfrac{1}{(1-x)^2} = \dfrac{3x^2}{(1-x)^4}$

$f''(x) = \dfrac{6x(1-x)^4 - 3x^2(4)(1-x)^3(-1)}{(1-x)^8} = \dfrac{6x(1+x)}{(1-x)^5}$

43. $2xf'(x^2 + 1)$ **45.** $2f(x)f'(x)$

47. $f'(x) = -4x(1+x^2)^{-3};$ (a) $x = 0$ (b) $x < 0$ (c) $x > 0$

49. $f'(x) = \dfrac{1-x^2}{(1+x^2)^2};$ (a) $x = \pm 1$ (b) $-1 < x < 1$ (c) $x < -1, \ x > 1$

51. $\dfrac{n!}{(1-x)^{n+1}}$

53. $n!\,b^n$

55. $y = (x^2+1)^3 + C$

57. $y = (x^3-2)^2 + C$

59. $L'(x) = \dfrac{1}{x^2+1}\cdot 2x = \dfrac{2x}{x^2+1}$

61. $T'(x) = 2f(x)\cdot f'(x) + 2g(x)\cdot g'(x) = 2f(x)\cdot g(x) - 2g(x)\cdot f(x) = 0$

63.   Suppose $P(x) = (x-a)^2 q(x)$, where $q(a)\neq 0$. Then

$$P'(x) = 2(x-a)q(x) + (x-a)^2 q'(x) \quad\text{and}\quad P''(x) = 2q(x) + 4(x-a)q'(x) + (x-a)^2 q''(x),$$

and it follows that $P(a) = P'(a) = 0$, and $P''(a)\neq 0$.

Now suppose that $P(a) = P'(a) = 0$ and $P''(a)\neq 0$.

$$P(a) = 0 \quad\Longrightarrow\quad P(x) = (x-a)g(x) \quad\text{for some polynomial } g.$$

Then $P'(x) = g(x) + (x-a)g'(x)$ and

$$P'(a) = 0 \quad\Longrightarrow\quad g(a) = 0 \text{ and so } g(x) = (x-a)q(x) \text{ for some polynomial } q.$$

Therefore, $P(x) = (x-a)^2 q(x)$. Finally, $P''(a)\neq 0$ implies $q(a)\neq 0$.

65.   Let P be a polynomial function of degree n. The number a is a root of P of multiplicity k, $(k<n)$ if and only if $P(a) = P'(a) = \cdots = P^{(k-1)}(a) = 0$ and $P^{(k)}(a)\neq 0$.

67. $V = \frac{4}{3}\pi r^3$ and $\dfrac{dr}{dt} = 2$ cm/sec. By the chain rule, $\dfrac{dV}{dt} = \dfrac{dV}{dr}\dfrac{dr}{dt} = 4\pi r^2\dfrac{dr}{dt} = 8\pi r^2.$

At the instant the radius is 10 centimeters, the volume is increasing at the rate

$$\dfrac{dV}{dt} = 8\pi(10)^2 = 800\pi \ \ \text{cm}^3/\text{sec}.$$

69.   (a) $\dfrac{dF}{dt} = \dfrac{dF}{dr}\cdot\dfrac{dr}{dt} = \dfrac{2k}{r^3}\cdot(49-9.8t) = \dfrac{2k}{(49t-4.9t^2)^3}(49-9.8t),\ \ 0\le t\le 10$

   (b) $\dfrac{dF}{dt}(3) = \dfrac{2k}{(102.9)^3}(19.6); \quad \dfrac{dF}{dt}(7) = -\dfrac{2k}{(102.9)^3}(19.6).$

71.   (a) $f(1) = \frac{1}{2},\ f'(1) = -\frac{1}{2};$ tangent $T: y = -\frac{1}{2}x + 1.$

   (b)

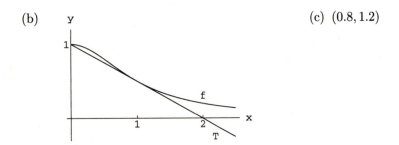

   (c) $(0.8, 1.2)$

73. (a) $\dfrac{d}{dx}\left[f\left(\dfrac{1}{x}\right)\right] = -\dfrac{f'(1/x)}{x^2}$ (b) $\dfrac{d}{dx}\left[f\left(\dfrac{x^2-1}{x^2+1}\right)\right] = \dfrac{4x\, f'\left[(x^2-1)/(x^2+1)\right]}{(1+x^2)^2}$

(c) $\dfrac{d}{dx}\left[\dfrac{f(x)}{1+f(x)}\right] = \dfrac{f'(x)}{[1+f(x)]^2}$

75. $\dfrac{d^2}{dx^2}\left[f(g(x))\right] = [g'(x)]^2\, f''[g(x)] + f'[g(x)]g''(x)$

SECTION 3.6

1. $\dfrac{dy}{dx} = -3\sin x - 4\sec x \tan x$ **3.** $\dfrac{dy}{dx} = 3x^2 \csc x - x^3 \csc x \cot x$

5. $\dfrac{dy}{dt} = -2\cos t \sin t$

7. $\dfrac{dy}{du} = 4\sin^3\sqrt{u}\,\dfrac{d}{du}\left(\sin\sqrt{u}\right) = 4\sin^3\sqrt{u}\,\cos\sqrt{u}\,\dfrac{d}{du}\left(\sqrt{u}\right) = 2u^{-1/2}\sin^3\sqrt{u}\,\cos\sqrt{u}$

9. $\dfrac{dy}{dx} = \sec^2 x^2\,\dfrac{d}{dx}\left(x^2\right) = 2x\sec^2 x^2$ **11.** $\dfrac{dy}{dx} = 4[x+\cot\pi x]^3[1-\pi\csc^2\pi x]$

13. $\dfrac{dy}{dx} = \cos x,\ \dfrac{d^2y}{dx^2} = -\sin x$

15. $\dfrac{dy}{dx} = \dfrac{(1+\sin x)(-\sin x) - \cos x\,(\cos x)}{(1+\sin x)^2} = \dfrac{-\sin x - (\sin^2 x + \cos^2 x)}{(1+\sin x)^2} = -(1+\sin x)^{-1}$

$\dfrac{d^2y}{dx^2} = (1+\sin x)^{-2}\,\dfrac{d}{dx}\,(1+\sin x) = \cos x\,(1+\sin x)^{-2}$

17. $\dfrac{dy}{du} = 3\cos^2 2u\,\dfrac{d}{du}\,(\cos 2u) = -6\cos^2 2u \sin 2u$

$\dfrac{d^2y}{du^2} = -6[\cos^2 2u\,\dfrac{d}{du}\,(\sin 2u) + \sin 2u\,\dfrac{d}{du}\,(\cos^2 2u)]$

$= -6[2\cos^3 2u + \sin 2u\,(-4\cos 2u \sin 2u)] = 12\cos 2u\,[2\sin^2 2u - \cos^2 2u]$

19. $\dfrac{dy}{dt} = 2\sec^2 2t,\quad \dfrac{d^2y}{dt^2} = 4\sec 2t\,\dfrac{d}{dt}\,(\sec 2t) = 8\sec^2 2t \tan 2t$

21. $\dfrac{dy}{dx} = x^2(3\cos 3x) + 2x\sin 3x$

$\dfrac{d^2y}{dx^2} = [x^2(-9\sin 3x) + 2x(3\cos 3x)] + [2x(3\cos 3x) + 2(\sin 3x)]$

$= (2 - 9x^2)\sin 3x + 12x\cos 3x$

23. $y = \sin^2 x + \cos^2 x = 1$ so $\dfrac{dy}{dx} = \dfrac{d^2y}{dx^2} = 0$

25. $\dfrac{d^4}{dx^4}\,(\sin x) = \dfrac{d^3}{dx^3}\,(\cos x) = \dfrac{d^2}{dx^2}\,(-\sin x) = \dfrac{d}{dx}\,(-\cos x) = \sin x$

27. $\dfrac{d}{dt}\left[t^2\dfrac{d^2}{dt^2}\,(t\cos 3t)\right] = \dfrac{d}{dt}\left[t^2\dfrac{d}{dt}\,(\cos 3t - 3t\sin 3t)\right]$

$$= \dfrac{d}{dt}[t^2(-3\sin 3t - 3\sin 3t - 9t\cos 3t)]$$

$$= \dfrac{d}{dt}[-6t^2\sin 3t - 9t^3\cos 3t]$$

$$= (-18t^2\cos 3t - 12t\sin 3t) + (27t^3\sin 3t - 27t^2\cos 3t)$$

$$= (27t^3 - 12t)\sin 3t - 45t^2\cos 3t$$

29. $\dfrac{d}{dx}[f(\sin 3x)] = f'(\sin 3x)\dfrac{d}{dx}(\sin 3x) = 3\cos 3x\,f'(\sin 3x)$

31. $\dfrac{dy}{dx} = \cos x;$ slope of tangent at $(0,0)$ is 1; tangent: $y = x.$

33. $\dfrac{dy}{dx} = -\csc^2 x;$ slope of tangent at $(\frac{\pi}{6}, \sqrt{3})$ is $-4,$ an equation for

tangent: $y - \sqrt{3} = -4\left(x - \dfrac{\pi}{6}\right).$

35. $\dfrac{dy}{dx} = \sec x\tan x,$ slope of tangent at $(\frac{\pi}{4}, \sqrt{2})$ is $\sqrt{2},$ an equation for

tangent is $y - \sqrt{2} = \sqrt{2}\left(x - \dfrac{\pi}{4}\right).$

37. $\dfrac{dy}{dx} = -\sin x;$ $x = \pi$

39. $\dfrac{dy}{dx} = \cos x - \sqrt{3}\sin x;$ $\dfrac{dy}{dx} = 0$ gives $\tan x = \dfrac{1}{\sqrt{3}};$ $x = \dfrac{\pi}{6}, \dfrac{7\pi}{6}$

41. $\dfrac{dy}{dx} = 2\sin x\cos x = \sin 2x;$ $x = \dfrac{\pi}{2}, \pi, \dfrac{3\pi}{2}$

43. $\dfrac{dy}{dx} = \sec^2 x - 2;$ $\dfrac{dy}{dx} = 0$ gives $\sec x = \pm\sqrt{2};$ $x = \dfrac{\pi}{4}, \dfrac{3\pi}{4}, \dfrac{5\pi}{4}, \dfrac{7\pi}{4}$

45. $\dfrac{dy}{dx} = 2\sec x\tan x + \sec^2 x;$ since $\sec x$ is never zero, $\dfrac{dy}{dx} = 0$ gives

$2\tan x + \sec x = 0$ so that $\sin x = -1/2;$ $x = \dfrac{7\pi}{6}, \dfrac{11\pi}{6}$

47. $f(x) = x + 2\cos x,$ $f'(x) = 1 - 2\sin x.$

(a) $xf'(x) = 0:$ $1 - 2\sin x = 0,$ $\sin x = 1/2,$ $x = \pi/6,\ 5\pi/6.$

$f':$
$$\begin{array}{ccccc} + & | & - & | & + \\ 0 & \frac{\pi}{6} & & \frac{5\pi}{6} & 2\pi \end{array}$$

(b) $f'(x) > 0$ on $(0, \pi/6) \cup (5\pi/6, 2\pi),$ (c) $f'(x) < 0$ on $(\pi/6, 5\pi/6).$

49. $f(x) = \sin x + \cos x, \quad f'(x) = \cos x - \sin x.$

(a) $f'(x) = 0: \quad \cos x - \sin x = 0, \quad \cos x = \sin x, \quad x = \pi/4, \; 5\pi/4.$

$f':$

```
      +        −        +
  ┝━━━━┿━━━━━━━━┿━━━━━┥
  0   π              5 π           2 π
      ─                ─
      4                4
```

(b) $f'(x) > 0$ on $(0, \pi/4) \cup (5\pi/4, 2\pi),$ (c) $f'(x) < 0$ on $(\pi/4, 5\pi/4).$

51. (a) $\dfrac{dy}{dt} = \dfrac{dy}{du} \dfrac{du}{dx} \dfrac{dx}{dt} = (2u)(\sec x \tan x)\pi = 2\pi \sec^2 \pi t \tan \pi t$

(b) $y = \sec^2 \pi t - 1, \quad \dfrac{dy}{dt} = 2 \sec \pi t \, (\sec \pi t \tan \pi t)\pi = 2\pi \sec^2 \pi t \tan \pi t$

53. (a) $\dfrac{dy}{dt} = \dfrac{dy}{du} \dfrac{du}{dx} \dfrac{dx}{dt} = 4\left[\dfrac{1}{2}(1-u)\right]^3 \left(-\dfrac{1}{2}\right)(-\sin x)(2) = 4\left[\dfrac{1}{2}(1-\cos 2t)\right]^3 \sin 2t$

$$= 4\sin^6 t \, (2\sin t \cos t) = 8\sin^7 t \cos t$$

(b) $y = \left[\dfrac{1}{2}(1 - \cos 2t)\right]^4 = \sin^8 t, \quad \dfrac{dy}{dt} = 8\sin^7 t \cos t$

55. $\dfrac{d^n}{dx^n}(\cos x) = \begin{cases} (-1)^{(n+1)/2} \sin x, & n \text{ odd} \\ (-1)^{n/2} \cos x, & n \text{ even} \end{cases}$

57. $\dfrac{d}{dx}(\cos x) = \dfrac{d}{dx}\left[\sin\left(\dfrac{\pi}{2} - x\right)\right] = -\cos\left(\dfrac{\pi}{2} - x\right) = -\sin x$

59. $f'(0) = \lim\limits_{h \to 0} \dfrac{\sin(0+h) - \sin 0}{h} = \lim\limits_{h \to 0} \dfrac{\sin h}{h} = \lim\limits_{x \to 0} \dfrac{\sin x}{x}$

61. $f(x) = 2\sin x + 3\cos x + C$ **63.** $f(x) = \sin 2x + \sec x + C$

65. $f(x) = \sin(x^2) + \cos 2x + C$

67. (a) $f'(x) = \sin(1/x) + x \, \cos(1/x)(-1/x^2)$

$$= \sin(1/x) - (1/x) \cos(1/x)$$

$$g'(x) = 2x \, \sin(1/x) + x^2 \cos(1/x)(-1/x^2)$$

$$= 2x \, \sin(1/x) - \cos(1/x)$$

(b) $\lim\limits_{x \to 0} g'(x) = \lim\limits_{x \to 0} \left[2x \, \sin(1/x) - \cos(1/x)\right] = -\lim\limits_{x \to 0} \cos(1/x)$ does not exist

69. (a) Continuity:

$$\lim_{x \to 2\pi/3^-} \sin x = \frac{\sqrt{3}}{2}, \quad \lim_{x \to 2\pi/3^+} (ax + b) = \frac{2\pi a}{3} + b; \quad \text{thus} \quad \frac{2\pi a}{3} + b = \frac{\sqrt{3}}{2}$$

Differentiability:

$$\lim_{x \to 2\pi/3^-} \cos x = -\frac{1}{2}, \quad \lim_{x \to 2\pi/3^+} (a) = a; \quad \text{thus} \quad a = -\frac{1}{2}$$

Therefore, f is differentiable at $2\pi/3$ if $a = -\frac{1}{2}$ and $b = \frac{1}{2}\sqrt{3} + \frac{1}{3}\pi$

(b)

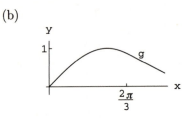

71. Let $y(t) = A \sin \omega t + B \cos \omega t$. Then

$$y'(t) = \omega A \cos \omega t - \omega B \sin \omega t \quad \text{and} \quad y''(t) = -\omega^2 A \sin \omega t - \omega^2 B \cos \omega t$$

Thus,

$$\frac{d^2 y}{dt^2} + \omega^2 y = 0.$$

73. $A = \frac{1}{2}c^2 \sin x; \quad \dfrac{dA}{dx} = \dfrac{1}{2}c^2 \cos x$

75. (a) $f^{(4p)}(x) = k^{4p} \cos kx, \quad f^{(4p+1)}(x) = -k^{4p+1} \sin kx, \quad f^{(4p+2)}(x) = -k^{4p+2} \cos kx,$

$f^{(4p+3)}(x) = k^{4p+3} \sin kx, \; p = 0, 1, 2, \ldots$

(b) $m = k^2, \; k = 1, 2, 3, \ldots$

77. f has horizontal at the points with x-coordinate $\frac{1}{2}\pi, \frac{3}{2}\pi, \cong 3.39, \cong 6.03$

79. $f(x) = \sin x, \; f'(x) = \cos x; \quad f(0) = 0, \; f'(0) = 1.$ Therefore an equation for the tangent line T is $y = x$.

The graphs of f and T are:

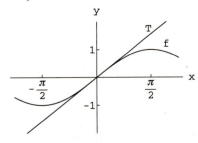

$|\sin x - x| < 0.01$ on $(-0.39, 0.39)$.

SECTION 3.7

1.
$$x^2 + y^2 = 4$$
$$2x + 2y\frac{dy}{dx} = 0$$
$$\frac{dy}{dx} = \frac{-x}{y}$$

3.
$$4x^2 + 9y^2 = 36$$
$$8x + 18y\frac{dy}{dx} = 0$$
$$\frac{dy}{dx} = \frac{-4x}{9y}$$

5.
$$x^4 + 4x^3y + y^4 = 1$$
$$4x^3 + 12x^2y + 4x^3\frac{dy}{dx} + 4y^3\frac{dy}{dx} = 0$$
$$\frac{dy}{dx} = -\frac{x^3 + 3x^2y}{x^3 + y^3}$$

7.
$$(x - y)^2 - y = 0$$
$$2(x - y)\left(1 - \frac{dy}{dx}\right) - \frac{dy}{dx} = 0$$
$$\frac{dy}{dx} = \frac{2(x - y)}{2(x - y) + 1}$$

9.
$$\sin(x + y) = xy$$
$$\cos(x + y)\left(1 + \frac{dy}{dx}\right) = x\frac{dy}{dx} + y$$
$$\frac{dy}{dx} = \frac{y - \cos(x + y)}{\cos(x + y) - x}$$

11.
$$y^2 + 2xy = 16$$
$$2y\frac{dy}{dx} + 2x\frac{dy}{dx} + 2y = 0$$
$$(x + y)\frac{dy}{dx} + y = 0.$$

Differentiating a second time, we have
$$(x + y)\frac{d^2y}{dx^2} + \frac{dy}{dx}\left(2 + \frac{dy}{dx}\right) = 0.$$

Substituting $\quad\dfrac{dy}{dx} = \dfrac{-y}{x + y},\quad$ we have
$$(x + y)\frac{d^2y}{dx^2} - \frac{y}{(x + y)}\left(\frac{2x + y}{x + y}\right) = 0, \quad \frac{d^2y}{dx^2} = \frac{2xy + y^2}{(x + y)^3} = \frac{16}{(x + y)^3}.$$

13.
$$y^2 + xy - x^2 = 9$$
$$2y\frac{dy}{dx} + x\frac{dy}{dx} + y - 2x = 0.$$

Differentiating a second time, we have
$$\left[2\left(\frac{dy}{dx}\right)^2 + 2y\frac{d^2y}{dx^2}\right] + \left[x\frac{d^2y}{dx^2} + \frac{dy}{dx}\right] + \frac{dy}{dx} - 2 = 0$$
$$(2y + x)\frac{d^2y}{dx^2} + 2\left[\left(\frac{dy}{dx}\right)^2 + \frac{dy}{dx} - 1\right] = 0.$$

Substituting $\quad\dfrac{dy}{dx} = \dfrac{2x - y}{2y + x},\quad$ we have
$$(2y + x)\frac{d^2y}{dx^2} + 2\left[\frac{(2x - y)^2 + (2x - y)(2y + x) - (2y + x)^2}{(2y + x)^2}\right] = 0$$
$$\frac{d^2y}{dx^2} = \frac{10(y^2 + xy - x^2)}{(2y + x)^3} = \frac{90}{(2y + x)^3}.$$

15.
$$4 \tan y = x^3$$

$$4 \sec^2 y \, \frac{dy}{dx} = 3x^2$$

$$\frac{dy}{dx} = \frac{3}{4} x^2 \cos^2 y$$

$$\frac{d^2 y}{dx^2} = \frac{3}{2} x \cos^2 y + \frac{3}{4} x^2 \left(2 \cos y (-\sin y) \frac{dy}{dx} \right)$$

$$= \frac{3}{2} x \cos^2 y - \frac{9}{8} x^4 \sin y \cos^3 y$$

17. $\quad x^2 - 4y^2 = 9, \qquad 2x - 8y \frac{dy}{dx} = 0.$

At $(5, 2)$, we get $\quad \dfrac{dy}{dx} = \dfrac{5}{8}.$ Then,

$$2 - 8 \left[y \frac{d^2 y}{dx^2} + \left(\frac{dy}{dx} \right)^2 \right] = 0.$$

At $(5, 2)$ we get

$$2 - 8 \left[2 \frac{d^2 y}{dx^2} + \frac{25}{64} \right] = 0 \quad \text{so that} \quad \frac{d^2 y}{dx^2} = -\frac{9}{128}.$$

19. $\quad \cos(x + 2y) = 0 \qquad -\sin(x + 2y) \left(1 + 2 \frac{dy}{dx} \right) = 0.$

At $(\pi/6, \pi/6)$, we get $\quad \dfrac{dy}{dx} = -1/2.$ Then,

$$-\cos(x + 2y) \left(1 + 2 \frac{dy}{dx} \right)^2 - \sin(x + 2y) \left(2 \frac{d^2 y}{dx^2} \right) = 0.$$

At $(\pi/6, \pi/6)$, we get

$$-\cos \frac{\pi}{2} (0)^2 - \sin \frac{\pi}{2} \left(2 \frac{d^2 y}{dx^2} \right) = 0 \quad \text{so that} \quad \frac{d^2 y}{dx^2} = 0.$$

21.
$$2x + 3y = 5$$

$$2 + 3 \frac{dy}{dx} = 0$$

slope of tangent at $(-2, 3)$: $\quad -2/3$

tangent: $\quad y - 3 = -\frac{2}{3}(x + 2)$

normal: $\quad y - 3 = \frac{3}{2}(x + 2)$

23.
$$x^2 + xy + 2y^2 = 28$$

$$2x + x \frac{dy}{dx} + y + 4y \frac{dy}{dx} = 0$$

slope of tangent at $(-2, -3)$: $\quad -1/2$

tangent: $\quad y + 3 = -\frac{1}{2}(x + 2)$

normal: $\quad y + 3 = 2(x + 2)$

25.
$$x = \cos y \, \frac{dy}{dx}$$

$$1 = -\sin y \, \frac{dy}{dx}$$

slope of tangent at $\left(\dfrac{1}{2}, \dfrac{\pi}{3} \right)$: $\quad \dfrac{-2}{\sqrt{3}}$

tangent: $\quad y - \dfrac{\pi}{3} = -\dfrac{2}{\sqrt{3}} \left(x - \dfrac{1}{2} \right)$

normal: $\quad y - \dfrac{\pi}{3} = \dfrac{\sqrt{3}}{2} \left(x - \dfrac{1}{2} \right)$

27. $\dfrac{dy}{dx} = \dfrac{1}{2}(x^3+1)^{-1/2}\dfrac{d}{dx}(x^3+1) = \dfrac{3}{2}x^2(x^3+1)^{-1/2}$

29. $\dfrac{dy}{dx} = \dfrac{1}{4}(2x^2+1)^{-3/4}\dfrac{d}{dx}(2x^2+1) = x(2x^2+1)^{-3/4}$

31. $\dfrac{dy}{dx} = \sqrt{2-x^2}\left[\dfrac{-x}{\sqrt{3-x^2}}\right] + \sqrt{3-x^2}\left[\dfrac{-x}{\sqrt{2-x^2}}\right] = \dfrac{x(2x^2-5)}{\sqrt{2-x^2}\,\sqrt{3-x^2}}$

33. $\dfrac{d}{dx}\left(\sqrt{x}+\dfrac{1}{\sqrt{x}}\right) = \dfrac{d}{dx}\left(x^{1/2}+x^{-1/2}\right) = \dfrac{1}{2}x^{-1/2}-\dfrac{1}{2}x^{-3/2} = \dfrac{1}{2}x^{-3/2}(x-1)$

35. $\dfrac{d}{dx}\left(\dfrac{x}{\sqrt{x^2+1}}\right) = \dfrac{d}{dx}\left(x(x^2+1)^{-1/2}\right)$

$\qquad\qquad = x\left(-\dfrac{1}{2}(x^2+1)^{-3/2}(2x)\right) + (x^2+1)^{-1/2} = (x^2+1)^{-3/2}$

37. (a) \qquad\qquad\qquad (b) \qquad\qquad\qquad (c)

39. $y = (a+bx)^{1/3};\quad \dfrac{dy}{dx} = \dfrac{b}{3}(a+bx)^{-2/3};\quad \dfrac{d^2y}{dx^2} = \dfrac{-2b^2}{9}(a+bx)^{-5/3}$

41. $y = \sqrt{x}\tan\sqrt{x};\quad \dfrac{dy}{dx} = \dfrac{1}{2\sqrt{x}}\tan\sqrt{x} + \sqrt{x}\sec^2\sqrt{x}\left(\dfrac{1}{2\sqrt{x}}\right) = \dfrac{1}{2\sqrt{x}}\tan\sqrt{x} + \dfrac{1}{2}\sec^2\sqrt{x}$

$\qquad \dfrac{d^2y}{dx^2} = \dfrac{2\sqrt{x}\sec^2\sqrt{x}\,(1/2\sqrt{x}) - \tan\sqrt{x}(1/\sqrt{x})}{4x} + \sec\sqrt{x}\sec\sqrt{x}\tan\sqrt{x}(1/2\sqrt{x})$

$\qquad\quad = \dfrac{\sqrt{x}\sec^2\sqrt{x} - \tan\sqrt{x} + 2x\sec^2\sqrt{x}\tan\sqrt{x}}{4x\sqrt{x}}$

43. Differentiation of $x^2+y^2=r^2$ gives $2x+2y\dfrac{dy}{dx}=0$ so that the slope of the normal line is

$$\dfrac{-1}{dy/dx} = \dfrac{y}{x} \quad (x\neq 0).$$

Let (x_0, y_0) be a point on the circle. Clearly, if $x_0 = 0$, the normal line, $x = 0$, passes through the origin. If $x_0 \neq 0$, the normal line is

$$y - y_0 = \dfrac{y_0}{x_0}(x - x_0), \quad \text{which simplifies to} \quad y = \dfrac{y_0}{x_0}x,$$

a line through the origin.

45. For the parabola $y^2 = 2px + p^2$, we have $2y\dfrac{dy}{dx} = 2p$ and the slope of a tangent is given by $m_1 = p/y$. For the parabola $y^2 = p^2 - 2px$, we obtain $m_2 = -p/y$ as the slope of a tangent. The parabolas intersect at the points $(0, \pm p)$. At each of these points $m_1 m_2 = -1$; the parabolas intersect at right angles.

47. For $y = x^2$ we have $m_1 = \dfrac{dy}{dx} = 2x$;   for $x = y^3$ we have $3y^2 \dfrac{dy}{dx} = 1$ or $m_2 = \dfrac{dy}{dx} = 1/3y^2$.
At $(1,1)$, $m_1 = 2$, $m_2 = 1/3$ and

$$\tan\alpha = \left| \frac{m_1 - m_2}{1 - m_1 m_2} \right| = \left| \frac{2 - (1/3)}{1 + 2(1/3)} \right| = 1 \;\Rightarrow\; \alpha = \frac{\pi}{4}$$

At $(0,0)$, $m_1 = 0$ and m_2 is undefined. Thus $\alpha = \pi/2$.

49. The hyperbola and the ellipse intersect at the points $(\pm 3, \pm 2)$. For the hyperbola, $\dfrac{dy}{dx} = \dfrac{x}{y}$ and for
the ellipse $\dfrac{dy}{dx} = -\dfrac{4x}{9y}$. The product of these slopes is $-\dfrac{4x^2}{9y^2}$. This product is -1 at each of the points
of intersection. Therefore the hyperbola and ellipse are orthogonal.

51. For the circles, $\dfrac{dy}{dx} = -\dfrac{x}{y}$, $y \neq 0$, and for the straight lines, $\dfrac{dy}{dx} = m = \dfrac{y}{x}$, $x \neq 0$. Since the product
of the slopes is -1, it follows that the two families are orthogonal trajectories.

53. The line $x + 2y + 3 = 0$ has slope $m = -1/2$. Thus, a line perpendicular to this line will have slope 2. A
tangent line to the ellipse $4x^2 + y^2 = 72$ has slope $m = \dfrac{dy}{dx} = -\dfrac{4x}{y}$. Setting $-\dfrac{4x}{y} = 2$ gives $y = -2x$.
Substituting into the equation for the ellipse, we have

$$4x^2 + 4x^2 = 72 \quad\Rightarrow\quad 8x^2 = 72 \quad\Rightarrow\quad x = \pm 3$$

It now follows that $y = \mp 6$ and the equations of the tangents are:

at $(3, -6):$ $y + 6 = 2(x - 3)$ or $y = 2x - 12$;

t   at $(-3, 6):$ $y - 6 = 2(x + 3)$ or $y = 2x + 12$.

55. Differentiate the equation $(x^2 + y^2)^2 = x^2 - y^2$ implicitly with respect to x :

$$2(x^2 + y^2)\left(2x + 2y\frac{dy}{dx}\right) = 2x - 2y\frac{dy}{dx}$$

Now set $dy/dx = 0$. This gives

$$2x(x^2 + y^2) = x$$
$$x^2 + y^2 = \frac{1}{2} \quad (x \neq 0)$$

Substituting this result into the original equation, we get

$$x^2 - y^2 = \frac{1}{4}$$

Now

$$\begin{array}{l} x^2 + y^2 = 1/2 \\ x^2 - y^2 = 1/4 \end{array} \;\Rightarrow\; x = \pm\frac{\sqrt{6}}{4}, \quad y = \pm\frac{\sqrt{2}}{4}$$

Thus, the points on the curve at which the tangent line is horizontal are:

$$(\sqrt{6}/4,\, \sqrt{2}/4), \quad (\sqrt{6}/4,\, -\sqrt{2}/4), \quad (-\sqrt{6}/4,\, \sqrt{2}/4), \quad (-\sqrt{6}/4,\, -\sqrt{2}/4).$$

57. Differentiate the equation $x^{1/2} + y^{1/2} = c^{1/2}$ implicitly with respect to x :

$$\frac{1}{2} x^{-1/2} + \frac{1}{2} y^{-1/2} \frac{dy}{dx} = 0 \quad \text{which implies} \quad \frac{dy}{dx} = - \left(\frac{y}{x}\right)^{1/2}$$

An equation for the tangent line to the graph at the point (x_0, y_0) is

$$y - y_0 = - \left(\frac{y_0}{x_0}\right)^{1/2} (x - x_0)$$

The x- and y-intercepts of this line are

$$a = (x_0 y_0)^{1/2} + x_0 \quad \text{and} \quad b = (x_0 y_0)^{1/2} + y_0 \quad \text{respectively.}$$

Now

$$a + b = 2(x_0 y_0)^{1/2} + x_0 + y_0 = \left(x_0^{1/2} + y_0^{1/2}\right)^2 = c.$$

59. (a) $d(h) = \dfrac{3 \sqrt[3]{h}}{h} = \dfrac{3}{h^{2/3}}$

(b) $d(h) \to \infty$ as $h \to 0^-$ and as $h \to 0^+$

(c) The graph of f has a vertical tangent at $(0, 0)$.

61. $f'(x) > 0$ on $(-\infty, \infty)$

63. $x = t, \quad y = \sqrt{4 - t^2}$ $\qquad\qquad x = t, \quad y = -\sqrt{4 - t^2}$

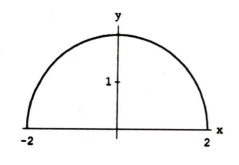   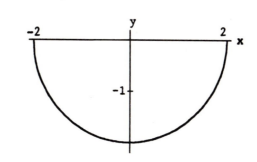

65. $\left.\dfrac{dy}{dx}\right|_{(3,4)} = 3$ $\qquad\qquad$ **67.** $\left.\dfrac{dy}{dx}\right|_{(1,3\sqrt{3})} = -\sqrt{3}$

69. (a) The graph of $x^4 = x^2 - y^2$ is: $\qquad$ (b) Differentiate the equation $x^4 = x^2 - y^2$ implicitly with respect to x :

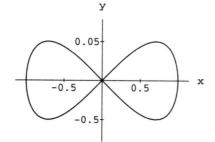

$$4x^3 = 2x - 2y \frac{dy}{dx}$$

Now set $dy/dx = 0$. This gives $4x^3 = 2x$ which implies $x = \pm \frac{\sqrt{2}}{2}$.

REVIEW EXERCISES

1.
$$f'(x) = \lim_{h \to 0} \frac{f(x+h) - f(x)}{h} = \lim_{h \to 0} \frac{[(x+h)^3 - 4(x+h) + 3] - [x^3 - 4x + 3]}{h}$$
$$= \lim_{h \to 0} \frac{3x^2 h + 3xh^2 + h^3 - 4h}{h} = \lim_{h \to 0} (3x^2 + 3xh + h^2 - 4) = 3x^2 - 4.$$

3.
$$g'(x) = \lim_{h \to 0} \frac{f(x+h) - f(x)}{h} = \lim_{h \to 0} \frac{\dfrac{1}{x+h-2} - \dfrac{1}{x-2}}{h}$$
$$= \lim_{h \to 0} \frac{(x-2) - (x+h-2)}{h(x+h-2)(x-2)} = \lim_{h \to 0} \frac{-1}{(x+h-2)(x-2)} = \frac{-1}{(x-2)^2}.$$

5. $y' = \frac{2}{3} x^{-1/3}$

7. $y' = \dfrac{(x^3 - 1)(2x + 2) - (1 + 2x + x^2)3x^2}{(x^3 - 1)^2} = -\dfrac{x^4 + 4x^3 + 3x^2 + 2x + 2}{(x^3 - 1)^2}.$

9. $f(x) = (a^2 - x^2)^{-1/2}; \quad f'(x) = -\frac{1}{2}(a^2 - x^2)^{-3/2}(-2x) = \dfrac{x}{(a^2 - x^2)^{3/2}}.$

11. $y' = 3\left(a + \dfrac{b}{x^2}\right)^2 \left(-\dfrac{2b}{x^3}\right) = -\dfrac{6b}{x^3}\left(a + \dfrac{b}{x^2}\right)^2.$

13. $y' = \sec^2 \sqrt{2x+1} \left[\frac{1}{2}(2x+1)^{-1/2}(2)\right] = \dfrac{\sec^2 \sqrt{2x+1}}{\sqrt{2x+1}}.$

15. $F'(x) = (x+2)^2 \frac{1}{2}(x^2 + 2)^{-1/2}(2x) + \sqrt{x^2+2}\,(2)(x+2) = 2(x+2)\sqrt{x^2+2} + \dfrac{x(x+2)^2}{\sqrt{x^2+2}}.$

17. $h'(t) = \sec t^2 + t(2t)\sec t^2 \tan t^2 + 6t^2 = \sec t^2 + 2t^2 \tan t^2 \sec t^2 + 6t^2.$

19. $s' = \dfrac{1}{3}\left(\dfrac{2-3t}{2+3t}\right)^{-2/3} \cdot \dfrac{(2+3t)(-3) - (2-3t)(3)}{(2+3t)^2} = -\dfrac{4}{(2+3t)^{4/3}(2-3t)^{2/3}}.$

21. $f'(\theta) = -3\csc^2(3\theta + \pi).$

23. $f'(x) = \frac{1}{3}x^{-2/3} + \frac{1}{2}x^{-1/2} = \dfrac{1}{3x^{2/3}} + \dfrac{1}{2x^{1/2}}; \quad f'(64) = \dfrac{1}{48} + \dfrac{1}{16} = \dfrac{1}{12}.$

25. $f'(x) = x^2(2)(\pi)\sin \pi x \cos \pi x + 2x \sin^2 \pi x = 2x \sin^2 \pi x + 2\pi x^2 \sin \pi x \cos \pi x; \quad f'(1/6) = \dfrac{1}{12} + \dfrac{\pi\sqrt{3}}{72}.$

27. $f(x) = 2x^3 - x^2 + 3, \; f'(x) = 6x^2 - 2x; \quad f'(1) = 4.$
Tangent line: $y - 4 = 4(x - 1)$ or $y = 4x$;   normal line: $y - 4 = -\frac{1}{4}(x - 1)$ or $y = -\frac{1}{4}x + \frac{17}{4}.$

29. $f(x) = (x+1)\sin 2x, \; f'(x) = 2(x+1)\cos 2x + \sin 2x; \quad f'(0) = 2.$
Tangent line: $y = 2x$;   normal line: $y = -\frac{1}{2}x.$

31. $f'(x) = -\sin(2-x)(-1) = \sin(2-x)$, $f''(x) = \cos(2-x)(-1) = -\cos(2-x)$.

33. $y' = x\cos x + \sin x$, $y'' = 2\cos x - x\sin x$.

35. $(-1)^n n! b^n$

37. $3x^2 y + x^3 y' + y^3 + 3xy^2 y' = 0$, $\quad y' = -\dfrac{y^3 + 3x^2 y}{x^3 + 3xy^2}$.

39. $6x^2 + 3\cos y - 3xy'\sin y = 2y + 2xy'$, $\quad y' = \dfrac{6x^2 + 3\cos y - 2y}{2x + 3x\sin y}$.

41. $2x + 2y + 2xy' - 6yy' = 0$, $y' = \dfrac{x+y}{3y-x}$; $\quad$ at $(3,2)$ $y' = \dfrac{5}{3}$.

Tangent line: $y - 2 = \frac{5}{3}(x-3)$ or $5x - 3y = 9$; normal line: $y - 2 = -\frac{3}{5}(x-3)$ or $3x + 5y = 19$.

43. $f'(x) = 3(x-4)(x-2)$.

f':

(a) $f'(x) = 0$ at $x = 2, 4$, $\qquad\qquad$ (b) $f'(x) > 0$ on $(-\infty, 2) \cup (4, \infty)$,

(c) $f'(x) < 0$ on $(2, 4)$.

45. $f'(x) = 1 + 2\cos 2x$.

f':

(a) $f'(x) = 0$ at $x = \pi/3, 2\pi/3, 4\pi/3, 5\pi/3$,

(b) $f'(x) > 0$ on $(0, \pi/3) \cup (2\pi/3, 4\pi/3) \cup (5\pi/3, 2\pi)$,

(c) $f'(x) < 0$ on $(\pi/3, 2\pi/3) \cup (4\pi/3, 5\pi/3)$.

47. $y' = \sqrt{x}$. $\qquad$ (a) 1, $\qquad$ (b) 3, $\qquad$ (c) $\frac{1}{3}$.

49. $y' = 3x^2 - 1$. The tangent line to the curve $y = x^3 - x$ at the point $(a, a^3 - a)$ has equation

$$y - (a^3 - a) = (3a^2 - 1)(x - a) \quad\text{or}\quad y = (3a^2 - 1)x - 2a^3.$$

Since this line must past through $(-2, 2)$, we have

$$2 = -6a^2 + 2 - 2a^3 \quad\text{which gives}\quad 6a^2 + 2a^3 = 0 \quad\text{and}\quad 2a^2(3 + a) = 0.$$

Thus, $a = 0, -3$. There are two tangent lines that pass through $(-2, 2)$: $y = -x$, $y = 26x + 54$.

51. The curve $y = Ax^3 + Bx^2 + Cx + D$ passes through the points $(1, 1)$ and $(-1, -9)$. This implies that
$A + B + C + D = 1$ and $-A + B - C + D = -9$.
$y' = 3Ax^2 + 2Bx + C$. The line $y = 5x - 4$ has slope 5; the line $y = 9x$ has slope 9. At
$x = 1$, $y' = 3A + 2B + C$; at $x = -1$, $y' = 3A - 2B + C$. Therefore we have $3A + 2B + C = 5$ and
$3A - 2B + C = 9$. The solution set of the system of equations

$$A + B + C + D = 1$$

$$-A + B - C + D = -9$$

$$3A + 2B + C = 5$$

$$3A - 2B + C = 9$$

is: $A = 1$, $B = -1$, $C = 4$, $D = -3$; $y = x^3 - x^2 + 4x - 3$.

53. Let $f(x) = x^2 - 2x + 1$; $f'(x) = 2x - 2$.
$$\lim_{h \to 0} \frac{(1+h)^2 - 2(1+h) + 1}{h} = \lim_{h \to 0} \frac{(1+h)^2 - 2(1+h) + 1 - 0}{h} = f'(1) = 0$$

55. Let $f(x) = \sin x$; $f'(x) = \cos x$.
$$\lim_{h \to 0} \frac{\sin(\frac{1}{6}\pi + h) - \frac{1}{2}}{h} = f'(\frac{1}{6}\pi) = \frac{\sqrt{3}}{2}$$

57. Let $f(x) = \sin x$; $f'(x) = \cos x$.
$$\lim_{x \to \pi} \frac{\sin x}{x - \pi} = f'(\pi) = -1$$

CHAPTER 4

SECTION 4.1

1. f is differentiable on $(0,1)$, continuous on $[0,1]$; and $f(0) = f(1) = 0$.

$$f'(c) = 3c^2 - 1; \quad 3c^2 - 1 = 0 \Longrightarrow c = \frac{\sqrt{3}}{3} \quad \left(-\frac{\sqrt{3}}{3} \notin (0,1)\right)$$

3. f is differentiable on $(0, 2\pi)$, continuous on $[0, 2\pi]$; and $f(0) = f(2\pi) = 0$.

$$f'(c) = 2\cos 2c; \quad 2\cos 2c = 0 \Longrightarrow 2c = \frac{\pi}{2} + n\pi, \quad \text{and} \quad c = \frac{\pi}{4} + \frac{n\pi}{2}, \quad n = 0, \pm 1, \pm 2 \ldots$$

Thus, $c = \dfrac{\pi}{4}, \dfrac{3\pi}{4}, \dfrac{5\pi}{4}, \dfrac{7\pi}{4}$

5. $f'(c) = 2c, \quad \dfrac{f(b) - f(a)}{b - a} = \dfrac{4 - 1}{2 - 1} = 3; \quad 2c = 3 \implies c = 3/2$

7. $f'(c) = 3c^2, \quad \dfrac{f(b) - f(a)}{b - a} = \dfrac{27 - 1}{3 - 1} = 13; \quad 3c^2 = 13 \implies c = \dfrac{1}{3}\sqrt{39} \quad \left(-\dfrac{1}{3}\sqrt{39} \text{ is not in } [a,b]\right)$

9. $f'(c) = \dfrac{-c}{\sqrt{1 - c^2}}, \quad \dfrac{f(b) - f(a)}{b - a} = \dfrac{0 - 1}{1 - 0} = -1; \quad \dfrac{-c}{\sqrt{1 - c^2}} = -1 \implies c = \dfrac{1}{2}\sqrt{2}$

$(-\dfrac{1}{2}\sqrt{2} \text{ is not in } [a,b])$

11. f is continuous on $[-1, 1]$, differentiable on $(-1, 1)$ and $f(-1) = f(1) = 0$.

$$f'(x) = \frac{-x(5 - x^2)}{(3 + x^2)^2\sqrt{1 - x^2}}, \quad f'(c) = 0 \text{ for } c \text{ in } (-1, 1) \text{ implies } c = 0.$$

13. No. By the mean-value theorem there exists at least one number $c \in (0, 2)$ such that

$$f'(c) = \frac{f(2) - f(0)}{2 - 0} = \frac{3}{2} > 1.$$

15. By the mean-value theorem there is a number $c \in (2, 6)$ such that

$$f(6) - f(2) = f'(c)(6 - 2) = f'(c)4.$$

Since $1 \le f'(x) \le 3$ for all $x \in (2, 6)$, it follows that

$$4 \le f(6) - f(2) \le 12.$$

17. f is everywhere continuous and everywhere differentiable except possibly at $x = -1$.

f is continuous at $x = -1$: as you can check,

$$\lim_{x \to -1^-} f(x) = 0, \quad \lim_{x \to -1^+} f(x) = 0, \quad \text{and} \quad f(-1) = 0.$$

f is differentiable at $x = -1$ and $f'(-1) = 2$: as you can check,

$$\lim_{h \to 0^-} \frac{f(-1 + h) - f(-1)}{h} = 2 \quad \text{and} \quad \lim_{h \to 0^+} \frac{f(-1 + h) - f(-1)}{h} = 2.$$

Thus f satisfies the conditions of the mean-value theorem on every closed interval $[a, b]$.

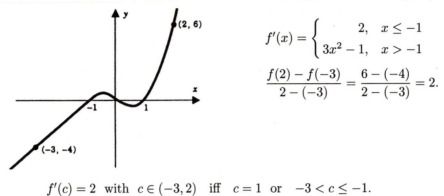

$$f'(x) = \begin{cases} 2, & x \le -1 \\ 3x^2 - 1, & x > -1 \end{cases}$$

$$\frac{f(2) - f(-3)}{2 - (-3)} = \frac{6 - (-4)}{2 - (-3)} = 2.$$

$f'(c) = 2$ with $c \in (-3, 2)$ iff $c = 1$ or $-3 < c \le -1$.

19. Let $f(x) = Ax^2 + Bx + C$. Then $f'(x) = 2Ax + B$. By the mean-value theorem

$$f'(c) = \frac{f(b) - f(a)}{b - a} = \frac{(Ab^2 + Bb + C) - (Aa^2 + Ba + C)}{b - a}$$

$$= \frac{A(b^2 - a^2) + B(b - a)}{b - a} = A(b + a) + B$$

Therefore, we have

$$2Ac + B = A(b + a) + B \Longrightarrow c = \frac{a + b}{2}$$

21. $\dfrac{f(1) - f(-1)}{1 - (-1)} = 0$ and $f'(x)$ is never zero. This result does not violate the mean-value theorem since f is not differentiable at 0; the theorem does not apply.

23. Set $P(x) = 6x^4 - 7x + 1$. If there existed three numbers $a < b < c$ at which $P(x) = 0$, then by Rolle's theorem $P'(x)$ would have to be zero for some x in (a, b) and also for some x in (b, c). This is not the case: $P'(x) = 24x^3 - 7$ is zero only at $x = (7/24)^{1/3}$.

25. Set $P(x) = x^3 + 9x^2 + 33x - 8$. Note that $P(0) < 0$ and $P(1) > 0$. Thus, by the intermediate-value theorem, there exists some number c between 0 and 1 at which $P(x) = 0$. If the equation $P(x) = 0$ had an additional real root, then by Rolle's theorem there would have to be some real number at which $P'(x) = 0$. This is not the case: $P'(x) = 3x^2 + 18x + 33$ is never zero since the discriminant $b^2 - 4ac = (18)^2 - 12(33) < 0$.

27. Let c and d be two consecutive roots of the equation $P'(x) = 0$. The equation $P(x) = 0$ cannot have two or more roots between c and d for then, by Rolle's theorem, $P'(x)$ would have to be zero somewhere between these two roots and thus between c and d. In this case c and d would no longer be consecutive roots of $P'(x) = 0$.

29. Suppose that f has two fixed points $a, b \in I$, with $a < b$. Let $g(x) = f(x) - x$. Then $g(a) = f(a) - a = 0$ and $g(b) = f(b) - b = 0$. Since f is differentiable on I, we can conclude that g is differentiable on (a, b) and continuous on $[a, b]$. By Rolle's theorem, there exists a number $c \in (a, b)$ such that $g'(c) = f'(c) - 1 = 0$ or $f'(c) = 1$. This contradicts the assumption that $f'(x) < 1$ on I.

31. (a) $f'(x) = 3x^2 - 3 < 0$ for all x in $(-1, 1)$. Also, f is differentiable on $(-1, 1)$ and continuous on $[-1, 1]$. Thus there cannot be a and b in $(-1, 1)$ such that $f(a) = f(b) = 0$, or they would contradict Rolle's theorem.

(b) When $f(x) = 0$, $b = 3x - x^3 = x(3 - x^2)$. When x is in $(-1, 1)$, then $|x(3 - x^2)| < 2$. Thus $|b| < 2$.

33. For $p(x) = x^n + ax + b$, $p'(x) = nx^{n-1} + a$, which has at most one real zero for n even $\left(x = -\frac{a}{n}^{\frac{1}{n-1}}\right)$. If there were more than two distinct real roots of $p(x)$, then by Rolle's theorem there would be more than one zero of $p'(x)$. Thus there are at most two distinct real roots of $p(x)$.

35. If $x_1 = x_2$, then $|f(x_1) - f(x_2)|$ and $|x_1 - x_2|$ are both 0 and the inequality holds. If $x_1 \neq x_2$, then by the mean-value theorem

$$\frac{f(x_1) - f(x_2)}{x_1 - x_2} = f'(c)$$

for some number c between x_1 and x_2. Since $|f'(c)| \leq 1$:

$$\left|\frac{f(x_1) - f(x_2)}{x_1 - x_2}\right| \leq 1 \quad \text{and thus} \quad |f(x_1) - f(x_2)| \leq |x_1 - x_2|.$$

37. Set, for instance, $f(x) = \begin{cases} 1, & a < x < b \\ 0, & x = a, b \end{cases}$

39. (a) By the mean-value theorem, there exists a number $c \in (a, b)$ such that $f(b) - f(a) = f'(c)(b - a)$. If $f'(x) \leq M$ for all $x \in (a, b)$, then it follows that

$$f(b) \leq f(a) + M(b - a)$$

(b) If $f'(x) \geq m$ for all $x \in (a, b)$, then it follows that

$$f(b) \geq f(a) + m(b - a)$$

(c) If $|f'(x)| \leq K$ on (a, b), then $-K \leq f'(x) \leq K$ on (a, b) and the result follows from parts (a) and (b).

41. We show first that the conditions on f and g imply that f and g cannot be simultaneously 0. Set $h(x) = f^2(x) + g(x)$. Then
$h'(x) = 2f(x)f'(x) + 2g(x)g'(x) = 2f(x)[g(x)] + 2g(x)[-f(x)] = 0 \implies f^2(x) + g^2(x) = C$ constant. If $C = 0$, then $f^2(x) = -g^2(x) \implies f(x) \equiv 0$, contradicting the assumptions on f. Therefore, $f^2(x) + g^2(x) = C$, $C > 0$. If $f(\alpha) = 0$, then $g(\alpha \neq 0$.

Assume that $f(a) = f(b) = 0$ and $f(x) \neq 0$ on (a, b). Suppose that $g(x) \neq 0$ on (a, b). We know also that $g(a) \neq 0$ and $g(b) \neq 0$. Set $h(x) = f(x)/g(x)$. Then h is continuous on the closed interval $[a, b]$ and differentiable on the open interval (a, b). Therefore, by Rolle's theorem, there exists at least one point $c \, \epsilon (a, b)$ such that $h'(c) = 0$. But,

$$h'(x) = \frac{g(x)f'(x) - f(x)g'(x)}{g^2(x)} = \frac{g^2(x) + f^2(x)}{g^2(x)} = \frac{C}{g^2(x)} > 0$$

for all $x \in (a,b)$, and we have a contradiction. Thus g must have at least one zero in (a,b). Since $g'(x) = -f(x) \neq 0$ in (a,b), g has exactly one zero in (a,b).

Simply reverse the roles of f and g to show that f has exactly one zero between two consecutive zeros of g.

43.
$$f'(x_0) = \lim_{h \to 0} \frac{f(x_0 + h) - f(x_0)}{h} = \lim_{h \to 0} \frac{f'(x_0 + \theta h)h}{h} = \lim_{h \to 0} f'(x_0 + \theta h)$$

$\underset{\text{(by the hint)}}{\uparrow}$

$$= \lim_{x \to x_0} f'(x) = L$$

$\underset{\text{(by 2.2.6)}}{\uparrow}$

45. Using the hint, F is continuous on $[a,b]$, differentiable on (a,b), and $F(a) = F(b)$. By Exercise 44, there is a number c in (a,b) such that $F'(c) = 0$.

Therefore $[f(b) - f(a)]g'(c) - [g(b) - g(a)]f'(c) = 0$ and $\dfrac{f(b) - f(a)}{g(b) - g(a)} = \dfrac{f'(c)}{g'(c)}$.

47. $f(x) = 1 - x^3 - \cos(\pi x/2)$ is differentiable on $(0,1)$, continuous on $[0,1]$, and $f(0) = f(1) = 0$.

$f'(x) = -3x^2 + \dfrac{\pi}{2} \sin(\pi x/2)$

$f'(c) = 0$ at $c \cong 0.676$

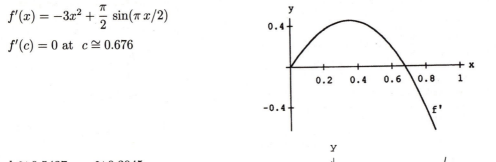

49. $b \cong 0.5437$, $c \cong 0.3045$,

$f(0.3045) \cong -0.1749$

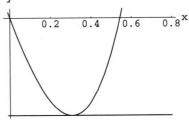

51. $0 \; c = 0$

53. $f(x) = x^4 - 7x^2 + 2$; $f'(x) = 4x^3 - 14x$

$g(x) = 4x^3 - 14x - \dfrac{f(3) - f(1)}{3 - 1} = 4x^3 - 14x - 12$

$g(c) = 0$ at $c \cong 2.205$

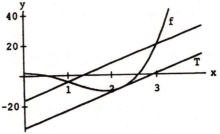

55. $f(x) = x^3 - x^2 + x - 1; \ f'(x) = 3x^2 - 2x + 1; \quad c = 8/3.$

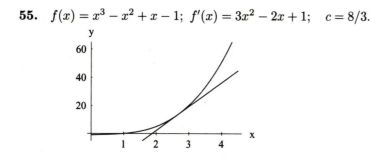

SECTION 4.2

1. $f'(x) = 3x^2 - 3 = 3\left(x^2 - 1\right) = 3(x + 1)(x - 1)$

f increases on $(-\infty, -1]$ and $[1, \infty)$, decreases on $[-1, 1]$

3. $f'(x) = 1 - \dfrac{1}{x^2} = \dfrac{x^2 - 1}{x^2} = \dfrac{(x + 1)(x - 1)}{x^2}$

f increases on $(-\infty, -1]$ and $[1, \infty)$, decreases on $[-1, 0)$ and $(0, 1]$ (f is not defined at 0)

5. $f'(x) = 3x^2 + 4x^3 = x^2(3 + 4x)$

f increases on $\left[-\frac{3}{4}, \infty\right)$,   decreases on $\left(-\infty, -\frac{3}{4}\right]$

7. $f'(x) = 4(x + 1)^3$

f increases on $[-1, \infty)$,   decreases on $(-\infty, -1]$

9. $f(x) = \begin{cases} \dfrac{1}{2 - x}, & x < 2 \\[2mm] \dfrac{1}{x - 2}, & x > 2 \end{cases} \qquad f'(x) = \begin{cases} \dfrac{1}{(2 - x)^2}, & x < 2 \\[2mm] \dfrac{-1}{(x - 2)^2}, & x > 2 \end{cases}$

f increases on $(-\infty, 2)$,   decreases on $(2, \infty)$ (f is not defined at 2)

11. $f'(x) = -\dfrac{4x}{\left(x^2 - 1\right)^2}$

f increases on $(-\infty, -1)$ and $(-1, 0]$, decreases on $[0, 1)$ and $(1, \infty)$ (f is not defined at ± 1)

13. $f(x) = \begin{cases} x^2 - 5, & x < -\sqrt{5} \\ -\left(x^2 - 5\right), & -\sqrt{5} \le x \le \sqrt{5} \\ x^2 - 5, & \sqrt{5} < x \end{cases} \qquad f'(x) = \begin{cases} 2x, & x < -\sqrt{5} \\ -2x, & -\sqrt{5} < x < \sqrt{5} \\ 2x, & \sqrt{5} < x \end{cases}$

f increases on $[-\sqrt{5}, 0]$ and $[\sqrt{5}, \infty)$, decreases on $(-\infty, -\sqrt{5}]$ and $[0, \sqrt{5}]$

15. $f'(x) = \dfrac{2}{(x + 1)^2};$ f increases on $(-\infty, -1)$ and $(-1, \infty)$ (f is not defined at -1)

17. $f'(x) = \dfrac{x}{(2 + x^2)^2}\sqrt{\dfrac{2 + x^2}{1 + x^2}}$ f increases on $[0, \infty)$,   decreases on $(-\infty, 0]$

19. $f'(x) = 1 + \sin x \geq 0$; f increases on $[0, 2\pi]$

21. $f'(x) = -2\sin 2x - 2\sin x = -2\sin x\,(2\cos x + 1)$; f increases on $\left[\frac{2}{3}\pi,\, \pi\right]$, decreases on $\left[0,\, \frac{2}{3}\pi\right]$

23. $f'(x) = \sqrt{3} + 2\sin 2x$; f increases on $\left[0,\, \frac{2}{3}\pi\right]$ and $\left[\frac{5}{6}\pi,\, \pi\right]$, decreases on $\left[\frac{2}{3}\pi,\, \frac{5}{6}\pi\right]$

25. $\dfrac{d}{dx}\left(\dfrac{x^3}{3} - x\right) = f'(x)$ $\implies$ $f(x) = \dfrac{x^3}{3} - x + C$

$f(1) = 2$ $\implies$ $2 = \frac{1}{3} - 1 + C$, so $C = \frac{8}{3}$. Thus, $f(x) = \frac{1}{3}x^3 - x + \frac{8}{3}$.

27. $\dfrac{d}{dx}\left(x^5 + x^4 + x^3 + x^2 + x\right) = f'(x)$ $\implies$ $f(x) = x^5 + x^4 + x^3 + x^2 + x + C$

$f(0) = 5$ $\implies$ $5 = 0 + C$, so $C = 5$. Thus, $f(x) = x^5 + x^4 + x^3 + x^2 + x + 5$.

29. $\dfrac{d}{dx}\left(\dfrac{3}{4}x^{4/3} - \dfrac{2}{3}x^{3/2}\right) = f'(x)$ $\implies$ $f(x) = \dfrac{3}{4}x^{4/3} - \dfrac{2}{3}x^{3/2} + C$

$f(0) = 1$ $\implies$ $1 = 0 + C$, so $C = 1$. Thus, $f(x) = \frac{3}{4}x^{4/3} - \frac{2}{3}x^{3/2} + 1,\ x \geq 0$.

31. $\dfrac{d}{dx}\left(2x - \cos x\right) = f'(x)$ $\implies$ $f(x) = 2x - \cos x + C$

$f(0) = 3$ $\implies$ $3 = 0 - 1 + C$, so $C = 4$. Thus, $f(x) = 2x - \cos x + 4$.

33. $f'(x) = \begin{cases} 1, & x < -3 \\ -1, & -3 < x < -1 \\ 1, & -1 < x < 1 \\ -2, & 1 < x \end{cases}$

f increases on $(-\infty, -3)$ and $[-1, 1]$;

decreases on $[-3, -1]$ and $[1, \infty)$

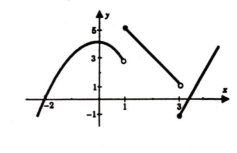

35. $f'(x) = \begin{cases} -2x, & x < 1 \\ -2, & 1 < x < 3 \\ 3, & 3 < x \end{cases}$

f increases on $(-\infty, 0]$ and $[3, \infty)$;

decreases on $[0, 1)$ and $[1, 3]$

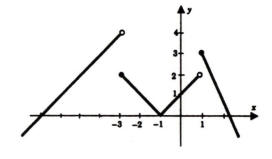

37.

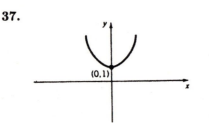

39.

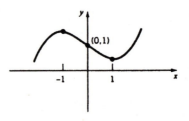

41.

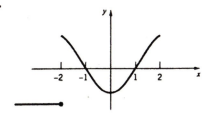

43.

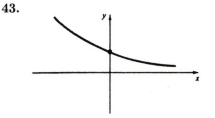

45. Not possible; f is increasing, so $f(2)$ must be greater than $f(-1)$.

47. (a) True. Let x_1, $x_2 \in [a, c]$, $x_1 < x_2$. If x_1, $x_2 \in [a, b]$, or if x_1, $x_2 \in [b, c]$, then $f(x_1) < f(x_2)$. If $x_1 \in [a, b)$ and $x_2 \in [b, c]$, then $f(x_1) < f(b) \le f(x_2)$. Therefore f increases on $[a, c]$

  (b) False. A slight modification of Example 6 is a counterexample. Let $g(x) = \begin{cases} \frac{1}{2}x + 2, & x \le 1 \\ x^3, & x > 1. \end{cases}$

49. (a) True. If $f'(c) < 0$ at some number $c \in (a, b)$, then there exists a number δ such that $f(x) > f(c) > f(z)$ for $x \in (c - \delta, c)$ and $z \in (c, c + \delta)$ (Theorem 4.1.2).

  (b) False. $f(x) = x^3$ increases on $(-1, 1)$ and $f'(0) = 0$.

51. Let $f(x) = x - \sin x$. Then $f'(x) = 1 - \cos x$.

  (a) $f'(x) \ge 0$ for all $x \in (-\infty, \infty)$ and $f'(x) = 0$ only at $x = \dfrac{\pi}{2} + n\pi$, $n = 0, \pm 1, \pm 2, \ldots$
    It follows from Theorem 4.2.3 that f is increasing on $(-\infty, \infty)$.

  (b) Since f is increasing on $(-\infty, \infty)$ and $f(0) = 0 - \sin 0 = 0$, we have:

$$f(x) > 0 \quad \text{for all} \quad x > 0 \Rightarrow x > \sin x \quad \text{on} \quad (0, \infty);$$

$$f(x) < 0 \quad \text{for all} \quad x < 0 \Rightarrow x < \sin x \quad \text{on} \quad (-\infty, 0).$$

53. $f'(x) = 2 \sec x (\sec x \tan x) = 2 \sec^2 x \tan x$ and $g'(x) = 2 \tan x \sec^2 x$.

  Therefore, $f'(x) = g'(x)$ for all $x \in I$.

55. Let f and g be functions such that $f'(x) = -g(x)$ and $g'(x) = f(x)$. Then:

  (a) Differentiating $f^2(x) + g^2(x)$ with respect to x, we have

$$2f(x)f'(x) + 2g(x)g'(x) = -2f(x)g(x) + 2g(x)f(x) = 0.$$

  Thus, $f^2(x) + g^2(x) = C$ (constant).

  (b) $f(0) = 0$ and $g(0) = 1$ implies $C = 1$.

  (c) The functions $f(x) = \sin x$, $g(x) = \cos x$ have these properties.

57. Let $f(x) = \tan x$ and $g(x) = x$ for $x \in [0, \pi/2)$. Then $f(0) = g(0) = 0$ and $f'(x) = \sec^2 x > g'(x) = 1$ for $x \in (0, \pi/2)$. Thus, $\tan x > x$ for $x \in (0, \pi/2)$ by Exercise 56(a).

59. Choose an integer $n > 1$. Let $f(x) = (1+x)^n$ and $g(x) = 1 + nx$, $x > 0$. Then, $f(0) = g(0) = 1$ and $f'(x) = n(1+x)^{n-1} > g'(x) = n$ since $(1+x)^{n-1} > 1$ for $x > 0$. The result follows from Exercise 56(a).

61. $4° \cong 0.06981$ radians. By Exercises 51 and 60,

$$0.6981 - \frac{(0.6981)^3}{6} = 0.06975 < \sin 4° < 0.6981$$

63. Let $f(x) = 3x^4 - 10x^3 - 4x^2 + 10x + 9$, $x \in [-2, 5]$. Then $f'(x) = 12x^3 - 30x^2 - 8x + 10$.

$f'(x) = 0$ at $x \cong -0.633$, 0.5, 2.633

f is decreasing on $[-2, -0.633]$

and $[0.5, 2.633]$

f is increasing on $[-0.633, 0.5]$

and $[2.633, 5]$

65. Let $f(x) = x \cos x - 3 \sin 2x$, $x \in [0, 6]$. Then $f'(x) = \cos x - x \sin x - 6 \cos 2x$.

$f'(x) = 0$ at $x \cong 0.770$, 2.155, 3.798, 5.812

f is decreasing on $[0, 0.770]$, $[2.155, 3.798]$

and $[5.812, 6]$

f is increasing on $[0.770, 2.155]$

and $[3.798, 5.812]$

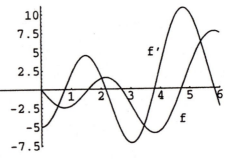

67. (a) $f'(x) = 0$ at $x = 0$, $\frac{\pi}{2}$, π, $\frac{3\pi}{2}$, 2π (b) $f'(x) > 0$ on $\left(\pi, \frac{3\pi}{2}\right) \cup \left(\frac{3\pi}{2}, 2\pi\right)$

(c) $f'(x) < 0$ on $\left(0, \frac{\pi}{2}\right) \cup \left(\frac{\pi}{2}, \pi\right)$

69. (a) $f'(x) = 0$ at $x = 0$ (b) $f'(x) > 0$ on $(0, \infty)$

(c) $f'(x) < 0$ on $(-\infty, 0)$

71. $f = C$, constant; $f'(x) \equiv 0$

SECTION 4.3

1. $f'(x) = 3x^2 + 3 > 0$;   no critical pts, no local extreme values

3. $f'(x) = 1 - \dfrac{1}{x^2}$; critical pts $-1, 1$

$f''(x) = \dfrac{2}{x^3}$, $f''(-1) = -2$, $f''(1) = 2$ $f(-1) = -2$ local max, $f(1) = 2$ local min

5. $f'(x) = 2x - 3x^2 = x(2 - 3x);$ critical pts $0, \frac{2}{3}$

$f''(x) = 2 - 6x;$ $f''(0) = 2,$ $f''(\frac{2}{3}) = -2$

$f(0) = 0$ local min, $f(\frac{2}{3}) = \frac{4}{27}$ local max

7. $f'(x) = \dfrac{2}{(1-x)^2}\,;$ no critical pts, no local extreme values

9. $f'(x) = -\dfrac{2(2x+1)}{x^2(x+1)^2}\,;$ critical pt $-\dfrac{1}{2}$

$f(-\frac{1}{2}) = -8$ local max

f':

```
+ + + + + +    + + + + + + + + 0 - - - - - - - -   - - - - - - -
―――――――○―――――――●―――――――○――――――→
       -1            -1             0              x
                      ――
                      2
                     max
```

11. $f'(x) = x^2(5x - 3)(x - 1);$ critical pts $0, \frac{3}{5}, 1$

f':

```
+ + + + + + 0 + + + + + + + + 0 - - - - - - 0 + + + + + +
―――――――●―――――――●―――――――●――――――→
       0               3         1              x
                       ―
                       5
                      max       min
```

$f\left(\dfrac{3}{5}\right) = \dfrac{2^2 3^3}{5^5}$ local max

$f(1) = 0$ local min

no local extreme at 0

13. $f'(x) = (5 - 8x)(x - 1)^2;$ critical pts $\frac{5}{8}, 1$

f':

```
+ + + + + + + + + + 0 - - - - - - - - - - - 0 - - - - - - - - - - -
――――――――●――――――――――●――――――→
        5                      1              x
        ―
        8                 no extreme
       max
```

$f\left(\dfrac{5}{8}\right) = \dfrac{27}{2048}$ local max

no local extreme at 1

15. $f'(x) = \dfrac{x(2 + x)}{(1 + x)^2};$ critical pts $-2, 0$

f':

```
+ + + + + + + 0 - - - - - - - -   - - - - - - - - 0 + + + + + + +
――――――●――――――――○――――――――●――――――→
      -2             -1             0              x
     max                           min
```

$f(-2) = -4$ local max

$f(0) = 0$ local min

17. $f'(x) = \frac{1}{3}x(7x + 12)(x + 2)^{-2/3};$ critical pts $-2, -\frac{12}{7}, 0$

f':

```
+ + + + + + + + dne + + + 0 - - - - - - - - - - - 0 + + + +
――――――――●―――――●――――――――――●―――――→
        -2         -12                0           x
    no extreme      ――              min
                    7
                   max
```

$f\left(-\dfrac{12}{7}\right) = \dfrac{144}{49}\left(\dfrac{2}{7}\right)^{1/3}$ local max

$f(0) = 0$ local min

19. $f(x) = \begin{cases} 2 - 3x, & x \le -\frac{1}{2} \\ x + 4, & -\frac{1}{2} < x < 3 \\ 3x - 2, & 3 \le x \end{cases}$ $\qquad f'(x) = \begin{cases} -3, & x < -\frac{1}{2} \\ 1, & -\frac{1}{2} < x < 3 \\ 3, & 3 < x \end{cases}$

critical pts $-\frac{1}{2}, 3$

f':

```
- - - - - - - dne + + + + + + + + dne + + + + + + + +
――――――●――――――――――●――――――→
      -1               3              x
      ――
      2            no extreme
     min
```

$f\left(-\dfrac{1}{2}\right) = \dfrac{7}{2}$ local min

no local extreme at 3

21. $f'(x) = \frac{2}{3}x^{-4/3}(x-1)$;   critical pt 1

f':

$f(1) = 3$ local min

no local extreme at 0

23. $f'(x) = \cos x - \sin x$;   critical pts $\frac{1}{4}\pi$, $\frac{5}{4}\pi$

$f''(x) = -\sin x - \cos x$, $f''\left(\frac{1}{4}\pi\right) = -\sqrt{2}$, $f''\left(\frac{5}{4}\pi\right) = \sqrt{2}$

$f(\frac{1}{4}\pi) = \sqrt{2}$ local max, $f(\frac{5}{4}\pi) = -\sqrt{2}$ local min

25. $f'(x) = \cos x\,(2\sin x - \sqrt{3}\,)$;   critical pts $\frac{1}{3}\pi$, $\frac{1}{2}\pi$, $\frac{2}{3}\pi$

f':

$f(\frac{1}{3}\pi) = f(\frac{2}{3}\pi) = -\frac{3}{4}$ local mins

$f(\frac{1}{2}\pi) = 1 - \sqrt{3}$ local max

27. $f'(x) = \cos^2 x - \sin^2 x - 3\cos x + 2 = (2\cos x - 1)(\cos x - 1)$ critical pts $\frac{1}{3}\pi$, $\frac{5}{3}\pi$

f':

$f(\frac{1}{3}\pi) = \frac{2}{3}\pi - \frac{5}{4}\sqrt{3}$ local min

$f(\frac{5}{3}\pi) = \frac{10}{3}\pi + \frac{5}{4}\sqrt{3}$ local max

29. (a) f increases on $[-2,0]$ and $[3,\infty)$; f decreases on $(-\infty,-2]$ and $[0,3]$.

(b) $f(-2)$ and $f(3)$ are local minima; $f(0) = 1$ is a local maximum.

31. Let $h(x) = f(x) - g(x)$. Then $h(x)$ gives the vertical separation between graphs of f and g at x. If h has a maximum at c, then $h'(c) = 0$. Since $h'(x) = f'(x) - g'(x)$, $h'(c) = 0$ implies $f'(c) = g'(c)$. Thus the lines tangent to the graphs of f and g are parallel at $x = c$.

33. Solving $f'(x) = 2ax + b = 0$ gives a critical point at $x = -\dfrac{b}{2a}$.   Since $f''(x) = 2a$, f has a local maximum at $-\dfrac{b}{2a}$ if $a < 0$ and a local minimum at $-\dfrac{b}{2a}$ if $a > 0$.

35.
$$P(x) = x^4 - 8x^3 + 22x^2 - 24x + 4$$
$$P'(x) = 4x^3 - 24x^2 + 44x - 24$$
$$P''(x) = 12x^2 - 48x + 44$$

Since $P'(1) = 0$, $P'(x)$ is divisible by $x - 1$. Division by $x - 1$ gives

$$P'(x) = (x-1)\left(4x^2 - 20x + 24\right) = 4(x-1)(x-2)(x-3).$$

The critical pts are 1, 2, 3. Since

$$P''(1) > 0, \quad P''(2) < 0, \quad P''(3) > 0,$$

$P(1) = -5$ is a local min, $P(2) = -4$ is a local max, and $P(3) = -5$ is a local min.

Since $P'(x) < 0$ for $x < 0$, P decreases on $(-\infty, 0]$. Since $P(0) > 0$, P does not take on the value 0 on $(-\infty, 0]$.

Since $P(0) > 0$ and $P(1) < 0$, P takes on the value 0 at least once on $(0, 1)$. Since $P'(x) < 0$ on $(0, 1)$, P decreases on $[0, 1]$. It follows that P takes on the value zero only once on $[0, 1]$.

Since $P'(x) > 0$ on $(1, 2)$ and $P'(x) < 0$ on $(2, 3)$, P increases on $[1, 2]$ and decreases on $[2, 3]$. Since $P(1)$, $P(2)$, $P(3)$ are all negative, P cannot take on the value 0 between 1 and 3.

Since $P(3) < 0$ and $P(100) > 0$, P takes on the value 0 at least once on $(3, 100)$. Since $P'(x) > 0$ on $(3, 100)$, P increases on $[3, 100]$. It follows that P takes on the value zero only once on $[3, 100]$.

Since $P'(x) > 0$ on $(100, \infty)$, P increases on $[100, \infty)$. Since $P(100) > 0$, P does not take on the value 0 on $[100, \infty)$.

37.   **(a)**   **(b)**

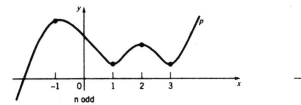
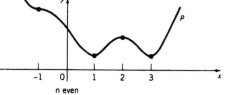

39.   Let $f(x) = \dfrac{ax}{x^2 + b^2}$. Then $f'(x) = \dfrac{a\left(b^2 - x^2\right)}{\left(b^2 + x^2\right)^2}$. Now

$$f'(0) = \frac{a}{b^2} = 1 \Rightarrow a = b^2 \quad \text{and} \quad f'(x) = \frac{b^2\left(b^2 - x^2\right)}{\left(b^2 + x^2\right)^2}$$

$$f'(-2) = \frac{b^2\left(b^2 - 4\right)}{\left(b^2 + 4\right)^2} = 0 \Rightarrow b = \pm 2$$

Thus, $a = 4$ and $b = \pm 2$.

41.   If p is a polynomial of degree n, then p' has degree $n - 1$. This implies that p' has at most $n - 1$ zeros, and it follows that p has at most $n - 1$ local extreme values.

43.   If $f(x) = x^4 - 7x^2 - 8x - 3$, then $f'(x) = 4x^3 - 14x - 8$ and $f''(x) = 12x^2 - 14$. Since $f'(2) = -4 < 0$ and $f'(3) = 58 > 0$, f' has at least one zero in $(2, 3)$. Since $f''(x) > 0$ for $x \in (2, 3)$, f' is increasing on this interval and so it has exactly one zero. Thus, f has exactly one critical point c in $(2, 3)$.

45. $f(x) = \dfrac{ax^2 + b}{cx^2 + d}$ and $f'(x) = \dfrac{2(ad - bc)x}{(cx^2 + d)^2}$; $x = 0$ is a critical number.

$f''(x) = \dfrac{2(ad - bc)(cx^2 - 4cx + d)}{(cx^2 + d)^3}$; $f''(0) = \dfrac{2(ad - bc)}{d^2}$

Therefore, $ad - bc > 0$ implies that $f(0)$ is a local minimum; $ad - bc < 0$ implies that $f(0)$ is a local maximum.

47. (a)

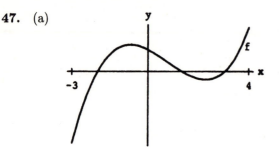

critical points: $x_1 \cong -0.692$, $x_2 \cong 2.248$

local extreme values: $f(-0.692) \cong 29.342$, $f(2.248) \cong -8.766$

(b) f is increasing on $[-3, -0.692]$, and $[2.248, 4]$; f is decreasing on $[-0.692, 2.248]$

49. (a)

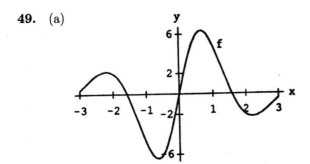

critical points: $x_1 \cong -2.201$, $x_2 \cong -0.654$, $x_3 \cong 0.654$, $x_4 \cong 2.201$

local extreme values: $f(-2.204) \cong 2.226$, $f(-0.654) \cong -6.634$, $f(0.654) \cong 6.634$,

$f(2.204) \cong -2.226$

(b) f is increasing on $[-3. - 2.204]$, $[-0.654, 0.654]$, and $[2.204, 3]$

f is decreasing on $[-2.204, -0.654]$, and $[0.654, 2.204]$

51. $f'(x) > 0$ on $\left(\frac{2}{3}, \infty\right)$; f has no local extrema.

53.

 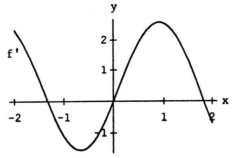

critical points of f : $x_1 \cong -1.326$, $x_2 = 0$, $x_3 \cong 1.816$

$f''(-1.326) \cong -4 < 0$ $\Rightarrow$ f has a local maximum at $x = -1.326$

$f''(0) = 4 > 0$ $\Rightarrow$ f has a local minimum at $x = 0$

$f''(1.816) \cong -4$ $\Rightarrow$ f has a local maximum at $x = 1.816$

SECTION 4.4

1. $f'(x) = \frac{1}{2}(x+2)^{-1/2}, \ x > -2;$

    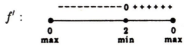

    $f(-2) = 0$ endpt and abs min;   as $x \to \infty$, $f(x) \to \infty$;   so no abs max

3. $f'(x) = 2x - 4, \ x \in (0,3);$

    critical pt. 2;

    $f(0) = 1$ endpt and abs max, $f(2) = -3$ local and abs min, $f(3) = -2$ endpt max

5. $f'(x) = 2x - \dfrac{1}{x^2} = \dfrac{2x^3 - 1}{x^2}, \ x \neq 0; \quad f'(x) = 0$ at $x = 2^{-1/3}$

    critical pt. $2^{-1/3}; \quad f''(x) = 2 + \dfrac{2}{x^3}, \quad f''\left(2^{-1/3}\right) = 6$

    $f\left(2^{-1/3}\right) = 2^{-2/3} + 2^{1/3} = 2^{-2/3} + 2 \cdot 2^{-2/3} = 3 \cdot 2^{-2/3}$ local min

7. $f'(x) = \dfrac{2x^3 - 1}{x^2}, \quad x \in \left(\dfrac{1}{10}, 2\right);$

    critical pt. $2^{-1/3};$

    $f\left(\frac{1}{10}\right) = 10\frac{1}{100}$ endpt and abs max, $f\left(2^{-1/3}\right) = 3 \cdot 2^{-2/3}$ local and abs min,

    $f(2) = 4\frac{1}{2}$ endpt max

9. $f'(x) = 2x - 3, \ x \in (0, 2);$

    critical pt. $\frac{3}{2};$

    $f(0) = 2$ endpt and abs max, $f\left(\frac{3}{2}\right) = -\frac{1}{4}$ local and abs min,

    $f(2) = 0$ endpt max

11. $f'(x) = \dfrac{(2-x)(2+x)}{(4+x^2)^2}, \ x \in (-3, 1);$

    critical pt. $-2;$

    $f(-3) = -\frac{3}{13}$ endpt max, $f(-2) = -\frac{1}{4}$ local and abs min,

    $f(1) = \frac{1}{5}$ endpt and abs max

13. $f'(x) = 2(x - \sqrt{x})\left(1 - \dfrac{1}{2\sqrt{x}}\right), \ x > 0;$

    critical pts. $\frac{1}{4}, 1;$

    $f(0) = 0$ endpt and abs min, $f\left(\frac{1}{4}\right) = \frac{1}{16}$ local max, $f(1) = 0$ local and abs min;

    as $x \to \infty$, $f(x) \to \infty$;   so no abs max

15. $f'(x) = \dfrac{3(2-x)}{2\sqrt{3-x}}, \quad x < 3$

critical pt. 2;

$f(2) = 2$ local and abs max, $f(3) = 0$ endpt min;

as $x \to -\infty$, $f(x) \to -\infty$; so no abs min

17. $f'(x) = -\frac{1}{3}(x-1)^{-2/3}, \quad x \neq 1;$

critical pt. 1;

no local extremes; $\left.\begin{array}{ll} \text{as} & x \to \infty, \quad f(x) \to -\infty \\ \text{as} & x \to -\infty, \quad f(x) \to \infty \end{array}\right\}$ no abs extremes

19. $f'(x) = \sin x \left(2\cos x + \sqrt{3}\right), x \in (0, \pi);$

critical pt. $\dfrac{5}{6}\pi$;

$f(0) = -\sqrt{3}$ endpt and abs min, $f\left(\frac{5}{6}\pi\right) = \dfrac{7}{4}$ local and abs max, $f(\pi) = \sqrt{3}$ endpt min

21. $f'(x) = -3\sin x \left(2\cos^2 x + 1\right) < 0, \quad x \in (0, \pi);$ no critical pts.

$f(0) = 5$ endpt and abs max, $f(\pi) = -5$ endpt and abs min

23. $f'(x) = \sec^2 x - 1 \geq 0, \quad x \in \left(-\frac{1}{3}\pi, \frac{1}{2}\pi\right);$ critical pt. 0;

$f\left(-\frac{1}{3}\pi\right) = \frac{1}{3}\pi - \sqrt{3}$ endpt and abs min, no abs max

25.

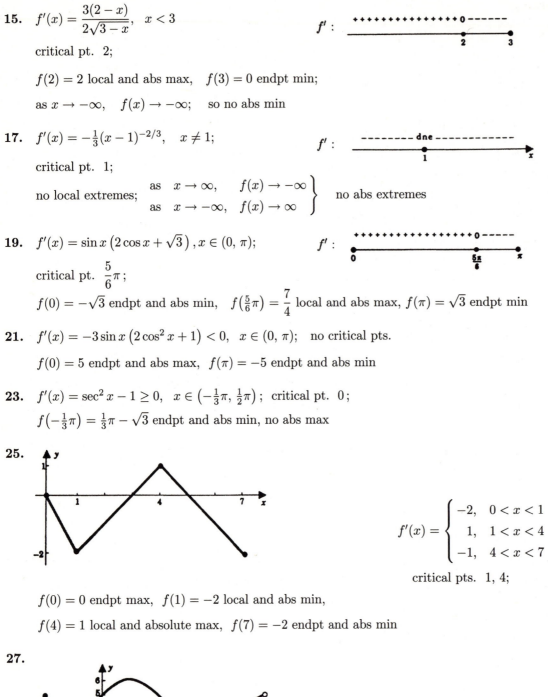

$f'(x) = \begin{cases} -2, & 0 < x < 1 \\ 1, & 1 < x < 4 \\ -1, & 4 < x < 7 \end{cases}$

critical pts. 1, 4;

$f(0) = 0$ endpt max, $f(1) = -2$ local and abs min,

$f(4) = 1$ local and absolute max, $f(7) = -2$ endpt and abs min

27.

$f'(x) = \begin{cases} 2x, & -2 < x < -1 \\ 2 - 2x, & -1 < x < 3 \\ 1, & 3 < x < 6 \end{cases}$

critical pts. $-1, 1, 3$

$f(-2) = 5$ endpt max, $f(-1) = 2$ local and abs min,

$f(1) = 6$ local and abs max, $f(3) = 2$ local and abs min

29.

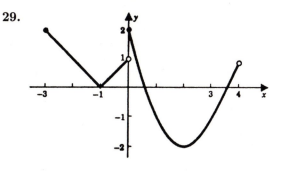

$$f'(x) = \begin{cases} -1, & -3 < x < -1 \\ 1, & -1 < x < 0 \\ 2x - 4, & 0 < x < 3 \\ 2, & 3 \le x < 4 \end{cases}$$

critical pts. $-1, 0, 2$

$f(-3) = 2$ endpt and abs max, $f(-1) = 0$ local min,

$f(0) = 2$ local and abs max, $f(2) = -2$ local and abs min

31.

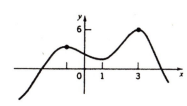

33. Not possible: $f(1) = f(3) = 0$ implies $f'(c) = 0$ for some $c \in (1,3)$ (Rolle's theorem).

35. Let $p(x) = x^3 + ax^2 + bx + c$. Then $p'(x) = 3x^2 + 2ax + b$ is a quadratic with discriminant $\Delta = 4a^2 - 12b = 4(a^2 - 3b)$. If $a^2 \le 3b$, then $\Delta \le 0$. This implies that $p'(x)$ does not change sign on $(-\infty, \infty)$. On the other hand, if $a^2 - 3b > 0$, then $\Delta > 0$ and p' has two real zeros, c_1 and c_2, from which it follows that p has extreme values at c_1 and c_2. Therefore, if p has no extreme values, then we must have $a^2 - 3b \le 0$.

37. By contradiction. If f is continuous at c, then, by the first-derivative test (4.3.4), $f(c)$ is not a local maximum.

39. If f is not differentiable on (a, b), then f has a critical point at each point c in (a, b) where $f'(c)$ does not exist. If f is differentiable on (a, b), then by the mean-value theorem there exists c in (a, b) where $f'(c) = [f(b) - f(a)]/(b - a) = 0$. This means c is a critical point of f.

41. Let $f(x) = \begin{cases} 1, & \text{if } x \text{ is a rational number} \\ 0, & \text{if } x \text{ is an irrational number} \end{cases}$

43. Let M be a positive number. Then

$$P(x) - M \ge a_n x^n - \left(|a_{n-1}| x^{n-1} + \cdots + |a_1| x + |a_0| + M \right) \quad \text{for} \quad x > 0$$

$$\ge a_n x^n - x^{n-1} \left(|a_{n-1}| + \cdots + |a_1| + |a_0| + M \right) \quad \text{for} \quad x > 1$$

$$\ge x^{n-1} [a_n x - (|a_{n-1}| + \cdots + |a_1| + |a_0| + M)]$$

It now follows that

$$P(x) - M \ge 0 \quad \text{for} \quad x \ge K = \frac{|a_{n-1}| + \cdots + |a_1| + |a_0| + M}{a_n}^{1/n}.$$

45. $f(0) = f(1) = 0$ and $f(x) > 0$ on $(0, 1)$.

$f'(x) = -qx^p(1-x)^{q-1} + px^{p-1}(1-x)^q; \quad f'(x) = 0$ implies $x = \dfrac{p}{p+q}$. The absolute maximum value

of f is $f(p/(p+q)) = \left(\dfrac{p}{p+q}\right)^p \cdot \left(\dfrac{q}{p+q}\right)^q$

47. Setting $R'(\theta) = \dfrac{v^2 \cos 2\theta}{16} = 0$, gives $\theta = \dfrac{\pi}{4}$.   Since $R''\left(\dfrac{\pi}{4}\right) = -\dfrac{v^2}{8} < 0$, $\theta = \dfrac{\pi}{4}$ is a maximum.

49.

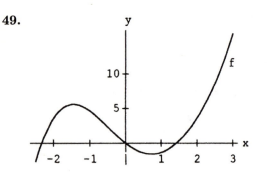

critical pts: $x_1 = -1.452$, $x_2 = 0.760$

$f(-1.452)$ local maximum

$f(0.727)$ local minimum

$f(3)$ absolute maximum

$f(-2.5)$ absolute minimum

51.

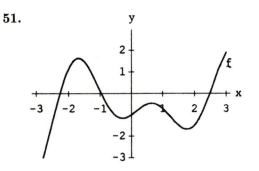

critical points: $x_1 = -1.683$, $x_2 = -0.284$,

$x_3 = 0.645$, $x_4 = 1.760$

$f(-1.683), f(0.645)$ local maxima

$f(-0.284), f(1.760)$ local minima

$f(\pi)$ absolute maximum

$f(-\pi)$ absolute minimum

53. Yes; $M = f(2) = 1$; $m = f(1) = f(3) = 0$

55. Yes; $M = f(6) = 2 + \sqrt{3}$; $m = f(1) = \dfrac{3}{2}$

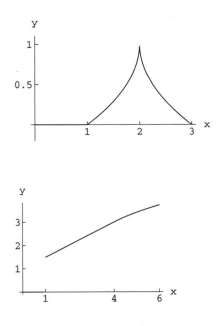

SECTION 4.5

1. Set $P = xy$ and $y = 40 - x$. We want to maximize

    $$P(x) = x(40 - x), \quad 0 \le x \le 40.$$

    $$P'(x) = 40 - 2x, \quad P'(x) = 0 \quad \Longrightarrow \quad x = 20.$$

    Since P increases on $(0, 20]$ and decreases on $[20, 40)$, the abs max of P occurs when $x = 20$. Then, $y = 20$ and $xy = 400$.

    The maximal value of xy is 400.

3.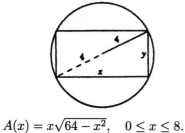

    <u>Minimize P</u>

    $P = x + 2y, \quad 200 = xy, \quad y = 200/x$

    $$P(x) = x + \frac{400}{x}, \quad x > 0.$$

    $$P'(x) = 1 - \frac{400}{x^2}, \quad P'(x) = 0 \quad \Longrightarrow \quad x = 20.$$

    Since P decreases on $(0, 20\,]$ and increases on $[20, \infty)$, the abs min of P occurs when $x = 20$.

    To minimize the fencing, make the garden 20 ft (parallel to barn) by 10 ft.

5. <u>Maximize A</u>

    $A = xy, \quad x^2 + y^2 = 8^2, \quad y = \sqrt{64 - x^2}$

    $$A(x) = x\sqrt{64 - x^2}, \quad 0 \le x \le 8.$$

    $$A'(x) = \sqrt{64 - x^2} + x\left(\frac{-x}{\sqrt{64 - x^2}}\right) = \frac{64 - 2x^2}{\sqrt{64 - x^2}}, \quad A'(x) = 0 \quad \Longrightarrow \quad x = 4\sqrt{2}.$$

    Since A increases on $(0, 4\sqrt{2}\,]$ and decreases on $[4\sqrt{2}, 8)$, the abs max of A occurs when $x = 4\sqrt{2}$. Then, $y = 4\sqrt{2}$ and $xy = 32$.

    The maximal area is 32.

7. <u>Minimize $\quad P = 3x + 2y$</u>

    $A = xy = 15{,}000, \quad y = \dfrac{15{,}000}{x}$

    $$P(x) = 3x + \frac{30{,}000}{x}, \quad 0 < x < \infty.$$

    $$P'(x) = 3 - \frac{30{,}000}{x^2} = \frac{3(x^2 - 10{,}000)}{x^2}; \quad P'(x) = 0 \quad \Longrightarrow \quad x = 100.$$

Since P decreases on $(0, 100]$ and increases on $[100, \infty)$, the abs min of P occurs when $x = 100$. When $x = 100$, $y = 150$; at least 600 feet of fencing is needed.

C has an abs min at $x = 61.24$. The dimensions that will minimize the cost are: $x = 61.24$, $y = 81.65$.

9.

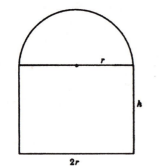

<u>Maximize L</u>

To account for the semi-circular portion admitting less light per square foot, we multiply its area by 1/3.

$$L = 2xy + \frac{1}{3}\left(\frac{\pi x^2}{2}\right),$$

$$2x + 2y + \pi x = p, \quad y = \tfrac{1}{2}(p - 2x - \pi x)$$

$$L = 2x\left(\frac{p - 2x - \pi x}{2}\right) + \frac{1}{6}\pi x^2$$

$$L(x) = px - \left(2 + \frac{5}{6}\pi\right)x^2, \quad 0 \le x \le \frac{p}{2 + \pi}.$$

$$L'(x) = p - \left(4 + \frac{5}{3}\pi\right)x; \quad L'(x) = 0 \implies x = \frac{3p}{12 + 5\pi}.$$

Since $L''(x) < 0$ for all x in the domain of L, the local max at $x = 3p/(12 + 5\pi)$ is the abs max.

For the window that admits the most light, take the radius of the semicircle as $\dfrac{3p}{12 + 5\pi}$ ft.

11.

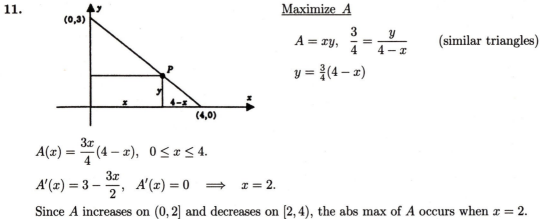

<u>Maximize A</u>

$$A = xy, \quad \frac{3}{4} = \frac{y}{4 - x} \quad \text{(similar triangles)}$$

$$y = \tfrac{3}{4}(4 - x)$$

$$A(x) = \frac{3x}{4}(4 - x), \quad 0 \le x \le 4.$$

$$A'(x) = 3 - \frac{3x}{2}, \quad A'(x) = 0 \implies x = 2.$$

Since A increases on $(0, 2]$ and decreases on $[2, 4)$, the abs max of A occurs when $x = 2$.

To maximize the area of the rectangle, take P as the point $\left(2, \frac{3}{2}\right)$.

13.

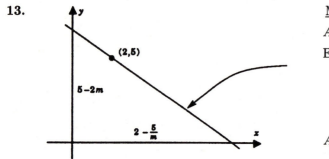

<u>Minimize A</u>

$A = \tfrac{1}{2}(x\text{-intercept})(y\text{-intercept})$

Equation of line: $y - 5 = m(x - 2)$

x-intercept: $2 - \dfrac{5}{m}$

y-intercept: $5 - 2m$

$$A = \frac{1}{2}\left(2 - \frac{5}{m}\right)(5 - 2m) = 10 - 2m - \frac{25}{2m}$$

$$A(m) = 10 - 2m - \frac{25}{2m}, \quad m < 0.$$

$$A'(m) = -2 + \frac{25}{2m^2}, \quad A'(m) = 0 \quad \Longrightarrow \quad m = -\frac{5}{2}.$$

Since $A''(m) = -25/m^3 > 0$ for $m < 0$, the local min at $m = -5/2$ is the abs min.

The triangle of minimal area is formed by the line of slope $-5/2$.

15.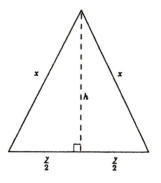

Maximize V

$$V = 2x^2 h, \quad 2\left(2x^2 + xh + 2xh\right) = 100, \quad h = \frac{50 - 2x^2}{3x}$$

$$V = 2x^2 \left(\frac{50 - 2x^2}{3x}\right)$$

$V(x) = \frac{100}{3}x - \frac{4}{3}x^3, \quad 0 \le x \le 5.$

$V'(x) = \frac{100}{3} - 4x^2, \quad V'(x) = 0 \quad \Longrightarrow \quad x = \frac{5}{3}\sqrt{3}.$

Since $V''(x) = -8x < 0$ on $(0,5)$, the local max at $x = \frac{5}{3}\sqrt{3}$ is the abs max.

The base of the box of greatest volume measures $\frac{5}{3}\sqrt{3}$ in. by $\frac{10}{3}\sqrt{3}$ in.

17.

Maximize A

$A = \frac{1}{2}hy$

$2x + y = 12 \quad \Longrightarrow \quad y = 12 - 2x$

Pythagorean Theorem:

$$h^2 + \left(\frac{y}{2}\right)^2 = x^2 \quad \Longrightarrow \quad h = \sqrt{x^2 - \left(\frac{y}{2}\right)^2}$$

Thus, $h = \sqrt{x^2 - (6-x)^2} = \sqrt{12x - 36}.$

$A(x) = (6-x)\sqrt{12x - 36}, \quad 3 \le x \le 6.$

$A'(x) = -\sqrt{12x - 36} + (6-x)\left(\frac{6}{\sqrt{12x - 36}}\right) = \frac{72 - 18x}{\sqrt{12x - 36}},$

$A'(x) = 0 \quad \Longrightarrow \quad x = 4.$

Since A increases on $(3,4]$ and decreases on $[4,6)$, the abs max of A occurs at $x = 4$.

The triangle of maximal area is equilateral with side of length 4.

19.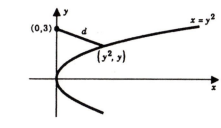

Minimize d

$$d = \sqrt{\left(y^2 - 0\right)^2 + (y - 3)^2}$$

The square-root function is increasing;

d is minimal when $D = d^2$ is minimal.

$D(y) = y^4 + (y - 3)^2$, y real.

$D'(y) = 4y^3 + 2(y - 3) = (y - 1)\left(4y^2 + 4y + 6\right)$, $D'(y) = 0$ at $y = 1$.

Since $D''(y) = 12y^2 + 2 > 0$, the local min at $y = 1$ is the abs min.

The point $(1, 1)$ is the point on the parabola closest to $(0, 3)$.

21.   The figure shows a rectangle inscribed in the ellipse $16x^2 + 9y^2 = 144$. The area of the rectangle is:

$A = (2x)(2y) = 4xy$. Solving the equation of the ellipse for y, we get $y = \frac{4}{3}\sqrt{9 - x^2}$.

<u>Maximize</u> $A = \frac{16}{3}x\sqrt{9 - x^2}$, $0 \le x \le 3$.

$A'(x) = \frac{16}{3}\left[\sqrt{9 - x^2} - \frac{x^2}{\sqrt{9 - x^2}}\right] = \frac{16}{3}\left(\frac{9 - 2x^2}{\sqrt{9 - x^2}}\right)$

$A'(x) = 0 \implies x = 3/\sqrt{2}$.

Since $A(0) = A(3) = 0$, we can conclude that $A(3/\sqrt{2})$

is the absolute maximum value of A.

At $x = 3/\sqrt{2}$, $y = 4/\sqrt{2}$ and $A = 24$; the maximum possible

area for a rectangle inscribed in the ellipse is 24 sq. units.

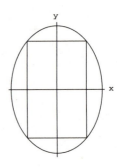

23.

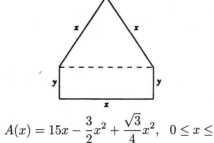

<u>Maximize</u> A

$$A = xy + \frac{\sqrt{3}}{4}x^2, \quad 30 = 3x + 2y, \quad y = \frac{30 - 3x}{2}$$

$A(x) = 15x - \frac{3}{2}x^2 + \frac{\sqrt{3}}{4}x^2$, $0 \le x \le 10$.

$A'(x) = 15 - 3x + \frac{\sqrt{3}}{2}x$, $A'(x) = 0 \implies x = \frac{30}{6 - \sqrt{3}} = \frac{10}{11}(6 + \sqrt{3})$.

Since $A''(x) = -3 + \frac{\sqrt{3}}{2} < 0$ on $(0, 10)$, the local max at $x = \frac{10}{11}\left(6 + \sqrt{3}\right)$ is the abs max.

The pentagon of greatest area is composed of an equilateral triangle with side $\frac{10}{11}\left(6 + \sqrt{3}\right) \cong 7.03$

in. and rectangle with height $\frac{15}{11}\left(5 - \sqrt{3}\right) \cong 4.46$ in.

25.

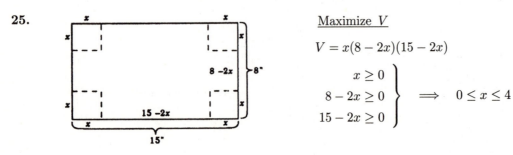

<u>Maximize</u> V

$$V = x(8 - 2x)(15 - 2x)$$

$$\left.\begin{array}{c} x \ge 0 \\ 8 - 2x \ge 0 \\ 15 - 2x \ge 0 \end{array}\right\} \implies 0 \le x \le 4$$

$V(x) = 120x - 46x^2 + 4x^3, \quad 0 \le x \le 4.$

$V'(x) = 120 - 92x + 12x^2 = 4(3x - 5)(x - 6), \quad V'(x) = 0 \quad \text{at} \quad x = \frac{5}{3}.$

Since V increases on $\left(0, \frac{5}{3}\right)$ and decreases on $\left[\frac{5}{3}, 4\right)$, the abs max of V occurs when $x = \frac{5}{3}$.

The box of maximal volume is made by cutting out squares 5/3 inches on a side.

27.

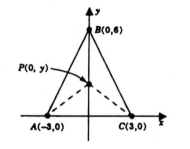

$\underline{\text{Minimize } \overline{AP} + \overline{BP} + \overline{CP} = S}$

length $AP = \sqrt{9 + y^2}$

length $BP = 6 - y$

length $CP = \sqrt{9 + y^2}$

$S(y) = 6 - y + 2\sqrt{9 + y^2}, \quad 0 \le y \le 6.$

$S'(y) = -1 + \dfrac{2y}{\sqrt{9 + y^2}}, \quad S'(y) = 0 \implies y = \sqrt{3}.$

Since

$$S(0) = 12, \quad S\left(\sqrt{3}\right) = 6 + 3\sqrt{3} \cong 11.2, \quad \text{and} \quad S(6) = 6\sqrt{5} \cong 13.4,$$

the abs min of S occurs when $y = \sqrt{3}$.

To minimize the sum of the distances, take P as the point $\left(0, \sqrt{3}\right)$.

29.

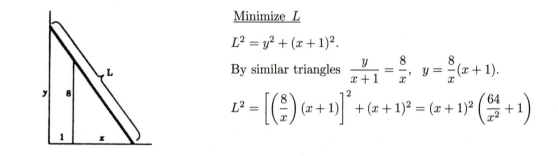

$\underline{\text{Minimize } L}$

$L^2 = y^2 + (x + 1)^2.$

By similar triangles $\dfrac{y}{x + 1} = \dfrac{8}{x}, \quad y = \dfrac{8}{x}(x + 1).$

$L^2 = \left[\left(\dfrac{8}{x}\right)(x + 1)\right]^2 + (x + 1)^2 = (x + 1)^2 \left(\dfrac{64}{x^2} + 1\right)$

Since L is minimal when L^2 is minimal, we consider the function

$$f(x) = (x + 1)^2 \left(\frac{64}{x^2} + 1\right), \quad x > 0.$$

$$f'(x) = 2(x + 1)\left(\frac{64}{x^2} + 1\right) + (x + 1)^2 \left(\frac{-128}{x^3}\right)$$

$$= \frac{2(x + 1)}{x^3}\left[x^3 - 64\right], \quad f'(x) = 0 \implies x = 4.$$

Since f decreases on $(0, 4]$ and increases on $[4, \infty)$, the abs min of f occurs when $x = 4$.

The shortest ladder is $5\sqrt{5}$ ft long.

31.

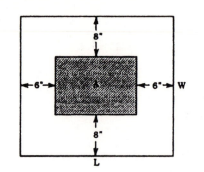

Maximize A

(We use feet rather than inches to reduce arithmetic.)

$$A = (L-1)\left(W - \tfrac{4}{3}\right)$$

$$LW = 27 \quad\Longrightarrow\quad W = \frac{27}{L}$$

$$A = (L-1)\left(\frac{27}{L} - \frac{4}{3}\right) = \frac{85}{3} - \frac{27}{L} - \frac{4}{3}L$$

$$A(L) = \frac{85}{3} - \frac{27}{L} - \frac{4}{3}L, \quad 1 \le L \le \frac{81}{4}.$$

$$A'(L) = \frac{27}{L^2} - \frac{4}{3}, \quad A'(L) = 0 \quad\Longrightarrow\quad L = \frac{9}{2}.$$

Since $A'(L) = -54/L^3 < 0$ for $1 < L < \frac{81}{4}$, the max at $L = \frac{9}{2}$ is the abs max.

The banner has length $9/2$ ft $= 54$ in. and height 6 ft $= 72$ in.

33.

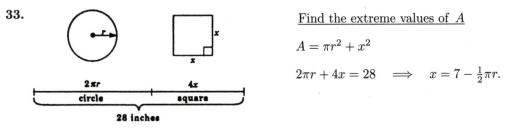

Find the extreme values of A

$$A = \pi r^2 + x^2$$

$$2\pi r + 4x = 28 \quad\Longrightarrow\quad x = 7 - \tfrac{1}{2}\pi r.$$

$$A(r) = \pi r^2 + \left(7 - \frac{1}{2}\pi r\right)^2, \quad 0 \le r \le \frac{14}{\pi}.$$

Note: the endpoints of the domain correspond to the instances when the string is not cut: $r = 0$ when no circle is formed, $r = 14/\pi$ when no square is formed.

$$A'(r) = 2\pi r - \pi\left(7 - \frac{1}{2}\pi r\right), \quad A'(r) = 0 \quad\Longrightarrow\quad r = \frac{14}{4 + \pi}.$$

Since $A''(r) = 2\pi + \pi^2/2 > 0$ on $(0, 14/\pi)$, the abs min of A occurs when $r = 14/(4 + \pi)$ and the abs max of A occurs at one of the endpts: $A(0) = 49$, $A(14/\pi) = 196/\pi > 49$.

(a) To maximize the sum of the two areas, use all of the string to form the circle.

(b) To minimize the sum of the two areas, use $2\pi r = 28\pi/(4 + \pi) \cong 12.32$ inches of string for the circle.

35.

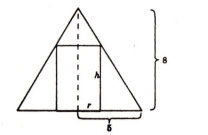

Maximize V

$$V = \pi r^2 h$$

By similar triangles

$$\frac{8}{5} = \frac{h}{5 - r} \quad \text{or} \quad h = \frac{8}{5}(5 - r).$$

$$V(r) = \frac{8\pi}{5}r^2(5 - r), \quad 0 \le r \le 5.$$

$$V'(r) = \frac{8\pi}{5}\left(10r - 3r^2\right), \quad V'(r) = 0 \implies r = 10/3.$$

Since V increases on $(0, 10/3]$ and decreases on $[10/3, 5)$, the abs max of V occurs when $r = 10/3$.

The cylinder with maximal volume has radius 10/3 and height 8/3.

37.

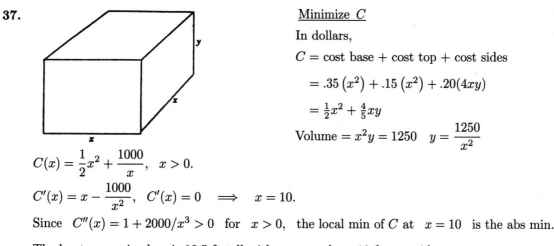

Minimize C

In dollars,

$C = $ cost base + cost top + cost sides

$$= .35\left(x^2\right) + .15\left(x^2\right) + .20(4xy)$$

$$= \tfrac{1}{2}x^2 + \tfrac{4}{5}xy$$

Volume $= x^2 y = 1250 \quad y = \dfrac{1250}{x^2}$

$$C(x) = \frac{1}{2}x^2 + \frac{1000}{x}, \quad x > 0.$$

$$C'(x) = x - \frac{1000}{x^2}, \quad C'(x) = 0 \implies x = 10.$$

Since $C''(x) = 1 + 2000/x^3 > 0$ for $x > 0$, the local min of C at $x = 10$ is the abs min.

The least expensive box is 12.5 ft tall with a square base 10 ft on a side.

39.

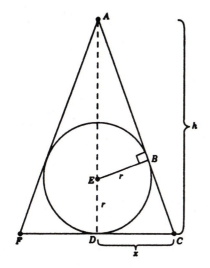

Minimize A

$$A = \tfrac{1}{2}(h)(2x) = hx$$

Triangles ADC and ABE are similar:

$$\frac{AD}{DC} = \frac{AB}{BE} \quad \text{or} \quad \frac{h}{x} = \frac{AB}{r}.$$

Pythagorean Theorem:

$$r^2 + (AB)^2 = (h - r)^2.$$

Thus

$$r^2 + \left(\frac{hr}{x}\right)^2 = (h - r)^2.$$

Solving this equation for h we find that

$$h = \frac{2x^2 r}{x^2 - r^2}.$$

$$A(x) = \frac{2x^3 r}{x^2 - r^2}, \quad x > r.$$

$$A'(x) = \frac{\left(x^2 - r^2\right)\left(6x^2 r\right) - 2x^3 r(2x)}{\left(x^2 - r^2\right)^2} = \frac{2x^2 r\left(x^2 - 3r^2\right)}{\left(x^2 - r^2\right)^2},$$

$$A'(x) = 0 \implies x = r\sqrt{3}.$$

Since A decreases on $\left(r, r\sqrt{3}\,\right]$ and increases on $\left[r\sqrt{3}, \infty\right)$, the local min at $x = r\sqrt{3}$ is the abs min of A. When $x = r\sqrt{3}$, we get $h = 3r$ so that $FC = 2r\sqrt{3}$ and $AF = FC = \sqrt{h^2 + x^2} = 2r\sqrt{3}$. The triangle of least area is equilateral with side of length $2r\sqrt{3}$.

41.

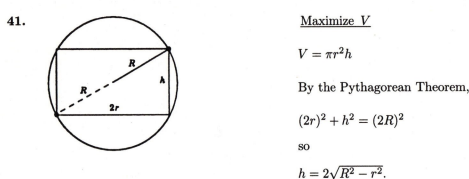

Maximize V

$$V = \pi r^2 h$$

By the Pythagorean Theorem,

$$(2r)^2 + h^2 = (2R)^2$$

so

$$h = 2\sqrt{R^2 - r^2}.$$

$V(r) = 2\pi r^2 \sqrt{R^2 - r^2}, \quad 0 \le r \le R.$

$V'(r) = 2\pi \left[2r\sqrt{R^2 - r^2} - \dfrac{r^3}{\sqrt{R^2 - r^2}} \right] = \dfrac{2\pi r \left(2R^2 - 3r^2\right)}{\sqrt{R^2 - r^2}}$

$V'(r) = 0 \implies r = \frac{1}{3}R\sqrt{6}.$

Since V increases on $\left(0, \frac{1}{3}R\sqrt{6}\,\right]$ and decreases on $\left[\frac{1}{3}R\sqrt{6}, R\right)$, the local max at $r = \frac{1}{3}R\sqrt{6}$ is the abs max.

The cylinder of maximal volume has base radius $\frac{1}{3}R\sqrt{6}$ and height $\frac{2}{3}R\sqrt{3}$.

43.

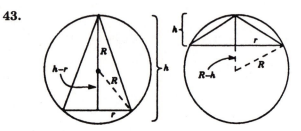

Case 1: $h \ge R$ Case 2: $h \le R$

Maximize V

$$V = \tfrac{1}{3}\pi r^2 h$$

Pythagorean Theorem

Case 1: $(h - R)^2 + r^2 = R^2$

Case 2: $(R - h)^2 + r^2 = R^2$

In both cases

$$r^2 = R^2 - (R - h)^2 = 2hR - h^2.$$

$V(h) = \frac{1}{3}\pi \left(2h^2 R - h^3\right), \quad 0 \le h \le 2R.$

$V'(h) = \frac{1}{3}\pi \left(4hR - 3h^2\right), \quad V'(h) = 0 \quad \text{at} \quad h = \dfrac{4R}{3}.$

Since V increases on $\left(0, \frac{4}{3}R\right]$ and decreases on $\left[\frac{4}{3}R, 2R\right)$, the local max at $h = \frac{4}{3}R$ is the abs max.

The cone of maximal volume has height $\frac{4}{3}R$ and radius $\frac{2}{3}R\sqrt{2}$.

45.

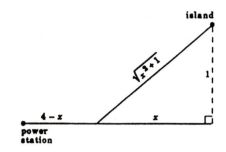

island

power
station

$4-x$ x $\sqrt{x^2+1}$ 1

<u>Minimize C</u>

In units of $10,000$,

$$C = \frac{\text{cost of cable}}{\text{underground}} + \frac{\text{cost of cable}}{\text{under water}}$$

$$\doteq 3(4-x) \qquad + \quad 5\sqrt{x^2+1}.$$

Clearly, the cost is unnecessarily high if

$$x > 4 \quad \text{or} \quad x < 0.$$

$$C(x) = 12 - 3x + 5\sqrt{x^2+1}, \quad 0 \le x \le 4.$$

$$C'(x) = -3 + \frac{5x}{\sqrt{x^2+1}}, \quad C'(x) = 0 \implies x = 3/4.$$

Since the domain of C is closed, the abs min can be identified by evaluating C at each critical point:

$$C(0) = 17, \quad C\left(\tfrac{3}{4}\right) = 16, \quad C(4) = 5\sqrt{17} \cong 20.6.$$

The minimum cost is $160,000$.

47. $P'(\theta) = \dfrac{-mW(m\cos\theta - \sin\theta)}{(m\sin\theta + \cos\theta)^2}; \quad P$ is minimized when $\tan\theta = m$.

49. Minimize $I = \dfrac{a}{x^2} + \dfrac{b}{(s-x)^2}$.

$$I'(x) = -\frac{2a}{x^3} + \frac{2b}{(s-x)^3}, \quad I'(x) = 0 \implies x = \frac{a^{\frac{1}{3}}s}{a^{\frac{1}{3}} + b^{\frac{1}{3}}}.$$

51. The slope of the line through (a,b) and $(x, f(x))$ is $\dfrac{f(x)-b}{x-a}$.

Let $D(x) = [x-a]^2 + [b - f(x)]^2$. Then $D'(x) = 0$

$$\implies \quad 2[x-a] - 2[b - f(x)]f'(x) = 0$$

$$\implies \quad f'(x) = \frac{x-a}{b - f(x)}.$$

53. Set $F(x) = 6x^4 - 16x^3 + 9x^2$. For integral values of x, $F(x) = f(x)$.

$$F'(x) = 24x^3 - 48x^2 + 18x = 6x(4x^2 - 8x + 3) = 6x(2x-1)(2x-3)$$

$F'(x) = 0$ at $x = 0, \ 1/2, \ 3/2$.

$F'(x) < 0$ on $(-\infty, 0) \cup (1/2, 3/2)$; $F'(x) > 0$ on $(0, 1/2) \cup (3/2, \infty)$. F has local minima at $x = 0$

and $x = 3/2$; F has a local maximum at $x = 1/2$. $F(0) = f(0) = 0$, $F(1) = f(1) = -1$, $F(2) =$

$f(2) = 4$; $n = 1$ minimizes $f(n)$.

55. Let x be the number of customers and P the net profit in dollars. Then $0 \leq x \leq 250$ and

$$P(x) = \begin{cases} 12x, & 0 \leq x \leq 50 \\ [12 - 0.06(x - 50)]\, x, & 50 < x \leq 250; \end{cases}$$

$$P'(x) = \begin{cases} 12, & 0 \leq x \leq 50 \\ 15 - 0.12x, & 50 < x \leq 250. \end{cases}$$

The critical points are: $x = 50$, $x = 125$. From $P(0) = 0$, $P(50) = 600$, $P(125) = 937.50$, and $P(250) = 0$, we conclude that the net profit is maximized by servicing 125 customers.

57. $y = mx - \dfrac{1}{400}(m^2 + 1)x^2$. When $y = 0$, $x = \dfrac{800m}{m^2 + 1}$.

Differentiating x with respect to m, $x' = \dfrac{800 - 800m^2}{(m^2 + 1)^2} = 0$

$\implies m = 1$.

59. Driving at ν mph, the trip takes $\dfrac{300}{\nu}$ hours and uses $\left(1 + \dfrac{1}{400}\nu^2\right)\dfrac{300}{\nu}$ gallons of fuel.

Thus, the expenses are:

$$E(\nu) = 2.60\left(1 + \frac{1}{400}\nu^2\right)\frac{300}{\nu} + 20\left(\frac{300}{\nu}\right) = 1.95\nu + \frac{6780}{\nu}, \quad 35 \leq x \leq 70.$$

Differentiating, we get

$$E'(\nu) = 1.95 - \frac{6780}{\nu^2}, \quad \text{and} \quad E'(\nu) = 0 \text{ at } \nu \cong 59$$

E is decreasing on $[35, 59]$ and increasing on $[59, 70]$; the minimal expenses occur when the truck is driven at 59 mph.

61. Minimize $SA = 2\pi rh + 2\pi r^2$, where $2r \leq h < 6$ and $\pi r^2 h = 16\pi$ (hence $h = \dfrac{16}{r^2}$).

Thus $SA = \dfrac{32\pi}{r} + 2\pi r^2$. Differentiating, $SA' = -\dfrac{32\pi}{r^2} + 4\pi r = 0$

$\implies r^3 = 8$, so $r = 2$ feet and $h = 4$ feet. Thus no minimum exists.

63. Swimming at 2 miles per hour, Maggie will reach point B in 1 hour; walking along the shore (a distance of π miles), she will reach point B in $\pi/5 \cong 0.63$ hours. Suppose that she swims to a point C and then walks to B. Let θ be the central angle (measured in radians) determined by the points B and C. See the figure. By the law of cosines, the square of the distance from A to C is given by

$$d^2 = 1^2 + 1^2 - 2(1)(1)\cos(\pi - \theta) = 2 + 2\cos\theta.$$

Therefore the distance d from A to C is $d = \sqrt{2 + 2\cos\theta}$. The distance from C to B is θ.

Now, the total length of time to swim to C and then walk to B is

$$T(\theta) = 2\sqrt{2 + 2\cos\theta} + 5\theta$$

Maggie wants to minimize T.

$$T'(\theta) = \frac{-2\sin\theta}{2\sqrt{2 + 2\cos\theta}} + 5.$$

Setting $T'(\theta) = 0$, we get

$$\frac{-2\sin\theta}{2\sqrt{2 + 2\cos\theta}} + 5 = 0$$

which reduces to $2\cos^2\theta + 25\cos\theta + 23 = 0$

and factors into $(\cos\theta + 1)(2\cos\theta + 23) = 0$.

This implies $\theta = \pi$. Maggie should walk the entire distance!

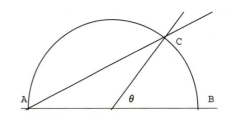

65. (b) The point on the graph of f that is closest to P is: $\left(1 + \sqrt{2}, 2 + \sqrt{2}\right)$

(c) l_{PQ}: $y - \left(2 + \sqrt{2}\right) = \dfrac{1 - \sqrt{2}}{3 - \sqrt{2}}\left(x - \left[1 + \sqrt{2}\right]\right)$

(d) and (e) $l_{PQ} = l_N$.

67. $D(x) = \sqrt{x^2 + (7 - 3x)^2}$;   the point on the line that is closest to the origin is: $\left(\frac{21}{10}, \frac{7}{10}\right)$

PROJECT 4.5

1. Distance over water: $\sqrt{36 + x^2}$.

Distance over land: $12 - x$.

Total energy: $E(x) = W\sqrt{36 + x^2} + L(12 - x)$, $0 \le x \le 12$.

3. (a) $W = kL$, $k > 1$, so $E(x) = kL\sqrt{36 + x^2} + L(12 - x)$, for $0 \le x \le 12$.

$E'(x) = \dfrac{kLx}{\sqrt{36 + x^2}} - L = 0 \implies x = \dfrac{6}{\sqrt{k^2 - 1}}$.

$E'(x) < 0$ on $\left(0, \dfrac{6}{\sqrt{k^2 - 1}}\right)$ and $E'(x) > 0$ on $\left(\dfrac{6}{\sqrt{k^2 - 1}}, 12\right)$, so E has an absolute minimum

at $\dfrac{6}{\sqrt{k^2 - 1}}$.

(b) As k increases, x decreases. As $k \to 1^+$, x increases.

(c) $x = 12 \implies k = \dfrac{\sqrt{5}}{2} \simeq 1.12$.

(d) No

SECTION 4.6

1. (a) f is increasing on $[a, b]$, $[d, n]$; f is decreasing on $[b, d]$, $[n, p]$.

 (b) The graph of f is concave up on (c, k), (l, m);

 The graph of f is concave down on (a, c), (k, l), (m, p).

 The x-coordinates of the points of inflection are: $x = c$, $x = k$, $x = l$, $x = m$.

3. (i) f', (ii) f, (iii) f''.

5. $f'(x) = -x^{-2}$, $f''(x) = 2x^{-3}$;

 concave down on $(-\infty, 0)$, concave up on $(0, \infty)$; no pts of inflection

7. $f'(x) = 3x^2 - 3$, $f''(x) = 6x$;

 concave down on $(-\infty, 0)$, concave up on $(0, \infty)$; pt of inflection $(0, 2)$

9. $f'(x) = x^3 - x$, $f''(x) = 3x^2 - 1$;

 concave up on $\left(-\infty, -\frac{1}{3}\sqrt{3}\right)$ and $\left(\frac{1}{3}\sqrt{3}, \infty\right)$, concave down on $\left(-\frac{1}{3}\sqrt{3}, \frac{1}{3}\sqrt{3}\right)$;

 pts of inflection $\left(-\frac{1}{3}\sqrt{3}, -\frac{5}{36}\right)$ and $\left(\frac{1}{3}\sqrt{3}, -\frac{5}{36}\right)$

11. $f'(x) = -\dfrac{x^2 + 1}{\left(x^2 - 1\right)^2}$, $f''(x) = \dfrac{2x\left(x^2 + 3\right)}{\left(x^2 - 1\right)^3}$;

 concave down on $(-\infty, -1)$ and $(0, 1)$, concave up on $(-1, 0)$ and on $(1, \infty)$;

 pts of inflection $(0, 0)$

13. $f'(x) = 4x^3 - 4x$, $f''(x) = 12x^2 - 4$;

 concave up on $\left(-\infty, -\frac{1}{3}\sqrt{3}\right)$ and $\left(\frac{1}{3}\sqrt{3}, \infty\right)$, concave down on $\left(-\frac{1}{3}\sqrt{3}, \frac{1}{3}\sqrt{3}\right)$;

 pts of inflection $\left(-\frac{1}{3}\sqrt{3}, \frac{4}{9}\right)$ and $\left(\frac{1}{3}\sqrt{3}, \frac{4}{9}\right)$

15. $f'(x) = \dfrac{-1}{\sqrt{x}\left(1 + \sqrt{x}\right)^2}$, $f''(x) = \dfrac{1 + 3\sqrt{x}}{2x\sqrt{x}\left(1 + \sqrt{x}\right)^3}$;

 concave up on $(0, \infty)$; no pts of inflection

17. $f'(x) = \frac{5}{3}(x + 2)^{2/3}$, $f''(x) = \frac{10}{9}(x + 2)^{-1/3}$;

 concave down on $(-\infty, -2)$, concave up on $(-2, \infty)$; pt of inflection $(-2, 0)$

19. $f'(x) = 2\sin x \cos x = \sin 2x$, $f''(x) = 2\cos 2x$;

 concave up on $\left(0, \frac{1}{4}\pi\right)$ and $\left(\frac{3}{4}\pi, \pi\right)$, concave down on $\left(\frac{1}{4}\pi, \frac{3}{4}\pi\right)$;

 pts of inflection $\left(\frac{1}{4}\pi, \frac{1}{2}\right)$ and $\left(\frac{3}{4}\pi, \frac{1}{2}\right)$

21. $f'(x) = 2x + 2\cos 2x, \quad f''(x) = 2 - 4\sin 2x;$

concave up on $\left(0, \frac{1}{12}\pi\right)$ and on $\left(\frac{5}{12}\pi, \pi\right)$, concave down on $\left(\frac{1}{12}\pi, \frac{5}{12}\pi\right)$;

pts of inflection $\left(\dfrac{1}{12}\pi, \dfrac{72 + \pi^2}{144}\right)$ and $\left(\dfrac{5}{12}\pi, \dfrac{72 + 25\pi^2}{144}\right)$

23. points of inflection: $(\pm 3.94822, 10.39228)$

25. points of inflection: $(-3, 0), \ (-2.11652, 2, 39953), \ (-0.28349, -18.43523)$

27. $f(x) = x^3 - 9x$

(a) $f'(x) = 3x^2 - 9 = 3(x^2 - 3)$

$f'(x) \geq 0 \Rightarrow x \leq -\sqrt{3} \ \text{ or } \ x \geq \sqrt{3};$

$f'(x) \leq 0 \Rightarrow -\sqrt{3} \leq x \leq \sqrt{3}.$

Thus, f is increasing on $(-\infty, -\sqrt{3}] \cup [\sqrt{3}, \infty)$

and decreasing on $[-\sqrt{3}, \sqrt{3}]$.

(b) $f(-\sqrt{3}) \cong 10.39$ is a local maximum;

$f(\sqrt{3}) \cong -10.39$ is a local minimum.

(c) $f''(x) = 6x;$

The graph of f is concave up on $(0, \infty)$ and concave down on $(-\infty, 0)$.

(d) point of inflection: $(0, 0)$

29. $f(x) = \dfrac{2x}{x^2 + 1}$

(a) $f'(x) = -\dfrac{2(x + 1)(x - 1)}{(x^2 + 1)^2}$

$f'(x) \geq 0 \Rightarrow -1 \leq x \leq 1;$

$f'(x) \leq 0 \Rightarrow x \leq -1 \ \text{ or } \ x \geq 1.$

v   Thus, f is increasing on $[-1, 1];$

and decreasing on $(-\infty, -1] \cup [1, \infty)$.

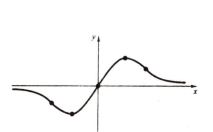

(b) $f(-1) = -1$ is a local minimum;

$f(1) = 1$ is a local maximum.

(c) $f''(x) = \dfrac{4x(x + \sqrt{3})(x - \sqrt{3})}{(x^2 + 1)^3}$

$f''(x) > 0 \Rightarrow x \leq -\sqrt{3} \ \text{ or } \ x \geq \sqrt{3};$

$f''(x) < 0 \Rightarrow -\sqrt{3} < x < \sqrt{3}.$

The graph of f is concave up on $(-\sqrt{3}, 0) \cup (\sqrt{3}, \infty)$ and concave down

on $(-\infty, -\sqrt{3}) \cup (0, \sqrt{3})$.

(d) points of inflection: $(-\sqrt{3}, -\sqrt{3}/2), \ (0, 0), \ (\sqrt{3}, \sqrt{3}/2)$

31. $f(x) = x + \sin x, \quad x \in [-\pi, \pi]$

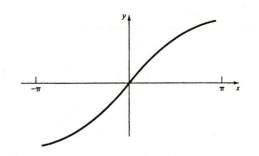

(a) $f'(x) = 1 + \cos x$

   $f'(x) > 0$ on $(-\pi, \pi)$

   Thus, f is increasing on $[-\pi, \pi]$.

(b) No local extrema

(c) $f''(x) = -\sin x$

   $f''(x) > 0$ for $x \in (-\pi, 0)$;

   $f''(x) < 0$ for $x \in (0, \pi)$.

   The graph of f is concave up on $(-\pi, 0)$ and concave down on $(0, \pi)$.

(d) point of inflection: $(0, 0)$

33. $f(x) = \begin{cases} x^3, & x < 1 \\ 3x - 2, & x \geq 1. \end{cases}$

(a) $f'(x) = \begin{cases} 3x^2, & x < 1 \\ 3, & x \geq 1; \end{cases}$

   $f'(x) > 0$ on $(-\infty, 0) \cup (0, \infty)$

   Thus, f is increasing on $(-\infty, \infty)$.

(b) No local extrema

(c) $f''(x) = \begin{cases} 6x, & x < 1 \\ 0, & x \geq 1; \end{cases}$

   $f''(x) > 0$ for $x \in (0, 1)$; $f''(x) < 0$ for $x \in (-\infty, 0)$.

   Thus, the graph of f is concave up on $(0, 1)$ and concave down on $(-\infty, 0)$.

   The graph of f is a straight line for $x \geq 1$.

(d) point of inflection: $(0, 0)$

35.

37.

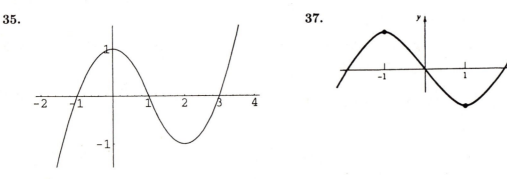

39. Since $f''(x) = 6x - 2(a + b + c)$, set $d = \frac{1}{3}(a + b + c)$. Note that $f''(d) = 0$ and that f is concave down on $(-\infty, d)$ and concave up on (d, ∞); $(d, f(d))$ is a point of inflection.

41. Since $(-1, 1)$ lies on the graph, $1 = -a + b$.

Since $f''(x)$ exists for all x and there is a pt of inflection at $x = \frac{1}{3}$, we must have $f''\left(\frac{1}{3}\right) = 0$. Therefore

$$0 = 2a + 2b.$$

Solving these two equations, we find $a = -\frac{1}{2}$ and $b = \frac{1}{2}$.

Verification: the function

$$f(x) = -\frac{1}{2}x^3 + \frac{1}{2}x^2$$

has second derivative $f''(x) = -3x + 1$. This does change sign at $x = \frac{1}{3}$.

43. First, we require that $\left(\frac{1}{6}\pi, 5\right)$ lie on the curve:

$$5 = \frac{1}{2}A + B.$$

Next we require that $\dfrac{d^2y}{dx^2} = -4A\cos 2x - 9B\sin 3x$ be zero (and change sign) at $x = \frac{1}{6}\pi$:

$$0 = -2A - 9B.$$

Solving these two equations, we find $A = 18$, $B = -4$.

Verification: the function

$$f(x) = 18\cos 2x - 4\sin 3x$$

has second derivative $f''(x) = -72\cos 2x + 36\sin 3x$. This does change sign at $x = \frac{1}{6}\pi$.

45. Let $f'(x) = 3x^2 - 6x + 3$. Then we must have $f(x) = x^3 - 3x^2 + 3x + c$ for some constant c. Note that $f''(x) = 6x - 6$ and $f''(1) = 0$. Since $(1, -2)$ is a point of inflection of the graph of f, $(1, -2)$ must lie on the graph. Therefore,

$$1^3 - 3(1)^2 + 3(1) + c = -2 \quad \text{which implies} \quad c = -3$$

and so $f(x) = x^3 - 3x^2 + 3x - 3$.

47. (a) $p''(x) = 6x + 2a$ is negative for $x < -a/3$, and positive for $x > -a/3$. Therefore, the graph of p has a point of inflection at $x = -a/3$.

(b) $p'(x) = 3x^2 + 2ax + b$. The discriminant of this quadratic is $4a^2 - 12b = 4(a^2 - 3b)$. Thus, p' has two real zeros iff $a^2 > 3b$.

(c) If $a^2 \leq 3b$, then $b \geq a^2/3$ and $p'(x) = 3x^2 + 2ax + b \geq 3x^2 + 2ax + a^2/3 = 3(x + \frac{1}{3}a)^2 \geq 0$ for all x; p is increasing in this case.

49. (a)

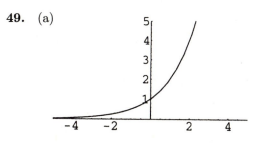

(b) No.   If $f''(x) < 0$ and $f'(x) < 0$ for all x, then

$f(x) < f'(0)x + f(0)$ on $(0, \infty)$, which implies that

$f(x) \to -\infty$ as $x \to \infty$.

51.

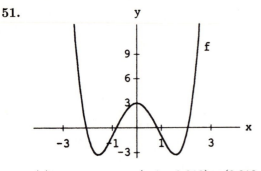

(a) concave up on $(-4, -0.913) \cup (0.913, 4)$

   concave down on $(-0.913, 0.913)$

(b) pts of inflection at $x = -0.913, \quad 0.913$

53.

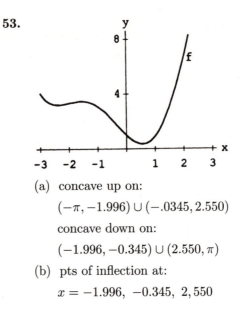

(a) concave up on:

   $(-\pi, -1.996) \cup (-.0345, 2.550)$

   concave down on:

   $(-1.996, -0.345) \cup (2.550, \pi)$

(b) pts of inflection at:

   $x = -1.996, \quad -0.345, \quad 2,550$

55. (a) $f''(x) = 0$ at $x = 0.68824, \; 2.27492, \; 4.00827, \; 5.59494$

   (b) $f''(x) > 0$ on $(0.68824, 2.27492) \cup (4.00827, 5.59494)$

   (c) $f''(x) < 0$ on $(0, 0.68824) \cup (2.27492, 4.00827) \cup (5.59494, 2\pi)$

57. (a) $f''(x) = 0$ at $x = \pm 1, \; 0, \; \pm 0.32654, \; \pm 0.71523$

   (b) $f''(x) > 0$ on $(-0.71523, -0.32654) \cup (0, 0.32654) \cup (0.71523, 1) \cup (1, \infty)$

   (c) $f''(x) < 0$ on $(-\infty, -1) \cup (-1, -0.71523) \cup (-0.32654, 0) \cup (0, 32654, 0.71523)$

SECTION 4.7

1. (a) ∞ (b) $-\infty$ (c) ∞ (d) 1

     (e) 0   (f) $x = -1, \; x = 1$ (g) $y = 0, \; y = 1$

3. vertical: $x = \frac{1}{3}$;   horizontal: $y = \frac{1}{3}$

5. vertical: $x = 2$;   horizontal: none

7. vertical: $x = \pm 3$;   horizontal: $y = 0$

9. vertical: $x = -\frac{4}{3}$;   horizontal: $y = \frac{4}{9}$

11. vertical: $x = \frac{5}{2}$;   horizontal: $y = 0$

13. vertical: none;   horizontal: $y = \pm \frac{3}{2}$

15. vertical: $x = 1$;   horizontal: $y = 0$

17. vertical: none;   horizontal: $y = 0$

19. vertical: $x = \left(2n + \frac{1}{2}\right)\pi$;   horizontal: none

21. $f'(x) = \frac{4}{3}(x + 3)^{1/3}$;   neither

23. $f'(x) = -\frac{4}{5}(2 - x)^{-1/5}$;   cusp

25. $f'(x) = \frac{6}{5}x^{-2/5}\left(1 - x^{3/5}\right)$;   tangent

27. $f(-2)$ undefined;   neither

29. $f'(x) = \begin{cases} \frac{1}{2}(x-1)^{-1/2}, & x > 1 \\ -\frac{1}{2}(1-x)^{-1/2}, & x < 1; \end{cases}$ cusp

31. $f'(x) = \begin{cases} \frac{1}{3}(x+8)^{-23}, & x > -8 \\ -\frac{1}{3}(x+8)^{-2/3}, & x < -8; \end{cases}$ cusp

33. f not continuous at 0; neither

35.

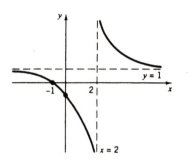

37.

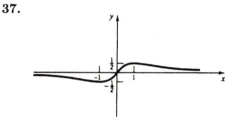

39. $f(x) = x - 3x^{1/3}$

    (a) $f'(x) = 1 - \dfrac{1}{x^{2/3}}$

       f is increasing on $(-\infty, -1] \cup [1, \infty)$

       f is decreasing on $[-1, 1]$

    (b) $f''(x) = \frac{2}{3} x^{-5/3}$

       concave up on $(0, \infty)$; concave down on $(-\infty, 0)$

       vertical tangent at $(0, 0)$

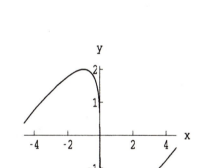

41. $f(x) = \frac{3}{5} x^{5/3} - 3x^{2/3}$

    (a) $f'(x) = x^{2/3} - 2x^{-1/3} = \dfrac{x-2}{x^{1/3}}$

       f is increasing on $(-\infty, 0] \cup [2, \infty)$

       f is decreasing on $[0, 2]$

    (b) $f''(x) = \frac{2}{3} x^{-1/3} + \frac{2}{3} x^{-4/3} = \dfrac{2x+2}{3x^{4/3}}$

       concave up on $(-1, 0) \cup (0, \infty)$; concave down on

         $(-\infty, -1)$

       vertical cusp at $(0, 0)$

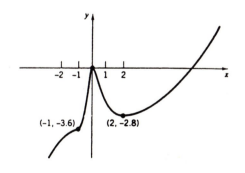

43.

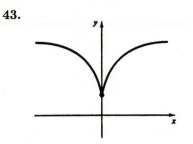

no asymptotes

vertical cusp at $(0,1)$

45.

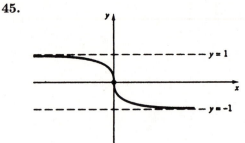

horizontal asymptotes: $y = -1, \; y = 1$

vertical tangent at $(0,0)$

47. (a) p odd; (b) p even.

49.

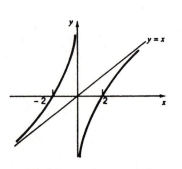

vertical asymptote: $x = 0$

oblique asymptote: $y = x$

51.

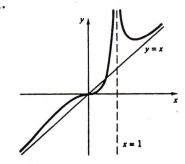

vertical asymptote: $x = 1$

oblique asymptote: $y = x$

53. oblique asymptote: $y = 3x - 4$

55. $y = 1$ is a horizontal asymptote.

$$f(x) = \sqrt{x^2 + 2x} - x = \left(\sqrt{x^2 + 2x} - x\right) \cdot \frac{\sqrt{x^2 + 2x} + x}{\sqrt{x^2 + 2x} + x} = \frac{2x}{\sqrt{x^2 + 2x} + x} \to 1.$$

SECTION 4.8

[Rough sketches; not scale drawings]

1. $f(x) = (x - 2)^2$

$f'(x) = 2(x - 2)$

$f''(x) = 2$

f':

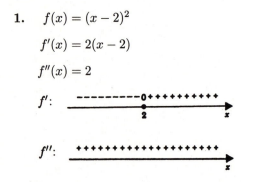

f'':

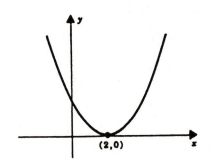

3. $f(x) = x^3 - 2x^2 + x + 1$

$f'(x) = (3x - 1)(x - 1)$

$f''(x) = 6x - 4$

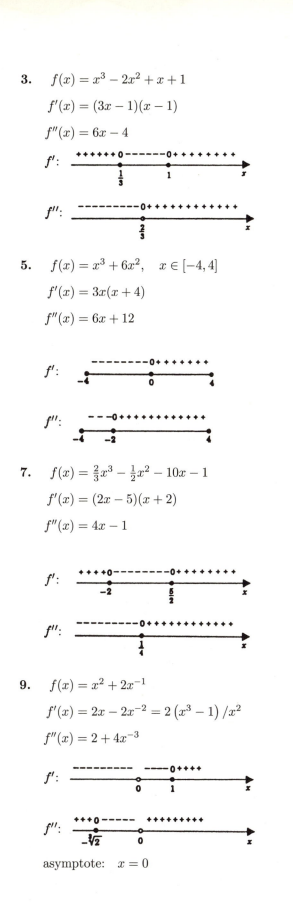

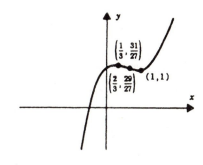

5. $f(x) = x^3 + 6x^2, \quad x \in [-4, 4]$

$f'(x) = 3x(x + 4)$

$f''(x) = 6x + 12$

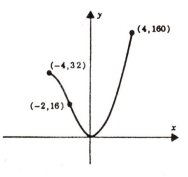

7. $f(x) = \frac{2}{3}x^3 - \frac{1}{2}x^2 - 10x - 1$

$f'(x) = (2x - 5)(x + 2)$

$f''(x) = 4x - 1$

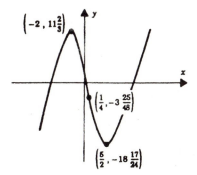

9. $f(x) = x^2 + 2x^{-1}$

$f'(x) = 2x - 2x^{-2} = 2\left(x^3 - 1\right)/x^2$

$f''(x) = 2 + 4x^{-3}$

asymptote: $x = 0$

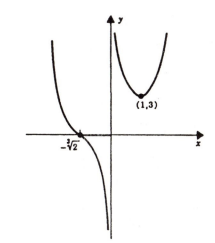

11. $f(x) = (x-4)/x^2$

$f'(x) = (8-x)/x^3$

$f''(x) = (2x-24)/x^4$

f':

$$--- \;+++++++0------$$
$$08$$

f'':

$$--- ------------0++++$$
$$012$$

asymptotes: $x = 0,\; y = 0$

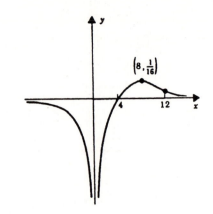

13. $f(x) = 2x^{1/2} - x,\quad x \in [0,4]$

$f'(x) = x^{-1/2}\left(1 - x^{1/2}\right)$

$f''(x) = -\frac{1}{2}x^{-3/2}$

f':

$$+++0-------------$$
$$014$$

f'':

$$------------------$$
$$04$$

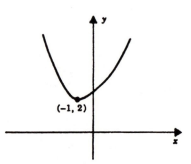

15. $f(x) = 2 + (x+1)^{6/5}$

$f'(x) = \frac{6}{5}(x+1)^{1/5}$

$f''(x) = \frac{6}{25}(x+1)^{-4/5}$

f':

$$----------0++++++++++++$$
$$-1$$

f'':

$$++++++++dne++++++++++$$
$$-1$$

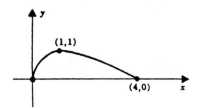

17. $f(x) = 3x^5 + 5x^3$

$f'(x) = 15x^2\left(x^2 + 1\right)$

$f''(x) = 30x\left(2x^2 + 1\right)$

f':

$$++++++++0++++++++++++$$
$$0$$

f'':

$$----------0++++++++++++$$
$$0$$

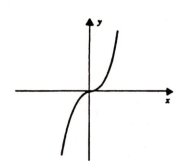

19. $f(x) = 1 + (x-2)^{5/3}$

$f'(x) = \frac{5}{3}(x-2)^{2/3}$

$f''(x) = \frac{10}{9}(x-2)^{-1/3}$

$f':$ ++++++++0++++++++++++
          2

$f'':$ ---------dne++++++++++
          2

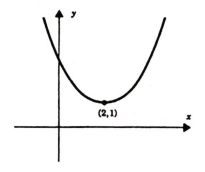

21. $f'(x) = \dfrac{8x}{(x^2+4)^2}$

$f''(x) = 48\dfrac{8(4-3x^2)}{(x^2+4)^3}$

$f'(x) < 0$ on $(-\infty, 0)$

$f'(x) > 0$ on $(0, \infty)$

$f''(x) < 0$ on $(-\infty, -2/\sqrt{3}) \cup (2/\sqrt{3}, \infty)$;

$f''(x) > 0$ on $(-2/\sqrt{3}, 2/\sqrt{3})$;

asymptote: $y = 1$

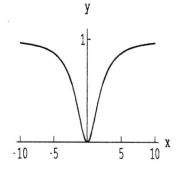

23. $f(x) = \dfrac{x}{(x+3)^2}$

$f'(x) = \dfrac{3-x}{(x+3)^3}$

$f''(x) = \dfrac{2x-12}{(x+3)^4}$

$f':$ --- +++++++0--------
          -3 3

$f'':$ --- ------------0+++
          -3 6

asymptotes: $x = -3,\ y = 0$

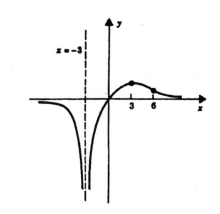

25. $f(x) = \dfrac{x^2}{x^2 - 4}$

$f'(x) = \dfrac{-8x}{(x^2-4)^2}$

$f''(x) = \dfrac{8(3x^2+4)}{(x^2-4)^3}$

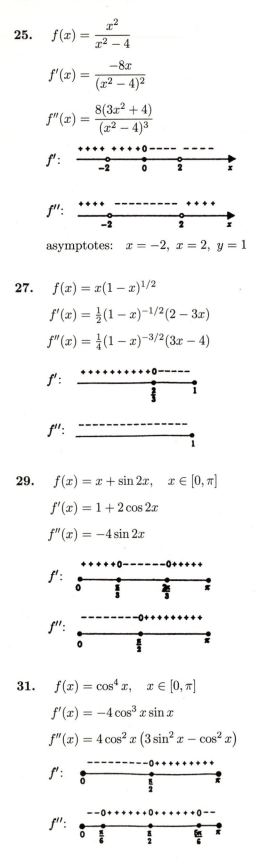

asymptotes: $x = -2,\ x = 2,\ y = 1$

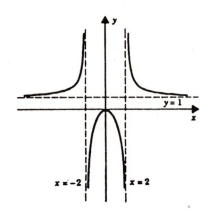

27. $f(x) = x(1-x)^{1/2}$

$f'(x) = \frac{1}{2}(1-x)^{-1/2}(2-3x)$

$f''(x) = \frac{1}{4}(1-x)^{-3/2}(3x-4)$

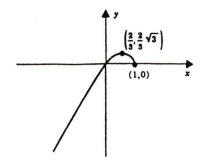

29. $f(x) = x + \sin 2x, \quad x \in [0, \pi]$

$f'(x) = 1 + 2\cos 2x$

$f''(x) = -4\sin 2x$

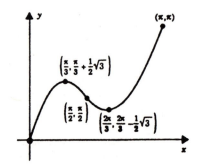

31. $f(x) = \cos^4 x, \quad x \in [0, \pi]$

$f'(x) = -4\cos^3 x \sin x$

$f''(x) = 4\cos^2 x \left(3\sin^2 x - \cos^2 x\right)$

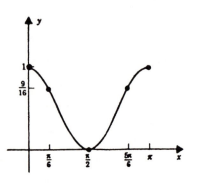

33. $f(x) = 2\sin^3 x + 3\sin x, \quad x \in [0, \pi]$

$f'(x) = 3\cos x \left(2\sin^2 x + 1\right)$

$f''(x) = 9\sin x \left(1 - 2\sin^2 x\right)$

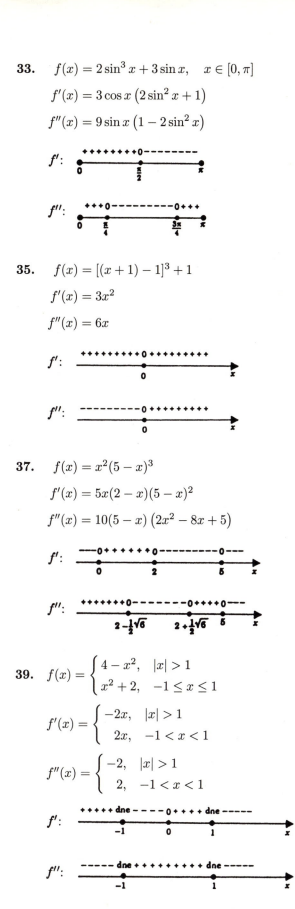

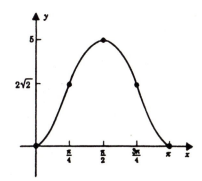

35. $f(x) = [(x+1) - 1]^3 + 1$

$f'(x) = 3x^2$

$f''(x) = 6x$

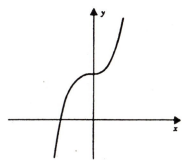

37. $f(x) = x^2(5-x)^3$

$f'(x) = 5x(2-x)(5-x)^2$

$f''(x) = 10(5-x)\left(2x^2 - 8x + 5\right)$

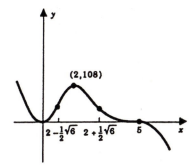

39. $f(x) = \begin{cases} 4 - x^2, & |x| > 1 \\ x^2 + 2, & -1 \le x \le 1 \end{cases}$

$f'(x) = \begin{cases} -2x, & |x| > 1 \\ 2x, & -1 < x < 1 \end{cases}$

$f''(x) = \begin{cases} -2, & |x| > 1 \\ 2, & -1 < x < 1 \end{cases}$

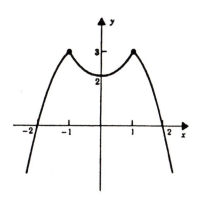

41. $f(x) = x(x-1)^{1/5}$

$f'(x) = \frac{1}{5}(x-1)^{-4/5}(6x-5)$

$f''(x) = \frac{2}{25}(x-1)^{-9/5}(3x-5)$

f':   -----0++++ dne ++++++++++++

          $\frac{5}{6}$ 1

f'':   ++++++++++ dne -------- 0++++

          1    $\frac{5}{3}$

vertical tangent at $(1,0)$

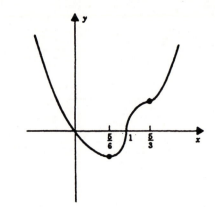

43. $f(x) = x^2 - 6x^{1/3}$

$f'(x) = 2x^{-2/3}\left(x^{5/3} - 1\right)$

$f''(x) = \frac{2}{3}x^{-5/3}\left(3x^{5/3} + 2\right)$

f':   ---------- dne ------ 0+++++

          0   1

f'':   +++0------ dne +++++++++++

      $-\left(\frac{2}{3}\right)^{3/5}$ 0

vertical tangent at $(0,0)$

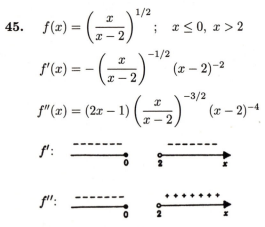

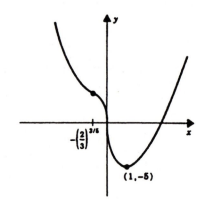

45. $f(x) = \left(\dfrac{x}{x-2}\right)^{1/2}; \quad x \le 0, \ x > 2$

$f'(x) = -\left(\dfrac{x}{x-2}\right)^{-1/2}(x-2)^{-2}$

$f''(x) = (2x-1)\left(\dfrac{x}{x-2}\right)^{-3/2}(x-2)^{-4}$

f':   ---------   --------

         0   2

f'':   --------   +++++++

         0   2

asymptotes: $x = 2, \ y = 1$

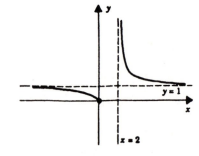

47. $f(x) = x^2 \left(x^2 - 2\right)^{-1/2}, \quad |x| > \sqrt{2}$

$f'(x) = x \left(x^2 - 4\right) \left(x^2 - 2\right)^{-3/2}$

$f''(x) = 2 \left(x^2 + 4\right) \left(x^2 - 2\right)^{-5/2}$

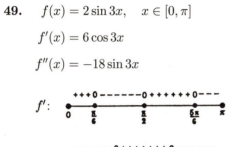

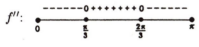

asymptotes: $x = -\sqrt{2}, \; x = \sqrt{2}$

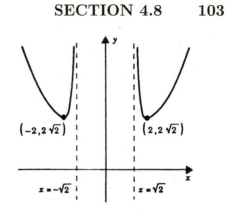

49. $f(x) = 2\sin 3x, \quad x \in [0, \pi]$

$f'(x) = 6\cos 3x$

$f''(x) = -18\sin 3x$

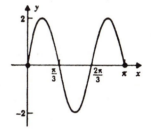

51. $f(x) = 2\tan x - \sec^2 x, \quad x \in (0, \pi/2)$

$\quad\quad = -(1 - \tan x)^2$

$f'(x) = 2\sec^2 x \left(1 - \tan x\right)$

$f''(x) = -2\sec^2 x \left(3\tan^2 x - 2\tan x + 1\right)$

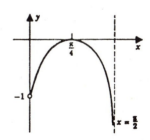

asymptote: $x = \frac{1}{2}\pi$

53. $f(x) = \dfrac{\sin x}{1 - \sin x}, \quad x \in (-\pi, \pi)$

$f'(x) = \dfrac{\cos x}{(1 - \sin x)^2}$

$f''(x) = \dfrac{1 - \sin x + \cos^2 x}{(1 - \sin x)^3}$

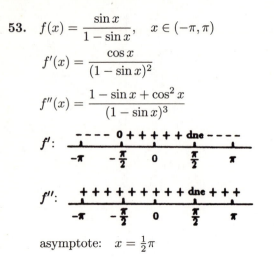

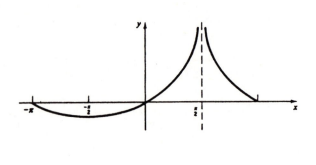

asymptote: $x = \frac{1}{2}\pi$

55. (a) f increases on $(-\infty, -1] \cup (0, 1] \cup [3, \infty)$;

f decreases on $[-1, 0) \cup [1, 3]$; critical points: $x = -1, 0, 1, 3$.

(b)

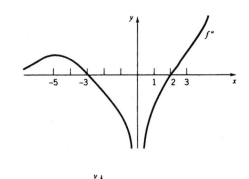

concave up on $(-\infty, -3) \cup (2, \infty)$

concave down on $(-3, 0) \cup (0, 2)$.

(c)

The graph does not necessarily have a horizontal asymptote.

57. $f(x) - x^2 = \dfrac{x^3 - x^{1/3}}{x} - x^2 = \dfrac{x^3 - x^{1/3} - x^3}{x} = -\dfrac{1}{x^{2/3}} \to 0$ as $x \to \pm\infty$.

The figure shows the graphs of f and $g(x) = x^2$ for $x > 0$. Since f and g are even functions, the graphs are symmetric about the y-axis.

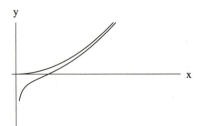

SECTION 4.9

1. $x(5) = -6$; $v(t) = 3 - 2t$ so $v(5) = -7$ and speed $= 7$; $a(t) = -2$ so $a(5) = -2$.

3. $x(1) = 6$; $v(t) = -18/(t+2)^2$ so $v(1) = -2$ and speed $= 2$,
   $a(t) = 36/(t+2)^3$ so $a(1) = 4/3$.

5. $x(1) = 0$, $v(t) = 4t^3 + 18t^2 + 6t - 10$ so $v(1) = 18$ and speed $= 18$,
   $a(t) = 12t^2 + 36t + 6$ so $a(1) = 54$.

7. $x(t) = 5t + 1$, $\quad x'(t) = v(t) = 5$, $\quad x''(t) = a(t) = 0$.
   $v(t) \neq 0$, $a(t) = 0$ for all t.

9. $x(t) = t^3 - 6t^2 + 9t - 1$, $\quad x'(t) = v(t) = 3t^2 - 12t + 9$, $\quad x''(t) = a(t) = 6t - 12$.
   $v(t) = 3(t-1)(t-3)$, $v(t) = 0$ at $t = 1, 3$; $\quad a(t) = 0$ at $t = 2$.

11. A 13. A 15. A and B 17. A 19. A and C

21. The object is moving right when $v(t) > 0$. Here,
    $v(t) = 4t^3 - 36t^2 + 56t = 4t(t-2)(t-7)$; $\quad v(t) > 0$ when $0 < t < 2$ and $7 < t$.

23. The object is speeding up when $v(t)$ and $a(t)$ have the same sign.

$v(t) = 5t^3(4-t)$	sign of $v(t)$:	
$a(t) = 20t^2(3-t)$	sign of $a(t)$:	

    Thus, $\quad 0 < t < 3 \quad$ and $\quad 4 < t$.

25. The object is moving left and slowing down when $v(t) < 0$ and $a(t) > 0$.

$v(t) = 3(t-5)(t+1)$	sign of $v(t)$:	
$a(t) = 6(t-2)$	sign of $a(t)$:	

    Thus, $\quad 2 < t < 5$.

27. The object is moving right and speeding up when $v(t) > 0$ and $a(t) > 0$.

$v(t) = 4t(t-2)(t-4)$	sign of $v(t)$:	
$a(t) = 4(3t^2 - 12t + 8)$	sign of $a(t)$:	

    Thus, $0 < t < 2 - \frac{2}{3}\sqrt{3} \quad$ and $\quad 4 < t$.

29. $x(t) = (t+1)^2(t-9)^3$, $\quad x'(t) = v(t) = 3(t+1)^2(t-9)^2 + 2(t+1)(t-9)^3 = 5(t+1)(t-9)^2(t-3)$.
    The object changes direction at time $t = 3$.

31. $x(t) = (t^3 - 12t)^4$, $\quad x/(t) = v(t) = 4(t^3-12t)^3(3t^2-12) = 12t^3(t+2\sqrt{3})^3(t-2\sqrt{3})^3(t+2)(t-2)$.
    The object changes direction at times $t = 2, 2\sqrt{3}$.

Exercises 33–37. The object is moving right with increasing speed when both $v(t)$ and $a(t)$ are positive.

33. $x(t) = \sin 3t$, $x'(t) = v(t) = 3\cos 3t$, $x''(t) = a(t) = -9\sin 3t$.

$v(t) > 0$ on $(0, \pi/6) \cup (\pi/2, 5\pi/6) \cup (7\pi/6, 3\pi/2) \cup (11\pi/6, 2\pi)$.

$a(t) > 0$ on $(\pi/3, 2\pi/3) \cup (\pi, 4\pi/3) \cup (5\pi/3, 2\pi)$.

$v(t)$ and $a(t)$ are positive on $(\pi/2, 2\pi/3) \cup (7\pi/6, 4\pi/3) \cup (11\pi/6, 2\pi)$.

35. $x(t) = \sin t - \cos t$, $x'(t) = v(t) = \cos t + \sin t$, $x''(t) = a(t) = -\sin t + \cos t$.

$v(t) > 0$ on $(0, 3\pi/4) \cup (7\pi/4, 2\pi)$; $a(t) > 0$ on $(0, \pi/4) \cup (5\pi/4, 2\pi)$.

$v(t)$ and $a(t)$ are positive on $(0, \pi/4) \cup (7\pi/4, 2\pi)$.

37. $x(t) = t + 2\cos t$, $x'(t) = v(t) = 1 - 2\sin t$, $x''(t) = a(t) = -2\cos t$.

$v(t) > 0$ on $(0, \pi/6) \cup (5\pi/6, 2\pi)$; $a(t) > 0$ on $(\pi/2, 3\pi/2)$.

$v(t)$ and $a(t)$ are positive on $(5\pi/6, 3\pi/2)$.

39. Since $v_0 = 0$ the equation of motion is

$$y(t) = -16t^2 + y_0.$$

We want to find y_0 so that $y(6) = 0$. From

$$0 = -16(6)^2 + y_0$$

we get $y_0 = 576$ feet.

41. The object's height and velocity at time t are given by

$$y(t) = -\frac{1}{2}gt^2 + v_0 t \qquad \text{and} \qquad v(t) = -gt + v_0$$

Since the object's velocity at its maximum height is 0, it takes v_0/g seconds to reach maximum height, and

$$y(v_0/g) = -\tfrac{1}{2}g(v_0/g)^2 + v_0(v_0/g) = v_0^2/2g \quad \text{or} \quad v_0^2/19.6 \ \ \text{(meters)}$$

43. At time t, the object's height is $y(t) = -\frac{1}{2}gt^2 + v_0 t + y_0$, and its velocity is $v(t) = -gt + v_0$. Suppose that $y(t_1) = y(t_2)$, $t_1 \neq t_2$. Then

$$-\tfrac{1}{2}gt_1^2 + v_0 t_1 + y_0 = -\tfrac{1}{2}gt_2^2 + v_0 t_2 + y_0$$

$$\tfrac{1}{2}g(t_2^2 - t_1^2) = v_0(t_2 - t_1)$$

$$gt_2 + gt_1 = 2v_0$$

From this equation, we get $-(-gt_1 + v_0) = -gt_2 + v_0$ and so $|v(t_1)| = |v(t_2)|$.

45. In the equation

$$y(t) = -16t^2 + v_0 t + y_0$$

we take $v_0 = -80$ and $y_0 = 224$. The ball first strikes the ground when

$$-16t^2 - 80t + 224 = 0;$$

that is, at $t = 2$. Since

$$v(t) = y'(t) = -32t - 80,$$

we have $v(2) = -144$ so that the speed of the ball the first time it strikes the ground is 144 ft/sec. Thus, the speed of the ball the third time it strikes the ground is $\dfrac{1}{4}\left[\dfrac{1}{4}(144)\right] = 9$ ft/sec.

47. The equation is $y(t) = -16t^2 + 32t$.   (Here $y_0 = 0$ and $v_0 = 32$.)

(a) We solve $y(t) = 0$ to find that the stone strikes the ground at $t = 2$ seconds.

(b) The stone attains its maximum height when $v(t) = 0$. Solving
$v(t) = -32t + 32 = 0,$ we get $t = 1$ and, thus, the maximum height is $y(1) = 16$ feet.

(c) We want to choose v_0 in

$$y(t) = -16t^2 + v_0 t$$

so that $y(t_0) = 36$ when $v(t_0) = 0$ for some time t_0.

From $v(t) = -32t + v_0 = 0$ we get $t_0 = v_0/32$ so that

$$-16\left(\frac{v_0}{32}\right)^2 + v_0\left(\frac{v_0}{32}\right) = 36, \quad \text{or} \quad \frac{v_0{}^2}{64} = 36.$$

Thus, $v_0 = 48$ ft/sec.

49. For all three parts of the problem the basic equation is

$$y(t) = -16t^2 + v_0 t + y_0$$

with

(∗) $y(t_0) = 100$ and $y(t_0 + 2) = 16$

for some time $t_0 > 0$.

We are asked to find y_0 for a given value of v_0.

From (∗) we get

$$16 - 100 = y(t_0 + 2) - y(t_0)$$

$$= [-16(t_0 + 2)^2 + v_0(t_0 + 2) + y_0] - [-16t_0{}^2 + v_0 t_0 + y_0]$$

$$= -64t_0 - 64 + 2v_0$$

so that

$$t_0 = \tfrac{1}{32}(v_0 + 10).$$

Substituting this result in the basic equation and noting that $y(t_0) = 100$, we have

$$-16\left(\frac{v_0 + 10}{32}\right)^2 + v_0\left(\frac{v_0 + 10}{32}\right) + y_0 = 100$$

and therefore

(**) $$y_0 = 100 - \frac{v_0{}^2}{64} + \frac{25}{16}.$$

We use (**) to find the answer to each part of the problem.

(a) $v_0 = 0$ so $y_0 = \frac{1625}{16}$ ft   (b) $v_0 = -5$ so $y_0 = \frac{6475}{64}$ ft   (c) $v_0 = 10$ so $y_0 = 100$ ft

51. Let $y_0 > 0$ be the initial height. The equation of motion becomes:
$0 = -16(8)^2 + 5(8) + y_0$,   so $y_0 = 984$ ft.

53. Let $f_1(t)$ and $f_2(t)$ be the positions of the horses at time t. Consider $f(t) = f_1(t) - f_2(t)$. Let T be the time the horses finish the race. Then $f(0) = f(T) = 0$. By the mean-value theorem there is a c in $(0, T)$ such that $f'(c) = 0$. Hence $f_1'(t) = f_2'(t)$, so the horses had the same speed at time c.

55. The driver must have exceeded the speed limit at some time during the trip. Let $s(t)$ denote the car's position at time t, with $s(0) = 0$ and $s(5/3) = 120$. Then, by the mean-value theorem, there exists at least one number (time) c such that

$$v(c) = s'(c) = \frac{s(5/3) - s(0)}{\frac{5}{3} - 0} = \frac{120}{\frac{5}{3}} = \frac{360}{5} = 72 \text{ mi./hr.}$$

57. If the speed $s(t)$ of the car is less than 60 mi/hr= 1 mi/min, then the distance traveled in 20 minutes is less than 20 miles. Therefore, the car must have gone at least 1 mi/min at some time $t < 20$. Let t_1 be the first instant the car's speed is 1 mi/min. Then the speed $s(t)$ is less than 1 mi/min on the interval $[0, t_1)$ and the distance r traveled in t_1 minutes is less than t_1 miles. Now, by the mean-value theorem, there is a time $c \in [t_1, 20]$ such that

$$s'(c) = \frac{20 - r}{20 - t_1} > \frac{20 - t_1}{20 - t_1} = 1 \; (= 60 \text{ mi/hr}).$$

59. $y(t) = A\sin(\omega t + \phi_0)$, $y'(t) = v(t) = \omega A\cos(\omega t + \phi_0)$, $y''(t) = a(t) = -\omega^2 A\sin(\omega t + \phi_0)$

(a) $y''(t) + \omega^2 y(t) = -\omega^2 A\sin(\omega t + \phi_0) + \omega^2 A\sin(\omega t + \phi_0) = 0$.

(b) $v'(t) = a(t) = -\omega^2 A\sin(\omega t + \phi_0) = 0$ when $\omega t + \phi_0 = n\pi; \; t = \dfrac{n\pi - \phi_0}{\omega}$.

The extreme values of v occur at these times. Now

$$y\left(\frac{n\pi - \phi_0}{\omega}\right) = A\sin\left(\omega\,\frac{n\pi - \phi_0}{\omega} + \phi_0\right) = A\sin(n\pi) = 0.$$

The bob attains maximum speed at the equilibrium position.

(c) $a'(t) - \omega^3 A \cos(\omega t + \phi_0) = 0$ when $\omega t + \phi_0 = (2n-1)\pi/2$; $t = \dfrac{(2n-1)\pi/2 - \phi_0}{\omega}$.

The extreme values of a occur at these times. Now

$$a\left(\frac{(2n-1)\pi/2 - \phi_0}{\omega}\right) = -\omega^2 A \sin\left(\frac{(2n-1)\pi/2 - \phi_0}{\omega} + \phi_0\right) = \pm \omega^2 A.$$

The bob attains these values at

$$y\left(\frac{(2n-1)\pi/2 - \phi_0}{\omega}\right) = A \sin\left(\omega \frac{(2n-1)\pi/2 - \phi_0}{\omega} + \phi_0\right) = A \sin(2n-1)\pi/2 = \pm A.$$

PROJECT 4.9A

1. length of arc $= r\theta$, speed $= \dfrac{d}{dt}[r\theta] = r\dfrac{d\theta}{dt} = r\omega$

3. We know that $d\theta/dt = \omega$ and, at time $t = 0$, $\theta = \theta_0$.   Therefore $\theta = \omega t + \theta_0$.   It follows that

$$x(t) = r\cos(\omega t + \theta_0) \quad and \quad y(t) = r\sin(\omega t + \theta_0).$$

$x(t) = r\cos(\omega t + \theta_0), \quad y(t) = r\sin(\omega t + \theta_0)$

$v(t) = x'(t) = -r\omega \sin(\omega t + \theta_0) = -\omega\, y(t)$

$a(t) = -r\omega^2 \cos(\omega t + \theta_0) = -\omega^2\, x(t)$

$v(t) = y'(t) = r\omega \cos(\omega t + \theta_0) = \omega\, x(t)$

$a(t) = -r\omega^2 \sin(\omega t + \theta_0) = -\omega^2\, y(t)$

5. From Exercise 4, $\dfrac{dA_T}{dt} = \frac{1}{2}r^2\omega \cos\theta$ and $\dfrac{dA_S}{dt} = \frac{1}{2}r^2\omega - \frac{1}{2}r^2\omega \cos\theta$
    Now,

$$\frac{1}{2}r^2\omega \cos\theta = \frac{1}{2}r^2\omega - \frac{1}{2}r^2\omega \cos\theta \quad \Longrightarrow \quad \cos\theta = \frac{1}{2} \quad \Longrightarrow \quad \theta = \frac{\pi}{3}.$$

PROJECT 4.9B

1.
$$\frac{d}{dt}\left[mgy + \tfrac{1}{2}mv^2\right] = mg\frac{dy}{dt} + \frac{1}{2}m\frac{d}{dt}(v^2)$$

$$= mgv + \frac{1}{2}m\left[2v\frac{dv}{dt}\right]$$

$$= mgv + mv\frac{dv}{dt}$$

$$= mgv + mv(-g) \quad (\text{since} \quad dv/dt = a = -g)$$

$$= mgv - mgv = 0$$

3. $y(t) = -\frac{1}{2}gt^2 + y_0 \quad \Longrightarrow \quad gt = \sqrt{2g(y_0 - y)}$
    $v(t) = y'(t) = -gt$ Therefore, $|v(t)| = \sqrt{2g(y_0 - y)}.$

SECTION 4.10

1. $x + 2y = 2$, $\quad \dfrac{dx}{dt} + 2\dfrac{dy}{dt} = 0$

   (a) If $\dfrac{dx}{dt} = 4$, then $\dfrac{dy}{dt} = -2$ units/sec.   (b) If $\dfrac{dy}{dt} = -2$, then $\dfrac{dx}{dt} = 4$ units/sec.

3. $y^2 = 4(x + 2)$, $\quad 2y\dfrac{dy}{dt} = 4\dfrac{dx}{dt}$ $\quad$ and $\quad \dfrac{dx}{dt} = \tfrac{1}{2}y\dfrac{dy}{dt}$

   At the point $(7, 6)$, $\quad \dfrac{dy}{dt} = 3$. Therefore $\quad \dfrac{dx}{dt} = \tfrac{1}{2} \cdot 6 \cdot 3 = 9$ units/sec.

5. Let $s = \sqrt{x^2 + y^2}$ denote the distance to the origin at time t. Since $x = 4\cos t$ and $y = 2\sin t$, we have

$$s(t) = \sqrt{16\cos^2 t + 4\sin^2 t} = \sqrt{12\cos^2 t + 4}$$

$$\frac{ds}{dt} = \frac{1}{2}(12\cos^2 t + 4)^{-1/2}(-24\cos t \sin t)$$

$$= \frac{-12\cos t \sin t}{\sqrt{12\cos^2 t + 4}}$$

   At $t = \pi/4$, $\quad \dfrac{ds}{dt} = \dfrac{-12\cos(\pi/4)\sin(\pi/4)}{\sqrt{12\cos^2(\pi/4) + 4}} = -\tfrac{3}{5}\sqrt{10}$.

7. Find $\quad \dfrac{dx}{dt}$ and $\dfrac{dS}{dt}$ $\quad$ when $\quad V = 27\text{m}^3$

   given that $\quad \dfrac{dV}{dt} = -2\text{m}^3/\text{min}$.

   $(*)$ $\quad V = x^3$, $\quad S = 6x^2$

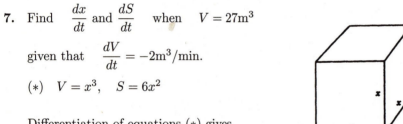

   Differentiation of equations $(*)$ gives

$$\frac{dV}{dt} = 3x^2\frac{dx}{dt} \quad and \quad \frac{dS}{dt} = 12x\frac{dx}{dt}.$$

   When $V = 27$, $x = 3$. Substituting $x = 3$ and $dV/dt = -2$, we get

$$-2 = 27\frac{dx}{dt} \quad \text{so that} \quad \frac{dx}{dt} = -2/27 \quad \text{and} \quad \frac{dS}{dt} = 12(3)\left(\frac{-2}{27}\right) = -8/3.$$

   The rate of change of an edge is $-2/27$ m/min; the rate of change of the surface area is $-8/3$ m^2/min.

9. The area of an equilateral triangle of side x is given by

$$A = \frac{1}{2}x\left(\frac{x\sqrt{3}}{2}\right) = \frac{\sqrt{3}}{4}x^2. \quad \text{Thus} \quad \frac{dA}{dt} = \frac{\sqrt{3}}{2}x\frac{dx}{dt}.$$

   When $x = \alpha$, $\dfrac{dx}{dt} = k$, and $\dfrac{dA}{dt} = \frac{\sqrt{3}}{2}\alpha k\,\text{cm}^2/\text{min}$.

11.

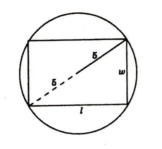

Find $\dfrac{dA}{dt}$ when $l = 6$ in.

given that $\dfrac{dl}{dt} = -2$ in./sec.

By the Pythagorean theorem

$$l^2 + w^2 = 100.$$

Also, $A = lw$. Thus, $A = l\sqrt{100 - l^2}$. Differentiation with respect to t gives

$$\frac{dA}{dt} = l\left(\frac{-l}{\sqrt{100 - l^2}}\right)\frac{dl}{dt} + \sqrt{100 - l^2}\,\frac{dl}{dt}.$$

Substituting $l = 6$ and $dl/dt = -2$, we get

$$\frac{dA}{dt} = 6\left(\frac{-6}{8}\right)(-2) + (8)(-2) = -7.$$

The area is decreasing at the rate of 7 in.^{2}/sec.

13.

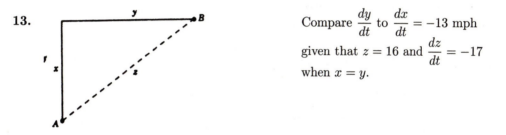

Compare $\dfrac{dy}{dt}$ to $\dfrac{dx}{dt} = -13$ mph

given that $z = 16$ and $\dfrac{dz}{dt} = -17$

when $x = y$.

By the Pythagorean theorem $x^2 + y^2 = z^2$. Thus,

$$2x\frac{dx}{dt} + 2y\frac{dy}{dt} = 2z\frac{dz}{dt}.$$

Since $x = y$ when $z = 16$, we have $x = y = 8\sqrt{2}$ and

$$2(8\sqrt{2})(-13) + 2(8\sqrt{2})\frac{dy}{dt} = 2(16)(-17).$$

Solving for dy/dt, we get

$$-13\sqrt{2} + \sqrt{2}\frac{dy}{dt} = -34 \quad\text{or}\quad \frac{dy}{dt} = \frac{1}{\sqrt{2}}(13\sqrt{2} - 34) \cong -11.$$

Thus, boat A wins the race.

15.   We want to find $\dfrac{dA}{dt}$ when $\dfrac{dx}{dt} = 2$ and $x = 12$.

$$A = \frac{1}{2}x\sqrt{169 - x^2}, \quad\text{so}\quad \frac{dA}{dt} = \left[\frac{1}{2}\sqrt{169 - x^2} - \frac{x^2}{2\sqrt{169 - x^2}}\right]\frac{dx}{dt}$$

When $\dfrac{dx}{dt} = 2$ and $x = 12$, $\dfrac{dA}{dt} = -\dfrac{119}{5}\,ft^2$/sec.

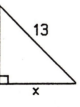

17. We want to find dV/dt when $V = 1000$ ft³ and $P = 5$ lb/in.² given that $dP/dt = -0.05$ lb/in.²/hr.

Differentiating $PV = C$ with respect to t, we get

$$P\frac{dV}{dt} + V\frac{dP}{dt} = 0 \quad \text{so that} \quad 5\frac{dV}{dt} + 1000(-0.05) = 0. \quad \text{Thus,} \frac{dV}{dt} = 10.$$

The volume increases at the rate of 10 ft³/hr.

19.

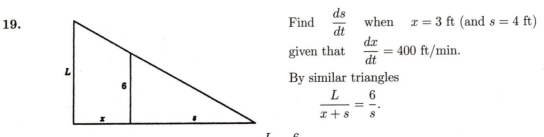

Find $\dfrac{ds}{dt}$ when $x = 3$ ft (and $s = 4$ ft)

given that $\dfrac{dx}{dt} = 400$ ft/min.

By similar triangles

$$\frac{L}{x+s} = \frac{6}{s}.$$

Substitution of $x = 3$ and $s = 4$ gives us $\dfrac{L}{7} = \dfrac{6}{4}$ so that the lamp post is $L = 10.5$ ft tall. Rewriting

$$\frac{10.5}{x+s} = \frac{6}{s} \quad \text{as} \quad s = \frac{4}{3}x$$

and differentiating with respect to t, we find that

$$\frac{ds}{dt} = \frac{4}{3}\frac{dx}{dt} = \frac{1600}{3}.$$

The shadow lengthens at the rate of 1600/3 ft/min.

If the tip of his shadow is at the point z, then

$$z = x + s \quad \text{and} \quad \frac{dz}{dt} = \frac{dx}{dt} + \frac{ds}{dt} = 400 + \frac{1600}{3} = \frac{2800}{3}.$$

The tip of his shadow is moving at the rate of 2800/3 ft./min.

21. Let $W(t) = 150\left(1 + \frac{1}{4000}r\right)^{-2}$. We want to find dW/dt when $r = 400$ given that $dr/dt = 10$ mi/sec. Differentiating with respect to t, we get

$$\frac{dW}{dt} = -300\left(1 + \frac{1}{4000}r\right)^{-3}\left(\frac{1}{4000}\right)\frac{dr}{dt}$$

Now set $r = 400$ and $dr/dt = 10$. Then

$$\frac{dW}{dt} = -300\left(1 + \frac{400}{4000}\right)^{-3}\frac{10}{4000} \cong -0.5634 \,\text{lbs/sec}$$

23.

Find $\dfrac{dh}{dt}$ when $h = 3$ in.

given that $\dfrac{dV}{dt} = -\dfrac{1}{2}$ cu in./min.

By similar triangles

$$r = \tfrac{1}{3}h.$$

Thus $V = \frac{1}{3}\pi r^2 h = \frac{1}{27}\pi h^3$. Differentiating with respect to t, we get

$$\frac{dV}{dt} = \frac{1}{9}\pi h^2\frac{dh}{dt}.$$

When $h = 3$,

$$-\frac{1}{2} = \frac{1}{9}\pi(9)\frac{dh}{dt} \quad \text{and} \quad \frac{dh}{dt} = -\frac{1}{2\pi}.$$

The water level is dropping at the rate of $1/2\pi$ inches per minute.

25. $\dfrac{dV}{dt} = 4\pi r^2 \dfrac{dr}{dt}$ and $\dfrac{dSA}{dt} = 8\pi r \dfrac{dr}{dt}$. Thus when $\dfrac{dSA}{dt} = 4$ and $\dfrac{dr}{dt} = 0.1$

we get $r = \dfrac{5}{\pi}$ and $\dfrac{dV}{dt} = \dfrac{10}{\pi}$ cm^3/min.

27.   Find $\dfrac{d\theta}{dt}$ when $x = 4$ ft

given that $\dfrac{dx}{dt} = 2$ in./min.

$$(*) \quad \tan\frac{\theta}{2} = \frac{3}{x}$$

Differentiation of $(*)$ with respect to t gives

$$\frac{1}{2}\sec^2\frac{\theta}{2}\frac{d\theta}{dt} = -\frac{3}{x^2}\frac{dx}{dt} \quad \text{or} \quad \frac{d\theta}{dt} = -\frac{6}{x^2}\cos^2\frac{\theta}{2}\frac{dx}{dt}.$$

Note that $dx/dt = 2$ in./min$=1/6$ ft/min. When $x = 4$, we have $\cos\theta/2 = 4/5$ and thus

$$\frac{d\theta}{dt} = -\frac{6}{16}\left(\frac{4}{5}\right)^2\left(\frac{1}{6}\right) = -\frac{1}{25}.$$

The vertex angle decreases at the rate of 0.04 rad/min.

29.   Find $\dfrac{dx}{dt}$ when $x = 1$ mi

given that $\dfrac{d\theta}{dt} = 2\pi$ rad/min.

$$(*) \quad \tan\theta = \frac{x}{1/2} = 2x$$

Differentiation of $(*)$ with respect to t gives

$$\sec^2\theta\frac{d\theta}{dt} = 2\frac{dx}{dt}.$$

When $x = 1$, we get $\sec\theta = \sqrt{5}$ and thus $\dfrac{dx}{dt} = 5\pi$.
The light is traveling at 5π mi/min.

31. We have $\tan\theta = \dfrac{x}{40}$, so $\sec^2\theta\dfrac{d\theta}{dt} = \dfrac{1}{40}\dfrac{dx}{dt}$, and $\dfrac{dx}{dt} = 4$.

At $t = 15, x = 60$ and $\sec\theta = \dfrac{\sqrt{5200}}{40}$, so $\dfrac{d\theta}{dt} = \dfrac{2}{65} \cong 0.031$ rad/sec.

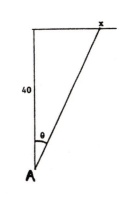

33. We have $\sin\theta = \dfrac{4}{5}$ so $\tan\theta = \dfrac{4}{3} = \dfrac{x}{h}$.   Thus $x = \dfrac{4}{3}h$. $V = 12\left(\dfrac{3+2x+3}{2}h\right) = 36h + 16h^2$.

Thus $\dfrac{dV}{dt} = (36 + 32h)\dfrac{dh}{dt}$, so at $\dfrac{dV}{dt} = 10$ and $h = 2$,

$\dfrac{dh}{dt} = \dfrac{1}{10}$ ft/min.

35. Find $\dfrac{d\theta}{dt}$ when $y = 4$ ft

given that $\dfrac{dx}{dt} = 3$ ft/sec.

$\tan\theta = \dfrac{16}{x}$, $x^2 + (16)^2 = (16 + y)^2$

Differentiating $\tan\theta = 16/x$ with respect to t, we obtain

$$\sec^2\theta\,\frac{d\theta}{dt} = \frac{-16}{x^2}\frac{dx}{dt} \quad \text{and thus} \quad \frac{d\theta}{dt} = \frac{-16}{x^2}\cos^2\theta\,\frac{dx}{dt}.$$

From $x^2 + (16)^2 = (16 + y)^2$ we conclude that $x = 12$, when $y = 4$. Thus

$$\cos\theta = \frac{x}{16 + y} = \frac{12}{20} = \frac{3}{5} \quad \text{and} \quad \frac{d\theta}{dt} = \frac{-16}{(12)^2}\left(\frac{3}{5}\right)^2(3) = \frac{-3}{25}.$$

The angle decreases at the rate of 0.12 rad/sec.

37. Find $\dfrac{d\theta}{dt}$ when $t = 6$ min.

$\tan\theta = \dfrac{100t}{500 + 75t} = \dfrac{4t}{20 + 3t}$

Differentiation with respect to t gives

$$\sec^2\theta\,\frac{d\theta}{dt} = \frac{(20 + 3t)4 - 4t(3)}{(20 + 3t)^2} = \frac{80}{(20 + 3t)^2}.$$

When $t = 6$

$$\tan\theta = \frac{24}{38} = \frac{12}{19} \quad \text{and} \quad \sec^2\theta = 1 + \left(\frac{12}{19}\right)^2 = \frac{505}{361}$$

so that

$$\frac{d\theta}{dt} = \frac{80}{(20 + 3t)^2} \cdot \frac{1}{\sec^2\theta} = \frac{80}{(38)^2} \cdot \frac{361}{505} = \frac{4}{101}.$$

The angle increases at the rate of 4/101 rad/min.

39. Let x be the distance from third base to the player. Then the distance from home plate to the player is given by: $y = \sqrt{(90)^2 + x^2}$. Differentiation with respect to t gives

$$\frac{dy}{dt} = \frac{x}{\sqrt{(90)^2 + x^2}}\frac{dx}{dt}$$

We are given that $dx/dt = -15$. Therefore,

$$\frac{dy}{dt} = -\frac{15x}{\sqrt{(90)^2 + x^2}} \quad \text{and} \quad \frac{dy}{dt}\bigg|_{x=10} = \frac{-15}{\sqrt{82}} \cong 1.66$$

The distance between home plate to the player is decreasing 1.66 ft./sec.

41. Let x be the position of the runner on the track, let y be the distance between the runner and the spectator, and let θ be the central angle indicated in the figure. The runner's speed is 5 meters per second so

$$50\frac{d\theta}{dt} = 5 \text{ and } \frac{d\theta}{dt} = \frac{1}{10}.$$

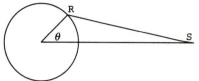

By the law of cosines:

$$y^2 = (50)^2 + (200)^2 - 2(50)(200)\cos\theta.$$

Differentiating with respect to t, we get

$$2y\frac{dy}{dt} = 20,000\sin\theta \frac{d\theta}{dt}$$

$$\frac{dy}{dt} = \frac{10,000\sin\theta}{y}\frac{1}{10}$$

$$\frac{dy}{dt} = \frac{1,000}{y}\sin\theta$$

Now, $\left.\dfrac{dy}{dt}\right|_{y=200} = 5\dfrac{\sqrt{(200)^2 - (25)^2}}{200} = 5\dfrac{\sqrt{39375}}{200} = 4.96.$

The runner is approaching the spectator at the rate of 4.96 meters per second.

43. $\dfrac{dx}{dt} = 4$

SECTION 4.11

1.
$$\begin{aligned}
\Delta V &= (x+h)^3 - x^3 \\
&= (x^3 + 3x^2h + 3xh^2 + h^3) - x^3 \\
&= 3x^2h + 3xh^2 + h^3, \\
dV &= 3x^2h, \\
\Delta V - dV &= 3xh^2 + h^3 \quad \text{(see figure)}
\end{aligned}$$

3. $f(x) = x^{1/3}, \quad x = 1000, \quad h = 2, \quad f'(x) = \frac{1}{3}x^{-2/3}$

$\sqrt[3]{1002} = f(x+h) \cong f(x) + hf'(x) = \sqrt[3]{1000} + 2\left[\frac{1}{3}(1000)^{-2/3}\right] = 10 + \frac{1}{150} \cong 10.0067$

5. $f(x) = x^{1/4}, \quad x = 16, \quad h = -0.5, \quad f'(x) = \frac{1}{4}x^{-3/4}$

$(15.5)^{1/4} = f(x+h) \cong f(x) + hf'(x) = (16)^{1/4} + (-0.5)\left[\frac{1}{4}(16)^{-3/4}\right] = 1\frac{63}{64} \cong 1.9844$

7. $f(x) = x^{3/5}, \quad x = 32, \quad h = 1, \quad f'(x) = \frac{3}{5}x^{-2/5}$

$(33)^{3/5} = f(x+h) \cong f(x) + hf'(x) = (32)^{3/5} + (1)\left[\frac{3}{5}(32)^{-2/5}\right] = 8.15$

9. $f(x) = \sin x, \quad x = \frac{\pi}{4}, \quad h = \frac{\pi}{180}, \quad f'(x) = \cos x$

$\sin 46° = f(x+h) \cong f(x) + hf'(x) = \sin\frac{\pi}{4} + \frac{\pi}{180}\cos\frac{\pi}{4} = \frac{\sqrt{2}}{2}\left(1 + \frac{\pi}{180}\right) \cong 0.719$

11. $f(x) = \tan x, \quad x = \dfrac{\pi}{6}, \quad h = \dfrac{-\pi}{90}, \quad f'(x) = \sec^2 x$

$\tan 28° = f(x+h) \cong f(x) + hf'(x) = \tan\dfrac{\pi}{6} + \left(\dfrac{-\pi}{90}\right)\left(\dfrac{4}{3}\right) = \dfrac{\sqrt{3}}{3} - \dfrac{2\pi}{135} \cong 0.531$

13. $f(2.8) \cong f(3) + (-0.2)f'(3) = 2 + (-0.2)(2) = 1.6$

15. $V(r) = \pi r^2 h; \quad \text{volume} = V(r+t) - V(r) \cong tV'(r) = 2\pi r h t$

17. $V(x) = x^3, \quad V'(x) = 3x^2, \quad \Delta V \cong dV = V'(10)h = 300h$

$|dV| \le 3 \quad \Longrightarrow \quad |300h| \le 3 \quad \Longrightarrow \quad |h| \le 0.01, \quad \text{error} \le 0.01 \text{ feet}$

19. $V(r) = \frac{2}{3}\pi r^3 \quad \text{and} \quad dr = 0.01.$

$V(r+0.01) - V(r) \cong V'(r)(0.01) = 2\pi r^2(0.01)$

$\qquad\qquad\qquad = 2\pi(600)^2(0.01) \quad (50 \text{ ft} = 600 \text{ in})$

$\qquad\qquad\qquad = 22619.5 \text{ in}^3 \quad \text{or} \quad 98 \text{ gallons (approx.)}$

21. $P = 2\pi\sqrt{\dfrac{L}{g}} \quad \text{implies} \quad P^2 = 4\pi^2\dfrac{L}{g}$

Differentiating with respect to t, we have

$\qquad 2P\dfrac{dP}{dt} = \dfrac{4\pi^2}{g}\cdot\dfrac{dL}{dt} = \dfrac{P^2}{L}\cdot\dfrac{dL}{dt} \quad \text{since} \quad \dfrac{P^2}{L} = \dfrac{4\pi^2}{g}.$

Thus $\quad \dfrac{dP}{P} = \dfrac{1}{2}\cdot\dfrac{dL}{L}$

23. $L = 3.26 \text{ ft}, \quad P = 2 \text{ sec}, \quad \text{and} \quad dL = 0.01 \text{ ft}$

$\dfrac{dP}{P} = \dfrac{1}{2}\cdot\dfrac{dL}{L}$

$dP = \dfrac{1}{2}\cdot\dfrac{dL}{L}\cdot P = \dfrac{1}{2}\cdot\dfrac{0.01}{3.26}\cdot 2 \quad dP \cong 0.00307 \text{ sec}$

25. $A(x) = \dfrac{1}{4}\pi x^2, \quad dA = \dfrac{1}{2}\pi x h, \quad \dfrac{dA}{A} = 2\dfrac{h}{x}$

$\dfrac{dA}{A} \le 0.01 \quad \Longleftrightarrow \quad 2\dfrac{h}{x} \le 0.01 \quad \Longleftrightarrow \quad \dfrac{h}{x} \le 0.005 \quad \text{within } \dfrac{1}{2}\%$

27. (a) and (b)

29. $\lim\limits_{h\to 0}\dfrac{g_1(h) + g_2(h)}{h} = \lim\limits_{h\to 0}\dfrac{g_1(h)}{h} + \lim\limits_{h\to 0}\dfrac{g_2(h)}{h} = 0 + 0 = 0$

$\lim\limits_{h\to 0}\dfrac{g_1(h)g_2(h)}{h} = \lim\limits_{h\to 0} h\,\dfrac{g_1(h)g_2(h)}{h^2} = \left(\lim\limits_{h\to 0} h\right)\left(\lim\limits_{h\to 0}\dfrac{g_1(h)}{h}\right)\left(\lim\limits_{h\to 0}\dfrac{g_2(h)}{h}\right) = (0)(0)(0) = 0$

31. Suppose that f is differentiable at x. Then there is a number m such that

$$\lim_{h \to 0} \frac{f(x+h) - f(x)}{h} = m \quad \text{or} \quad \lim_{h \to 0} \frac{f(x+h) - f(x)}{h} - m = 0.$$

Let $g(h) = f(x+h) - f(x) - mh$. Then

$$\frac{g(h)}{h} = \frac{f(x+h) - f(x)}{h} - m \quad \text{and} \quad \lim_{h \to 0} \frac{g(h)}{h} = \lim_{h \to 0} \frac{f(x+h) - f(x)}{h} - m = 0.$$

Now suppose that $f(x+h) - f(x) - mh = g(h)$ where $g(h) = o(h)$. Then

$$\lim_{h \to 0} \left(\frac{f(x+h) - f(x)}{h} - m \right) = \lim_{h \to 0} \frac{g(h)}{h} = 0.$$

Therefore

$$\lim_{h \to 0} \frac{f(x+h) - f(x)}{h} = m$$

and f is differentiable at x. $m = f'(x)$.

PROJECT 4.11

1. $C(x) = 2000 + 50x - \dfrac{x^2}{20}$.

Marginal cost: $C'(x) = 50 - \dfrac{x}{10}$; $C'(20) = 48$

Exact cost of the 21st component: $C(21) - C(20) = 3027.95 - 2980 = 47.95$.

3. (a) Profit function: $P(x) = R(x) - C(x) = 20x - \dfrac{x^2}{50} - (4x + 1400) = 16x - \dfrac{x^2}{50} - 1400$.

Break-even points: $16x - \dfrac{x^2}{50} - 1400 = 0$ or $x^2 - 800x + 70{,}000 = 0$

Thus $x = 100$, or $x = 700$ units.

(b) $P'(x) = 16 - \dfrac{x}{25}$; $P'(x) = 0$ at $x = 400$ units.

(c)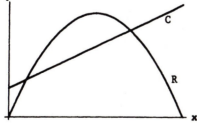

5. (a)

(b) $P(x) = \dfrac{10x}{1 + 0.25x^2} - 4 - 0.75x$

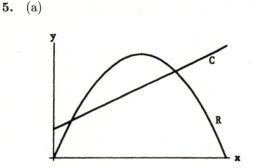

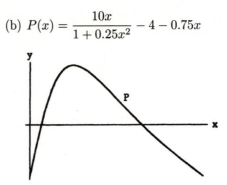

Breakeven points at $x \cong 0.46$, and $x \cong 4.53$

Maximum profit at 175 units

7. $B(x) = \dfrac{R(x)}{x}, \quad B'(x) = \dfrac{xP'(x) - P(x)}{x^2}.$

$B'(x) = 0 \quad$ implies $\quad xP'(x) - P(x) = 0 \quad$ or $\quad P'(x) = \dfrac{P(x)}{x} = B(x).$ The critical points of B are the points where $P'(x) = P(x)/x.$

$B''(x) = \dfrac{x^2[xP''(x)] - [xP'(x) - P(x)](2x)}{x^4}$

At the points where $xP'(x) - P(x) = 0, \quad B''(x) = \dfrac{P''(x)}{x}.$ If $P''(x) < 0$ at such a point, then $B(x)$ is a maximum.

SECTION 4.12

1. (a) $x_{n+1} = \dfrac{1}{2}x_n + 12\left(\dfrac{1}{x_n}\right)$

(b) $x_4 \cong 4.89898$

3. (a) $x_{n+1} = \dfrac{2}{3}x_n + \dfrac{25}{3}\left(\dfrac{1}{x_n}\right)^2$

(b) $x_4 \cong 2.92402$

5. (a) $x_{n+1} = \dfrac{x_n \sin x_n + \cos x_n}{\sin x_n + 1}$

(b) $x_4 \cong 0.73909$

7. (a) $x_{n+1} = \dfrac{6 + x_n}{2\sqrt{x_n + 3} - 1}$

(b) $x_4 \cong 2.30278$

9. Let $f(x) = x^{1/3}$. Then $f'(x) = \frac{1}{3}x^{-2/3}$. The Newton-Raphson method applied to this function gives:

$$x_{n+1} = x_n - \frac{f(x_n)}{f'(x_n)} = x_n - \frac{x_n^{1/3}}{\frac{1}{3}x_n^{-2/3}} = -2x_n.$$

Choose any $x_1 \neq 0$. Then $x_2 = -2x_1, \; x_3 = -2x_2 = 4x_1, \; \cdots,$
$x_n = -2x_{n-1} = (-1)^{n-1}2^n \, x_1, \; \cdots.$

11. (a) f is a continuous function, and $f(1) = -2 < 0, \; f(2) = 3 > 0.$ Thus, f has a root in $(1, 2)$.

   (b) $f'(x) = 6x^2 - 6x$ and $f'(1) = 0.$ Therefore, $x_1 = 1$ will fail to generate values that will approach the root in $(1, 2)$.

   (c) $x_{n+1} = x_n - \dfrac{2x_n^3 - 3x_n^2 - 1}{6x_n^2 - 6x_n};$

    $x_1 = 2, \; x_2 = 1.75, \; x_3 = 1.68254, \; x_4 = 1.67768, \; f(x_4) \cong 0.00020.$

13. (a) $f(x) = x^2 - a$; $f'(x) = 2x$. Substituting into (4.12.1), we have

$$x_{n+1} = x_n - \frac{x_n^2 - a}{2x_n} = \frac{x_n^2 + a}{2x_n} = \frac{1}{2}\left(x_n + \frac{a}{x_n}\right), \quad n \geq 1$$

(b) Let $a = 5$, $x_1 = 2$, and $x_{n+1} = \frac{1}{2}\left(x_n + \frac{a}{x_n}\right)$, $n \geq 1$.

Then $x_2 = 2.25$, $x_3 = 2.23611$, $x_4 = 2.23607$, and $f(x_4) \cong 0.000009045$.

15. (a) Let $f(x) = \frac{1}{x} - a$. Then $f'(x) = -\frac{1}{x^2}$. The Newton-Raphson method applied to this function gives:

$$x_{n+1} = x_n - \frac{\frac{1}{x_n} - a}{-\frac{1}{x_n^2}} = x_n + x_n - ax_n^2 = 2x_n - ax_n^2$$

(b) Let $a = 2.7153$, and $x_1 = 0.3$. Then

$x_2 = 0.35562$, $x_3 = 0.36785$, $x_4 = x_5 = 0.36828$,

Thus $\frac{1}{2.7153} \simeq 0.36828$.

17. (a) $f(x) = \sin x + \frac{1}{2}x^2 - 2x$, $f'(x) = \cos x + x - 2$, $f''(x) = -\sin x + 1$.

$f'(2) = \cos 2 \cong -0.4161 < 0$ and $f'(3) = \cos 3 + 1 \cong 0.0100 > 0$; f' has a zero in $(2, 3)$.

$f''(x) = -\sin x + 1 > 0$ on $(2, 3)$. Therefore, f' has exactly one zero in this interval.

(b) $x_{n+1} = x_n - \frac{\cos x_n + x_n - 2}{1 - \sin x_n} = \frac{2 - x_n \sin x_n - \cos x_n}{1 - \sin x_n}$; $x_3 \cong 2.9883$. Since $f''(x_3) > 0$, f has a local minimum at c.

19. (a) $x_{n+1} = x_n - \frac{x_n + \tan x_n}{1 + \sec^2 x_n}$, $x_1 = 2\pi/3$; $r_1 \cong 2.029$

(b) $x_1 = 5\pi/3$; $r_2 \cong 4.913$

REVIEW EXERCISES

1. f is differentiable on $(-1, 1)$ and continuous on $[-1, 1]$; $f(1) = f(-1) = 0$.

$f'(x) = 3x^2 - 1$; $3c^2 - 1 = 0 \implies c = \pm\frac{\sqrt{3}}{3}$.

3. f is differentiable on $(-2, 3)$ and continuous on $[-2, 3]$.

$$f'(c) = 3c^2 - 2 = \frac{f(3) - f(-2)}{5} = 5 \implies 3c^2 = 7 \implies c = \pm\sqrt{\frac{7}{3}}$$

5. f is differentiable on $(2, 4)$ and continuous on $[2, 4]$.

$$f'(c) = -\frac{2}{(c-1)^2} = \frac{f(4) - f(2)}{2} = -\frac{2}{3} \implies c = 1 + \sqrt{3}.$$

7. $f'(x) = \frac{1 + x^{2/3}}{3x^{2/3}} \neq 0$ for all $x \in (-1, 1)$. $f'(0)$ does not exist. Therefore f is not differentiable on $(-1, 1)$.

9. No. Reason: If such a function did exist, then, by the mean-value theorem, there is a number $c \in (1, 4)$ such that $f'(c) = -\frac{4}{3} < -1$.

11. $f'(x) = 6x(x+1)$.

f':

```
        +        −        +
─────────────────────────────
        −1       0
```

f increases on $(-\infty, -1] \cup [0, \infty)$ and decreases on $[-1, 0]$.

The critical points are $x = -1, 0$; $f(-1) = 2$ is a local max and $f(0) = 1$ is a local min.

13. $f' = (x+2)(x-1)^2(5x+4)$.

f':

```
      +        −        +      +
──────────────────────────────────
      −2       − 4/5    1
```

f increases on $(-\infty, -2] \cup [-\frac{4}{5}, \infty)$ and decreases on $[-2, -\frac{4}{5}]$.

The critical points are $x = -2, 1, -\frac{4}{5}$. $f(-2) = 0$ is a local max and $f(-\frac{4}{5}) \cong -8.981$ is a local min.

15. $f'(x) = \dfrac{1-x^2}{(1+x^2)^2}$.

f':

```
        −        +        −
─────────────────────────────
        −1       1
```

f increases on $[-1, 1]$ and decreases on $(-\infty, -1] \cup [1, \infty)$. The critical points are $x = -1, 1$. $f(-1) = -\frac{1}{2}$ is a local min and $f(1) = \frac{1}{2}$ is a local max.

17. $f'(x) = 3x^2 + 4x + 1 = (3x+1)(x+1)$; critical points: $x = -\frac{1}{3}, -1$.

f':

```
      +        −        +
──────────────────────────────────
   −2      −1     − 1/3        1
```

$f(-2) = -1$ endpoint and abs. min; $f(-1) = 1$ local max; $f(-\frac{1}{3}) = \frac{23}{27}$ local min; $f(1) = 5$ endpoint and abs. max.

19. $f'(x) = 2x - \dfrac{8}{x^3} = \dfrac{2(x^4-8)}{x^3}$; critical point: $x = 4^{1/4} = \sqrt{2}$.

f':

```
     −          +
──────────────────────────────
   1  √2                    4
```

$f(1) = 5$ endpoint max; $f(\sqrt{2}) = 4$ local and abs. min; $f(4) = \frac{65}{4}$ endpoint and abs. max.

21. $f'(x) = \dfrac{-x}{2\sqrt{1-x}} + \sqrt{1-x} = \dfrac{2-3x}{2\sqrt{1-x}}$; critical point: $x = \frac{2}{3}$.

$f(1) = 0$ endpoint min; $f(\frac{2}{3}) = \frac{2\sqrt{3}}{9}$ local and abs. max.

23. $f(x) = \dfrac{3x(x-3)}{(x-4)(x+3)}$. Vertical asymptotes: $x = -3$ and $x = 4$; horizontal asymptote: $y = 3$.

25. $f(x) = x - \dfrac{x}{(x-1)(x^2+x+1)}$. Vertical asymptote: $x = 1$; oblique asymptote: $y = x$.

27. $f'(x) = x^{2/5} - 2x^{-3/5} = \dfrac{x-2}{x^{3/5}}$, vertical cusp.

29. $f(x) = 6 + 4x^3 - 3x^4$, domain: $(-\infty, \infty)$

$f'(x) = 12x^2(1-x)$

critical pts. $x = 0, 1$

$f''(x) = 12x(2 - 3x)$

$f'(x) > 0$ on $(-\infty, 1)$,

$f'(x) < 0$ on $(1, \infty)$

$f''(x) > 0$ on $(0, 2/3)$

$f''(x) < 0$ on $(-\infty, 0) \cup (2/3, \infty)$

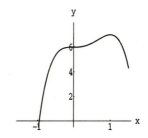

31. $f(x) = \dfrac{2x}{x^2 + 4}$, domain: $(-\infty, \infty)$

symmetric with respect to the origin.

$f'(x) = \dfrac{2\left(4 - x^2\right)}{\left(x^2 + 4\right)^2}$

critical pts. $x = 2$, $x = -2$

$f''(x) = \dfrac{4x\left(x^2 - 12\right)}{\left(x^2 + 4\right)^3}$

$f'(x) > 0$ on $(-2, 2)$,

$f'(x) < 0$ on $(-\infty, 2) \cup (2, \infty)$

$f''(x) > 0$ on $\left(-\sqrt{12}, 0\right) \cup \left(\sqrt{12}, \infty\right)$

$f''(x) < 0$ on $\left(-\infty, -\sqrt{12}\right) \cup \left(0, \sqrt{12}\right)$

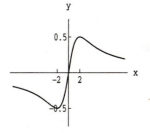

33. $f(x) = x\sqrt{4 - x}$, domain: $(-\infty, 4]$

$f'(x) = \dfrac{8 - 3x}{2\sqrt{4 - x}}$

$f''(x) = \dfrac{3x - 16}{4(4 - x)^{3/2}}$

$f'(x) > 0$ on $\left(-\infty, \frac{8}{3}\right]$,

$f'(x) < 0$ on $\left[\frac{8}{3}, 4\right)$

$f''(x) < 0$ on $(-\infty, 4)$

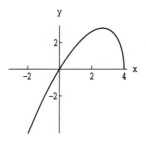

35. $f(x) = \sin x + \sqrt{3}\cos x$, domain $[0, 2\pi]$

$f'(x) = \cos x - \sqrt{3}\sin x$

critical pts. $x = \frac{1}{6}\pi$, $x = \frac{7}{6}\pi$

$f''(x) = -\sin x - \sqrt{3}\cos x$

$f'(x) > 0$ on $[0, \pi/6) \cup (7\pi/6, 2\pi]$

$f'(x) < 0$ on $(\pi/6, 7\pi/6)$

$f''(x) > 0$ on $(2\pi/3, 5\pi/3)$

$f''(x) < 0$ on $[0, \pi/3) \cup (5\pi/3, 2\pi]$

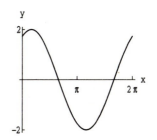

37.

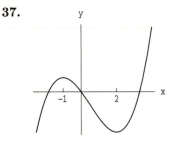

39. Let x be the side length of the square base. Then, since the volume is 27, the height $y = 27/x^2$. The conditions: $x^2 \leq 18$, $0 \leq y \leq 4$ imply $\frac{3\sqrt{3}}{2} \leq x \leq 3\sqrt{2}$. Thus, the surface area S of the box is given by

$$S(x) = 2x^2 + \frac{108}{x}, \quad \frac{3\sqrt{3}}{2} \leq x \leq 3\sqrt{2}$$

$$S'(x) = 4x - \frac{108}{x^2}; \quad S'(x) = 0 \implies 4x^3 = 27 \implies x = 3$$

Now,

$S\left(\frac{3\sqrt{3}}{2}\right) = \frac{27}{2} + 24\sqrt{3} \cong 55.07,$

$S(3) = 54$

$S(3\sqrt{2}) = 36 + 18\sqrt{2} \cong 61.46$

(a) The minimal surface area is 54 sq. ft. at $x = 3$.

(b) The maximal surface area is 61.46 sq. ft. (approx.) at $x = 3\sqrt{2}$.

41. Introduce a rectangular coordinate system so that the rectangle is in the first quadrant and the base and left side lie on the coordinate axes. Let x denote the length of the base, and y the height. Let $P = 2x + 2y$ be the given perimeter. If the rectangle is rotated about the y-axis, then the volume V is given by:

$$V(x) = \frac{\pi P}{2} x^2 - \pi x^3, \ 0 \leq x \leq P/2.$$

Now,

$$V'(x) = \pi P x - 3\pi x^2 \quad \text{and} \quad V'(x) = 0 \implies x = 0 \text{ or } x = \tfrac{1}{3}P.$$

At $x = 0, V = 0$; at $x = P/3$, $y = P/6$, $V = \pi P^3/54$; at $x = P/2$, $V = 0$. The dimensions that generate the cylinder of maximum volume are $x = P/3$, $y = P/6$.

43. $x'(t) = 1 - 2\sin t, \quad x''(t) = -2\cos t.$

The object is slowing down when $x'(t)x''(t) < 0 \implies t \in [0, \tfrac{1}{6}\pi) \cup (\tfrac{1}{2}\pi, \tfrac{5}{6}\pi) \cup (\tfrac{3}{2}\pi, 2\pi].$

45. (a) $x(t) = \sqrt{t+1}$

$x'(t) = v(t) = \dfrac{1}{2\sqrt{t+1}}$

$x''(t) = a(t) = -\dfrac{1}{4(t+1)^{3/2}} = -2\left[v(t)\right]^3,$

$x'''(t) = a'(t) = \dfrac{3}{8(t+1)^{5/2}}$

(b) Position: $x(17) = x(15+2) \cong x(15) + 2x'(15) = 4.25;$

Velocity: $v(17) = v(15+2) \cong v(15) + 2v'(15) = \dfrac{15}{128} \cong 0.1172;$

Acceleration: $a(17) = a(15+2) \cong a(15) + 2a'(15) \cong 0.0032.$

47. The height at time t is: $y(t) = -16t^2 + 8t + y_0$. $y(10) = 0 = -1600 + 80 + y_0 \implies y_0 = 1520$ ft.

49. Let x be the position of the boy and let y be the length of his shadow. Then

$$\frac{dx}{dt} = 168 \quad \text{and} \quad \frac{x+y}{12} = \frac{y}{5} \implies 7y = 5x \quad \text{and} \quad y = \frac{5}{7}x$$

Differentiating implicitly with respect to t, we get

$$7\frac{dy}{dt} = 5\frac{dx}{dt} \implies \frac{dy}{dt} = \left(\frac{5}{7}\right)(168) = 120$$

The shadow is lengthening at the rate of 120 ft/min.

51. Let x be the position of the locomotive at time t and let y be the position of the car. By the law of cosines the distance between the locomotive and the car is given by

$$s^2 = x^2 + y^2 - 2xy \cos 60° = x^2 + y^2 - xy.$$

Differentiating with respect to t, we get

$$2s\frac{ds}{dt} = 2x\frac{dx}{dt} + 2y\frac{dy}{dt} - x\frac{dy}{dt} - y\frac{dx}{dt}.$$

Setting $dx/dt = 60$, $dy/dt = -30$ and converting to feet, gives

$$2s\frac{ds}{dt} = 2x(60)(5280) - 2y(30)(5280) + x(30)(5280) - y(60)(5280) = 5280(150x - 120y).$$

When $x = y = 500$, $s = 500$ and

$$2(500)\frac{ds}{dt} = 5280(30)(500) \implies \frac{ds}{dt} = 15(5280).$$

The distance between the locomotive and the car is increasing at the rate of 15 miles per hour.

53. The cross-section of the water at time t is an isosceles triangle with base x and height y where x and y are related by $\dfrac{\frac{1}{2}x}{y} = \frac{1}{2}$ (similar triangles). Thus $x = y$.

The volume of water in the trough at time t is

$$V = 12(\tfrac{1}{2}xy) = 6y^2.$$

Differentiating with respect to t gives

$$\frac{dV}{dt} = 12y\frac{dy}{dt} \implies \frac{dy}{dt} = \frac{1}{12y}\frac{dV}{dt}.$$

Since $dV/dt = -3$, $\dfrac{dy}{dt} = -\dfrac{1}{4y}$ and $\dfrac{dy}{dt}\bigg|_{t=1.5} = -\dfrac{1}{6}$. The water level is falling at the rate of 1/6 feet per minute.

55. $f(x) = \sqrt{x} + 1/\sqrt{x}, \quad f'(x) = \frac{1}{2}x^{-1/2} - \frac{1}{2}x^{-3/2}$

$f(4.2) \cong f(4) + f'(4)(0.2) = \frac{5}{2} + \frac{3}{16}(0.2) = 2.5375.$

57. $f(x) = \tan x, \quad f'(x) = \sec^2 x$

$f(43°) = f(\frac{1}{4}\pi - \frac{1}{90}\pi) \cong f(\pi/4) + f'(\pi/4)(-\pi/90) = 1 - 2(\pi/90) = 1 - (\pi/45) \cong 0.9302.$

59. $f(x) = x^3 - 10$

(a) $x_{n+1} = x_n - \dfrac{x_n^3 - 10}{3x_n^2} = \dfrac{2x_n^3 + 10}{3x_n^2}$

(b) $x_4 \cong 2.15443; \quad f(2.15443) \cong -0.00007$

CHAPTER 5

SECTION 5.2

1. $L_f(P) = 0(\frac{1}{4}) + \frac{1}{2}(\frac{1}{4}) + 1(\frac{1}{2}) = \frac{5}{8}$, $U_f(P) = \frac{1}{2}(\frac{1}{4}) + 1(\frac{1}{4}) + 2(\frac{1}{2}) = \frac{11}{8}$

3. $L_f(P) = \frac{1}{4}(\frac{1}{2}) + \frac{1}{16}(\frac{1}{4}) + 0(\frac{1}{4}) = \frac{9}{64}$, $U_f(P) = 1(\frac{1}{2}) + \frac{1}{4}(\frac{1}{4}) + \frac{1}{16}(\frac{1}{4}) = \frac{37}{64}$

5. $L_f(P) = 1(\frac{1}{2}) + \frac{9}{8}(\frac{1}{2}) = \frac{17}{16}$, $U_f(P) = \frac{9}{8}(\frac{1}{2}) + 2(\frac{1}{2}) = \frac{25}{16}$

7. $L_f(P) = \frac{1}{16}(\frac{3}{4}) + 0(\frac{1}{2}) + \frac{1}{16}(\frac{1}{4}) + \frac{1}{4}(\frac{1}{2}) = \frac{3}{16}$, $U_f(P) = 1(\frac{3}{4}) + \frac{1}{16}(\frac{1}{2}) + \frac{1}{4}(\frac{1}{4}) + 1(\frac{1}{2}) = \frac{43}{32}$

9. $L_f(P) = 0(\frac{\pi}{6}) + \frac{1}{2}(\frac{\pi}{3}) + 0(\frac{\pi}{2}) = \frac{\pi}{6}$, $U_f(P) = \frac{1}{2}(\frac{\pi}{6}) + 1(\frac{\pi}{3}) + 1(\frac{\pi}{2}) = \frac{11\pi}{12}$

11. (a) $L_f(P) \le U_f(P)$ but $3 \not\le 2$.

(b) $L_f(P) \le \displaystyle\int_{-1}^{1} f(x)\,dx \le U_f(P)$ but $3 \not\le 2 \le 6$.

(c) $L_f(P) \le \displaystyle\int_{-1}^{1} f(x)\,dx \le U_f(P)$ but $3 \le 10 \not\le 6$.

13. (a) $L_f(P) = -3x_1(x_1 - x_0) - 3x_2(x_2 - x_1) - \cdots - 3x_n(x_n - x_{n-1})$,

$U_f(P) = -3x_0(x_1 - x_0) - 3x_1(x_2 - x_1) - \cdots - 3x_{n-1}(x_n - x_{n-1})$

(b) For each index i

$$-3x_i \le -\tfrac{3}{2}(x_i + x_{i-1}) \le -3x_{i-1}.$$

Multiplying by $\Delta x_i = x_i - x_{i-1}$ gives

$$-3x_i\,\Delta x_i \le -\tfrac{3}{2}(x_i^2 - x_{i-1}^2) \le -3x_{i-1}\,\Delta x_i.$$

Summing from $i = 1$ to $i = n$, we find that

$$L_f(P) \le -\tfrac{3}{2}(x_1^2 - x_0^2) - \cdots - \tfrac{3}{2}(x_n^2 - x_{n-1}^2) \le U_f(P).$$

The middle sum collapses to

$$-\tfrac{3}{2}(x_n^2 - x_0^2) = -\tfrac{3}{2}(b^2 - a^2).$$

Thus

$$L_f(P) \le -\frac{3}{2}(b^2 - a^2) \le U_f(P) \quad \text{so that} \quad \int_a^b -3x\,dx = -\frac{3}{2}(b^2 - a^2).$$

15. $\displaystyle\int_{-1}^{2} (x^2 + 2x - 3)\,dx$

17. $\displaystyle\int_{0}^{2\pi} t^2 \sin(2t + 1)\,dt$

19.

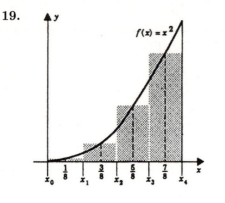

21. $\Delta x_1 = \Delta x_2 = \frac{1}{8}, \quad \Delta x_3 = \Delta x_4 = \Delta x_5 = \frac{1}{4}$

$m_1 = 0, \quad m_2 = \frac{1}{4}, \quad m_3 = \frac{1}{2}, \quad m_4 = 1, \quad m_5 = \frac{3}{2}$

$f(x_1^*) = \frac{1}{8}, \quad f(x_2^*) = \frac{3}{8}, \quad f(x_3^*) = \frac{3}{4}, \quad f(x_4^*) = \frac{5}{4}, \quad f(x_5^*) = \frac{3}{2}$

$M_1 = \frac{1}{4}, \quad M_2 = \frac{1}{2}, \quad M_3 = 1, \quad M_4 = \frac{3}{2}, \quad M_5 = 2$

(a) $L_f(P) = \frac{25}{32}$ (b) $S^*(P) = \frac{15}{16}$ (c) $U_f(P) = \frac{39}{32}$

23. $L_f(P) = x_0{}^3(x_1 - x_0) + x_1{}^3(x_2 - x_1) + \cdots + x_{n-1}^3(x_n - x_{n-1})$

$U_f(P) = x_1{}^3(x_1 - x_0) + x_2{}^3(x_2 - x_1) + \cdots + x_n{}^3(x_n - x_{n-1})$

For each index i

$$x_{i-1}^3 \le \tfrac{1}{4}\left(x_i{}^3 + x_i{}^2 x_{i-1} + x_i x_{i-1}^2 + x_{i-1}^3\right) \le x_i{}^3$$

and thus by the hint

$$x_{i-1}^3(x_i - x_{i-1}) \le \tfrac{1}{4}\left(x_i{}^4 - x_{i-1}^4\right) \le x_i{}^3(x_i - x_{i-1}).$$

Adding up these inequalities, we find that

$$L_f(P) \le \tfrac{1}{4}\left(x_n{}^4 - x_0{}^4\right) \le U_f(P).$$

Since $x_n = 1$ and $x_0 = 0$, the middle term is $\dfrac{1}{4}$: $\displaystyle\int_0^1 x^3\,dx = \dfrac{1}{4}.$

25. Necessarily holds: $L_g(P) \le \int_a^b g(x)\,dx < \int_a^b f(x)\,dx \le U_f(P).$

27. Necessarily holds: $L_g(P) \le \int_a^b g(x)\,dx < \int_a^b f(x)\,dx$

29. Necessarily holds: $U_f(P) \ge \int_a^b f(x)\,dx > \int_a^b g(x)\,dx$

31. Let $P = \{x_0, x_1, x_2, \ldots, x_n\}$ be a regular partition of $[a, b]$ and let $\Delta x = (b - a)/n$. Since f is increasing on $[a, b]$,

$$L_f(P) = f(x_0)\Delta x + f(x_1)\Delta x + \cdots + f(x_{n-1})\Delta x$$

and

$$U_f(P) = f(x_1)\Delta x + f(x_2)\Delta x + \cdots + f(x_n)\Delta x.$$

Now,

$$U_f(P) - L_f(P) = [f(x_n) - f(x_0)]\Delta x = [f(b) - f(a)]\Delta x.$$

33. (a) $f'(x) = \dfrac{x}{\sqrt{1+x^2}} > 0$ for $x \in [0, 2]$. Thus, f is increasing on $[0, 2]$.

(b) Let $P = \{x_0, x_1, \ldots, x_n\}$ be a regular partition of $[0, 2]$ and let $\Delta x = 2/n$

By Exercise 30,

$$\int_0^2 f(x)\,dx - L_f(P) \le |f(2) - f(0)|\frac{2}{n} = \frac{2(\sqrt{5}-1)}{n} \cong \frac{2.47}{n}$$

It now follows that $\int_0^2 f(x)\,dx - L_f(P) < 0.1$ if $n > 25$.

(c) $\displaystyle\int_0^2 f(x)\,dx \cong 2.96$

35. Let S be the set of positive integers for which the statement is true. Since $\dfrac{1(2)}{2} = 1, 1 \in S$. Assume that $k \in S$. Then

$$1 + 2 + \cdots + k + k + 1 = (1 + 2 + \cdots + k) + k + 1 = \frac{k(k+1)}{2} + k + 1$$

$$= \frac{(k+1)(k+2)}{2}$$

Thus, $k + 1 \in S$ and so S is the set of positive integers.

37. Let $f(x) = x$ and let $P = \{x_0, x_1, x_2, \ldots, x_n\}$ be a regular partition of $[0, b]$. Then $\Delta x = b/n$ and $x_i = \dfrac{ib}{n}$, $i = 0, 1, 2, \ldots, n$.

(a) Since f is increasing on $[0, b]$,

$$L_f(P) = \left[f(0) + f\left(\frac{b}{n}\right) + f\left(\frac{2b}{n}\right) + \cdots + f\left(\frac{(n-1)b}{n}\right)\right]\frac{b}{n}$$

$$= \left[0 + \frac{b}{n} + \frac{2b}{n} + \cdots + \frac{(n-1)b}{n}\right]\frac{b}{n}$$

$$= \frac{b^2}{n^2}[1 + 2 + \cdots + (n-1)]$$

(b)

$$U_f(P) = \left[f\left(\frac{b}{n}\right) + f\left(\frac{2b}{n}\right) + \cdots + f\left(\frac{(n-1)b}{n}\right) + f(b)\right]\frac{b}{n}$$

$$= \left[\frac{b}{n} + \frac{2b}{n} + \cdots + \frac{(n-1)b}{n} + b\right]\frac{b}{n}$$

$$= \frac{b^2}{n^2}[1 + 2 + \cdots + (n-1) + n]$$

(c) By Exercise 35,

$$L_f(P) = \frac{b^2}{n^2} \cdot \frac{(n-1)n}{2} = \frac{1}{2}b^2\left(\frac{n^2-n}{n^2}\right) = \frac{1}{2}b^2\left(1 - \frac{1}{n}\right) = \frac{1}{2}b^2(1 - ||P||)$$

$$L_f(P) = \frac{b^2}{n^2} \cdot \frac{n(n+1)}{2} = \frac{1}{2}b^2\left(\frac{n^2+n}{n^2}\right) = \frac{1}{2}b^2\left(1 + \frac{1}{n}\right) = \frac{1}{2}b^2(1 + ||P||)$$

(d) For any partition $P, L_f(P) \leq \S^*(P) \leq U_f(P)$. Since

$$\lim_{\|P\|\to 0} L_f(P) = \lim_{\|P\|\to 0} U_f(P) = \frac{1}{2}b^2,$$

$\lim_{\|P\|\to 0} S^*(P) = \frac{1}{2}b^2$ by the pinching theorem.

39. Choose each x_i^* so that $f(x_i^*) = m_i$. Then $S_i^*(P) = L_f(P)$.

Similarly, choosing each x_i^* so that $f(x_i^*) = M_i$ gives $S_i^*(P) = U_f(P)$.

Also, choosing each x_i^* so that $f(x_i^*) = \frac{1}{2}(m_i + M_i)$ (they exist by the intermediate value theorem)

gives

$$S_i^*(P) = \frac{1}{2}(m_1 + M_1)\Delta x_1 + \cdots + \frac{1}{2}(m_n + M_n)\Delta x_n$$

$$= \frac{1}{2}[m_1 \Delta x_1 + \cdots + m_n \Delta x_n + M_1 \Delta x_1 + \cdots + M_n \Delta x_n]$$

$$= \frac{1}{2}[L_f(P) + U_f(P)].$$

41. (a) $L_f(P) \cong 0.6105, \quad U_f(P) \cong 0.7105$

(b) $\frac{1}{2}[L_f(P) + U_f(P)] \cong 0.6605$ (c) $S^*(P) \cong 0.6684$

43. (a) $L_f(P) \cong 0.53138, \quad U_f(P) \cong 0.73138$

(b) $\frac{1}{2}[L_f(P) + U_f(P)] \cong 0.63138$ (c) $S^*(P) \cong 0.63926$

SECTION 5.3

1. (a) $\int_0^5 f(x)\, dx = \int_0^2 f(x)\, dx + \int_2^5 f(x)\, dx = 4 + 1 = 5$

(b) $\int_1^2 f(x)\, dx = \int_0^2 f(x)\, dx - \int_0^1 f(x)\, dx = 4 - 6 = -2$

(c) $\int_1^5 f(x)\, dx = \int_0^5 f(x)\, dx - \int_0^1 f(x)\, dx = 5 - 6 = -1$

(d) 0   (e) $\int_2^0 f(x)\, dx = -\int_0^2 f(x)\, dx = -4$

(f) $\int_5^1 f(x)\, dx = -\int_1^5 f(x)\, dx = 1$

3. With $P = \left\{1, \frac{3}{2}, 2\right\}$ and $f(x) = \frac{1}{x}$, we have

$$0.5 \leq \frac{7}{12} = L_f(P) \leq \int_1^2 \frac{dx}{x} \leq U_f(P) = \frac{5}{6} < 1.$$

5. (a) $F(0) = 0$ (b) $F'(x) = x\sqrt{x+1}$ (c) $F'(2) = 2\sqrt{3}$

     (d) $F(2) = \int_0^2 t\sqrt{t+1}\,dt$ (e) $-F(x) = \int_x^0 t\sqrt{t+1}\,dt$

7. $F'(x) = \dfrac{1}{x^2+9}$;   (a) $\dfrac{1}{10}$ (b) $\dfrac{1}{9}$ (c) $\dfrac{4}{37}$ (d) $\dfrac{-2x}{(x^2+9)^2}$

9. $F'(x) = -x\sqrt{x^2+1}$;   (a) $\sqrt{2}$ (b) 0 (c) $-\frac{1}{4}\sqrt{5}$ (d) $-\left(\sqrt{x^2+1} + \dfrac{x^2}{\sqrt{x^2+1}}\right)$

11. $F'(x) = \cos \pi x$;   (a) -1 (b) 1 (c) 0 (d) $-\pi \sin \pi x$

13. (a) Since $P_1 \subseteq P_2$, $U_f(P_2) \leq U_f(P_1)$ but $5 \not\leq 4$.

     (b) Since $P_1 \subseteq P_2$, $L_f(P_1) \leq L_f(P_2)$ but $5 \not\leq 4$.

15. constant functions

17. $F'(x) = \dfrac{x-1}{1+x^2} = 0 \implies x = 1$ is a critical number.

     $F''(x) = \dfrac{(1+x^2) - 2x(x-1)}{(1+x^2)^2}$, so $F''(1) = \dfrac{1}{2} > 0$ means $x = 1$ is a local minimum.

19. (a) $F'(x) = \dfrac{1}{x} > 0$ for $x > 0$.   (b) $F''(x) = -\dfrac{1}{x^2} < 0$ for $x > 0$.

      Thus, F is increasing on $(0, \infty)$;   The graph of F is concave down on $(0, \infty)$;

      there are no critical numbers.   there are no points of inflection.

     (c)

21. (a) F is differentiable, therefore continuous   (b) $F'(x) = f(x)$ f is differentiable; $F''(x) = f'(x)$

     (c) $F'(1) = f(1) = 0$ (d) $F''(1) = f'(1) > 0$

     (e) $f(1) = 0$ and f increasing $(f' > 0)$ implies $f < 0$ on $(0, 1)$ and $f > 0$ on $(1, \infty)$.

       Since $F' = f$, F is decreasing on $(0, 1)$ and increasing on $(1, \infty)$;

       $F(0) = 0$ implies $F(1) < 0$.

23. (a)

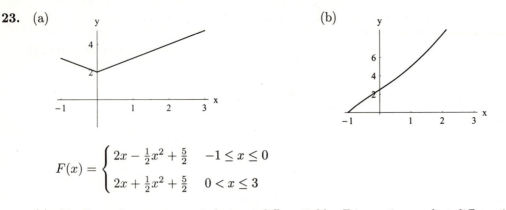

(b)

$$F(x) = \begin{cases} 2x - \frac{1}{2}x^2 + \frac{5}{2} & -1 \leq x \leq 0 \\ 2x + \frac{1}{2}x^2 + \frac{5}{2} & 0 < x \leq 3 \end{cases}$$

(c) f is discontinuous at $x = 0$, but not differentiable; F is continuous but differentiable at $x = 0$.

25. Let $u = x^3$. Then $F(u) = \int_1^u t \cos t \, dt$ and

$$\frac{dF}{dx} = \frac{dF}{du} \frac{du}{dx} = u \cos u \, (3x^2) = 3x^5 \cos x^3.$$

27. $F(x) = \int_{x^2}^1 (t - \sin^2 t) \, dt = -\int_1^{x^2} (t - \sin^2 t) \, dt$. Let $u = x^2$. Then

$$\frac{dF}{dx} = \frac{dF}{du} \frac{du}{dx} = -(u - \sin^2 u)(2x) = 2x \left[\sin^2(x^2) - x^2 \right].$$

29. (a) $F(0) = 0$

(b) $F'(0) = 2 + \dfrac{\sin 2(0)}{1 + 0^2} = 2$

(c) $F''(0) = \dfrac{(1+0)^2 2 \cos 2(0) - \sin 2(0)(2)(0)}{(1+0)^2} = 2$

31. $f(x) = \dfrac{d}{dx} \left(\dfrac{2x}{4 + x^2} \right) = \dfrac{8 - 2x^2}{(4 + x^2)^2}$

(a) $f(0) = \dfrac{1}{2}$ (b) $f(x) = 0$ at $x = -2, \, 2$

33. By the hint $\dfrac{F(b) - F(a)}{b - a} = F'(c)$ for some c in (a, b). The result follows by observing that

$$F(b) = \int_a^b f(t) \, dt, \quad F(a) = 0, \quad \text{and} \quad F'(c) = f(c).$$

33. Set $G(x) = \int_a^x f(t) \, dt$. Then $F(x) = \int_c^a f(t) \, dt + G(x)$. First, note that $\int_c^a f(t) \, dt$

is a constant. By (5.3.5) G, and thus F, is continuous on $[a, b]$, is differentiable

on (a, b), and $F'(x) = G'(x) = f(x)$ for all x in (a, b).

35. (a) $F'(x) = f(x) = G'(x)$, on $[a, b]$. Therefore, by Theorem 4.2.4, F and G differ by a constant.

(b) $F(x) = -\int_a^c f(t) \, dt + \int_a^x f(t) \, dt$ and $G(x) = -\int_a^d f(t) \, dt + \int_a^x f(t) \, dt$.

Thus $F(x) - G(x) = -\int_a^c f(t) \, dt + \int_a^d f(t) \, dt = \int_c^d f(t) \, dt$, a constant

37. (a) $F'(x) = 0$ at $x = -1, 4$; F is increasing on $(-\infty, -1]$, $[4, \infty)$; F is decreasing on $[-1, 4]$

(b) $F''(x) = 0$ at $x = \frac{3}{2}$;   the graph of F is concave up on $\left(\frac{3}{2}, \infty\right)$; concave down on $\left(-\infty, \frac{3}{2}\right)$

39. (a) $F'(x) = 0$ at $x = 0, \frac{\pi}{2}, \pi, \frac{3\pi}{2}, 2\pi$

F is increasing on $[\frac{\pi}{2}, \pi], [\frac{3\pi}{2}, 2\pi]$; F is decreasing on $[0, \frac{\pi}{2}], [\pi, \frac{3\pi}{2}]$

(b) $F''(x) = 0$ at $x = \frac{\pi}{4}, \frac{3\pi}{4}, \frac{5\pi}{4}, \frac{7\pi}{4}$;

the graph of F is concave up on $\left(\frac{\pi}{4}, \frac{3\pi}{4}\right), \left(\frac{5\pi}{4}, \frac{7\pi}{4}\right)$; concave down on $\left(0, \frac{\pi}{4}\right)$, $\left(\frac{3\pi}{4}, \frac{5\pi}{4}\right)$, $\left(\frac{7\pi}{4}, 2\pi\right)$

SECTION 5.4

1. $\displaystyle\int_0^1 (2x - 3)\, dx = [x^2 - 3x]_0^1 = (-2) - (0) = -2$

3. $\displaystyle\int_{-1}^0 5x^4\, dx = [x^5]_{-1}^0 = (0) - (-1) = 1$

5. $\displaystyle\int_1^4 2\sqrt{x}\, dx = 2\int_1^4 x^{1/2}\, dx = 2\left[\frac{2}{3}x^{3/2}\right]_1^4 = \frac{4}{3}\left[x^{3/2}\right]_1^4 = \frac{4}{3}(8 - 1) = \frac{28}{3}$

7. $\displaystyle\int_1^5 2\sqrt{x - 1}\, dx = \int_1^5 2(x - 1)^{1/2}\, dx = \left[\frac{4}{3}(x - 1)^{3/2}\right]_1^5 = \frac{4}{3}[4^{3/2} - 0] = \frac{32}{3}$

9. $\displaystyle\int_{-2}^0 (x + 1)(x - 2)\, dx = \int_{-2}^0 (x^2 - x - 2)\, dx = \left[\frac{x^3}{3} - \frac{x^2}{2} - 2x\right]_{-2}^0 = \left[0 - \left(\frac{-8}{3} - 2 + 4\right)\right] = \frac{2}{3}$

11. $\displaystyle\int_1^2 \left(3t + \frac{4}{t^2}\right) dt = \int_1^2 (3t + 4t^{-2})\, dt = \left[\frac{3}{2}t^2 - 4t^{-1}\right]_1^2 = \left[(6 - 2) - \left(\frac{3}{2} - 4\right)\right] = \frac{13}{2}$

13. $\displaystyle\int_0^1 (x^{3/2} - x^{1/2})\, dx = \left[\frac{2}{5}x^{5/2} - \frac{2}{3}x^{3/2}\right]_0^1 = \left[\left(\frac{2}{5} - \frac{2}{3}\right) - 0\right] = -\frac{4}{15}$

15. $\displaystyle\int_0^1 (x + 1)^{17}\, dx = \left[\frac{1}{18}(x + 1)^{18}\right]_0^1 = \frac{1}{18}(2^{18} - 1)$

17. $\displaystyle\int_0^a (\sqrt{a} - \sqrt{x})^2\, dx = \int_0^a (a - 2\sqrt{a}\, x^{1/2} + x)\, dx = \left[ax - \frac{4}{3}\sqrt{a}\, x^{3/2} + \frac{x^2}{2}\right]_0^a = a^2 - \frac{4}{3}a^2 + \frac{a^2}{2} = \frac{1}{6}a^2$

19. $\displaystyle\int_1^2 \frac{6 - t}{t^3}\, dt = \int_1^2 (6t^{-3} - t^{-2})\, dt = \left[-3t^{-2} + t^{-1}\right]_1^2 = \left[\frac{-3}{4} + \frac{1}{2}\right] - [-3 + 1] = \frac{7}{4}$

21. $\displaystyle\int_1^2 2x(x^2 + 1)\, dx = \int_1^2 (2x^3 + 2x)\, dx = \left[\frac{x^4}{2} + x^2\right]_1^2 = 12 - \frac{3}{2} = \frac{21}{2}$

23. $\displaystyle\int_0^{\pi/2} \cos x\, dx = [\sin x]_0^{\pi/2} = 1$

25. $\displaystyle\int_0^{\pi/4} 2\sec^2 x\,dx = 2\left[\tan x\right]_0^{\pi/4} = 2$

27. $\displaystyle\int_{\pi/6}^{\pi/4} \csc u\,\cot u\,dx = \left[-\csc u\right]_{\pi/6}^{\pi/4} = -\sqrt{2} - (-2) = 2 - \sqrt{2}$

29. $\displaystyle\int_0^{2\pi} \sin x\,dx = \left[-\cos x\right]_0^{2\pi} = -1 - (-1) = 0$

31. $\displaystyle\int_0^{\pi/3} \left(\frac{2}{\pi}x - 2\sec^2 x\right) dx = \left[\frac{1}{\pi}x^2 - 2\tan x\right]_0^{\pi/3} = \frac{\pi}{9} - 2\sqrt{3}$

33. $\displaystyle\int_0^3 \left[\frac{d}{dx}\left(\sqrt{4+x^2}\right)\right] dx = \left[\sqrt{4+x^2}\right]_0^3 = \sqrt{13} - 2$

35. (a) $F(x) = \displaystyle\int_1^x (t+2)^2\,dt \quad\Longrightarrow\quad F'(x) = (x+2)^2$

(b) $\displaystyle\int_1^x (t+2)^2\,dt = \left[\frac{t^3}{3} + 2t^2 + 4t\right]_1^x = \frac{x^3}{3} + 2x^2 + 4x - 6\frac{1}{3}$

$\Longrightarrow\quad F'(x) = x^2 + 4x + 4 = (x+2)^2$

37. (a) $F(x) = \displaystyle\int_1^{2x+1} \frac{1}{2}\sec u\tan u\,du \quad\Rightarrow\quad F'(x) = \sec(2x+1)\,\tan(2x+1)$

(b) $\displaystyle\int_1^{2x+1} \frac{1}{2}\sec u\,\tan u\,du = \left[\frac{1}{2}\sec u\right]_1^{2x+1} = \frac{1}{2}\sec(2x+1) - \frac{1}{2}\sec 1$

$\Longrightarrow\quad F'(x) = \sec(2x+1)\,\tan(2x+1)$

39. (a) $F(x) = \displaystyle\int_2^x \frac{dt}{t}$ \qquad\qquad\qquad (b) $F(x) = -3 + \displaystyle\int_2^x \frac{dt}{t}$

41. Area $= \displaystyle\int_0^4 (4x - x^2)\,dx = \left[2x^2 - \frac{x^3}{3}\right]_0^4 = \frac{32}{3}$

43. Area $= \displaystyle\int_{-\pi/2}^{\pi/4} 2\cos x\,dx = 2\left[\sin x\right]_{-\pi/2}^{\pi/4} = \sqrt{2} + 2$

45. (a) $\displaystyle\int_2^5 (x-3)\,dx = \left[\frac{x^2}{2} - 3x\right]_2^5 = \frac{3}{2}$ \qquad (b) $\displaystyle\int_2^5 |x-3|\,dx = \int_2^3 (3-x)\,dx + \int_3^5 (x-3)\,dx$

$= \left[3x - \frac{x^2}{2}\right]_2^3 + \left[\frac{x^2}{2} - 3x\right]_3^5 = \frac{5}{2}$

47. (a) $\displaystyle\int_{-2}^2 (x^2 - 1)\,dx = \left[\frac{x^3}{3} - x\right]_{-2}^2 = \frac{4}{3}$

(b) $\displaystyle\int_{-2}^2 |x^2 - 1|\,dx = \int_{-2}^{-1} (x^2 - 1)\,dx + \int_{-1}^1 (1 - x^2)\,dx + \int_1^2 (x^2 - 1)\,dx$

$= \left[\frac{x^3}{3} - x\right]_{-2}^{-1} + \left[x - \frac{x^3}{3}\right]_{-1}^1 + \left[\frac{x^3}{3} - x\right]_1^2 = 4$

49. valid

51. not valid; $1/x^3$ is not defined at $x = 0$

53. (a) $x(t) = \displaystyle\int_0^t (10u - u^2)\, du = \left[5u^2 - \frac{u^3}{3}\right]_0^t = 5t^2 - \frac{t^3}{3}, \quad 0 \le t \le 10$

(b) $v'(t) = 10 - 2t;$ v has an absolute maximum at $t = 5$. The object's position at $t = 5$ is

$x(5) = \dfrac{250}{3}.$

55. $\displaystyle\int_0^4 f(x)\, dx = \int_0^1 (2x + 1)\, dx + \int_1^4 (4 - x)\, dx = \left[x^2 + x\right]_0^1 + \left[4x - \frac{x^2}{2}\right]_1^4 = \frac{13}{2}$

57. $\displaystyle\int_{-\pi/2}^{\pi} f(x)\, dx = \int_{-\pi/2}^{\pi/3} (1 + 2\cos x\, dx + \int_{\pi/3}^{\pi} \left[\frac{3}{\pi}x + 1\right]\, dx = \left[x + 2\sin x\right]_{-\pi/2}^{\pi/3} + \left[\frac{3x^2}{2\pi} + x\right]_{\pi/3}^{\pi}$

$= 2 + \sqrt{3} + \tfrac{17}{6}\pi$

59. (a) f is continuous on $[-2, 2]$.

For $x \in [-2, 0],$ $g(x) = \displaystyle\int_{-2}^x (t + 2)\, dt = \left[\frac{1}{2}t^2 + 2t\right]_{-2}^x = \frac{1}{2}x^2 + 2x + 2.$

For $x \in [0, 1],$ $g(x) = \displaystyle\int_{-2}^0 (t + 2)\, dt + \int_0^x 2\, dt = 2 + [2t]_0^x = 2 + 2x.$

For $x \in [1, 2],$ $g(x) = \displaystyle\int_{-2}^0 (t + 2)\, dt + \int_0^1 2\, dt + \int_1^x (4 - 2t)\, dt = 2 + 2 + \left[4t - t^2\right]_1^x = 1 + 4x - x^2.$

Thus $g(x) = \begin{cases} \frac{1}{2}x^2 + 2x + 2, & -2 \le x \le 0 \\ 2x + 2, & 0 \le x \le 1 \\ 1 + 4x - x^2, & 1 \le x \le 2 \end{cases}$

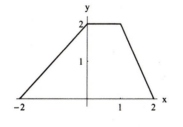

(b)

(c) f is continuous on $[-2, 2]$; f is differentiable on $(-2, 0)$, $(0, 1)$, and $(1, 2)$.

g is differentiable on $(-2, 2)$.

61. Follows from Theorem 5.3.2 since $f(x)$ is an antiderivative of $f'(x)$.

63. $\dfrac{d}{dx}\left[\displaystyle\int_a^x f(t)\,dt\right] = f(x);\qquad \displaystyle\int_a^x \dfrac{d}{dt}\,[f(t)]\,dt = f(x) - f(a)$

SECTION 5.5

1. $A = \displaystyle\int_0^1 (2 + x^3)\,dx = \left[2x + \dfrac{x^4}{4}\right]_0^1 = \dfrac{9}{4}$

3. $A = \displaystyle\int_3^8 \sqrt{x+1}\,dx = \int_3^8 (x+1)^{1/2}\,dx = \left[\dfrac{2}{3}(x+1)^{3/2}\right]_3^8 = \dfrac{2}{3}[27 - 8] = \dfrac{38}{3}$

5. $A = \displaystyle\int_0^1 (2x^2 + 1)^2\,dx = \int_0^1 (4x^4 + 4x^2 + 1)\,dx = \left[\dfrac{4}{5}x^5 + \dfrac{4}{3}x^3 + x\right]_0^1 = \dfrac{47}{15}$

7. $A = \displaystyle\int_1^2 [0 - (x^2 - 4)]\,dx = \int_1^2 (4 - x^2)\,dx = \left[4x - \dfrac{x^3}{3}\right]_1^2 = \left[8 - \dfrac{8}{3}\right] - \left[4 - \dfrac{1}{3}\right] = \dfrac{5}{3}$

9. $A = \displaystyle\int_{\pi/3}^{\pi/2} \sin x\,dx = [-\cos x]_{\pi/3}^{\pi/2} = (0) - \left(-\dfrac{1}{2}\right) = \dfrac{1}{2}$

11.

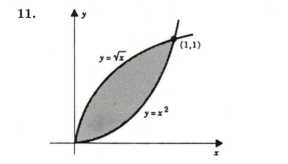

$A = \displaystyle\int_0^1 [x^{1/2} - x^2]\,dx$

$\qquad = \left[\tfrac{2}{3}x^{3/2} - \tfrac{1}{3}x^3\right]_0^1 = \tfrac{1}{3}$

13.

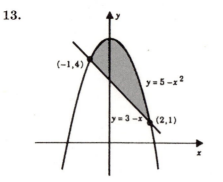

$A = \displaystyle\int_{-1}^2 [(5 - x^2) - (3 - x)]\,dx$

$\qquad = \displaystyle\int_{-1}^2 (2 + x - x^2)\,dx$

$\qquad = \left[2x + \dfrac{x^2}{2} - \dfrac{x^3}{3}\right]_{-1}^2$

$\qquad = \left[4 + 2 - \tfrac{8}{3}\right] - \left[-2 + \tfrac{1}{2} + \tfrac{1}{3}\right] = \dfrac{9}{2}$

15.

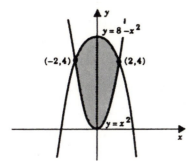

$$A = \int_{-2}^{2} [(8 - x^2) - (x^2)]\, dx$$

$$= \int_{-2}^{2} (8 - 2x^2)\, dx$$

$$= \left[8x - \tfrac{2}{3}x^3\right]_{-2}^{2}$$

$$= \left[16 - \tfrac{16}{3}\right] - \left[-16 + \tfrac{16}{3}\right] = \tfrac{64}{3}$$

17.

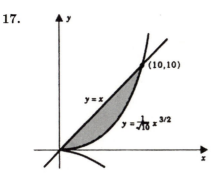

$$A = \int_{0}^{10} \left[x - \frac{1}{\sqrt{10}}\, x^{3/2}\right] dx$$

$$= \left[\frac{x^2}{2} - \frac{2\sqrt{10}}{50}\, x^{5/2}\right]_{0}^{10}$$

$$= 50 - \frac{2\sqrt{10}}{50}(10)^{5/2} = 10$$

19.

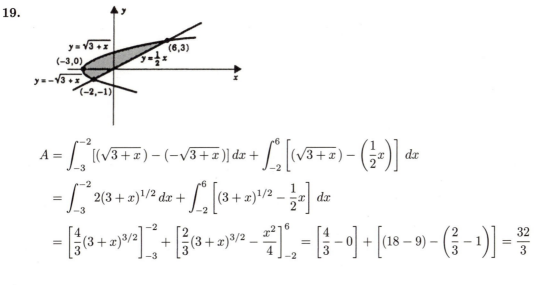

$$A = \int_{-3}^{-2} [(\sqrt{3+x}) - (-\sqrt{3+x})]\, dx + \int_{-2}^{6} \left[(\sqrt{3+x}) - \left(\frac{1}{2}x\right)\right] dx$$

$$= \int_{-3}^{-2} 2(3+x)^{1/2}\, dx + \int_{-2}^{6} \left[(3+x)^{1/2} - \frac{1}{2}x\right] dx$$

$$= \left[\frac{4}{3}(3+x)^{3/2}\right]_{-3}^{-2} + \left[\frac{2}{3}(3+x)^{3/2} - \frac{x^2}{4}\right]_{-2}^{6} = \left[\frac{4}{3} - 0\right] + \left[(18 - 9) - \left(\frac{2}{3} - 1\right)\right] = \frac{32}{3}$$

21.

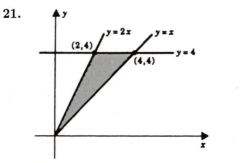

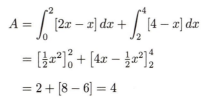

$$A = \int_{0}^{2} [2x - x]\, dx + \int_{2}^{4} [4 - x]\, dx$$

$$= \left[\tfrac{1}{2}x^2\right]_{0}^{2} + \left[4x - \tfrac{1}{2}x^2\right]_{2}^{4}$$

$$= 2 + [8 - 6] = 4$$

23.

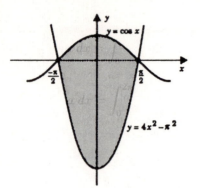

$$A = \int_{-\pi/2}^{\pi/2} [\cos x - (4x^2 - \pi^2)] \, dx$$

$$= \left[\sin x - \tfrac{4}{3}x^3 + \pi^2 x\right]_{-\pi/2}^{\pi/2}$$

$$= [1 - \tfrac{1}{6}\pi^3 + \tfrac{1}{2}\pi^3] - [-1 + \tfrac{1}{6}\pi^3 - \tfrac{1}{2}\pi^3]$$

$$= 2 + \tfrac{2}{3}\pi^3$$

25.

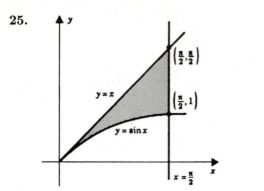

$$A = \int_0^{\pi/2} [x - \sin x] \, dx$$

$$= \left[\frac{x^2}{2} + \cos x\right]_0^{\pi/2}$$

$$= \frac{\pi^2}{8} - 1$$

27. (a) $\displaystyle\int_{-3}^4 (x^2 - x - 6)\, dx = \left[\frac{1}{3}x^3 - \frac{1}{2}x^2 - 6x\right]_{-3}^4 = -\frac{91}{6};$

the area of the region bounded by the graph of f and the x-axis for $x \in [-3, -2] \cup [3, 4]$

minus the area of the region bounded the graph of f and the x-axis for $x \in [-2, 3]$.

(b) $\displaystyle A = \int_{-3}^{-2} (x^2 - x - 6)\, dx + \int_{-2}^3 (-x^2 + x + 6)\, dx + \int_3^4 (x^2 - x - 6)\, dx$

$$= \left[\tfrac{1}{3}x^3 - \tfrac{1}{2}x^2 - 6x\right]_{-3}^{-2} + \left[-\tfrac{1}{3}x^3 + \tfrac{1}{2}x^2 + 6x\right]_{-2}^3 + \left[\tfrac{1}{3}x^3 - \tfrac{1}{2}x^2 - 6x\right]_3^4 = \frac{17}{6} + \frac{125}{6} + \frac{17}{6} = \frac{53}{2}$$

(c) $\displaystyle A = -\int_{-2}^3 (x^2 - x - 6)\, dx = \frac{125}{6}$

29. (a) $\displaystyle\int_{-2}^2 (x^3 - x)\, dx = \left[\frac{1}{4}x^4 - \frac{1}{2}x^2\right]_{-2}^2 = 0$

(b)

$$A = 2\left[-\int_0^1 (x^3 - x)\, dx + \int_1^2 (x^3 - x)\, dx\right]$$

$$= -2[\tfrac{1}{4}x^4 - \tfrac{1}{2}x^2]_0^1 + 2[\tfrac{1}{4}x^4 - \tfrac{1}{2}x^2]_1^2$$

$$= \frac{1}{2} + \frac{9}{2} = 5$$

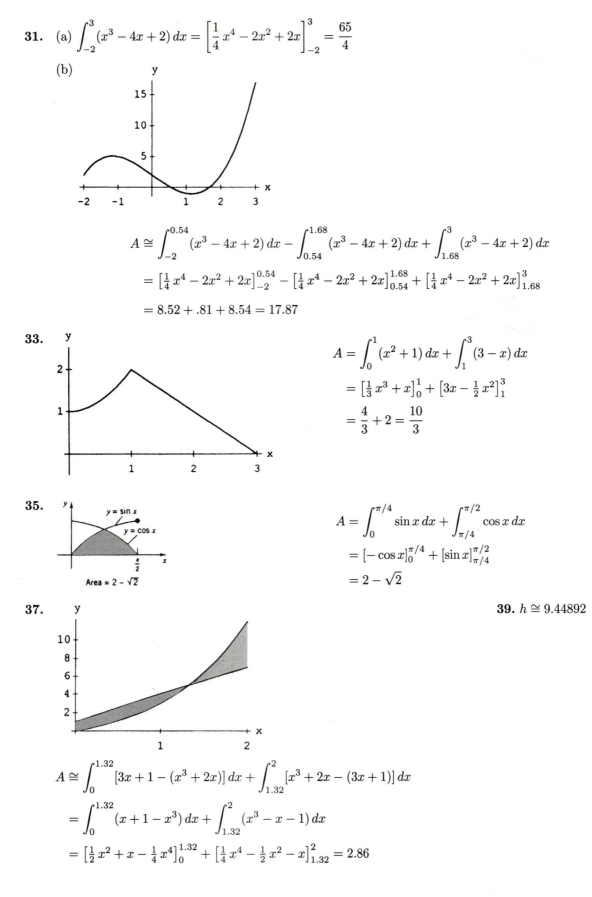

31. (a) $\displaystyle\int_{-2}^{3}(x^3-4x+2)\,dx=\left[\frac{1}{4}\,x^4-2x^2+2x\right]_{-2}^{3}=\frac{65}{4}$

(b)

$$A\cong\int_{-2}^{0.54}(x^3-4x+2)\,dx-\int_{0.54}^{1.68}(x^3-4x+2)\,dx+\int_{1.68}^{3}(x^3-4x+2)\,dx$$

$$=\left[\tfrac{1}{4}\,x^4-2x^2+2x\right]_{-2}^{0.54}-\left[\tfrac{1}{4}\,x^4-2x^2+2x\right]_{0.54}^{1.68}+\left[\tfrac{1}{4}\,x^4-2x^2+2x\right]_{1.68}^{3}$$

$$=8.52+.81+8.54=17.87$$

33.

$$A=\int_{0}^{1}(x^2+1)\,dx+\int_{1}^{3}(3-x)\,dx$$

$$=\left[\tfrac{1}{3}\,x^3+x\right]_{0}^{1}+\left[3x-\tfrac{1}{2}\,x^2\right]_{1}^{3}$$

$$=\frac{4}{3}+2=\frac{10}{3}$$

35.

$$A=\int_{0}^{\pi/4}\sin x\,dx+\int_{\pi/4}^{\pi/2}\cos x\,dx$$

$$=\left[-\cos x\right]_{0}^{\pi/4}+\left[\sin x\right]_{\pi/4}^{\pi/2}$$

$$=2-\sqrt{2}$$

37.

39. $h\cong 9.44892$

$$A\cong\int_{0}^{1.32}[3x+1-(x^3+2x)]\,dx+\int_{1.32}^{2}[x^3+2x-(3x+1)]\,dx$$

$$=\int_{0}^{1.32}(x+1-x^3)\,dx+\int_{1.32}^{2}(x^3-x-1)\,dx$$

$$=\left[\tfrac{1}{2}\,x^2+x-\tfrac{1}{4}\,x^4\right]_{0}^{1.32}+\left[\tfrac{1}{4}\,x^4-\tfrac{1}{2}\,x^2-x\right]_{1.32}^{2}=2.86$$

PROJECT 5.5

1. (a) $g(x) = \begin{cases} -1, & x = 0 \\ 0, & 0 < x \le 1 \\ 1, & 1 < x \le 2 \\ \vdots & \\ 4, & 4 < x \le 5. \end{cases}$

$$\int_0^5 g(x)\,dx = \int_0^1 g(x)\,dx + \int_1^2 g(x)\,dx + \cdots + \int_4^5 g(x)\,dx = 0 + 1 + 2 + 3 + 4 = 10.$$

(b) $g(x) = \begin{cases} 0, & 0 \le x < 1 \\ 1, & 1 \le x < 2 \\ \vdots & \\ 4, & 4 \le< x < 5 \\ 5 & x = 5. \end{cases}$

$$\int_0^5 g(x)\,dx = \int_0^1 g(x)\,dx + \int_1^2 g(x)\,dx + \cdots + \int_4^5 g(x)\,dx = 0 + 1 + 2 + 3 + 4 = 10.$$

(c) $g(x) = \begin{cases} 1 & x = 1, 2, \cdots 5 \\ 0 & \text{otherwise} \end{cases}$; $\displaystyle\int_0^5 g(x)\,dx = 0.$

3. (a) For $0 \le x < 1$, $G(x) = \int_0^x (2-t)\,dt = \left[2t - \tfrac{1}{2}t^2\right]_0^x = 2x - \tfrac{1}{2}x^2$

For $1 \le x \le 2$, $G(x) = \int_0^1 (2-t)\,dt + \int_1^x (2+t)\,dt = \tfrac{3}{2} + \left[2t + \tfrac{1}{2}t^2\right]_1^x = 2x + \tfrac{1}{2}x^2 - 1$

Thus, $G(x) = \begin{cases} 2x - \tfrac{1}{2}x^2, & 0 \le x < 1 \\ 2x + \tfrac{1}{2}x^2 - 1, & 1 \le x \le 2 \end{cases}$.

$$\lim_{h \to 0^-} \frac{G(1+h) - G(1)}{h} = \lim_{h \to 0^-} \frac{2(1+h) - \tfrac{1}{2}(1+h)^2 - \tfrac{3}{2}}{h} = \lim_{h \to 0^-} \frac{h - \tfrac{1}{2}h^2}{h} = 1$$

$$\lim_{h \to 0^+} \frac{G(1+h) - G(1)}{h} = \lim_{h \to 0^+} \frac{2(1+h) + \tfrac{1}{2}(1+h)^2 - 1 - \tfrac{3}{2}}{h} = \lim_{h \to 0^+} \frac{3h + \tfrac{1}{2}h^2}{h} = 3$$

Thus, G is not differentiable at $x = 1$.

(b) $G(x) = \begin{cases} \frac{1}{3}x^3, & 0 \le x < 2 \\ \frac{1}{2}x^2 + \frac{2}{3}, & 2 \le x \le 5 \end{cases}$.

$$\lim_{h \to 0^-} \frac{G(2+h) - G(2)}{h} = \lim_{h \to 0^-} \frac{\frac{1}{3}(2+h)^3 - \frac{8}{3}}{h} = \lim_{h \to 0^-} \frac{4h + 2h^2 + \frac{1}{3}h^3}{h} = 4$$

$$\lim_{h \to 0^+} \frac{G(2+h) - G(2)}{h} = \lim_{h \to 0^+} \frac{\frac{1}{2}(2+h)^2 + \frac{2}{3} - \frac{8}{3}}{h} = \lim_{h \to 0^+} \frac{2h + \frac{1}{2}h^2}{h} = 2$$

Thus, G is not differentiable at $x = 2$.

(c) $G(x) = \begin{cases} \sin x, & 0 \le x < \pi/2 \\ 1 - \cos x, & \pi/2 \le x < \pi \\ 2 + \frac{1}{2}x - \pi/2, & \pi \le x \le 2\pi \end{cases}$.

G is not differentiable at $x = \pi/2, \ \pi$;
see the graph of G:

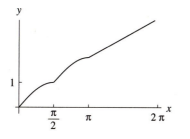

SECTION 5.6

1. $\displaystyle \int \frac{dx}{x^4} = \int x^{-4}\, dx = -\frac{1}{3}x^{-3} + C$

3. $\displaystyle \int (ax + b)\, dx = \frac{1}{2}ax^2 + bx + C$

5. $\displaystyle \int \frac{dx}{\sqrt{1+x}} = \int (1+x)^{-1/2}\, dx = 2(1+x)^{1/2} + C$

7. $\displaystyle \int \left(\frac{x^3 - 1}{x^2} \right) dx = \int (x - x^{-2})\, dx = \frac{1}{2}x^2 + x^{-1} + C$

9. $\displaystyle \int (t-a)(t-b)\, dt = \int [t^2 - (a+b)t + ab]\, dt = \frac{1}{3}t^3 - \frac{a+b}{2}t^2 + abt + C$

11. $\displaystyle \int \frac{(t^2 - a)(t^2 - b)}{\sqrt{t}}\, dt = \int [t^{7/2} - (a+b)t^{3/2} + abt^{-1/2}]\, dt$

$$= \tfrac{2}{9}t^{9/2} - \tfrac{2}{5}(a+b)t^{5/2} + 2abt^{1/2} + C$$

13. $\int g(x)g'(x)\,dx = \frac{1}{2}[g(x)]^2 + C$

15. $\int \tan x \sec^2 x\,dx = \int \sec x \frac{d}{dx}[\sec x]\,dx = \frac{1}{2}\sec^2 x + C$

$\int \tan x \sec^2 x\,dx = \int \tan x \frac{d}{dx}[\tan x]\,dx = \frac{1}{2}\tan^2 x + C$

17. $\int \frac{4}{(4x+1)^2}\,dx = \int 4(4x+1)^{-2}\,dx = -(4x+1)^{-1} + C$

19. $f(x) = \int f'(x)\,dx = \int (2x-1)\,dx = x^2 - x + C.$

Since $f(3) = 4$, we get $4 = 9 - 3 + C$ so that $C = -2$ and

$$f(x) = x^2 - x - 2.$$

21. $f(x) = \int f'(x)\,dx = \int (ax+b)\,dx = \frac{1}{2}ax^2 + bx + C.$

Since $f(2) = 0$, we get $0 = 2a + 2b + C$ so that $C = -2a - 2b$ and

$$f(x) = \tfrac{1}{2}ax^2 + bx - 2a - 2b.$$

23. $f(x) = \int f'(x)\,dx = \int \sin x\,dx = -\cos x + C.$

Since $f(0) = 2$, we get $2 = -1 + C$ so that $C = 3$ and

$$f(x) = 3 - \cos x.$$

25. First,

$$f'(x) = \int f''(x)\,dx = \int (6x-2)\,dx = 3x^2 - 2x + C.$$

Since $f'(0) = 1$, we get $1 = 0 + C$ so that $C = 1$ and

$$f'(x) = 3x^2 - 2x + 1.$$

Next,

$$f(x) = \int f'(x)\,dx = \int (3x^2 - 2x + 1)\,dx = x^3 - x^2 + x + K.$$

Since $f(0) = 2$, we get $2 = 0 + K$ so that $K = 2$ and

$$f(x) = x^3 - x^2 + x + 2.$$

27. First,

$$f'(x) = \int f''(x)\,dx = \int (x^2 - x)\,dx = \frac{1}{3}x^3 - \frac{1}{2}x^2 + C.$$

Since $f'(1) = 0$, we get $0 = \frac{1}{3} - \frac{1}{2} + C$ so that $C = \frac{1}{6}$ and

$$f'(x) = \tfrac{1}{3}x^3 - \tfrac{1}{2}x^2 + \tfrac{1}{6}.$$

Next,

$$f(x) = \int f'(x)\,dx = \int \left(\frac{1}{3}x^3 - \frac{1}{2}x^2 + \frac{1}{6}\right)dx = \frac{1}{12}x^4 - \frac{1}{6}x^3 + \frac{1}{6}x + K.$$

Since $f(1) = 2$, we get $2 = \frac{1}{12} - \frac{1}{6} + \frac{1}{6} + K$ so that $K = \frac{23}{12}$ and

$$f(x) = \frac{x^4}{12} - \frac{x^3}{6} + \frac{x}{6} + \frac{23}{12} = \frac{1}{12}(x^4 - 2x^3 + 2x + 23).$$

29. First,

$$f'(x) = \int f''(x)\,dx = \int \cos x\,dx = \sin x + C.$$

Since $f'(0) = 1$, we get $1 = 0 + C$ so that $C = 1$ and

$$f'(x) = \sin x + 1.$$

Next,

$$f(x) = \int f'(x)\,dx = \int (\sin x + 1)\,dx = -\cos x + x + K.$$

Since $f(0) = 2$, we get $2 = -1 + 0 + K$ so that $K = 3$ and

$$f(x) = -\cos x + x + 3.$$

31. First,

$$f'(x) = \int f''(x)\,dx = \int (2x - 3)\,dx = x^2 - 3x + C.$$

Then,

$$f(x) = \int f'(x)\,dx = \int (x^2 - 3x + C)\,dx = \frac{1}{3}x^3 - \frac{3}{2}x^2 + Cx + K.$$

Since $f(2) = -1$, we get

$$(1) \qquad\qquad\qquad -1 = \tfrac{8}{3} - 6 + 2C + K;$$

and, from $f(0) = 3$, we conclude that

$$(2) \qquad\qquad\qquad 3 = 0 + K.$$

Solving (1) and (2) simultaneously, we get $K = 3$ and $C = -\frac{1}{3}$ so that

$$f(x) = \tfrac{1}{3}x^3 - \tfrac{3}{2}x^2 - \tfrac{1}{3}x + 3.$$

33. $\dfrac{d}{dx}\left[\int f(x)\,dx\right] = f(x); \quad \int \dfrac{d}{dx}[f(x)]\,dx = f(x) + C$

35. (a) $x(t) = \displaystyle\int v(t)\,dt = \int (6t^2 - 6)\,dt = 2t^3 - 6t + C.$

Since $x(0) = -2$, we get $-2 = 0 + C$ so that $C = -2$ and

$x(t) = 2t^3 - 6t - 2.$ Therefore $x(3) = 34.$

Three seconds later the object is 34 units to the right of the origin.

(b) $s = \int_0^3 |v(t)|\, dt = \int_0^3 |6t^2 - 6|\, dt = \int_0^1 (6 - 6t^2)\, dt + \int_1^3 (6t^2 - 6)\, dt$

$\qquad = [6t - 2t^3]_0^1 + [2t^3 - 6t]_1^3 = 4 + [36 - (-4)] = 44.$

The object traveled 44 units.

37. (a) $v(t) = \int a(t)\, dt = \int (t+1)^{-1/2}\, dt = 2(t+1)^{1/2} + C.$

Since $v(0) = 1$, we get $1 = 2 + C$ so that $C = -1$ and

$$v(t) = 2(t+1)^{1/2} - 1.$$

(b) We know $v(t)$ by part (a). Therefore,

$$x(t) = \int v(t)\, dt = \int [2(t+1)^{1/2} - 1]\, dt = \frac{4}{3}(t+1)^{3/2} - t + C.$$

Since $x(0) = 0$, we get $0 = \frac{4}{3} - 0 + C$ so that

$C = -\frac{4}{3}$ and $x(t) = \frac{4}{3}(t+1)^{3/2} - t - \frac{4}{3}.$

39. (a) $v_0 = 60$ mph $= 88$ feet per second. In general, $v(t) = at + v_0$. Here, in feet and

seconds, $v(t) = -20t + 88$. Thus $v(t) = 0$ at $t = 4.4$ seconds.

(b) In general, $x(t) = \frac{1}{2}at^2 + v_0 t + x_0$. Here we take $x_0 = 0$. In feet and seconds

$$x(t) = -10(4.4)^2 + 88(4.4) = 10(4.4)^2 = 193.6 \text{ ft}.$$

41. $\qquad [v(t)]^2 = (at + v_0)^2 = a^2 t^2 + 2av_0 t + v_0{}^2$

$\qquad\qquad\qquad = v_0{}^2 + a(at^2 + 2v_0 t)$

$\qquad\qquad\qquad = v_0{}^2 + 2a(\frac{1}{2}at^2 + v_0 t)$

(set $x(t) = \frac{1}{2}at^2 + v_0 t + x_0$)

$\qquad\qquad\qquad = v_0{}^2 + 2a\,[x(t) - x_0]$

43. The car can accelerate to 60 mph (88 ft/sec) in 20 seconds thereby covering a distance of 880 ft. It can decelerate from 88 ft/sec to 0 ft/sec in 4 seconds thereby covering a distance of 176 ft. At full speed, 88 ft/sec, it must cover a distance of

$$\frac{5280}{2} - 880 - 176 = 1584 \text{ ft}.$$

This takes $\dfrac{1584}{88} = 18$ seconds. The run takes at least $20 + 18 + 4 = 42$ seconds.

45. $v(t) = \int a(t)\, dt = \int (2A + 6Bt)\, dt = 2At + 3Bt^2 + C.$

Since $v(0) = v_0$, we have $v_0 = 0 + C$ so that $v(t) = 2At + 3Bt^2 + v_0$.

$$x(t) = \int v(t)\, dt = \int (2At + 3Bt^2 + v_0)\, dt = At^2 + Bt^3 + v_0 t + K.$$

Since $x(0) = x_0$, we have $x_0 = 0 + K$ so that $K = x_0$ and

$$x(t) = x_0 + v_0 t + At^2 + Bt^3.$$

47. $x'(t) = t^2 - 5,$ $y'(t) = 3t,$

          $x(t) = \frac{1}{3}t^3 - 5t + C.$ $y(t) = \frac{3}{2}t^2 + K.$

When $t = 2$, the particle is at $(4, 2)$. Thus, $x(2) = 4$ and $y(2) = 2$.

$$4 = \tfrac{8}{3} - 10 + C \quad \Longrightarrow \quad C = \tfrac{34}{3}. \qquad\qquad 2 = 6 + K \quad \Longrightarrow \quad K = -4.$$

$$x(t) = \tfrac{1}{3}t^3 - 5t + \tfrac{34}{3}, \qquad\qquad y(t) = \tfrac{3}{2}t^2 - 4.$$

Four seconds later the particle is at $(x(6), y(6)) = (\frac{160}{3}, 50)$.

49.   Since $v(0) = 2$, we have $2 = A \cdot 0 + B$ so that $B = 2$. Therefore

$$x(t) = \int v(t)\, dt = \int (At + 2)\, dt = \frac{1}{2}At^2 + 2t + C.$$

Since $x(2) = x(0) - 1$, we have

$$2A + 4 + C = C - 1 \quad \text{so that} \quad A = -\tfrac{5}{2}.$$

51. $x(t) = \displaystyle\int v(t)\, dt = \int \sin t\, dt = -\cos t + C$

     Since $x(\pi/6) = 0$, we have $0 = -\dfrac{\sqrt{3}}{2} + C$ so that $C = \dfrac{\sqrt{3}}{2}$ and $x(t) = \frac{\sqrt{3}}{2} - \cos t.$

    (a)   At $t = 11\pi/6$ sec.   (b) We want to find the smallest $t_0 > \pi/6$ for

                                             which $x(t_0) = 0$ and $v(t_0) > 0$. We get

                                             $t_0 = 13\pi/6$ seconds.

53.   The mean-value theorem. With obvious notation

$$\frac{x(1/12) - x(0)}{1/12} = \frac{4}{1/12} = 48.$$

By the mean-value theorem there exists some time t_0 at which

$$x'(t_0) = \frac{x(1/12) - x(0)}{1/12}.$$

55. $\dfrac{v'(t)}{[v(t)]^2} = 2 \quad \Longrightarrow \quad -[v(t)]^{-1} = 2t - v_0^{-1}.$

        $\Longrightarrow \quad [v(t)]^{-1} = v_0^{-1} - 2t \quad \Longrightarrow \quad v(t) = \dfrac{1}{v_0^{-1} - 2t} = \dfrac{v_0}{1 - 2tv_0}.$

57. $\int (\cos x - 2 \sin x)\, dx = \sin x + 2 \cos x + C$ and so

$$\frac{d}{dx}\left(\int (\cos x - 2 \sin x)\, dx\right) = \frac{d}{dx}\left[\sin x + 2 \cos x\right] = \cos x - 2 \sin x;$$

$\dfrac{d}{dx}\left[\cos x - 2 \sin x\right] = -\sin x - 2 \cos x$ and so

$$\int \frac{d}{dx}\left[\cos x - 2 \sin x\right] dx = \int (-\sin x - 2 \cos x)\, dx = \cos x - 2 \sin x + C$$

59. $f(x) = \sin x + 2 \cos x + 1$ **61.** $\frac{1}{12} x^4 - \frac{1}{2} x^3 + \frac{5}{2} x^2 + 4x - 3$

SECTION 5.7

1. $\left\{ \begin{array}{l} u = 2 - 3x \\ du = -3\, dx \end{array} \right\};$ $\displaystyle\int \frac{dx}{(2-3x)^2} = \int (2-3x)^{-2}\, dx = -\frac{1}{3}\int u^{-2}\, du = \frac{1}{3} u^{-1} + C$

$$= \tfrac{1}{3}(2-3x)^{-1} + C$$

3. $\left\{ \begin{array}{l} u = 2x + 1 \\ du = 2\, dx \end{array} \right\};$ $\displaystyle\int \sqrt{2x+1}\, dx = \int (2x+1)^{1/2}\, dx = \frac{1}{2}\int u^{1/2}\, du = \frac{1}{3} u^{3/2} + C$

$$= \frac{1}{3}(2x+1)^{3/2} + C$$

5. $\left\{ \begin{array}{l} u = ax + b \\ du = a\, dx \end{array} \right\};$ $\displaystyle\int (ax+b)^{3/4}\, dx = \frac{1}{a}\int u^{3/4}\, du = \frac{4}{7a} u^{7/4} + C$

$$= \frac{4}{7a}(ax+b)^{7/4} + C$$

7. $\left\{ \begin{array}{l} u = 4t^2 + 9 \\ du = 8t\, dt \end{array} \right\};$ $\displaystyle\int \frac{t}{(4t^2+9)^2}\, dt = \frac{1}{8}\int \frac{du}{u^2} = -\frac{1}{8} u^{-1} + C = -\frac{1}{8}(4t^2+9)^{-1} + C$

9. $\left\{ \begin{array}{l} u = 1 + x^3 \\ du = 3x^2\, dx \end{array} \right\};$ $\displaystyle\int x^2(1+x^3)^{1/4}\, dx = \frac{1}{3}\int u^{1/4}\, du = \frac{4}{15} u^{5/4} + C = \frac{4}{15}(1+x^3)^{5/4} + C$

11. $\left\{ \begin{array}{l} u = 1 + s^2 \\ du = 2s\, ds \end{array} \right\};$ $\displaystyle\int \frac{s}{(1+s^2)^3}\, ds = \frac{1}{2}\int \frac{du}{u^3} = -\frac{1}{4} u^{-2} + C = -\frac{1}{4}(1+s^2)^{-2} + C$

13. $\left\{ \begin{array}{l} u = x^2 + 1 \\ du = 2x\, dx \end{array} \right\};$ $\displaystyle\int \frac{x}{\sqrt{x^2+1}}\, dx = \int (x^2+1)^{-1/2}\, x\, dx = \frac{1}{2}\int u^{-1/2}\, du = u^{1/2} + C = \sqrt{x^2+1} + C$

15. $\left\{ \begin{array}{l} u = x^2 + 1 \\ du = 2x \end{array} \right\};$ $\displaystyle\int 5x\left(x^2+1\right)^{-3}\, dx = \frac{5}{2}\int u^{-3}\, du = -\frac{5}{4} u^{-2} + C = -\frac{5}{4}(x^2+1)^{-2} + C$

17. $\left\{\begin{array}{l} u = x^{1/4} + 1 \\ du = \frac{1}{4}x^{-3/4}\,dx \end{array}\right\}$; $\quad \int x^{-3/4}\left(x^{1/4} + 1\right)^{-2} dx = 4 \int u^{-2}\,du = -4u^{-1} + C = -4(x^{1/4} + 1)^{-1} + C$

19. $\left\{\begin{array}{l} u = 1 - a^4 x^4 \\ du = -4a^4 x^3\,dx \end{array}\right\}$;

$$\int \frac{b^3 x^3}{\sqrt{1 - a^4 x^4}}\,dx = -\frac{b^3}{4a^4}\int u^{-1/2}\,du = -\frac{b^3}{2a^4}u^{1/2} + C = -\frac{b^3}{2a^4}\sqrt{1 - a^4 x^4} + C$$

21. $\left\{\begin{array}{l} u = x^2 + 1 \\ du = 2x\,dx \end{array}\right\}$; $\quad \int x\left(x^2 + 1\right)^3 dx = \frac{1}{2}\int_1^2 u^3\,du = \frac{1}{8}u^4 + C = \frac{1}{8}(x^2 + 1)^4 + C$

$$\int_0^1 x(x^2 + 1)\,dx = \left[\frac{1}{8}(x^2 + 1)^4\right]_0^1 = \frac{1}{8}[16 - 1] = \frac{15}{8}$$

23. 0; the integrand is an odd function

25. $\left\{\begin{array}{l} u = a^2 - y^2 \\ du = -2y\,dy \end{array}\right\}$; $\quad \int y\sqrt{a^2 - y^2}\,dy = -\frac{1}{2}\int u^{1/2}\,du = -\frac{1}{3}u^{3/2} + C = -\frac{1}{3}(a^2 - y^2)^{3/2} + C$

$$\int_0^a y\sqrt{a^2 - y^2}\,dy = -\frac{1}{3}\left[(a^2 - y^2)^{3/2}\right]_0^a = \frac{1}{3}(a^2)^{3/2} = \frac{1}{3}|a|^3$$

27. $\left\{\begin{array}{l} u = 2x^2 + 1 \\ du = 4x\,dx \end{array}\right\}$; $\quad \int x\sqrt{2x^2 + 1}\,dx = \frac{1}{4}\int u^{1/2}\,du\,\frac{1}{6}u^{3/2} + C = \frac{1}{6}(2x^2 + 1)^{3/2} + C$

$$\int_0^2 x\sqrt{2x^2 + 1}\,dx = \left[\frac{1}{6}(2x^2 + 1)^{3/2}\right]_0^2 = \frac{13}{3}$$

29. $\left\{\begin{array}{l} u = 1 + x^{-2} \\ du = -2x^{-3}\,dx \end{array}\right\}$; $\quad \int x^{-3}(1 + x^{-2})^{-3}\,dx = -\frac{1}{2}\int u^{-3}\,du = \frac{1}{4}u^{-2} + C = \frac{1}{4}(1 + x^{-2})^{-2} + C$

$$\int_0^2 x^{-3}(1 + x^{-2})^{-3}\,dx = \left[\frac{1}{4}(1 + x^{-2})^{-2}\right]_1^2 = \frac{39}{400}$$

31. $\left\{\begin{array}{l} u = x + 1 \\ du = dx \end{array}\right\}$; $\quad \int x\sqrt{x + 1}\,dx = \int (u - 1)\sqrt{u}\,du = \int (u^{3/2} - u^{1/2})\,du$

$$= \tfrac{2}{5}u^{5/2} - \tfrac{2}{3}u^{3/2} + C = \tfrac{2}{5}(x + 1)^{5/2} - \tfrac{2}{3}(x + 1)^{3/2} + C$$

33. $\left\{\begin{array}{l} u = 2x - 1 \\ du = dx \end{array}\right\}$; $\quad \displaystyle\int x\sqrt{2x-1}\,dx = \frac{1}{2}\int \frac{(u-1)}{2}\sqrt{u}\,du = \frac{1}{4}\int (u^{3/2} + u^{1/2})\,du$

$$= \frac{1}{10}u^{5/2} + \frac{1}{6}u^{3/2} + C = \frac{1}{10}(2x-1)^{5/2} + \frac{1}{6}(2x-1)^{3/2} + C$$

35. $\displaystyle\int \frac{1}{\sqrt{x\sqrt{x}+x}}\,dx = \int \frac{1}{\sqrt{x}\sqrt{\sqrt{x}+1}}\,dx$

$\left\{\begin{array}{l} u = \sqrt{x} + 1 \\ du = dx/2\sqrt{x} \end{array}\right\}$; $\quad \displaystyle\int \frac{1}{\sqrt{x}\sqrt{\sqrt{x}+1}}\,dx = 2\int u^{-1/2}\,du = 4\sqrt{u} + C = 4\sqrt{\sqrt{x}+1} + C$

37. $\left\{\begin{array}{l|l} u = x + 1 & x = 0 \;\Rightarrow\; u = 1 \\ du = dx & x = 1 \;\Rightarrow\; u = 2 \end{array}\right\}$;

$$\int_0^1 \frac{x+3}{\sqrt{x+1}}\,dx = \int_1^2 \frac{u+2}{\sqrt{u}}\,du = \int_1^2 (u^{1/2} + 2u^{-1/2})\,du = \left[\frac{2}{3}u^{3/2} + 4u^{1/2}\right]_1^2 = \frac{16}{3}\sqrt{2} - \frac{14}{3}$$

39. $\left\{\begin{array}{l} u = x^2 + 1 \\ du = 2x\,dx \end{array}\right\}$; $\quad \displaystyle\int x\sqrt{x^2+1}\,dx = \frac{1}{2}\int \sqrt{u}\,du = \frac{1}{3}u^{3/2} + C = \frac{1}{3}(x^2+1)^{3/2} + C.$

Also, $1 = \dfrac{1}{3}(0^2 + 1) + C \implies C = \dfrac{2}{3}.$ Thus $y = \dfrac{1}{3}(x^2+1)^{\frac{3}{2}} + \dfrac{2}{3}.$

41. $\displaystyle\int \cos(3x+1)\,dx = -\frac{1}{3}\sin(3x+1) + C$ **43.** $\displaystyle\int \csc^2 \pi x\,dx = -\frac{1}{\pi}\cot \pi x + C$

45. $\left\{\begin{array}{l} u = 3 - 2x \\ du = -2\,dx \end{array}\right\}$; $\quad \displaystyle\int \sin(3-2x)\,dx = \int -\tfrac{1}{2}\sin u\,du = \tfrac{1}{2}\cos u + C = \tfrac{1}{2}\cos(3-2x) + C$

47. $\left\{\begin{array}{l} u = \cos x \\ du = -\sin x\,dx \end{array}\right\}$; $\quad \displaystyle\int \cos^4 x \sin x\,dx = \int -u^4\,du = -\frac{1}{5}u^5 + C = -\frac{1}{5}\cos^5 x + C$

49. $\left\{\begin{array}{l} u = x^{1/2} \\ du = \frac{1}{2}x^{-1/2}\,dx \end{array}\right\}$; $\quad \displaystyle\int x^{-1/2}\sin x^{1/2}\,dx = \int 2\sin u\,du = -2\cos u + C = -2\cos x^{1/2} + C$

51. $\left\{\begin{array}{l} u = 1 + \sin x \\ du = \cos x\,dx \end{array}\right\}$; $\quad \displaystyle\int \sqrt{1+\sin x}\,\cos x\,dx = \int u^{1/2}\,du = \frac{2}{3}u^{3/2} + C = \tfrac{2}{3}(1+\sin x)^{3/2} + C$

53. $\left\{\begin{array}{l} u = \sin \pi x \\ du = \pi \cos \pi x\,dx \end{array}\right\}$; $\quad \displaystyle\int \sin \pi x \cos \pi x\,dx = \frac{1}{\pi}\int u\,du = \frac{1}{2\pi}u^2 + C = \frac{1}{2\pi}\sin^2 \pi x + C$

55. $\left\{\begin{array}{l} u = \cos \pi x \\ du = -\pi \cos \pi x\, dx \end{array}\right\}$; $\quad \displaystyle\int \cos^2 \pi x \sin \pi x\, dx = -\frac{1}{\pi}\int u^2\, du = -\frac{1}{3\pi} u^3 + C = -\frac{1}{3\pi}\cos^3 \pi x + C$

57. $\left\{\begin{array}{l} u = \sin x^2 \\ du = 2x \cos x^2\, dx \end{array}\right\}$; $\quad \displaystyle\int x \sin^3 x^2 \cos x^2\, dx = \frac{1}{2}\int u^3\, du = \frac{1}{8}u^4 + C = \frac{1}{8}\sin^4 x^2 + C$

59. $\left\{\begin{array}{l} u = 1 + \tan x \\ du = \sec^2 x\, dx \end{array}\right\}$; $\quad \displaystyle\int \frac{\sec^2 x}{\sqrt{1 + \tan x}}\, dx = \int u^{-1/2}\, du = 2u^{1/2} + C = 2(1 + \tan x)^{1/2} + C$

61. $\left\{\begin{array}{l} u = 1/x \\ du = -1/x^2\, dx \end{array}\right\}$; $\quad \displaystyle\int \frac{\cos(1/x)}{x^2}\, dx = -\int \cos u\, du = -\sin u + C = -\sin(1/x) + C.$

63. $\left\{\begin{array}{l} u = \tan(x^3 + \pi) \\ du = 3x^2 \sec^2(x^3 + \pi)\, dx \end{array}\right\}$;

$\displaystyle\int x^2 \tan(x^3 + \pi) \sec^2(x^3 + \pi)\, dx = \frac{1}{3}\int u\, du = \frac{1}{6}u^2 + C = \frac{1}{6}\tan^2(x^3 + \pi) + C$

65. $\left\{\begin{array}{l|l} u = \sin x & x = -\pi \ \Rightarrow\ u = 0 \\ du = \cos x\, dx & x = \pi \ \Rightarrow\ u = 0 \end{array}\right\}$; $\displaystyle\int_{-\pi}^{\pi} \sin^4 x \cos x\, dx = \int_0^0 u^4\, du = 0.$

67. $\displaystyle\int_{1/4}^{1/3} \sec^2 \pi x\, dx = \frac{1}{\pi}\left[\tan \pi x\right]_{1/4}^{1/3} = \frac{1}{\pi}(\sqrt{3} - 1)$

69. $\left\{\begin{array}{l|l} u = \cos x & x = 0 \ \Rightarrow\ u = 1 \\ du = -\sin x\, dx & x = \pi/2 \ \Rightarrow\ u = 0 \end{array}\right\}$;

$\displaystyle\int_0^{\pi/2} \sin x \cos^3 x\, dx = -\int_1^0 u^3\, du = \int_0^1 u^3\, du = \left[\frac{u^4}{4}\right]_0^1 = \frac{1}{4}.$

71. $\displaystyle\int \sin^2 x\, dx = \int \frac{1 - \cos 2x}{2}\, dx = \frac{1}{2}x - \frac{1}{4}\sin 2x + C$

73. $\displaystyle\int \cos^2 5x\, dx = \int \frac{1 + \cos 10x}{2}\, dx = \frac{1}{2}x + \frac{1}{20}\sin 10x + C$

75. $\displaystyle\int_0^{\pi/2} \cos^2 2x\, dx = \int_0^{\pi/2} \frac{1 + \cos 4x}{2}\, dx = \left[\frac{1}{2}x + \frac{1}{8}\sin 4x\right]_0^{\pi/2} = \frac{\pi}{4}$

77. $A = \displaystyle\int_0^{\frac{\pi}{2}} [\cos x - (-\sin x)]\, dx = [\sin x - \cos x]_0^{\frac{\pi}{2}} = 2$

79. $A = \displaystyle\int_0^{1/4} (\cos^2 \pi x - \sin^2 \pi x)\, dx = \int_0^{1/4} \cos 2\pi x\, dx = \frac{1}{2\pi}[\sin 2\pi x]_0^{1/4} = \frac{1}{2\pi}$

81. $A = \displaystyle\int_{1/6}^{1/4} (\csc^2 \pi x - \sec^2 \pi x)\, dx = \left[\dfrac{1}{\pi}(-\cot \pi x - \tan \pi x)\right]_{1/6}^{1/4}$

$$= \dfrac{1}{\pi}\left(-2 + \cot \dfrac{\pi}{6} + \tan \dfrac{\pi}{6}\right)$$

$$= \dfrac{1}{\pi}\left(-2 + \sqrt{3} + \dfrac{1}{\sqrt{3}}\right) = \dfrac{1}{3\pi}(4\sqrt{3} - 6)$$

83. (a) $\left\{\begin{array}{l} u = \sec x \\ du = \sec x \tan x\, dx \end{array}\right\}$; $\displaystyle\int \sec^2 x \tan x\, dx = \int u\, du = \tfrac{1}{2}u^2 + C = \tfrac{1}{2}\sec^2 x + C$

(b) $\left\{\begin{array}{l} u = \tan x \\ du = \sec^2 x\, dx \end{array}\right\}$; $\displaystyle\int \sec^2 x \tan x\, dx = \int u\, du = \tfrac{1}{2}u^2 + C' = \tfrac{1}{2}\tan^2 x + C'$

(c) $C' = \tfrac{1}{2} + C$

85. $A = 4 \displaystyle\int_0^r \sqrt{r^2 - x^2}\, dx = 4 \int_0^{\pi/2} \sqrt{r^2 - r^2 \sin^2 u}\,(r \cos u)\, du$ $(x = r \sin u)$

$$= 4 \int_0^{\pi/2} r^2 \cos^2 u\, du = 4r^2 \left[\dfrac{1}{2}u + \dfrac{1}{4}\sin 2u\right]_0^{\pi/2} = \pi r^2$$

SECTION 5.8

1. Yes; $\displaystyle\int_a^b [f(x) - g(x)]\, dx = \int_a^b f(x)\, dx - \int_a^b g(x)\, dx > 0.$

3. Yes;   otherwise we would have $f(x) \le g(x)$ for all $x \in [a, b]$ and it would follow that

$$\int_a^b f(x)\, dx \le \int_a^b g(x)\, dx.$$

5. No;   take $f(x) = 0$, $g(x) = -1$ on $[0,1]$.

7. No;   take, for example, any odd function on an interval of the form $[-c, c]$.

9. No; $\displaystyle\int_{-1}^1 x\, dx = 0$ but $\displaystyle\int_{-1}^1 |x|\, dx \ne 0.$ **11.** Yes; $U_f(P) \ge \displaystyle\int_a^b f(x)\, dx = 0.$

13. No; $L_f(P) \le \displaystyle\int_a^b f(x)\, dx = 0.$

15. Yes; $\displaystyle\int_a^b [f(x) + 1]\, dx = \int_a^b f(x)\, dx + \int_a^b 1\, dx = 0 + b - a = b - a.$

17. $\dfrac{d}{dx}\left[\displaystyle\int_0^{1+x^2} \dfrac{dt}{\sqrt{2t + 5}}\right] = \dfrac{1}{\sqrt{2(1 + x^2) + 5}}\dfrac{d}{dx}(1 + x^2) = \dfrac{2x}{\sqrt{2x^2 + 7}}$

19. $\dfrac{d}{dx}\left[\displaystyle\int_x^a f(t)\,dt\right] = \dfrac{d}{dx}\left[-\displaystyle\int_a^x f(t)\,dt\right] = -f(x)$

21. $\dfrac{d}{dx}\left[\displaystyle\int_{x^2}^3 \dfrac{\sin t}{t}\,dt\right] = -\dfrac{d}{dx}\left[\displaystyle\int_3^{x^2} \dfrac{\sin t}{t}\,dt\right] = -\dfrac{\sin(x^2)}{x^2}(2x) = -\dfrac{2\sin(x^2)}{x}$

23. $\dfrac{d}{dx}\left[\displaystyle\int_1^{\sqrt{x}} \dfrac{t^2}{1+t^2}\,dt\right] = \dfrac{x}{1+x}\cdot\dfrac{1}{2\sqrt{x}} = \dfrac{\sqrt{x}}{2(1+x)}$

25. $\dfrac{d}{dx}\left[\displaystyle\int_x^{x^2} \dfrac{dt}{t}\right] = \dfrac{1}{x^2}\dfrac{d}{dx}\left(x^2\right) - \dfrac{1}{x}\dfrac{d}{dx}\left(x\right) = \dfrac{2x}{x^2} - \dfrac{1}{x} = \dfrac{1}{x}$

27. $\dfrac{d}{dx}\left[\displaystyle\int_{\tan x}^{2x} t\sqrt{1+t^2}\,dt\right] = 2x\sqrt{1+(2x)^2}\,(2) - \tan x\,\sqrt{1+\tan^2 x}\,(\sec^2 x)$

$$= 4x\sqrt{1+4x^2} - \tan x\,\sec^2 x\,|\sec x|$$

29. Set $h(x) = g(x) - f(x)$ and apply (5.8.2) to h.

31. $H(x) = \displaystyle\int_{2x}^{x^3-4} \dfrac{x\,dt}{1+\sqrt{t}} = x\int_{2x}^{x^3-4} \dfrac{dt}{1+\sqrt{t}}$,

$H'(x) = x\cdot\left[\dfrac{3x^2}{1+\sqrt{x^3-4}} - \dfrac{2}{1+\sqrt{2x}}\right] + 1\cdot\displaystyle\int_{2x}^{x^3-4} \dfrac{dt}{1+\sqrt{t}}$,

$H'(2) = 2\left[\dfrac{12}{3} - \dfrac{2}{3}\right] + \underbrace{\displaystyle\int_4^4 \dfrac{dt}{1+\sqrt{t}}}_{=\,0} = \dfrac{20}{3}$

33. (a) Let $u = -x$. Then $du = -dx$; and $u = 0$ when $x = 0$, $u = a$ when $x = -a$.

$$\int_{-a}^0 f(x)\,dx = -\int_a^0 f(-u)\,du = \int_0^a f(-u)\,du = \int_0^a f(-x)\,dx$$

(b) $\displaystyle\int_{-a}^a f(x)\,dx = \int_{-a}^0 f(x)\,dx + \int_0^a f(x)\,dx = -\int_a^0 f(u)\,du + \int_0^a f(x)\,dx$

$\qquad\qquad\qquad\qquad\qquad\qquad\qquad \overset{\textstyle\uparrow}{\underset{\displaystyle u = -x,\ du = -dx}{}}$

$$= \int_0^a f(u)\,du + \int_0^a f(x)\,dx = \int_0^a [f(x) + f(-x)]\,dx$$

35. $\displaystyle\int_{-\pi/4}^{\pi/4} (x + \sin 2x)\,dx = 0$ since $f(x) = x + \sin 2x$ is an odd function.

37. $\displaystyle\int_{-\pi/3}^{\pi/3} (1+x^2 - \cos x)\, dx = 2\int_0^{\pi/3} (1+x^2 - \cos x)\, dx$ since $f(x) = 1+x^2 - \cos x$ is an even function.

$$2\int_0^{\pi/3} (1+x^2 - \cos x)\, dx = 2\left[x + \frac{1}{3}x^3 - \sin x\right]_0^{\pi/3} = \frac{2}{3}\pi + \frac{2}{81}\pi^3 - \sqrt{3}$$

SECTION 5.9

1. A.V. $= \displaystyle\frac{1}{c}\int_0^c (mx+b)\, dx = \frac{1}{c}\left[\frac{m}{2}x^2 + bx\right]_0^c = \frac{mc}{2} + b;$ at $x = c/2$

3. A.V. $= \frac{1}{2}\displaystyle\int_{-1}^1 x^3\, dx = 0$ since the integrand is odd; at $x = 0$

5. A.V. $= \displaystyle\frac{1}{4}\int_{-2}^2 |x|\, dx = \frac{1}{2}\int_0^2 |x|\, dx = \frac{1}{2}\int_0^2 x\, dx = \frac{1}{2}\left[\frac{x^2}{2}\right]_0^2 = 1;$ at $x = \pm 1$

7. A.V. $= \displaystyle\frac{1}{2}\int_0^2 (2x - x^2)\, dx = \frac{1}{2}\left[x^2 - \frac{x^3}{3}\right]_0^2 = \frac{2}{3};$ at $x = 1 \pm \frac{1}{3}\sqrt{3}$

9. A.V. $= \displaystyle\frac{1}{9}\int_0^9 \sqrt{x}\, dx = \frac{1}{9}\left[\frac{2}{3}x^{3/2}\right]_0^9 = 2;$ at $x = 4$

11. A.V. $= \displaystyle\frac{1}{2\pi}\int_0^{2\pi} \sin x\, dx = \frac{1}{2\pi}\left[-\cos x\right]_0^{2\pi} = 0;$ at $x = \pi$

13. A.V. $= \displaystyle\frac{1}{b-a}\int_a^b x^n\, dx = \frac{1}{b-a}\left[\frac{x^{n+1}}{n+1}\right]_a^b = \frac{b^{n+1} - a^{n+1}}{(n+1)(b-a)}.$

15. Average of f' on $[a, b] = \displaystyle\frac{1}{b-a}\int_a^b f'(x)\, dx = \frac{1}{b-a}\,[f(x)]_a^b = \frac{f(b) - f(a)}{b-a}.$

17. Distance from (x, y) to the origin: $\sqrt{x^2 + y^2}$. Since $y = x^2$, $D(x) = \sqrt{x^2 + x^4}$.

On $[0, \sqrt{3}]$,   A.V.$= \displaystyle\frac{1}{\sqrt{3}}\int_0^{\sqrt{3}} x\sqrt{1+x^2}\, dx = \frac{1}{\sqrt{3}}\left[\frac{1}{3}(1+x^2)^{3/2}\right]_0^{\sqrt{3}} = \frac{1}{3\sqrt{3}}\,7 = \frac{7}{9}\sqrt{3}.$

19. The distance the stone has fallen after t seconds is given by $s(t) = 16t^2$.

    (a)   The terminal velocity after x seconds is $s'(x) = 32x$. The average velocity
        is $\displaystyle\frac{s(x) - s(0)}{x - 0} = 16x$. Thus the terminal velocity is twice the average velocity.

    (b)   For the first $\frac{1}{2}x$ seconds, aver. vel. $= \displaystyle\frac{s\left(\frac{1}{2}x\right) - s(0)}{\frac{1}{2}x - 0} = 8x.$

        For the next $\frac{1}{2}x$ seconds, aver. vel. $= \displaystyle\frac{s(x) - s\left(\frac{1}{2}x\right)}{x - \frac{1}{2}x} = 24x.$

        Thus, for the first $\frac{1}{2}x$ seconds the average velocity is one-third of the average velocity
        during the next $\frac{1}{2}x$ seconds.

21. Suppose $f(x) \neq 0$ for all x in (a, b). Then, since f is continuous, either
$$f(x) > 0 \text{ on } (a, b) \quad \text{or} \quad f(x) > 0 \text{ on } (a, b).$$
In either case, $\displaystyle\int_a^b f(x)\,dx \neq 0.$

23. (a) $v(t) - v(0) = \displaystyle\int_0^t a\,du; \quad v(0) = 0.$ Thus $v(t) = at.$

$x(t) - x(0) = \displaystyle\int_0^t v(u)\,du; \quad x(0) = x_0.$ Thus $x(t) = \displaystyle\int_0^t au\,du + x_0 = \frac{1}{2}at^2 + x_0.$

(b) $v_{avg} = \dfrac{1}{t_2 - t_1} \displaystyle\int_{t_1}^{t_2} at\,dt = \dfrac{1}{t_2 - t_1} \left[\dfrac{1}{2}at^2\right]_{t_1}^{t_2}$

$$= \frac{at_2^2 - at_1^2}{2(t_2 - t_1)} = \frac{v(t_1) + v(t_2)}{2}$$

25. (a) $M = \displaystyle\int_0^6 \frac{12}{\sqrt{x+1}}\,dx = 12\int_0^6 \frac{1}{\sqrt{x+1}}\,dx = 24\left[\sqrt{x+1}\right]_0^6 = 24(\sqrt{7} - 1)$

$x_M M = \displaystyle\int_0^6 \frac{12x}{\sqrt{x+1}}\,dx = 12\int_1^7 \left(u^{1/2} - u^{-1/2}\right)du$

$\underset{\uparrow}{}$

$\quad\quad u = x+1, \quad du = dx, \quad x = u - 1$

$\quad\quad = 12\left[\frac{2}{3}u^{3/2} - 2u^{1/2}\right]_1^7 = 16 + 32\sqrt{7};$

$x_M = \dfrac{16 + 32\sqrt{7}}{24(\sqrt{7} - 1)} = \dfrac{4\sqrt{7} + 2}{3\sqrt{7} - 3}$

(b) A.V.$= \dfrac{1}{6}\displaystyle\int_0^6 \frac{12}{\sqrt{x+1}}\,dx = \frac{1}{6}[24(\sqrt{7} - 1)] = 4(\sqrt{7} - 1)$

27. (a) $M = \displaystyle\int_0^L k\sqrt{x}\,dx = k\left[\frac{2}{3}x^{3/2}\right]_0^L = \frac{2}{3}kL^{3/2}$

$x_M M = \displaystyle\int_0^L x\left(k\sqrt{x}\right)dx = \int_0^L kx^{3/2}\,dx = \left[\frac{2}{5}kx^{5/2}\right]_0^L = \frac{2}{5}kL^{5/2}$

$x_M = \left(\frac{2}{5}kL^{5/2}\right) / \left(\frac{2}{3}kL^{3/2}\right) = \frac{3}{5}L$

(b) $M = \displaystyle\int_0^L k(L - x)^2\,dx = \left[-\frac{1}{3}k(L-x)^3\right]_0^L = \frac{1}{3}kL^3$

$x_M M = \displaystyle\int_0^L x\left[k(L-x)^2\right]dx = \int_0^L k\left(L^2x - 2Lx^2 + x^3\right)dx$

$\quad\quad = k\left[\frac{1}{2}L^2x^2 - \frac{2}{3}Lx^3 + \frac{1}{4}x^4\right]_0^L = \frac{1}{12}kL^4$

$x_M = \left(\frac{1}{12}kL^4\right) / \left(\frac{1}{3}kL^3\right) = \frac{1}{4}L$

29. $\frac{1}{4}LM = \frac{1}{8}LM_1 + x_{M_2}M_2$

$x_{M_2} = \dfrac{1}{M_2}\left(\dfrac{1}{4}LM - \dfrac{1}{8}LM_1\right) = \dfrac{L}{8M_2}(2M - M_1)$

31. Let $M = \int_a^{a+L} kx\,dx$, where a is the point of the first cut.

Thus $M = \left[\dfrac{kx^2}{2}\right]_a^{a+L} = \dfrac{k}{2}(2aL + L^2)$. Hence $a = \dfrac{2M - kL^2}{2kL}$, and $a + L = \dfrac{2M + kL^2}{2kL}$.

33. If f is continuous on $[a, b]$, then, by Theorem 5.2.5, F satisfies the conditions of the mean-value theorem of differential calculus (Theorem 4.1.1). Therefore, by that theorem, there is at least one number c in (a, b) for which

$$F'(c) = \frac{F(b) - F(a)}{b - a}.$$

Then

$$\int_a^b f(x)\,dx = F(b) - F(a) = F'(c)(b - a) = f(c)(b - a).$$

35. If f and g take on the same average value on every interval $[a, x]$, then

$$\frac{1}{x - a}\int_a^x f(t)\,dt = \frac{1}{x - a}\int_a^x g(t)\,dt.$$

Multiplication by $(x - a)$ gives

$$\int_a^x f(t)\,dt = \int_a^x g(t)\,dt.$$

Differentiation with respect to x gives $f(x) = g(x)$. This shows that, if the averages are everywhere the same, then the functions are everywhere the same.

37. Let $P = \{a = x_0, x_1, x_2, \ldots, x_n = b\}$ be a partition of the interval $[a, b]$. Then

$$\int_a^b f(x)\,dx = \int_{x_0}^{x_1} f(x)\,dx + \int_{x_1}^{x_2} f(x)\,dx + \cdots + \int_{x_{n-1}}^{x_n} f(x)\,dx$$

By the mean-value theorem for integrals, there exists a number $x_i^* \in (x_{i-1}, x_i)$ such that

$$\int_{x_{i-1}}^{x_i} f(x)\,dx = f(x_i^*)(x_i - x_{i-1}) = f(x_i^*)\,\Delta x_i, \qquad i = 1, 2, \ldots, n$$

Thus

$$\int_a^b f(x)\,dx = f(x_1^*)\,\Delta x_1 + f(x_2^*)\,\Delta x_2 + \cdots + f(x_n^*)\,\Delta x_n$$

39. (a) A.V. $= \dfrac{1}{\pi}\displaystyle\int_0^\pi \sin x\,dx = \dfrac{1}{\pi}\left[-\cos x\right]_0^\pi = \dfrac{2}{\pi}$

(b) $\sin x = \dfrac{2}{\pi} \implies x = 0.690$

(c)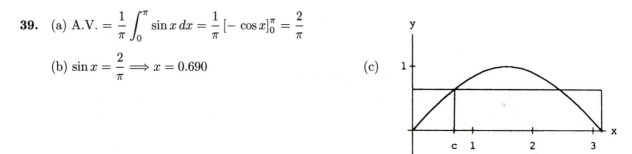

41. (a) $f(x) = 0$ at $a \cong -3.4743$ and $b \cong 3.4743$.

(b)

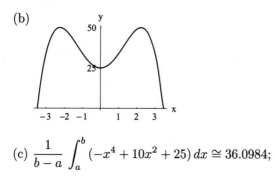

(c) $\dfrac{1}{b-a}\displaystyle\int_a^b (-x^4 + 10x^2 + 25)\,dx \cong 36.0984;$

solving $f(c) = 36.0984$ for c we get $c \cong \pm 2.9545,\ \pm 1.1274$

REVIEW EXERCISES

1. $\displaystyle\int \frac{x^3 - 2x + 1}{\sqrt{x}}\,dx = \int (x^{5/2} - 2\sqrt{x} + x^{-1/2})\,dx = \frac{2}{7}x^{\frac{7}{2}} - \frac{4}{3}x^{\frac{3}{2}} + 2x^{\frac{1}{2}} + C$

3. $\left\{ \begin{array}{c} u = 1 + t^3 \\ du = 3t^2\,dt \end{array} \right\};\qquad \displaystyle\int t^2(1+t^3)^{10}\,dt = \frac{1}{3}\int u^{10}\,du = \frac{1}{33}u^{11} + C = \frac{1}{33}(1+t^3)^{11} + C$

5. $\left\{ \begin{array}{c} u = t^{2/3} - 1 \\ du = \dfrac{3}{3t^{1/3}}\,dt \end{array} \right\};\ \displaystyle\int \frac{\left(t^{2/3} - 1\right)^2}{t^{1/3}}\,dt = \frac{3}{2}\int u^2\,du = \frac{1}{2}u^3 + C = \frac{1}{2}(t^{2/3} - 1)^3 + C$

7. Set $u = 2 - x$. Then $du = -dx$ and $x = 2 - u$.

$$\int x\sqrt{2-x}\,dx = -\int (2-u)u^{1/2}\,du = \int \left(u^{3/2} - 2u^{1/2}\right)du$$

$$= \frac{2}{5}u^{5/2} - \frac{4}{3}u^{3/2} + C = \frac{2}{5}(2-x)^{5/2} - \frac{4}{3}(2-x)^{3/2} + C$$

9. $\left\{ \begin{array}{c} u = 1 + \sqrt{x} \\ du = \dfrac{1}{\sqrt{x}}\,dx \end{array} \right\};\qquad \displaystyle\int \frac{(1+\sqrt{x})^5}{\sqrt{x}}\,dx = \int 2u^5\,du = \frac{1}{3}u^6 + C = \frac{1}{3}\left(1+\sqrt{x}\right)^6 + C$

11. $\left\{ \begin{array}{c} u = 1 + \sin x \\ du = \cos x\,dx \end{array} \right\};\qquad \displaystyle\int \frac{\cos x}{\sqrt{1+\sin x}}\,dx = \int \frac{1}{\sqrt{u}}\,du = 2\sqrt{u} + C = 2\sqrt{1+\sin x} + C$

13. $\displaystyle\int (\tan 3\theta - \cot 3\theta)^2\,d\theta = \int (\tan^2 3\theta + \cot^2 3\theta - 2)\,d\theta = \int (\sec^2 3\theta + \csc^2 3\theta - 4)\,d\theta$

$$= \frac{1}{3}\tan 3\theta - \frac{1}{3}\cot 3\theta - 4\theta + C$$

15. $\displaystyle\int \frac{1}{1 + \cos 2x}\,dx = \int \frac{1}{1 + 2\cos^2 x - 1}\,dx = \frac{1}{2}\int \frac{1}{\cos^2 x}\,dx = \frac{1}{2}\int \sec^2 x\,dx = \frac{1}{2}\tan x + C$

17. $\begin{Bmatrix} u = \sec \pi x \\ du = \pi \sec \pi x \tan \pi x \, dx \end{Bmatrix};$

$$\int \sec^3 \pi x \tan \pi x \, dx = \int \sec^2 \pi x (\sec \pi x \tan \pi x) \, dx = \int \frac{1}{\pi} u^2 \, du = \frac{1}{3\pi} u^3 + C = \frac{1}{3\pi} (\sec \pi x)^3 + C$$

19. Set $u = 1 + bx$. Then $du = b \, dx$ and $x = \frac{1}{b}(u - 1)$.

$$\int ax\sqrt{1 + bx} \, dx = \frac{a}{b^2} \int u^{1/2}(u - 1) \, du = \frac{a}{b^2} \int \left(u^{3/2} - u^{1/2} \right) du$$

$$= \frac{a}{b^2} \left[\frac{2}{5} u^{5/2} - \frac{2}{3} u^{3/2} \right] + C = \frac{2a}{5b^2} (1 + bx)^{5/2} - \frac{2a}{3b^2} (1 + bx)^{3/2} + C$$

21. $\begin{Bmatrix} u = 1 + g^2(x) \\ du = 2g(x)g'(x)dx \end{Bmatrix};$ $\int \frac{g(x)g'(x)}{\sqrt{1 + g^2(x)}} \, dx = \frac{1}{2} \int \frac{1}{\sqrt{u}} \, du = \sqrt{u} + C = \sqrt{1 + g^2(x)} + C$

23. $\int_{-1}^{2} (x^2 - 2x + 3) \, dx = \int_{-1}^{2} x^2 \, dx - \int_{-1}^{2} 2x \, dx + \int_{-1}^{2} 3 \, dx = \frac{1}{3} \left[x^3 \right]_{-1}^{2} - \left[x^2 \right]_{-1}^{2} + 3 \left[x \right]_{-1}^{2} = 9$

25. Set $u = \sin 2x$. Then $du = 2 \cos 2x \, dx$, $u(0) = 0$, $u(\pi/4) = 1$.

$$\int_0^{\pi/4} \sin^3 2x \cos 2x dx = \frac{1}{2} \int_0^1 u^3 \, du = \frac{1}{8} \left[u^4 \right]_0^1 = \frac{1}{8}$$

27. Set $u = x^3 + 3x - 6$. Then $du = 3(x^2 + 1) \, dx$, $u(0) = -6$, $u(2) = 8$.

$$\int_0^2 (x^2 + 1)(x^3 + 3x - 6)^{1/3} \, dx = \frac{1}{3} \int_{-6}^8 u^{1/3} \, du = \frac{1}{4} \left[u^{4/3} \right]_{-6}^8 = 4 - \frac{1}{4}(6)^{4/3}$$

29. (a) $\int_2^3 f(x) \, dx = \int_0^3 f(x) \, dx - \int_0^2 f(x) \, dx = -2$

(b) $\int_3^5 f(x) \, dx = \int_0^3 f(x) \, dx + \int_3^5 f(x) \, dx - \int_0^2 f(x) \, dx = 6;$

(c) Mean-value theorem: there exists a $c \in (3, 5)$ such that $f(c) = \frac{1}{2} \int_3^5 f(x) \, dx = 4.$

(d) If $f(x) \geq 0$ on $[2, 3]$, then $\int_2^3 f \geq 0$. But $\int_2^3 f = -2 < 0$.

31. $A = \int_{-2}^1 \left[(4 - x^2) - (x + 2) \right] dx = \int_{-2}^1 (-x^2 - x + 2) dx = \frac{9}{2}$

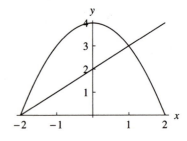

33. $A = \int_0^3 (3y - y^2)\, dy = \dfrac{9}{2}$

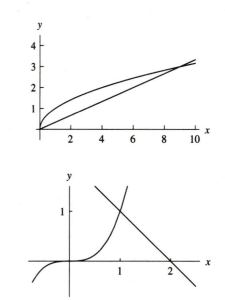

35. $A = \int_0^1 [x^3]\, dx + \int_1^2 (2 - x)\, dx = \dfrac{3}{4}$

$A = \int_0^1 (2 - y - y^{1/3}]\, dy = \dfrac{3}{4}$

37. $\dfrac{d}{dx}\left(\int_0^x \dfrac{dt}{1 + t^2}\right) = \dfrac{1}{1 + x^2}$

39. Fix a number a. Then

$$\frac{d}{dx}\left(\int_x^{x^2} \frac{dt}{1 + t^2}\right) = \frac{d}{dx}\left[\int_a^{x^2} \frac{dt}{1 + t^2} - \int_a^x \frac{dt}{1 + t^2}\right] = \frac{2x}{1 + x^4} - \frac{1}{1 + x^2}$$

41. $\dfrac{d}{dx}\left(\int_0^{\cos x} \dfrac{dt}{1 - t^2}\right) = \dfrac{1}{1 - \cos^2 x}(-\sin x) = -\dfrac{1}{\sin x} = -\csc x$

43. (a) Yes, at $x = 0$.

(b) $F'(x) = \dfrac{1}{x^2 + 2x + 2} > 0 \Longrightarrow F$ increases on $(-\infty, \infty)$.

(c) $F''(x) = \dfrac{-2(x + 1)}{x^2 + 2x + 2}$. The graph of F is concave up on $(-\infty, -1)$ and concave down on $(-1, \infty)$

(d)

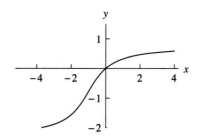

45. $f_{avg} = \dfrac{1}{4} \displaystyle\int_0^4 \dfrac{x}{\sqrt{x^2+9}}\, dx = \dfrac{1}{4}\left[\sqrt{x^2+9}\right]_0^4 = \dfrac{1}{2}$

47. $f_{avg} = \dfrac{1}{2\pi} \displaystyle\int_a^{a+2\pi} \cos x\, dx = \left[\sin x\right]_a^{a+2\pi} = \sin(a+2\pi) - \sin a = 0$

49. $\displaystyle\int_\alpha^\beta |f(x)|dx$

51. $\dfrac{1}{2}\left[\displaystyle\int_\alpha^\beta |f(x)|dx - \int_\alpha^\beta f(x)dx\right]$

53. $\lambda(x) = k\left(\frac{1}{4}a - x\right)$ for $0 \le x \le \frac{1}{4}a$ and $\lambda(x) = k(x - \frac{1}{4}a)$ for $\frac{1}{4}a \le x \le a$; $k > 0$.

$$M = \int_0^{a/4} k\left(\tfrac{1}{4}a - x\right) dx + \int_{a/4}^a k\left(x - \tfrac{1}{4}a\right) dx = \tfrac{5}{16}ka^2$$

$$x_M M = \int_0^{a/4} kx\left(\tfrac{1}{4}a - x\right) dx + \int_{a/4}^a kx\left(x - \tfrac{1}{4}a\right) dx = \frac{41}{192}ka^3; \qquad x_M = \frac{\frac{41}{192}ka^3}{\frac{5}{16}ka^2} = \frac{41a}{60}$$

CHAPTER 6

SECTION 6.1

1.

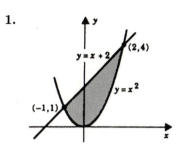

(a) $\displaystyle\int_{-1}^{2} [(x+2)-x^2]\,dx$

(b) $\displaystyle\int_{0}^{1} [(\sqrt{y})-(-\sqrt{y})]\,dy + \int_{1}^{4} [(\sqrt{y})-(y-2)]\,dy$

3.

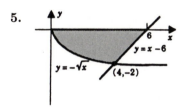

(a) $\displaystyle\int_{0}^{2} \left[(2x^2)-(x^3)\right]\,dx$

(b) $\displaystyle\int_{0}^{8} \left[\left(y^{1/3}\right)-\left(\frac{1}{2}y\right)^{1/2}\right]\,dy$

5.

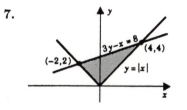

(a) $\displaystyle\int_{0}^{4} \left[(0)-(-\sqrt{x})\right]\,dx + \int_{4}^{6} [(0)-(x-6)]\,dx$

(b) $\displaystyle\int_{-2}^{0} \left[(y+6)-(y^2)\right]\,dy$

7.

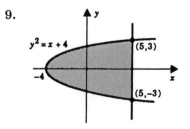

(a) $\displaystyle\int_{-2}^{0} \left[\left(\frac{8+x}{3}\right)-(-x)\right]\,dx + \int_{0}^{4} \left[\left(\frac{8+x}{3}\right)-(x)\right]\,dx$

(b) $\displaystyle\int_{0}^{2} [(y)-(-y)]\,dy + \int_{2}^{4} [(y)-(3y-8)]\,dy$

9.

(a) $\displaystyle\int_{-4}^{5} \left[(\sqrt{4+x})-(-\sqrt{4+x})\right]\,dx$

(b) $\displaystyle\int_{-3}^{3} \left[(5)-(y^2-4)\right]\,dy$

11.

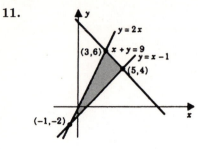

(a) $\displaystyle\int_{-1}^{3} [(2x) - (x-1)]\, dx + \int_{3}^{5} [(9-x) - (x-1)]\, dx$

(b) $\displaystyle\int_{-2}^{4} \left[(y+1) - \left(\tfrac{1}{2}y\right)\right] dy + \int_{4}^{6} \left[(9-y) - \left(\tfrac{1}{2}y\right)\right] dy$

13.

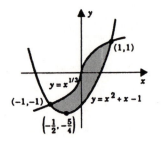

(a) $\displaystyle\int_{-1}^{1} \left[\left(x^{1/3}\right) - (x^2 + x - 1)\right] dx$

(b) $\displaystyle\int_{-5/4}^{-1} \left[\left(-\tfrac{1}{2} + \tfrac{1}{2}\sqrt{4y+5}\right) - \left(-\tfrac{1}{2} - \tfrac{1}{2}\sqrt{4y+5}\right)\right] dy$
$\displaystyle + \int_{-1}^{1} \left[\left(-\tfrac{1}{2} + \tfrac{1}{2}\sqrt{4y+5}\right) - \left(y^3\right)\right] dy$

15.

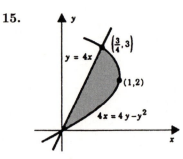

$\displaystyle A = \int_{0}^{3} \left[\left(\frac{4y - y^2}{4}\right) - \left(\frac{y}{4}\right)\right] dy$

$\displaystyle = \int_{0}^{3} \left(\frac{3}{4}y - \frac{1}{4}y^2\right) dy$

$\displaystyle = \left[\tfrac{3}{8}y^2 - \tfrac{1}{12}y^3\right]_0^3 = \tfrac{9}{8}$

17.

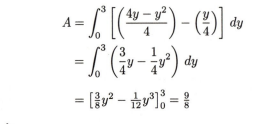

$\displaystyle A = 2\int_{0}^{2} \left[(12 - 2y^2) - (y^2)\right] dy$

$\displaystyle = 2\int_{0}^{2} (12 - 3y^2)\, dy$

$\displaystyle = 2\left[12y - y^3\right]_0^2 = 2\,(16) = 32$

19.

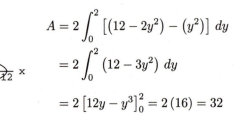

$\displaystyle A = \int_{-2}^{0} \left[(y^3 - y) - (y - y^2)\right] dy + \int_{0}^{1} \left[(y - y^2) - (y^3 - y)\right] dy$

$\displaystyle = \int_{-2}^{0} (y^3 + y^2 - 2y)\, dy + \int_{0}^{1} (2y - y^2 - y^3)\, dy$

$\displaystyle = \left[\tfrac{1}{4}y^4 + \tfrac{1}{3}y^3 - y^2\right]_{-2}^{0} + \left[y^2 - \tfrac{1}{3}y^3 - \tfrac{1}{4}y^4\right]_{0}^{1} = \tfrac{8}{3} + \tfrac{5}{12} = \tfrac{37}{12}$

21.

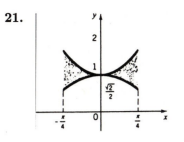

$$A = \int_{-\pi/4}^{\pi/4} \left[\sec^2 x - \cos x\right]\, dx$$

$$= 2\int_0^{\pi/4} \left[\sec^2 x - \cos x\right]\, dx$$

$$= 2\left[\tan x + \sin x\right]_0^{\pi/4} = 2\left[1 + \sqrt{2}/2\right] = 2 + \sqrt{2}$$

23.

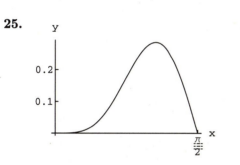

$$A = \int_{-\pi}^{-\pi/2} \left[\sin 2x - 2\cos x\right]\, dx + \int_{-\pi/2}^{\pi/2} \left[2\cos x - \sin 2x\right]\, dx$$

$$+ \int_{\pi/2}^{\pi} \left[\sin 2x - 2\cos x\right]\, dx$$

$$= \left[-\tfrac{1}{2}\cos 2x - 2\sin x\right]_{-\pi}^{-\pi/2} + \left[2\sin x + \tfrac{1}{2}\cos 2x\right]_{-\pi/2}^{\pi/2}$$

$$+ \left[-\tfrac{1}{2}\cos 2x - 2\sin x\right]_{\pi/2}^{\pi} = 8$$

25.

$$A = \int_0^{\pi/2} \left(\sin^4 x \cos x\right)\, dx$$

$$= \int_0^1 u^4\, du, \quad (u = \sin x)$$

$$= \left[\frac{u^5}{5}\right]_0^1 = \frac{1}{5}$$

27. $\quad A = \int_0^1 \left[3x - \dfrac{1}{3}x\right]\, dx + \int_1^3 \left[-x + 4 - \dfrac{1}{3}x\right]\, dx = \left[\dfrac{4}{3}x^2\right]_0^1 + \left[-\dfrac{2}{3}x^2 + 4x\right]_1^3 = 4$

29. $\quad A = \int_{-2}^1 \left[x - (-2)\right]\, dx + \int_1^5 \left[1 - (-2)\right]\, dx + \int_5^7 \left[-\dfrac{3}{2}x + \dfrac{17}{2} - (-2)\right]\, dx$

$$= \left[\tfrac{1}{2}x^2 + 2x\right]_{-2}^1 + \left[3x\right]_1^5 + \left[-\tfrac{3}{4}x^2 + \tfrac{21}{2}x\right]_5^7 = \frac{39}{2}$$

31.

$$A = \int_{-3}^0 \left[6 - x^2 - x\right]\, dx + \int_0^3 \left[6 - x^2 - (-x)\right]\, dx$$

$$= \left[6x - \frac{1}{3}x^3 - \frac{1}{2}x^2\right]_{-3}^0 + \left[6x - \frac{1}{3}x^3 + \frac{1}{2}x^2\right]_0^3 = 27$$

33.

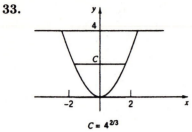

$$\int_0^{\sqrt{c}} \left[c - x^2\right] dx = \frac{1}{2} \int_0^2 \left[4 - x^2\right] dx$$

$$\left[cx - \frac{1}{3}x^3\right]_0^{\sqrt{c}} = \frac{1}{2}\left[4x - \frac{1}{3}x^3\right]_0^2$$

$$\frac{2}{3}c^{3/2} = \frac{8}{3} \quad \text{and} \quad c = 4^{2/3}$$

36. $A = \int_0^1 \sqrt{3}\,dx + \int_1^2 \sqrt{4 - x^2}\,dx;$

$$A = \int_0^{\sqrt{3}} \left(\sqrt{4 - y^2} - \frac{y}{\sqrt{3}}\right) dy$$

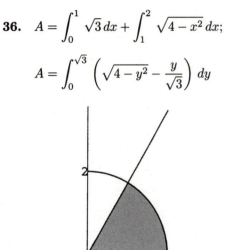

37. $A = \int_0^2 \left[\sqrt{4 - x^2} - (2 - \sqrt{4x - x^2})\right] dx$

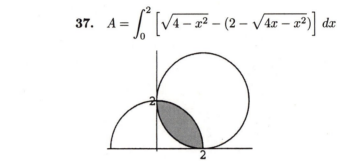

39.

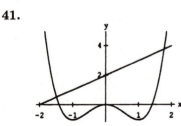

The area under the curve is $A_c = \int_0^a bx^n\,dx = \dfrac{ba^{n+1}}{n+1}$.

For the rectangle, $A_r = ba^{n+1}$.   Thus the ratio is $\dfrac{1}{n+1}$.

41.

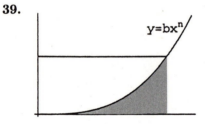

$$A \cong \int_{-1.49}^{1.79} \left[x + 2 - (x^4 - 2x^2)\right] dx$$

$$= \left[\frac{1}{2}x^2 - 2x - \frac{1}{5}x^5 + \frac{2}{3}x^3\right]_{-1.49}^{1.78} \cong 7.93$$

43.

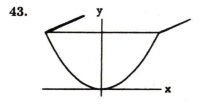

$$V = 8 \cdot 12 \int_{-3}^{3} \left[4 - \frac{4}{9} x^2 \right] dx$$

$$= 96 \cdot 2 \int_{0}^{3} \left[4 - \frac{4}{9} x^2 \right] dx$$

$$= 192 \left[4x - \tfrac{4}{27} x^3 \right]_{0}^{3}$$

$$= 1536 \text{ cu. in.} \cong 0.89 \text{ cu. ft.}$$

45.

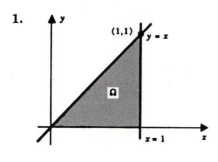

(a) $A = \int_{1}^{b} \frac{1}{\sqrt{x}} dx = \left[2\sqrt{x} \right]_{1}^{b} = 2\sqrt{b} - 2.$

(b) $2\sqrt{b} - 2 \to \infty$ as $b \to \infty.$

SECTION 6.2

1.

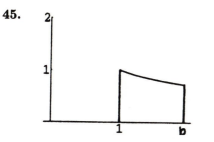

$$V = \int_{0}^{1} \pi \left[(x)^2 - (0)^2 \right] dx = \pi \left[\frac{x^3}{3} \right]_{0}^{1} = \frac{\pi}{3}$$

3.

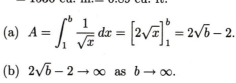

$$V = \int_{-3}^{3} \pi \left[(9)^2 - \left(x^2 \right)^2 \right] dx = 2 \int_{0}^{3} \pi \left(81 - x^4 \right) dx$$

$$= 2\pi \left[81x - \frac{x^5}{5} \right]_{0}^{3} = \frac{1944\pi}{5}$$

5.

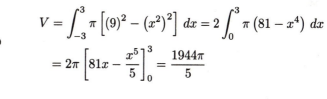

$$V = \int_{0}^{1} \pi \left[\left(\sqrt{x} \right)^2 - \left(x^3 \right)^2 \right] dx$$

$$= \int_{0}^{1} \pi \left(x - x^6 \right) dx$$

$$= \pi \left[\frac{1}{2} x^2 - \frac{1}{7} x^7 \right]_{0}^{1} = \frac{5\pi}{14}$$

7.

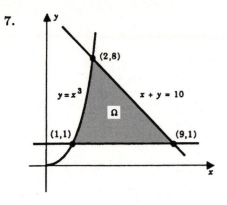

$$V = \int_1^2 \pi \left[(x^3)^2 - (1)^2 \right] dx + \int_2^9 \pi \left[(10-x)^2 - (1)^2 \right] dx$$

$$= \int_1^2 \pi \left(x^6 - 1 \right) dx + \int_2^9 \pi \left(99 - 20x + x^2 \right) dx$$

$$= \pi \left[\frac{1}{7}x^7 - x \right]_1^2 + \pi \left[99x - 10x^2 + \frac{1}{3}x^3 \right]_2^9 = \frac{3790\pi}{21}$$

9.

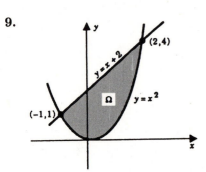

$$V = \int_{-1}^2 \pi \left[(x+2)^2 - (x^2)^2 \right] dx$$

$$= \int_{-1}^2 \pi \left(x^2 + 4x + 4 - x^4 \right) dx$$

$$= \pi \left[\tfrac{1}{3}x^3 + 2x^2 + 4x - \tfrac{1}{5}x^5 \right]_{-1}^2 = \frac{72}{5}\pi$$

11.

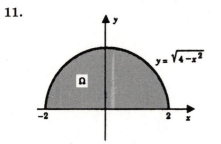

$$V = \int_{-2}^2 \pi \left[\sqrt{4-x^2} \right]^2 dx = 2 \int_0^2 \pi \left(4 - x^2 \right) dx$$

$$= 2\pi \left[4x - \frac{x^3}{3} \right]_0^2 = \frac{32}{3}\pi$$

13.

$$V = \int_0^{\pi/4} \pi \sec^2 x \, dx = \pi \left[\tan x \right]_0^{\pi/4} = \pi$$

15.

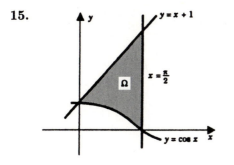

$$V = \int_0^{\pi/2} \pi \left[(x+1)^2 - (\cos x)^2 \right] \, dx$$

$$= \int_0^{\pi/2} \pi \left[(x+1)^2 - \left(\frac{1}{2} + \frac{1}{2} \cos 2x \right) \right] \, dx$$

$$= \pi \left[\frac{1}{3}(x+1)^3 - \frac{1}{2}x - \frac{1}{4}\sin 2x \right]_0^{\pi/2}$$

$$= \frac{\pi^2}{24} \left(\pi^2 + 6\pi + 6 \right)$$

17.

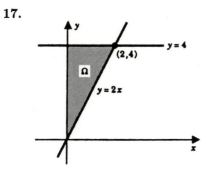

$$V = \int_0^4 \pi \left(\frac{y}{2} \right)^2 \, dy = \frac{\pi}{12} \left[y^3 \right]_0^4 = \frac{16\pi}{3}$$

19.

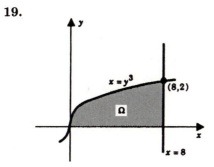

$$V = \int_0^2 \pi \left[(8)^2 - \left(y^3 \right)^2 \right] \, dy$$

$$= \int_0^2 \pi \left(64 - y^6 \right) \, dy$$

$$= \pi \left[64y - \frac{1}{7}y^7 \right]_0^2 = \frac{768}{7}\pi$$

21.

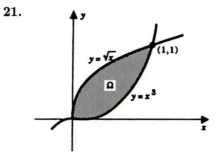

$$V = \int_0^1 \pi \left[\left(y^{1/3} \right)^2 - \left(y^2 \right)^2 \right] \, dy$$

$$= \int_0^1 \pi \left[y^{2/3} - y^4 \right] \, dy$$

$$= \pi \left[\frac{3}{5}y^{5/3} - \frac{1}{5}y^5 \right]_0^1 = \frac{2}{5}\pi$$

23.

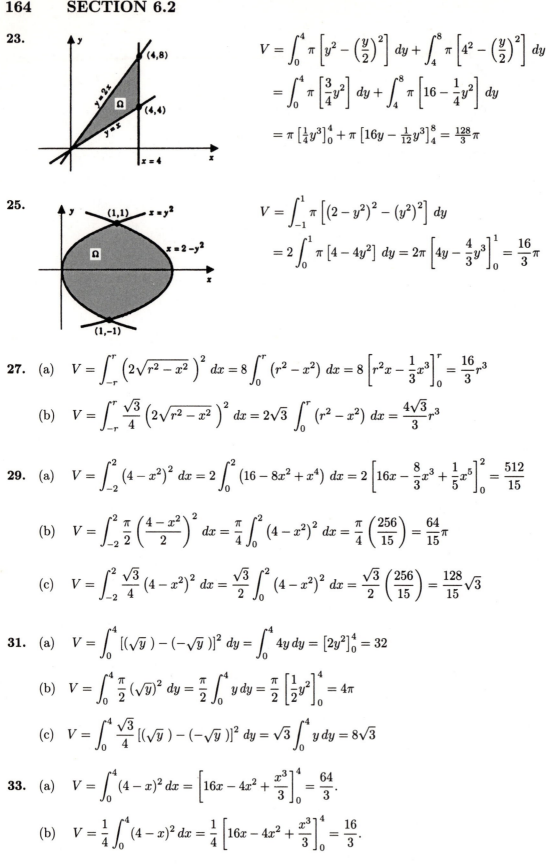

$$V = \int_0^4 \pi \left[y^2 - \left(\frac{y}{2} \right)^2 \right] dy + \int_4^8 \pi \left[4^2 - \left(\frac{y}{2} \right)^2 \right] dy$$

$$= \int_0^4 \pi \left[\frac{3}{4} y^2 \right] dy + \int_4^8 \pi \left[16 - \frac{1}{4} y^2 \right] dy$$

$$= \pi \left[\frac{1}{4} y^3 \right]_0^4 + \pi \left[16y - \frac{1}{12} y^3 \right]_4^8 = \frac{128}{3} \pi$$

25.

$$V = \int_{-1}^1 \pi \left[(2 - y^2)^2 - (y^2)^2 \right] dy$$

$$= 2 \int_0^1 \pi \left[4 - 4y^2 \right] dy = 2\pi \left[4y - \frac{4}{3} y^3 \right]_0^1 = \frac{16}{3} \pi$$

27. (a) $\quad V = \int_{-r}^r \left(2\sqrt{r^2 - x^2} \right)^2 dx = 8 \int_0^r (r^2 - x^2) \, dx = 8 \left[r^2 x - \frac{1}{3} x^3 \right]_0^r = \frac{16}{3} r^3$

(b) $\quad V = \int_{-r}^r \frac{\sqrt{3}}{4} \left(2\sqrt{r^2 - x^2} \right)^2 dx = 2\sqrt{3} \int_0^r (r^2 - x^2) \, dx = \frac{4\sqrt{3}}{3} r^3$

29. (a) $\quad V = \int_{-2}^2 \left(4 - x^2 \right)^2 dx = 2 \int_0^2 \left(16 - 8x^2 + x^4 \right) dx = 2 \left[16x - \frac{8}{3} x^3 + \frac{1}{5} x^5 \right]_0^2 = \frac{512}{15}$

(b) $\quad V = \int_{-2}^2 \frac{\pi}{2} \left(\frac{4 - x^2}{2} \right)^2 dx = \frac{\pi}{4} \int_0^2 \left(4 - x^2 \right)^2 dx = \frac{\pi}{4} \left(\frac{256}{15} \right) = \frac{64}{15} \pi$

(c) $\quad V = \int_{-2}^2 \frac{\sqrt{3}}{4} \left(4 - x^2 \right)^2 dx = \frac{\sqrt{3}}{2} \int_0^2 \left(4 - x^2 \right)^2 dx = \frac{\sqrt{3}}{2} \left(\frac{256}{15} \right) = \frac{128}{15} \sqrt{3}$

31. (a) $\quad V = \int_0^4 \left[(\sqrt{y}) - (-\sqrt{y}) \right]^2 dy = \int_0^4 4y \, dy = \left[2y^2 \right]_0^4 = 32$

(b) $\quad V = \int_0^4 \frac{\pi}{2} (\sqrt{y})^2 dy = \frac{\pi}{2} \int_0^4 y \, dy = \frac{\pi}{2} \left[\frac{1}{2} y^2 \right]_0^4 = 4\pi$

(c) $\quad V = \int_0^4 \frac{\sqrt{3}}{4} \left[(\sqrt{y}) - (-\sqrt{y}) \right]^2 dy = \sqrt{3} \int_0^4 y \, dy = 8\sqrt{3}$

33. (a) $\quad V = \int_0^4 (4 - x)^2 dx = \left[16x - 4x^2 + \frac{x^3}{3} \right]_0^4 = \frac{64}{3}.$

(b) $\quad V = \frac{1}{4} \int_0^4 (4 - x)^2 dx = \frac{1}{4} \left[16x - 4x^2 + \frac{x^3}{3} \right]_0^4 = \frac{16}{3}.$

35. (a) $V = \int_0^{\pi/2} \sqrt{3} \sin x \, dx = -\sqrt{3} \left[\cos x\right]_0^{\pi/2} = \sqrt{3}$

   (b) $V = \int_0^{\pi/2} 4 \sin x \, dx = -4 \left[\cos x\right]_0^{\pi/2} = 4$

37. $V = \int_{-a}^{a} \pi \left(b\sqrt{1 - \dfrac{x^2}{a^2}}\right)^2 dx = \dfrac{2\pi b^2}{a^2} \int_0^a \pi \left(a^2 - x^2\right) dx = \dfrac{2\pi b^2}{a^2} \pi \left[a^2 x - \dfrac{1}{3}x^3\right]_0^a$

$$= \dfrac{2\pi b^2}{a^2} \left(\dfrac{2}{3}a^3\right) = \dfrac{4}{3}\pi ab^2$$

39.

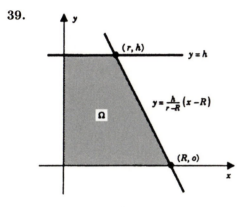

The specified frustum is generated by revolving the region Ω about the y-axis.

$$V = \int_0^h \pi \left[\dfrac{r - R}{h}y + R\right]^2 dy$$

$$= \pi \left[\dfrac{h}{3(r - R)} \left(\dfrac{r - R}{h}y + R\right)^3\right]_0^h$$

$$= \dfrac{\pi h}{3(r - R)} \left(r^3 - R^3\right) = \dfrac{\pi h}{3} \left(r^2 + rR + R^2\right)$$

41. Capacity of basin $= \dfrac{1}{2}\left(\dfrac{4}{3}\pi r^3\right) = \dfrac{2}{3}\pi r^3$.

   (a) Volume of water $= \int_{r/2}^{r} \pi \left[\sqrt{r^2 - x^2}\right]^2 dx$

$$= \pi \int_{r/2}^{r} \left(r^2 - x^2\right) dx = \pi \left[r^2 x - \dfrac{1}{3}x^3\right]_{r/2}^{r} = \dfrac{5}{24}\pi r^3.$$

   The basin is $\left(\dfrac{5}{24}\pi r^3\right)(100) / \left(\dfrac{2}{3}\pi r^3\right) = 31\dfrac{1}{4}\%$ full.

   (b) Volume of water $= \int_{2r/3}^{r} \pi \left[\sqrt{r^2 - x^2}\right]^2 dx = \pi \int_{2r/3}^{r} \left(r^2 - x^2\right) dx = \dfrac{8}{81}\pi r^3.$

   The basin is $\left(\dfrac{8}{81}\pi r^3\right)(100) / \left(\dfrac{2}{3}\pi r^3\right) = 14\dfrac{22}{27}\%$ full.

43. $V = \int_h^r \pi(r^2 - y^2)\, dy = \pi \left[r^2 y - \dfrac{y^3}{3}\right]_h^r = \dfrac{\pi}{3}\left[2r^3 - 3r^2 h + h^3\right].$

45. (a)

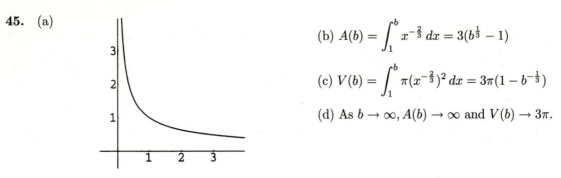

(b) $A(b) = \displaystyle\int_1^b x^{-\frac{2}{3}}\,dx = 3(b^{\frac{1}{3}} - 1)$

(c) $V(b) = \displaystyle\int_1^b \pi(x^{-\frac{2}{3}})^2\,dx = 3\pi(1 - b^{-\frac{1}{3}})$

(d) As $b \to \infty$, $A(b) \to \infty$ and $V(b) \to 3\pi$.

47. If the depth of the liquid in the container is h feet, then the volume of the liquid is:

$$V(h) = \int_0^h \pi \left(\sqrt{y+1}\right)^2\,dy = \int_0^h \pi\,[y+1]\,dy.$$

Differentiation with respect to t gives

$$\frac{dV}{dt} = \frac{dV}{dh}\cdot\frac{dh}{dt} = \pi(h+1)\frac{dh}{dt}.$$

Now, since $\dfrac{dV}{dt} = 2$, it follows that $\dfrac{dh}{dt} = \dfrac{2}{\pi(h+1)}$. Thus

$$\frac{dh}{dt}\bigg|_{h=1} = \frac{2}{2\pi} = \frac{1}{\pi}\ \text{ft/min} \quad\text{and}\quad \frac{dh}{dt}\bigg|_{h=2} = \frac{2}{3\pi}\ \text{ft/min}.$$

49. (b) The x-coordinates of the points of intersection are: $x = 0$, $x = 2^{1/4} \cong 1.1892$.

(c) $A \cong 2\displaystyle\int_0^{1.1892} \left(2x - x^5\right)\,dx \cong 2(0.9428) = 1.8856$

(d) $V = \displaystyle\int_0^{1.1892} \pi\left(4x^2 - x^{10}\right)\,dx \cong 5.1234$

51. $V = \displaystyle\int_0^4 \pi\left[4 - \left(2 - \sqrt{x}\right)^2\right]\,dx = \int_0^4 \pi\left[4\sqrt{x} - x\right]\,dx = \pi\left[\frac{8}{3}x^{3/2} - \frac{1}{2}x^2\right]_0^4 = \frac{40\pi}{3}$

53. $V = \displaystyle\int_0^\pi \pi\left[2\sin x - \sin^2 x\right]\,dx = \pi\left[-2\cos x\right]_0^\pi - \frac{\pi}{2}\int_0^\pi (1 - \cos 2x)\,dx$

$= 4\pi - \dfrac{\pi}{2}\left[x - \dfrac{1}{2}\sin 2x\right]_0^\pi = 4\pi - \dfrac{1}{2}\pi^2$

55.

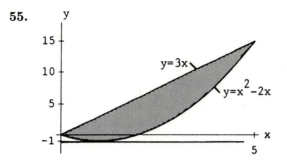

$$V = \int_0^5 \pi \left([3x - (-1)]^2 - [x^2 - 2x - (-1)]^2 \right) dx$$

$$= \pi \int_0^5 [3x + 1]^2 \, dx - \int_0^5 [x - 1]^4 \, dx$$

$$= \pi \left[\frac{1}{9}(3x+1)^3 \right]_0^5 - \left[\frac{(x-1)^5}{5} \right]_0^5$$

$$= 250\pi$$

57.

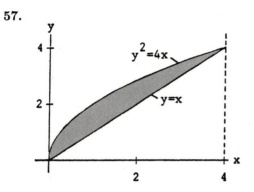

(a) $V = \int_0^4 \pi \left[\left(\sqrt{4x} \right)^2 - x^2 \right] dx$

$$= \pi \int_0^4 [4x - x^2] \, dx$$

$$= \pi \left[2x^2 - \frac{1}{3} x^3 \right]_0^4 = \frac{32\pi}{3}$$

(b) $V = \int_0^4 \pi \left[\left(\frac{1}{4} y^2 - 4 \right)^2 - (y - 4)^2 \right] dy$

$$= \pi \int_0^4 \left[\frac{1}{16} y^4 - 3y^2 + 8y \right] dy$$

$$= \pi \left[\frac{1}{80} y^5 - y^3 + 4y^2 \right]_0^4 = \frac{64\pi}{5}$$

59. (a) $V = \int_0^4 \pi \left(x^{3/2} \right)^2 dx = \pi \int_0^4 x^3 \, dx = \pi \left[\frac{1}{4} x^4 \right]_0^4 = 64\pi$

(b) $V = \int_0^8 \pi \left(4 - y^{2/3} \right)^2 dy = \pi \int_0^8 \left(16 - 8y^{2/3} + y^{4/3} \right) dy$

$$= \pi \left[16y - \frac{24}{5} y^{5/3} + \frac{3}{7} y^{7/3} \right]_0^8 = \frac{1024}{35} \pi$$

(c) $V = \int_0^4 \pi \left[(8)^2 - \left(8 - x^{3/2} \right)^2 \right] dx = \pi \int_0^4 \left(16x^{3/2} - x^3 \right) dx$

$$= \pi \left[\frac{32}{5} x^{5/2} - \frac{1}{4} x^4 \right]_0^4 = \frac{704}{5} \pi$$

(d) $V = \int_0^8 \pi \left[(4)^2 - \left(y^{2/3} \right)^2 \right] dy = \pi \int_0^8 \left(16 - y^{4/3} \right) dy = \pi \left[16y - \frac{3}{7} y^{7/3} \right]_0^8 = \frac{512}{7} \pi$

SECTION 6.3

1.

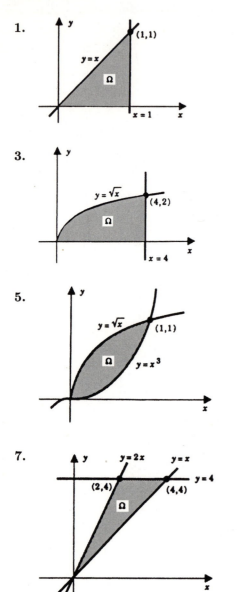

$$V = \int_0^1 2\pi x \,[x - 0]\, dx = 2\pi \int_0^1 x^2 \, dx$$

$$= 2\pi \left[\frac{1}{3}x^3\right]_0^1 = \frac{2\pi}{3}$$

3.

$$V = \int_0^4 2\pi x \left[\sqrt{x} - 0\right] dx = 2\pi \int_0^4 x^{3/2} \, dx$$

$$= 2\pi \left[\frac{2}{5}x^{5/2}\right]_0^4 = \frac{128}{5}\pi$$

5.

$$V = \int_0^1 2\pi x \left[\sqrt{x} - x^3\right] dx$$

$$= 2\pi \int_0^1 \left(x^{3/2} - x^4\right) dx$$

$$= 2\pi \left[\frac{2}{5}x^{5/2} - \frac{1}{5}x^5\right]_0^1 = \frac{2\pi}{5}$$

7.

$$V = \int_0^2 2\pi x \,[2x - x]\, dx + \int_2^4 2\pi x \,[4 - x]\, dx$$

$$= 2\pi \int_0^2 x^2 \, dx + 2\pi \int_2^4 \left(4x - x^2\right) dx$$

$$= 2\pi \left[\frac{1}{3}x^3\right]_0^2 + 2\pi \left[2x^2 - \frac{1}{3}x^3\right]_2^4 = 16\pi$$

9.

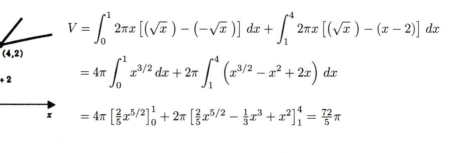

$$V = \int_0^1 2\pi x \left[\left(\sqrt{x}\right) - \left(-\sqrt{x}\right)\right] dx + \int_1^4 2\pi x \left[\left(\sqrt{x}\right) - (x - 2)\right] dx$$

$$= 4\pi \int_0^1 x^{3/2} \, dx + 2\pi \int_1^4 \left(x^{3/2} - x^2 + 2x\right) dx$$

$$= 4\pi \left[\frac{2}{5}x^{5/2}\right]_0^1 + 2\pi \left[\frac{2}{5}x^{5/2} - \frac{1}{3}x^3 + x^2\right]_1^4 = \frac{72}{5}\pi$$

11.

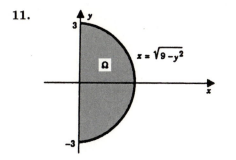

$$V = \int_0^3 2\pi x \left[\sqrt{9 - x^2} - \left(-\sqrt{9 - x^2} \right) \right] dx$$

$$= 4\pi \int_0^3 x \left(9 - x^2 \right)^{1/2} dx$$

$$= 4\pi \left[-\tfrac{1}{3} \left(9 - x^2 \right)^{3/2} \right]_0^3 = 36\pi$$

13.

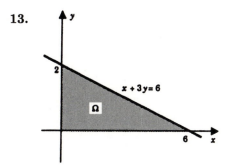

$$V = \int_0^2 2\pi y \left[6 - 3y \right] dy$$

$$= 6\pi \int_0^2 \left(2y - y^2 \right) dy$$

$$= 6\pi \left[y^2 - \tfrac{1}{3} y^3 \right]_0^2 = 8\pi$$

15.

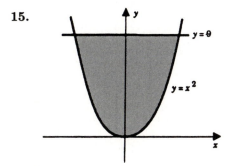

$$V = \int_0^9 2\pi y \left[\left(\sqrt{y} \right) - \left(-\sqrt{y} \right) \right] dy$$

$$= 4\pi \int_0^9 y^{3/2} dy$$

$$= 4\pi \left[\tfrac{2}{5} y^{5/2} \right]_0^9 = \tfrac{1944}{5} \pi$$

17.

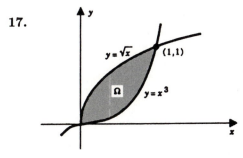
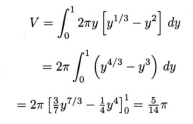

$$V = \int_0^1 2\pi y \left[y^{1/3} - y^2 \right] dy$$

$$= 2\pi \int_0^1 \left(y^{4/3} - y^3 \right) dy$$

$$= 2\pi \left[\tfrac{3}{7} y^{7/3} - \tfrac{1}{4} y^4 \right]_0^1 = \tfrac{5}{14} \pi$$

19.

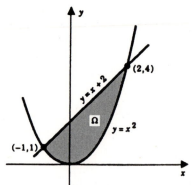

$$V = \int_0^1 2\pi y \left[(\sqrt{y}) - (-\sqrt{y}) \right] dy + \int_1^4 2\pi y \left[(\sqrt{y}) - (y - 2) \right] dy$$

$$= 4\pi \int_0^1 y^{3/2} \, dy + 2\pi \int_1^4 \left(y^{3/2} - y^2 + 2y \right) dy$$

$$= 4\pi \left[\tfrac{2}{5} y^{5/2} \right]_0^1 + 2\pi \left[\tfrac{2}{5} y^{5/2} - \tfrac{1}{3} y^3 + y^2 \right]_1^4 = \tfrac{72}{5}\pi$$

21.

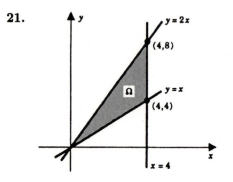

$$V = \int_0^4 2\pi y \left[y - \frac{y}{2} \right] dy + \int_4^8 2\pi y \left[4 - \frac{y}{2} \right] dy$$

$$= \pi \int_0^4 y^2 \, dy + \pi \int_4^8 \left(8y - y^2 \right) dy$$

$$= \pi \left[\tfrac{1}{3} y^3 \right]_0^4 + \pi \left[4y^2 - \tfrac{1}{3} y^3 \right]_4^8 = 64\pi$$

23.

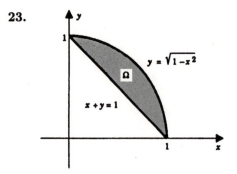

$$V = \int_0^1 2\pi y \left[\sqrt{1 - y^2} - (1 - y) \right] dy$$

$$= 2\pi \int_0^1 \left[y \left(1 - y^2 \right)^{1/2} - y + y^2 \right] dy$$

$$= 2\pi \left[-\frac{1}{3} \left(1 - y^2 \right)^{3/2} - \frac{1}{2} y^2 + \frac{1}{3} y^3 \right]_0^1 = \frac{\pi}{3}$$

25. (a) $V = \int_0^1 2\pi x \left[1 - \sqrt{x} \right] dx$ (b) $V = \int_0^1 \pi y^4 \, dy$

$$= \pi \left[\tfrac{1}{5} y^5 \right]_0^1 = \tfrac{1}{5}\pi$$

27. (a) $V = \int_0^1 \pi \left(x - x^4 \right) dx$ (b) $V = \int_0^1 2\pi y \left(\sqrt{y} - y^2 \right) dy$

$$= \pi \left[\tfrac{1}{2} x^2 - \tfrac{1}{5} x^5 \right]_0^1 = \frac{3\pi}{10}$$

29. (a) $V = \int_0^1 2\pi x \cdot x^2 \, dx = 2\pi \int_0^1 x^3 \, dx$ (b) $V = \int_0^1 \pi (1 - y) \, dy$

$$= 2\pi \left[\tfrac{1}{4} x^4 \right]_0^1 = \frac{\pi}{2}$$

31. $V = \int_0^a 2\pi x \left[2b\sqrt{1 - \dfrac{x^2}{a^2}} \right] dx = \dfrac{4\pi b}{a} \int_0^a x \left(a^2 - x^2\right)^{1/2} dx = \dfrac{4\pi b}{a} \left[-\dfrac{1}{3} \left(a^2 - x^2\right)^{3/2} \right]_0^a = \dfrac{4}{3}\pi a^2 b$

33.

By the shell method

$V = \int_0^{a/2} 2\pi x \left(\sqrt{3}\,x\right) dx + \int_{a/2}^a 2\pi x \left[\sqrt{3}\,(a - x)\right] dx$

$= 2\pi\sqrt{3} \int_0^{a/2} x^2\, dx + 2\pi\sqrt{3} \int_{a/2}^a \left(ax - x^2\right) dx$

$= 2\pi\sqrt{3} \left[\dfrac{1}{3}x^3\right]_0^{a/2} + 2\pi\sqrt{3} \left[\dfrac{a}{2}x^2 - \dfrac{1}{3}x^3\right]_{a/2}^a = \dfrac{\sqrt{3}}{4}a^3\pi$

35. (a) $V = \int_0^8 2\pi y \left[4 - y^{2/3}\right] dy = 2\pi \int_0^8 \left(4y - y^{5/3}\right) dy = 2\pi \left[2y^2 - \dfrac{3}{8}y^{8/3}\right]_0^8 = 64\pi$

(b) $V = \int_0^4 2\pi (4 - x) \left[x^{3/2}\right] dx = 2\pi \int_0^4 \left(4x^{3/2} - x^{5/2}\right) dx$

$= 2\pi \left[\dfrac{8}{5}x^{5/2} - \dfrac{2}{7}x^{7/2}\right]_0^4 = \dfrac{1024}{35}\pi$

(c) $V = \int_0^8 2\pi (8 - y) \left[4 - y^{2/3}\right] dy = 2\pi \int_0^8 \left(32 - 4y - 8y^{2/3} + y^{5/3}\right) dy$

$= 2\pi \left[32y - 2y^2 - \dfrac{24}{5}y^{5/3} + \dfrac{3}{8}y^{8/3}\right]_0^8 = \dfrac{704}{5}\pi$

(d) $V = \int_0^4 2\pi x \left[x^{3/2}\right] dx = 2\pi \int_0^4 x^{5/2}\, dx = 2\pi \left[\dfrac{2}{7}x^{7/2}\right]_0^4 = \dfrac{512}{7}\pi$

37. (a) $F'(x) = \sin x + x \cos x - \sin x = x \cos x = f(x).$

(b) $V = \int_0^{\pi/2} 2\pi x \cdot \cos x\, dx = 2\pi \left[x \sin x + \cos x\right]_0^{\pi/2} = \pi^2 - 2\pi$

39. (a) $V = \int_0^1 2\sqrt{3}\pi x^2\, dx + \int_1^2 2\pi x \sqrt{4 - x^2}\, dx$ (b) $V = \int_0^{\sqrt{3}} \pi \left[4 - \dfrac{4}{3}y^2\right] dy$

(c) $V = \int_0^{\sqrt{3}} \pi \left[4 - \dfrac{4}{3}y^2\right] dy = \pi \left[4y - \dfrac{4}{9}y^3\right]_0^{\sqrt{3}} = \dfrac{8\pi\sqrt{3}}{3}$

41. (a) $V = \int_0^1 2\sqrt{3}\,\pi x(2-x)\,dx + \int_1^2 2\pi(2-x)\sqrt{4-x^2}\,dx$

   (b) $V = \int_0^{\sqrt{3}} \pi \left[\left(2 - \frac{y}{\sqrt{3}}\right)^2 - \left(2 - \sqrt{4-y^2}\right)^2 \right] dy$

43. (a) $V = 2\int_{b-a}^{b+a} 2\pi x \sqrt{a^2 - (x-b)^2}\,dx$

   (b) $V = \int_{-a}^{a} \pi \left[\left(b + \sqrt{a^2 - y^2}\right)^2 - \left(b - \sqrt{a^2 - y^2}\right)^2 \right] dy$

45. $V = \int_0^r 2\pi x\left(h - \frac{h}{r}x\right)dx = 2\pi h \left[\frac{x^2}{2} - \frac{x^3}{3r}\right]_0^r = \frac{\pi r^2 h}{3}.$

47. (a) $V = \int_a^r 2\pi x(r^2 - x^2)\,dx = 2\pi \int_a^r (r^2 x - x^3)\,dx = 2\pi\left[\frac{1}{2}r^2 x^2 - \frac{1}{4}x^4\right]_a^r = \frac{1}{2}\pi\left(r^2 - a^2\right)^2$

   (b) $V = \int_0^{r^2-a^2} \pi\left[\left(\sqrt{r^2 - y}\right)^2 - a^2\right]dy = \pi \int_0^{r^2-a^2}(r^2 - y - a^2)\,dy = \pi\left[(r^2 - a^2)y - \frac{1}{2}y^2\right]_0^{r^2-a^2}$

   $= \frac{1}{2}\pi(r^2 - a^2)^2$

49. (a)

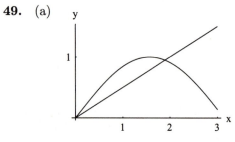

   (b) points of intersection: $x = 0,\ x \cong 1.8955$

   (c) $A \cong \int_0^{1.8955}\left(\sin x - \frac{1}{2}x\right)dx \cong 0.4208$

   (d) $V \cong \int_0^{1.8955} 2\pi\left(\sin x - \frac{1}{2}x\right)dx \cong 2.6226$

SECTION 6.4

1.

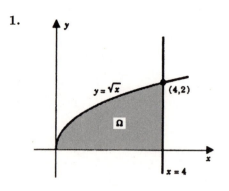

$A = \int_0^4 \sqrt{x}\,dx = \frac{16}{3}$

$\bar{x}A = \int_0^4 x\sqrt{x}\,dx = \frac{64}{5}, \quad \bar{x} = \frac{12}{5}$

$\bar{y}A = \int_0^4 \frac{1}{2}\left(\sqrt{x}\right)^2 dx = 4, \quad \bar{y} = \frac{3}{4}$

$V_x = 2\pi\bar{y}A = 8\pi, \quad V_y = 2\pi\bar{x}A = \frac{128}{5}\pi$

3.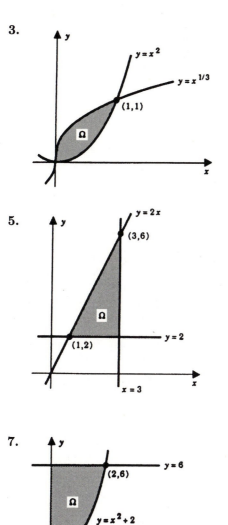

$$A = \int_0^1 \left(x^{1/3} - x^2 \right) dx = \frac{5}{12}$$

$$\overline{x}A = \int_0^1 x \left(x^{1/3} - x^2 \right) dx = \frac{5}{28}, \quad \overline{x} = \frac{3}{7}$$

$$\overline{y}A = \int_0^1 \frac{1}{2} \left[\left(x^{1/3} \right)^2 - \left(x^2 \right)^2 \right] dx = \frac{1}{5}, \quad \overline{y} = \frac{12}{25}$$

$$V_x = 2\pi\overline{y}A = \tfrac{2}{5}\pi, \quad V_y = 2\pi\overline{x}A = \tfrac{5}{14}\pi$$

5.

$$A = \int_1^3 (2x - 2) \, dx = 4$$

$$\overline{x}A = \int_1^3 x (2x - 2) \, dx = \frac{28}{3}, \quad \overline{x} = \frac{7}{3}$$

$$\overline{y}A = \int_1^3 \frac{1}{2} \left[(2x)^2 - (2)^2 \right] dx = \frac{40}{3}, \quad \overline{y} = \frac{10}{3}$$

$$V_x = 2\pi\overline{y}A = \tfrac{80}{3}\pi, \quad V_y = 2\pi\overline{x}A = \tfrac{56}{3}\pi$$

7.

$$A = \int_0^2 \left[6 - \left(x^2 + 2x \right) \right] dx = \frac{16}{3}$$

$$\overline{x}A = \int_0^2 x \left[6 - \left(x^2 + 2 \right) \right] dx = 4, \quad \overline{x} = \frac{3}{4}$$

$$\overline{y}A = \int_0^2 \frac{1}{2} \left[(6)^2 - \left(x^2 + 2 \right)^2 \right] dx = \frac{352}{15}, \quad \overline{y} = \frac{22}{5}$$

$$V_x = 2\pi\overline{y}A = \tfrac{704}{15}\pi, \quad V_y = 2\pi\overline{x}A = 8\pi$$

9.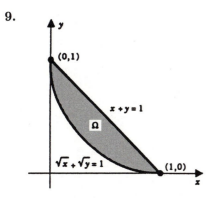

$$A = \int_0^1 \left[(1 - x) - \left(1 - \sqrt{x} \right)^2 \right] dx = \frac{1}{3}$$

$$\overline{x}A = \int_0^1 x \left[(1 - x) - \left(1 - \sqrt{x} \right)^2 \right] dx = \frac{2}{15}, \quad \overline{x} = \frac{2}{5}$$

$$\overline{y} = \tfrac{2}{5} \qquad \text{by symmetry}$$

$$V_x = 2\pi\overline{y}A = \tfrac{4}{15}\pi, \quad V_y = \tfrac{4}{15}\pi \qquad \text{by symmetry}$$

11.

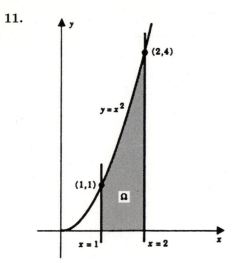

$$A = \int_1^2 x^2 \, dx = \frac{7}{3}$$

$$\overline{x}A = \int_1^2 x \left(x^2 \right) \, dx = \frac{15}{4}, \quad \overline{x} = \frac{45}{28}$$

$$\overline{y}A = \int_1^2 \frac{1}{2} \left(x^2 \right)^2 \, dx = \frac{31}{10}, \quad \overline{y} = \frac{93}{70}$$

$$V_x = 2\pi \overline{y}A = \frac{31}{5}\pi, \quad V_y = 2\pi \overline{x}A = \frac{15}{2}\pi$$

13.

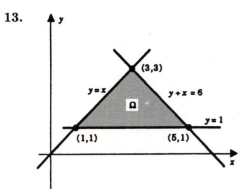

$$A = \tfrac{1}{2}bh = 4; \qquad \text{by symmetry,} \quad \overline{x} = 3$$

$$\overline{y}A = \int_1^3 y\left[(6-y) - y \right] \, dy = \frac{20}{3}, \quad \overline{y} = \frac{5}{3}$$

$$V_x = 2\pi \overline{y}A = \frac{40}{3}\pi, \quad V_y = 2\pi \overline{x}A = 24\pi$$

15. $\left(\frac{5}{2}, 5 \right)$

16. $$A = \int_{-1}^3 (4x - x^2 - 2x + 3) \, dx = \frac{32}{3}$$

$$\overline{x}A = \int_{-1}^3 x(2x - x^2 + 3) \, dx = \frac{32}{3} \implies \overline{x} = 1$$

$$\overline{y}A = \int_{-1}^3 \frac{1}{2} \left[(4x - x^2)^2 - (2x - 3)^2 \right] \, dx = \frac{32}{5} \implies \overline{y} = \frac{3}{5}$$

19. $\left(\frac{10}{3}, \frac{40}{21} \right)$

21. $(2, 4)$

23. $\left(-\frac{3}{5}, 0 \right)$

25. (a) $(0,0)$ by symmetry

(b) Ω_1 smaller quarter disc, Ω_2 the larger quarter disc

$$A_1 = \frac{1}{16}\pi, \quad A_2 = \pi; \quad \bar{x}_1 = \bar{y}_1 = \frac{2}{3\pi}, \quad \bar{x}_2 = \bar{y}_2 = \frac{8}{3\pi} \quad \text{(Example 1)}$$

$$\bar{x}A = \left(\frac{8}{3\pi}\right)(\pi) - \frac{2}{3\pi}\left(\frac{1}{16}\pi\right) = \frac{63}{24}, \quad A = \frac{15}{16}\pi$$

$$\bar{x} = \left(\frac{63}{24}\right) \bigg/ \left(\frac{15\pi}{16}\right) = \frac{14}{5\pi}, \quad \bar{y} = \bar{x} = \frac{14}{5\pi} \quad \text{(symmetry)}$$

(c) $\bar{x} = 0, \quad \bar{y} = \dfrac{14}{5\pi}$

27. Use theorem of Pappus. Centroid of rectangle is located

$$c + \sqrt{\left(\frac{a}{2}\right)^2 + \left(\frac{b}{2}\right)^2} \text{ units}$$

from line l. The area of the rectangle is ab. Thus,

$$\text{volume} = 2\pi \left[c + \sqrt{\left(\frac{a}{2}\right)^2 + \left(\frac{b}{2}\right)^2}\right](ab) = \pi ab\left(2c + \sqrt{a^2 + b^2}\right).$$

29. (a) $\left(\frac{2}{3}a, \frac{1}{3}h\right)$ (b) $\left(\frac{2}{3}a + \frac{1}{3}b, \frac{1}{3}h\right)$ (c) $\left(\frac{1}{3}a + \frac{1}{3}b, \frac{1}{3}h\right)$

31. (a) $V = \frac{2}{3}\pi R^3 \sin^3\theta + \frac{1}{3}\pi R^3 \sin^2\theta\cos\theta = \frac{1}{3}\pi R^3 \sin^2\theta\,(2\sin\theta + \cos\theta)$

(b) $\bar{x} = \dfrac{V}{2\pi A} = \dfrac{\frac{1}{3}\pi R^3 \sin^2\theta\,(2\sin\theta + \cos\theta)}{2\pi\left(\frac{1}{2}R^2 \sin\theta\cos\theta + \frac{1}{4}\pi R^2 \sin^2\theta\right)} = \dfrac{2R\sin\theta\,(2\sin\theta + \cos\theta)}{3\,(\pi\sin\theta + 2\cos\theta)}$

33. An annular region; see Exercise 25(a).

35. (a) $A = \frac{1}{2}$ (b) $\left(\frac{16}{35}, \frac{16}{35}\right)$ (c) $V = \frac{16}{35}\pi$ (d) $V = \frac{16}{35}\pi$

37. (a) $A = \frac{250}{3}$ (b) $\left(-\frac{9}{8}, \frac{290}{21}\right) \cong (-1.125, 13.8095)$

PROJECT 6.4

1. Let $P = \{x_0, x_1, \ldots, x_n\}$ be a partition of $[a, b]$. P breaks up $[a, b]$ into n subintervals $[x_{i-1}, x_i]$. Choose x_i^* as the midpoint of $[x_{i-1}, x_i]$. By revolving the ith midpoint rectangle about x-axis, we obtain a solid cylinder of volume $V_i = \pi\,[f\,(x_i^*)]^2\,\Delta x_i$ and centroid (center) on the x-axis at $x = x_i^*$. The union of all these cylinders has centroid at $x = \bar{x}_P$ where

$$x_p V_p = \pi x_1^*\,[f\,(x_1^*)]^2\,\Delta x_1 + \cdots + \pi x_n^*\,[f\,(x_n^*)]^2\,\Delta x_n$$

(Here V_P represents the union of the n cylinders.) As $\|P\| \to 0$, the union of the cylinders tends to the shape of S and the equation just derived tends to the given formula.

3. (a)

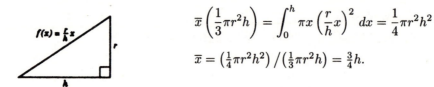

$$\bar{x}\left(\frac{1}{3}\pi r^2 h\right) = \int_0^h \pi x \left(\frac{r}{h}x\right)^2 dx = \frac{1}{4}\pi r^2 h^2$$

$$\bar{x} = \left(\frac{1}{4}\pi r^2 h^2\right) / \left(\frac{1}{3}\pi r^2 h\right) = \frac{3}{4}h.$$

The centroid of the cone lies on the axis of the cone at a distance $\frac{3}{4}h$ from the vertex.

(b) The ball is obtained by rotating $f(x) = \sqrt{r^2 - x^2}, x \in [-r, r]$, around the x-axis.

$$V_x = \frac{2}{3}\pi r^3; \quad \bar{x}V_x = \int_{-r}^r \pi x(r^2 - x^2)\, dx = \pi\left[r^2\frac{x^2}{2} - \frac{x^4}{4}\right]_{-r}^r = 0$$

$$\implies \bar{x} = 0; \text{ no surprise here.}$$

(c) $V_x = \int_0^a \pi \frac{b^2}{a^2}\left(a^2 - x^2\right) dx = \frac{2}{3}\pi ab^2, \quad \bar{x}V_x = \int_0^a \pi x \frac{b^2}{a^2}\left(a^2 - x^2\right) dx = \frac{1}{4}\pi a^2 b^2$

$\bar{x} = \left(\frac{1}{4}\pi a^2 b^2\right) / \left(\frac{2}{3}\pi ab^2\right) = \frac{3}{8}a; \quad$ centroid $\left(\frac{3}{8}a, 0\right)$

(d) (i) $V_x = \int_0^1 \pi\left(\sqrt{x}\right)^2 dx = \frac{1}{2}\pi, \quad \bar{x}V_x = \int_0^1 \pi x\left(\sqrt{x}\right)^2 dx = \frac{1}{3}\pi$

$\bar{x} = \left(\frac{1}{3}\pi\right) / \left(\frac{1}{2}\pi\right) = \frac{2}{3}; \quad$ centroid $\left(\frac{2}{3}, 0\right)$

(ii) $V_y = \int_0^1 2\pi x\sqrt{x}\, dx = \frac{4}{5}\pi, \quad \bar{y}V_y = \int_0^1 \pi x\left(\sqrt{x}\right)^2 dx = \frac{1}{3}\pi$

$\bar{y} = \left(\frac{1}{3}\pi\right) / \left(\frac{4}{5}\pi\right) = \frac{5}{12}; \quad$ centroid $\left(0, \frac{5}{12}\right)$

(e) (i) $V_x = \int_0^2 \pi(4 - x^2)^2 dx = \frac{256}{15}\pi$

$\bar{x}V_x = \int_0^2 \pi x(4 - x^2)^2 dx = \pi\left[-\frac{(4 - x^2)^3}{6}\right]_0^2 = \frac{32\pi}{3} \implies \bar{x} = \frac{5}{8}; \quad \bar{y} = 0$

(ii) $V_y = \int_0^2 2\pi x(4 - x^2)\, dx = 8\pi$

$\bar{y}V_y = \int_0^2 \pi x(4 - x^2)^2 dx = \frac{32}{3}\pi \implies \bar{y} = \frac{4}{3}; \quad \bar{x} = 0$

SECTION 6.5

1. $W = \int_1^4 x\left(x^2 + 1\right)^2 dx = \frac{1}{6}\left[(x^2 + 1)^3\right]_1^4 = 817.5$ ft-lb

3. $W = \int_1^3 x\sqrt{x^2 + 7}\, dx = \frac{1}{3}\left[(x^2 + 7)^{\frac{3}{2}}\right]_0^3 = \frac{1}{3}(64 - 7^{\frac{3}{2}})$ newton-meters

5. $W = \int_{\pi/6}^\pi (x + \sin 2x)\, dx = \left[\frac{1}{2}x^2 - \frac{1}{2}\cos 2x\right]_{\pi/6}^\pi = \frac{35}{72}\pi^2 - \frac{1}{4}$ newton-meters

7. By Hooke's law, we have $600 = -k(-1)$. Therefore $k = 600$.

    The work required to compress the spring to 5 inches is given by

    $$W = \int_{10}^{5} 600(x - 10)\, dx = 600 \left[\frac{1}{2}x^2 - 10x \right]_{10}^{5}$$

    $$= 7500 \text{ in-lb, or } 625 \text{ ft-lb}$$

9. To counteract the restoring force of the spring we must apply a force $F(x) = kx$.
    Since $F(4) = 200$, we see that $k = 50$ and therefore $F(x) = 50x$.

    (a) $W = \int_{0}^{1} 50x\, dx = 25$ ft-lb   (b) $W = \int_{0}^{3/2} 50x\, dx = \dfrac{225}{4}$ ft-lb

11. Let L be the natural length of the spring.

    $$\int_{2-L}^{2.1-L} kx\, dx = \frac{1}{2} \int_{2.1-L}^{2.2-L} kx\, dx$$

    $$\left[\tfrac{1}{2} kx^2 \right]_{2-L}^{2.1-L} = \tfrac{1}{2} \left[\tfrac{1}{2} kx^2 \right]_{2.1-L}^{2.2-L}$$

    $$(2.1 - L)^2 - (2 - L)^2 = \tfrac{1}{2} \left[(2.2 - L)^2 - (2.1 - L)^2 \right].$$

    Solve this equation for L and you will find that $L = 1.95$.   Answer: 1.95 ft

13.

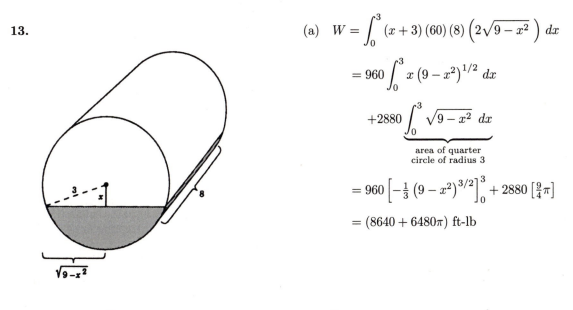

(a) $W = \int_{0}^{3} (x + 3)\,(60)\,(8) \left(2\sqrt{9 - x^2}\, \right)\, dx$

$$= 960 \int_{0}^{3} x \left(9 - x^2 \right)^{1/2}\, dx$$

$$+ 2880 \underbrace{\int_{0}^{3} \sqrt{9 - x^2}\, dx}_{\substack{\text{area of quarter} \\ \text{circle of radius 3}}}$$

$$= 960 \left[-\tfrac{1}{3} \left(9 - x^2 \right)^{3/2} \right]_{0}^{3} + 2880 \left[\tfrac{9}{4}\pi \right]$$

$$= (8640 + 6480\pi) \text{ ft-lb}$$

(b) $W = \int_{0}^{3} (x + 7)(60)(8) \left(2\sqrt{9 - x^2}\, \right)\, dx = 960 \int_{0}^{3} x \left(9 - x^2 \right)^{1/2}\, dx + 6720 \int_{0}^{3} \sqrt{9 - x^2}\, dx$

$$= (8640 + 15120\pi) \text{ ft-lb}$$

15. Set $w(x) =$ the weight of the chain from height x to the ground. (Weight measured in pounds, height measured in feet.) Then $w(x) = 1.5x$ and

$$W = \int_0^{50} 1.5x\,dx = 1.5 \left[\frac{x^2}{2}\right]_0^{50} = 1875 \text{ ft-lb}$$

17.

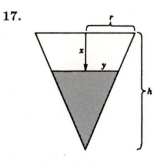

By similar triangles

$$\frac{h}{r} = \frac{h-x}{y} \quad \text{so that} \quad y = \frac{r}{h}(h-x).$$

Thus, the area of a cross section of the fluid at a depth of x feet is

$$\pi y^2 = \pi \frac{r^2}{h^2}(h-x)^2.$$

(a) $\quad W = \int_0^{h/2} x\sigma \left[\pi\frac{r^2}{h^2}(h-x)^2\right] dx = \frac{\sigma\pi r^2}{h^2} \int_0^{h/2} \left(h^2 x - 2hx^2 + x^3\right) dx = \frac{11}{192}\sigma\pi r^2 h^2 \text{ ft-lb}$

(b) $\quad W = \int_0^{h/2} (x+k)\sigma \left[\pi\frac{r^2}{h^2}(h-x)^2\right] dx = \frac{11}{192}\pi r^2 h^2\sigma + \frac{7}{24}\pi r^2 hk\sigma \text{ ft-lb}$

19. $y = \frac{3}{4}x^2,\ 0 \le x \le 4$

(a) $W = \int_0^{12} \sigma(12-y)\pi x^2\,dy = \frac{4}{3}\pi\sigma \int_0^{12} (12y - y^2)\,dy$

$$= \frac{4}{3}\pi\sigma \left[6y^2 - \frac{y^3}{3}\right]_0^{12} = \frac{4}{3}\pi\sigma(288) = 384\pi\sigma \text{ newton-meters.}$$

(b) $W = \int_0^{12} \sigma(13-y)\pi x^2\,dy = \frac{4}{3}\pi\sigma \int_0^{12} (13y - y^2)\,dy$

$$= \frac{4}{3}\pi\sigma \left[\frac{13y^2}{2} - \frac{y^3}{3}\right]_0^{12} = \frac{4}{3}\pi\sigma(360) = 480\pi\sigma \text{ newton-meters.}$$

21. $W = \int_0^{80} (80-x)15\,dx = 15\left[80x - \frac{1}{2}x^2\right]_0^{80} = 48,000 \text{ ft-lb}$

23. (a) $\quad W = 200 \cdot 100 = 20,000 \text{ ft-lb}$ \qquad (b) $\quad W = \int_0^{100} [(100-x)2 + 200]\,dx$

$$= \int_0^{100} (400 - 2x)\,dx$$

$$= \left[400x - x^2\right]_0^{100} = 30,000 \text{ ft-lb}$$

25. The bag is raised 8 feet and loses a total of 1 pound at a constant rate. Thus, the bag loses sand at the rate of 1/8 lb/ft. After the bag has been raised x feet it weighs $100 - \dfrac{x}{8}$ pounds.

$$W = \int_0^8 \left(100 - \frac{x}{8}\right) dx = \left[100x - \frac{x^2}{16}\right]_0^8 = 796 \text{ ft-lb.}$$

27. (a) $W = \displaystyle\int_0^l x\sigma \, dx = \frac{1}{2}\sigma l^2$ ft-lb

    (b) $W = \displaystyle\int_0^l (x+l)\,\sigma \, dx = \frac{3}{2}\sigma l^2$ ft-lb

29. Thirty feet of cable and the steel beam weighing a total of: $800 + 30(6) = 980$ lbs. are raised 20 feet. The work required is: $(20)(980) = 19,600$ ft-lb.

Next, the remaining 20 feet of cable is raised a varying distance and wound onto the steel drum. Thus the total work is given by

$$W = 19,600 + \int_0^{20} 6x\, dx = 19,600 + 1,200 = 20,800 \text{ ft-lb.}$$

31. Let $\lambda(x)$ be the mass density of the chain at the point x units above the ground. Let g be the gravitational constant. the work done to pull the chain to the top of the building is given by:

$$W = \int_0^H (H-x)\,g\,\lambda(x)\,dx = Hg \int_0^H \lambda(x)\,dx - g \int_0^H x\,\lambda(x)\,dx$$

$$= HgM - g\,\overline{x}\,M = (H - \overline{x})\,gM$$

$$= \text{(weight of chain)} \times \text{(distance from the center of mass to top of building)}$$

33. The height of the object at time t is $y(t) = -\frac{1}{2}gt^2 + h$; time at impact is $t = \sqrt{2h/g}$; velocity at time t is $v(t) = -gt$; velocity at impact is $v\left(\sqrt{2h/g}\right) = -g\sqrt{2h/g} = -\sqrt{2gh}$.

35. Assume that $v_a = 0$ and $v_b = 95$ mph.

mass of ball: $\dfrac{5/16}{32} = \dfrac{1}{512}$;   speed in feet per second: $\dfrac{(95)(5280)}{3600}$

$$W = \frac{1}{2}\frac{1}{512}\left[\frac{(95)(5280)}{3600}\right]^2 \cong 18.96 \text{ ft-lbs.}$$

37. Assume that $v_a = 0$ and $v_b = 17,000$ mph.

mass of satellite: $\dfrac{1000}{32} = 31.25$;   speed in feet per second: $\dfrac{(17,000)(5280)}{3600}$

$$W = \frac{1}{2}(31.25)\left[\frac{(17,000)(5280)}{3600}\right]^2 \cong 9.714 \times 10^9 \text{ ft-lbs.}$$

39. (a) The work required to pump the water out of the tank is given by

$$W = \int_5^{10} (62.5)\pi \, 5^2 x \, dx = 1562.5\pi \left[\frac{1}{2}x^2\right]_5^{10} \cong 184{,}078 \text{ ft-lb}$$

A $\frac{1}{2}$-horsepower pump can do 275 ft-lb of work per second. Therefore it will take

$$\frac{184{,}078}{275} \cong 669 \text{ seconds} \cong 11 \text{ min, } 10 \text{ sec, to empty the tank.}$$

(b) The work required to pump the water to a point 5 feet above the top of the tank is given by

$$W = \int_5^{10} (62.5)\pi \, 5^2 (x+5) \, dx = \int_5^{10} (62.5)\pi \, 5^2 x \, dx + \int_5^{10} (62.5)\pi \, 5^3 \, dx \cong 306796 \text{ ft-lb}$$

It will take a $\frac{1}{2}$-horsepower pump approximately $1{,}116$ sec, or 18 min, 36 sec, in this case.

41. $KE = \frac{1}{2}mv^2; \quad \dfrac{d\,KE}{dt} = mv\dfrac{dv}{dt} = mav = Fv = P.$

SECTION 6.6

1. $F = \displaystyle\int_0^6 (62.5) \cdot x \cdot 8 \, dx = 250 \left[x^2\right]_0^6 = 9{,}000 \text{ lbs}$

3. The width of the plate x meters below the surface is given by $w(x) = 60 + 2(20 - x) = 100 - 2x$ (see the figure). The force against the dam is

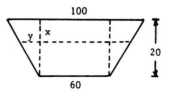

$$F = \int_0^{20} 9800x(100 - 2x)\,dx$$

$$= 9800 \int_0^{20} (100x - 2x^2)\,dx$$

$$= 9800 \left[50x^2 - \tfrac{2}{3}x^3\right]_0^{20}$$

$$\cong 1.437 \times 10^8 \text{ newtons}$$

5. The width of the gate x meters below its top is given by $w(x) = 4 + \frac{2}{3}x$ (see the figure). The force of the water against the gate is

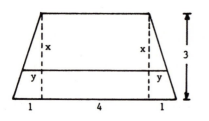

$$F = \int_0^3 9800(10 + x)\left(4 + \frac{2}{3}x\right)dx$$

$$= 9800 \int_0^3 \left[\frac{2}{3}x^2 + \frac{32}{3}x + 40\right]dx$$

$$= 9800 \left[\tfrac{2}{9}x^3 + \tfrac{16}{3}x^2 + 40x\right]_0^3$$

$$\cong 1.7052 \times 10^6 \text{ Newtons}$$

7.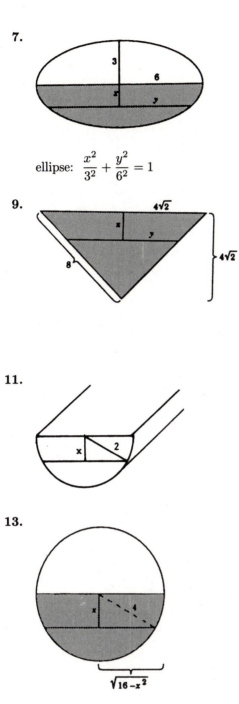

ellipse: $\dfrac{x^2}{3^2} + \dfrac{y^2}{6^2} = 1$

$$F = \int_0^3 (60)\, x \left[12\sqrt{1 - \frac{x^2}{9}} \,\right] dx$$

$$= 240 \int_0^3 x \left(9 - x^2\right)^{1/2} dx$$

$$= 240 \left[-\tfrac{1}{3} \left(9 - x^2\right)^{3/2} \right]_0^3 = 2160 \text{ lb}$$

9.

By similar triangles

$$\frac{4\sqrt{2}}{4\sqrt{2}} = \frac{y}{4\sqrt{2} - x} \quad \text{so} \quad y = 4\sqrt{2} - x.$$

$$F = \int_0^{4\sqrt{2}} (62.5)\, x \left[2\left(4\sqrt{2} - x\right) \right] dx$$

$$= 125 \int_0^{4\sqrt{2}} \left(4\sqrt{2}\, x - x^2\right) dx = \frac{8000}{3}\sqrt{2} \text{ lb}$$

11.

$$F = \int_0^2 (62.5) \cdot x \cdot 2\sqrt{4 - x^2}\, dx$$

$$= 125 \int_0^2 x\sqrt{4 - x^2}\, dx$$

$$= -\tfrac{125}{3} \left[\left(4 - x^2\right)^{3/2} \right]_0^2 = 333.33 \text{ lb}$$

13.

$$F = \int_0^4 60x \left(2\sqrt{16 - x^2}\, \right) dx$$

$$= 120 \int_0^4 x \left(16 - x^2\right)^{1/2} dx$$

$$= 120 \left[-\tfrac{1}{3} \left(16 - x^2\right)^{3/2} \right]_0^4 = 2560 \text{ lb}$$

15. (a) The width of the plate is 10 feet and the depth of the plate ranges from 8 feet to 14 feet. Thus

$$F = \int_8^{14} 62.5x\,(10)\, dx = 41{,}250 \text{ lb}.$$

(b) The width of the plate is 6 feet and the depth of the plate ranges from 6 feet to 16 feet. Thus

$$F = \int_6^{16} 62.5x\,(6)\, dx = 41{,}250 \text{ lb}.$$

17. (a) Force on the sides:

$$F = \int_0^1 (9800)\, x\, 14\, dx + \int_0^2 (9800)(1+x)\, 7(2-x)\, dx$$

$$= 68,600 \left[x^2\right]_0^1 + 68,600 \int_0^2 \left[2 + x - x^2\right] dx$$

$$= 68,600 + 68,600 \left[2x + \tfrac{1}{2}x^2 - \tfrac{1}{3}x^3\right]_0^2$$

$$\cong 297,267 \text{ Newtons}$$

(b) Force at the shallow end:

$$F = \int_0^1 (9800) \cdot x \cdot 8\, dx = 39,200 \left[x^2\right]_0^1 = 39,200 \text{ Newtons}$$

Force at the deep end:

$$F = \int_0^3 (9800) \cdot x \cdot 8\, dx = 39,200 \left[x^2\right]_0^3 = 352,800 \text{ Newtons}$$

19. From Exercise 18, $\quad F_1 = \sigma \bar{x}_1 A_1 = \sigma h_1 A_1, \quad F_2 = \sigma h_2 A_2. \quad$ Since $\quad A_1 = A_2, \quad F_2 = \dfrac{h_2}{h_1} F_1$

21. $\quad F = \displaystyle\int_0^{14} (9800) \left(1 + \frac{1}{7}x\right) \cdot 8 \frac{5\sqrt{2}}{7}\, dx = 392,000 \frac{\sqrt{2}}{7} \left[x + \frac{1}{14}x^2\right]_0^{14} \cong 2.217 \times 10^6 \text{ Newtons}$

REVIEW EXERCISES

1.

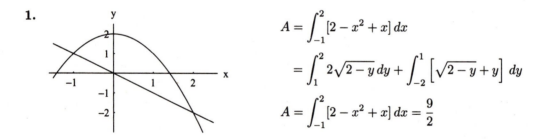

$$A = \int_{-1}^2 \left[2 - x^2 + x\right] dx$$

$$= \int_1^2 2\sqrt{2-y}\, dy + \int_{-2}^1 \left[\sqrt{2-y} + y\right] dy$$

$$A = \int_{-1}^2 \left[2 - x^2 + x\right] dx = \frac{9}{2}$$

3.

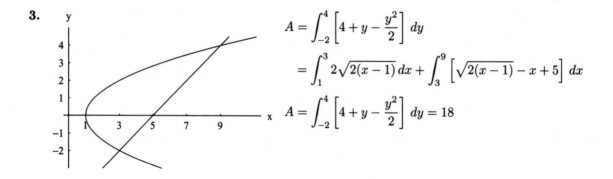

$$A = \int_{-2}^4 \left[4 + y - \frac{y^2}{2}\right] dy$$

$$= \int_1^3 2\sqrt{2(x-1)}\, dx + \int_3^9 \left[\sqrt{2(x-1)} - x + 5\right] dx$$

$$A = \int_{-2}^4 \left[4 + y - \frac{y^2}{2}\right] dy = 18$$

5. Consecutive intersections of $\sin x$ and $\cos x$ occur at $x = \frac{\pi}{4}$, $x = \frac{5\pi}{4}$.

$$A = \int_{\pi/4}^{5\pi/4} (\sin x - \cos x)\, dx = \Big[-\cos x - \sin x \Big]_{\pi/4}^{5\pi/4} = 2\sqrt{2}$$

7. By symmetry, $A = 2 \int_0^1 (1-x)\sqrt{x}\, dx = 2 \int_0^1 \left(x^{1/2} - x^{3/2} \right) dx = 2\Big[\frac{2}{3}x^{3/2} - \frac{2}{5}x^{5/2} \Big]_0^1 = \frac{8}{15}$

9. (a) $V = \int_{-r}^{r} A(x)\, dx = \int_{-r}^{r} \frac{1}{2}\pi \left(\sqrt{r^2 - x^2} \right)^2 dx = \int_{-r}^{r} \frac{\pi}{2}(r^2 - x^2)\, dx = \frac{\pi}{2}\Big[r^2 x - \frac{1}{3}\pi x^3 \Big]_{-r}^{r} = \frac{2}{3}\pi r^3$

    (b) $V = \int_{-r}^{r} A(x)\, dx = \int_{-r}^{r} \frac{1}{2}(2)\sqrt{r^2 - x^2}\sqrt{r^2 - x^2}\, dx = \int_{-r}^{r} (r^2 - x^2)\, dx = \Big[r^2 x - \frac{1}{3}x^3 \Big]_{-r}^{r} = \frac{4}{3}r^3$

11. $V = \int_0^3 \frac{1}{2}\pi \left(\frac{3-x}{3} \right)^2 dx = \frac{1}{2}\pi \int_0^3 \left(1 - \frac{2}{3}x + \frac{1}{9}x^2 \right) dx = \frac{1}{2}\pi \Big[x - \frac{1}{3}x^2 + \frac{1}{27}x^3 \Big]_0^3 = \frac{1}{2}\pi$

13. $V = \int_0^2 \pi \left[(x/2)^2 - (x^2/4)^2 \right] dx = \pi \int_0^2 \left(\frac{x^2}{4} - \frac{x^4}{16} \right) dx = \pi \Big[\frac{x^3}{12} - \frac{x^5}{80} \Big]_0^2 = \frac{4\pi}{15}$

15. $V = \int_0^1 \pi[(1)^2 - (x^3)^2]\, dx = \pi \int_0^1 (1 - x^6)\, dx = \frac{6\pi}{7}$

17. $V = \int_0^{\pi/4} \pi \sec^2 x\, dx = \pi \Big[\tan x \Big]_0^{\pi/4} = \pi$

19. $V = \int_0^{\sqrt{\pi}} 2\pi x \sin x^2\, dx = \pi \int_0^{\sqrt{\pi}} 2x \sin x^2\, dx = \pi \Big[-\cos x^2 \Big]_0^{\sqrt{\pi}} = 2\pi$

21. By symmetry, $V = 2 \int_0^3 2\pi x(3x - x^2)\, dx = 4\pi \int_0^3 (3x^2 - x^3)\, dx = 4\pi \Big[x^3 - \frac{1}{4}x^4 \Big]_0^3 = 27\pi$

23. $V = \int_0^3 \pi[(x+1)^2 - (x-1)^4]\, dx = \pi \Big[\frac{1}{3}(x+1)^3 - \frac{1}{5}(x-1)^5 \Big]_0^3 = \frac{72}{5}\pi$

25. With respect to x: $V = \int_0^4 \pi \left[(2^2 - (\sqrt{x})^2 \right] dx = \pi \int_0^4 (4 - x)\, dx$;

    with respect to y: $V = \int_0^2 2\pi y(y^2)\, dy = 2\pi \int_0^2 y^3\, dy = 8\pi$

27. With respect to x: $V = \int_0^4 2\pi(x+1)\left(\sqrt{x} - \frac{1}{2}x \right) dx = 2\pi \int_0^4 \left(x^{3/2} + x^{1/2} - \frac{1}{2}x^2 - \frac{1}{2}x \right) dx = \frac{104}{15}\pi$

    with respect to y: $V = \int_0^2 \pi\{[2y+1]^2 - [y^2+1]^2\}\, dy$

29. With respect to x: $V = \int_0^4 2\pi x (\frac{1}{2}x)\, dx = \pi \int_0^4 x^2\, dx = \frac{64}{3}\pi$;

with respect to y: $V = \int_0^2 \pi [4^2 - (2y)^2]\, dy$

31. Since the region is symmetric about the y-axis, $\bar{x} = 0$;

$A = \int_{-2}^2 (4 - x^2)\, dx = \frac{32}{3}$;

$\bar{y}A = \int_{-2}^2 \frac{1}{2}(4 - x^2)^2\, dx = \frac{1}{2}\int_{-2}^2 (16 - 8x^2 + x^4)\, dx = \frac{256}{15};\quad \bar{y} = \frac{8}{5}$

33. $A = \int_{-1}^2 \left[(2x - x^2) - (x^2 - 4)\right] dx = \int_{-1}^2 (2x - 2x^2 + 4)\, dx = 9$;

$\bar{x}A = \int_{-1}^2 x\,(2x - 2x^2 + 4)\, dx = \int_{-1}^2 (2x^2 - 2x^3 + 4x)\, dx = \frac{9}{2}$;

$\bar{y}A = \int_{-1}^2 \frac{1}{2}\left[(2x - x^2) - (x^2 - 4)\right] dx = \frac{1}{2}\int_{-1}^2 [12x^2 - 4x^3 - 16]\, dx = -\frac{27}{2}$;

$\bar{x} = \frac{1}{2};\quad \bar{y} = -\frac{3}{2}$

35. $A = \int_0^1 \left[(2 - x^2) - x\right] dx = \int_0^1 (2 - x^2 - x)\, dx = \frac{7}{6}$;

$\bar{x}A = \int_0^1 x(2 - x^2 - x)dx = \int_0^1 (2x - x^3 - x^2)\, dx = \frac{5}{12};\quad \bar{x} = \frac{5}{14}$

$\bar{y}A = \int_0^1 \frac{1}{2}\left[(2 - x^2)^2 - x^2\right] dx = \frac{1}{2}\int_0^1 [4 - 5x^2 + x^4]\, dx = \frac{19}{15};\quad \bar{y} = \frac{38}{35}$

around x-axis: $V = 2\pi \left(\frac{38}{35}\right)\left(\frac{7}{6}\right) = \frac{38\pi}{15}$

around y-axis: $V = 2\pi \left(\frac{5}{14}\right)\left(\frac{7}{6}\right) = \frac{5\pi}{6}$

37. $W = \int_0^3 F(x)dx = \int_0^3 x\sqrt{7 + x^2}\, dx = \left[\frac{1}{3}(7 + x^2)^{3/2}\right]_0^3 = \frac{1}{3}(64 - 7^{3/2})$ ft-lbs

39. Assume the natural length is x_0. Then,

$$\int_9^{10} k(x - x_0)\, dx = \frac{3}{2}\int_8^9 k(x - x_0)\, dx$$

The solution of this equation is $x_0 = \frac{13}{2}$ inches.

41. $W = \int_0^{25} 4(25 - x)\, dx = \left[100x - 2x^2\right]_0^{25} = 1250$ ft-lbs

43. $W = \displaystyle\int_0^{10} 60(20-x)\pi(20x - x^2)\,dx = 60\pi \int_0^{10} (400x - 40x^2 + x^3)\,dx$

$\qquad = 60\pi\left[200x^2 - \tfrac{40}{3}x^3 + \tfrac{1}{4}x^4\right]_0^{10} = 550,000\pi \text{ ft-lbs.}$

45. (a) $F = \displaystyle\int_0^{50} 9800x(100 - 2x)\,dx = 9800 \int_0^{50} (100x - 2x^2)\,dx$

$\qquad = 9800\left[50x^2 - \tfrac{2}{3}x^3\right]_0^{50} = \tfrac{1225}{3} \times 10^6 \text{ newtons.}$

(b) $F = \displaystyle\int_{10}^{50} 9800x(100 - 2x)\,dx = \dfrac{10976}{3} \times 10^5 \text{ newtons.}$

CHAPTER 7

SECTION 7.1

1. Suppose $f(x_1) = f(x_2)$ $x_1 \neq x_2$. Then

$5x_1 + 3 = 5x_2 + 3 \Rightarrow x_1 = x_2$;

f is one-to-one

$$f(y) = x$$
$$5y + 3 = x$$
$$5y = x - 3$$
$$y = \tfrac{1}{5}(x - 3)$$
$$f^{-1}(x) = \tfrac{1}{5}(x - 3)$$
$$\text{dom } f^{-1} = (-\infty, \infty)$$

3. f is not one-to-one;   e.g. $f(1) = f(-1)$

5. $f'(x) = 5x^4 \geq 0$ on $(-\infty, \infty)$ and $f'(x) = 0$ only at $x = 0$; f is increasing. Therefore, f is one-to-one.

$$f(y) = x$$
$$y^5 + 1 = x$$
$$y^5 = x - 1$$
$$y = (x - 1)^{1/5}$$
$$f^{-1}(x) = (x - 1)^{1/5}$$
$$\text{dom } f^{-1} = (-\infty, \infty)$$

7. $f'(x) = 9x^2 \geq 0$ on $(-\infty, \infty)$ and $f'(x) = 0$ only at $x = 0$; f is increasing. Therefore, f is one-to-one.

$$f(y) = x$$
$$1 + 3y^3 = x$$
$$y^3 = \tfrac{1}{3}(x - 1)$$
$$y = \left[\tfrac{1}{3}(x - 1)\right]^{1/3}$$
$$f^{-1}(x) = \left[\tfrac{1}{3}(x - 1)\right]^{1/3}$$
$$\text{dom } f^{-1} = (-\infty, \infty)$$

9. $f'(x) = 3(1 - x)^2 \geq 0$ on $(-\infty, \infty)$ and $f'(x) = 0$ only at $x = 1$; f is increasing. Therefore, f is one-to-one.

$$f(y) = x$$
$$(1 - y)^3 = x$$
$$1 - y = x^{1/3}$$
$$y = 1 - x^{1/3}$$
$$f^{-1}(x) = 1 - x^{1/3}$$
$$\text{dom } f^{-1} = (-\infty, \infty)$$

11. $f'(x) = 3(x + 1)^2 \geq 0$ on $(-\infty, \infty)$ and $f'(x) = 0$ only at $x = -1$; f is increasing. Therefore, f is one-to-one.

$$f(y) = x$$
$$(y + 1)^3 + 2 = x$$
$$(y + 1)^3 = x - 2$$
$$y + 1 = (x - 2)^{1/3}$$
$$y = (x - 2)^{1/3} - 1$$
$$f^{-1}(x) = (x - 2)^{1/3} - 1$$
$$\text{dom } f^{-1} = (-\infty, \infty)$$

13. $f'(x) = \dfrac{3}{5x^{2/5}} > 0$ for all $x \neq 0$;

f is increasing on $(-\infty, \infty)$

$$f(y) = x$$
$$y^{3/5} = x$$
$$y = x^{5/3}$$
$$f^{-1}(x) = x^{5/3}$$
$$\text{dom } f^{-1} = (-\infty, \infty)$$

15. $f'(x) = -3(2 - 3x)^2 \leq 0$ for all x and

$f'(x) = 0$ only at $x = 2/3$; f is decreasing

$$f(y) = x$$
$$(2 - 3y)^3 = x$$
$$2 - 3y = x^{1/3}$$
$$3y = 2 - x^{1/3}$$
$$y = \tfrac{1}{3}(2 - x^{1/3})$$
$$f^{-1}(x) = \tfrac{1}{3}(2 - x^{1/3})$$
$$\text{dom } f^{-1} = (-\infty, \infty)$$

17. $f'(x) = \cos x \geq 0$ on $[-\pi/2, \pi/2]$; and $f'(x) = 0$ only at $x = -\pi/2, \pi/2$. Therefore f is increasing on $[-\pi/2, \pi/2]$ and f has an inverse. The inverse is denoted by $\arcsin x$; this function will be studied in Section 7.7. The domain of $\arcsin x$ is $[-1, 1] = $ range of $\sin x$ on $[-\pi/2, \pi/2]$.

19. $f'(x) = -\dfrac{1}{x^2} < 0$ for all $x \neq 0$;

f is decreasing on $(-\infty, 0) \cup (0, \infty)$

$$f(y) = x \implies \frac{1}{y} = x \implies y = \frac{1}{x}$$
$$f^{-1}(x) = \frac{1}{x} \quad \text{dom } f^{-1}: x \neq 0$$

21. f is not one-to-one; e.g. $f\left(\tfrac{1}{2}\right) = f(2)$

23. $f'(x) = -\dfrac{3x^2}{(x^3 + 1)^2} \leq 0$ for all $x \neq -1$;

f is decreasing on $(-\infty, -1) \cup (-1, \infty)$

$$f(y) = x$$
$$\frac{1}{y^3 + 1} = x$$
$$y^3 + 1 = \frac{1}{x}$$
$$y^3 = \frac{1}{x} - 1$$
$$y = \left(\frac{1}{x} - 1\right)^{1/3}$$
$$f^{-1}(x) = \left(\frac{1}{x} - 1\right)^{1/3}$$

$\text{dom } f^{-1}: x \neq 0$

25. $f'(x) = \dfrac{-1}{(x+1)^2} < 0$ for all $x \neq -1$;

f is decreasing on $(-\infty, -1) \cup (-1, \infty)$

$$f(y) = x$$
$$\frac{y + 2}{y + 1} = x$$
$$y + 2 = xy + x$$
$$y(1 - x) = x - 2$$
$$y = \frac{x - 2}{1 - x}$$
$$f^{-1}(x) = \frac{x - 2}{1 - x}$$

$\text{dom } f^{-1}: x \neq 1$

27. They are equal.

29.

31.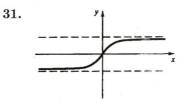

33. (a) $f'(x) = x^2 + 2x + k$: f will be increasing on $(-\infty, \infty)$ if f' does not change sign. This will occur if the discriminant of f', namely $4 - 4k$ is non-positive.

$$4 - 4k \leq 0 \quad \Longrightarrow \quad k \geq 1$$

(b) $g'(x) = 3x^2 + 2kx + 1$: g will be increasing on $(-\infty, \infty)$ if g' does not change sign. This will occur if the discriminant of g', namely $4k^2 - 12$ is non-positive.

$$4k^2 - 12 \leq 0 \quad \Longrightarrow \quad k^2 \leq 3 \quad \Longrightarrow \quad -\sqrt{3} \leq k \leq \sqrt{3}$$

35. $f'(x) = 3x^2 \geq 0$ on $I = (-\infty, \infty)$ and $f'(x) = 0$ only at $x = 0$; f is increasing on I and so it has an inverse.

$$f(2) = 9 \text{ and } f'(2) = 12; \quad (f^{-1})'(9) = \frac{1}{f'(2)} = \frac{1}{12}$$

37. $f'(x) = 1 + \dfrac{1}{\sqrt{x}} > 0$ on $I = (0, \infty)$; f is increasing on I and so it has an inverse.

$$f(4) = 8 \text{ and } f'(4) = 1 + \frac{1}{2} = \frac{3}{2}; \quad (f^{-1})'(8) = \frac{1}{f'(4)} = \frac{1}{3/2} = \frac{2}{3}$$

39. $f'(x) = 2 - \sin x > 0$ on $I = (-\infty, \infty)$; f is increasing on I and so it has an inverse.

$$f(\pi/2) = \pi \text{ and } f'(\pi/2) = 1; \quad (f^{-1})'(\pi) = \frac{1}{f'(\pi/2)} = 1$$

41. $f'(x) = \sec^2 x > 0$ on $I = (-\pi/2, \pi/2)$; f is increasing on I and so it has an inverse

$$f(\pi/3) = \sqrt{3} \text{ and } f'(\pi/3) = 4; \quad (f^{-1})'(\sqrt{3}) = \frac{1}{f'(\pi/3)} = \frac{1}{4}$$

43. $f'(x) = 3 + \dfrac{3}{x^4} > 0$ on $I = (0, \infty)$; f is increasing on I and so it has an inverse.

$$f(1) = 2 \text{ and } f'(1) = 6; \quad f^{-1'}(2) = \frac{1}{f'(1)} = \frac{1}{6}.$$

45. Let $x \in \text{dom}\,(f^{-1})$ and let $f(z) = x$. Then

$$(f^{-1})'(x) = \frac{1}{f'(z)} = \frac{1}{f(z)} = \frac{1}{x}$$

47. Let $x \in \text{dom}\,(f^{-1})$ and let $f(z) = x$. Then

$$(f^{-1})'(x) = \frac{1}{f'(z)} = \frac{1}{\sqrt{1 - [f(z)]^2}} = \frac{1}{\sqrt{1 - x^2}}$$

49. (a)

$$f'(x) = \frac{(cx + d)a - (ax + b)c}{(cx + d)^2} = \frac{ad - bc}{(cx + d)^2}, \quad x \neq -d/c$$

Thus, $f'(x) \neq 0$ iff $ad - bc \neq 0$.

(b)

$$\frac{at + b}{ct + d} = x$$

$$at + b = ctx + dx$$

$$(a - cx)t = dx - b$$

$$t = \frac{dx - b}{a - cx}; \quad f^{-1}(x) = \frac{dx - b}{a - cx}$$

51. (a) $f'(x) = \sqrt{1 + x^2} > 0$, so f is always increasing, hence one-to-one.

(b) $(f^{-1})'(0) = \dfrac{1}{f'(f^{-1}(0))} = \dfrac{1}{f'(2)} = \dfrac{1}{\sqrt{5}}$.

53. (a) $g'(x) = \dfrac{1}{f'[g(x)]}$; $g''(x) = -\dfrac{1}{(f'[g(x)])^2} f''[g(x)]g'(x) = -\dfrac{f''[g(x)]}{(f'[g(x)])^3}$

(b) If f' is increasing, then the graphs of f and g have opposite concavity;

if f' is decreasing then the graphs of f and g have the same concavity.

55. Let $f(x) = \sin x$ and let $y = f^{-1}(x)$. Then

$$\sin y = x$$

$$\cos y \,\frac{dy}{dx} = 1$$

$$\frac{dy}{dx} = \frac{1}{\cos y} \quad (y \neq \pm\pi/2)$$

$$= \frac{1}{\sqrt{1 - \sin^2 y}} = \frac{1}{\sqrt{1 - x^2}} \quad (x \neq \pm 1)$$

57. $f^{-1}(x) = \dfrac{x^2 - 8x + 25}{9}, \quad x \geq 4$

59. $f^{-1}(x) = 16 - 12x + 6x^2 - x^3$

61. $f'(x) = 3x^2 + 3 > 0$ for all x;

f is increasing on $(-\infty, \infty)$

63. $f'(x) = 8 \cos 2x > 0, \ x \in (-\pi/4, \pi/4)$;

f is increasing on $[-\pi/4, \pi/4]$

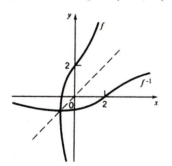

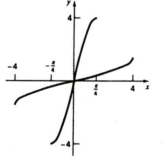

SECTION 7.2

1. $\ln 20 = \ln 2 + \ln 10 \cong 2.99$

3. $\ln 1.6 = \ln \frac{16}{10} = 2\ln 4 - \ln 10 \cong 0.48$

5. $\ln 0.1 = \ln \frac{1}{10} = \ln 1 - \ln 10 \cong -2.30$

7. $\ln 7.2 = \ln \frac{72}{10} = \ln 8 + \ln 9 - \ln 10 \cong 1.98$

9. $\ln \sqrt{2} = \frac{1}{2} \ln 2 \cong 0.35$

11. For any positive integer k: $\displaystyle\int_k^{2k} \frac{1}{x} \, dx = [\ln x]_k^{2k} = \ln 2k - \ln k = \ln\left(\frac{2k}{k}\right) = \ln 2.$

13. $\frac{1}{2}\left[L_f(P) + U_f(P)\right] = \frac{1}{2}\left[\frac{763}{1980} + \frac{1691}{3960}\right] \cong 0.406$

15. (a) $\ln 5.2 \cong \ln 5 + \frac{1}{5}(0.2) \cong 1.65$

   (b) $\ln 4.8 \cong \ln 5 - \frac{1}{5}(0.2) \cong 1.57$ (b)

   (c) $\ln 5.5 \cong \ln 5 + \frac{1}{5}(0.5) \cong 1.71$

17. $x = e^2$

19. $2 - \ln x = 0$ or $\ln x = 0$. Thus $x = e^2$ or $x = 1$.

21.
$$\ln[(2x + 1)(x + 2)] = 2\ln(x + 2)$$
$$\ln[(2x + 1)(x + 2)] = \ln[(x + 2)^2]$$
$$(2x + 1)(x + 2) = (x + 2)^2$$
$$x^2 + x - 2 = 0$$
$$(x + 2)(x - 1) = 0$$
$$x = -2, 1$$

We disregard the solution $x = -2$ since it does not satisfy the initial equation.

Thus, the only solution is $x = 1$.

23. See Exercises 3.1, (3.1.6).

$$\lim_{x \to 1} \frac{\ln x}{x - 1} = \frac{d}{dx}(\ln x)\bigg|_{x=1} = \frac{1}{x}\bigg|_{x=1} = 1$$

25. Continuing Exercise 24, the upper sum

$$\frac{1}{2} + \frac{1}{3} + \cdots + \frac{1}{n - 1} > \ln n.$$

Therefore, $k = n - 1$.

27. (a) Let $G(x) = \ln x - \sin x$. Then $G(2) = \ln 2 - \sin 2 \cong -0.22 < 0$ and $G(3) = \ln 3 - \sin 3 \cong 0.96 > 0$. Thus, G has at least one zero on $[2, 3]$ which implies that there is at least one number $r \in [2, 3]$ such that $\sin r = \ln r$.

(b) $r \cong 2.2191$

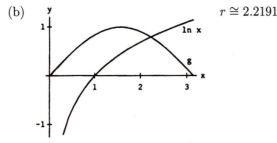

29. $L = 1$

SECTION 7.3

1. $\text{dom}(f) = (0, \infty)$, $f'(x) = \dfrac{1}{4x}(4) = \dfrac{1}{x}$

3. $\text{dom}(f) = (-1, \infty)$, $f'(x) = \dfrac{1}{x^3 + 1}\dfrac{d}{dx}(x^3 + 1) = \dfrac{3x^2}{x^3 + 1}$

5. $\operatorname{dom}(f) = (-\infty, \infty)$, $f(x) = \dfrac{1}{2}\ln(1 + x^2)$ so $f'(x) = \dfrac{1}{2}\left[\dfrac{1}{1 + x^2}(2x)\right] = \dfrac{x}{1 + x^2}$

7. $\operatorname{dom}(f) = \{x \mid x \neq \pm 1\}$, $f'(x) = \dfrac{1}{x^4 - 1}\dfrac{d}{dx}(x^4 - 1) = \dfrac{4x^3}{x^4 - 1}$

9. $\operatorname{dom}(f) = \left(-\dfrac{1}{2}, \infty\right)$,

$$f'(x) = (2x + 1)^2 \dfrac{d}{dx}\left[\ln(2x + 1)\right] + 2(2x + 1)(2)\ln(2x + 1)$$

$$= (2x + 1)^2 \dfrac{2}{2x + 1} + 4(2x + 1)\ln(2x + 1)$$

$$= 2(2x + 1) + 4(2x + 1)\ln(2x + 1) = 2(2x + 1)\left[1 + 2\ln(2x + 1)\right]$$

11. $\operatorname{dom}(f) = (0, 1) \cup (1, \infty)$, $f(x) = (\ln x)^{-1}$ so $f'(x) = -(\ln x)^{-2}\dfrac{d}{dx}(\ln x) = -\dfrac{1}{x(\ln x)^2}$

13. $\operatorname{dom}(f) = (0, \infty)$, $f'(x) = \cos(\ln x)\left(\dfrac{1}{x}\right) = \dfrac{\cos(\ln x)}{x}$

15. $\displaystyle\int \dfrac{dx}{x + 1} = \ln|x + 1| + C$

17. $\left\{\begin{array}{l} u = 3 - x^2 \\ du = -2x\,dx \end{array}\right\}$; $\displaystyle\int \dfrac{x}{3 - x^2}\,dx = -\dfrac{1}{2}\int \dfrac{du}{u} = -\dfrac{1}{2}\ln|u| + C = -\dfrac{1}{2}\ln|3 - x^2| + C$

19. $\left\{\begin{array}{l} u = 3x \\ du = 3\,dx \end{array}\right\}$; $\displaystyle\int \tan 3x\,dx = \dfrac{1}{3}\int \tan u\,du = \dfrac{1}{3}\ln|\sec u| + C = \dfrac{1}{3}\ln|\sec 3x| + C$

21. $\left\{\begin{array}{l} u = x^2 \\ du = 2x\,dx \end{array}\right\}$; $\displaystyle\int x\sec x^2\,dx = \dfrac{1}{2}\int \sec u\,du = \dfrac{1}{2}\ln|\sec u + \tan u| + C$

$$= \tfrac{1}{2}\ln|\sec x^2 + \tan x^2| + C$$

23. $\left\{\begin{array}{l} u = 3 - x^2 \\ du = -2x\,dx \end{array}\right\}$; $\displaystyle\int \dfrac{x}{(3 - x^2)^2}\,dx = -\dfrac{1}{2}\int \dfrac{du}{u^2} = \dfrac{1}{2u} + C = \dfrac{1}{2(3 - x^2)} + C$

25. $\left\{\begin{array}{l} u = 2 + \cos x \\ du = -\sin x\,dx \end{array}\right\}$; $\displaystyle\int \dfrac{\sin x}{2 + \cos x}\,dx = -\int \dfrac{1}{u}\,du = -\ln|u| + C = -\ln|2 + \cos x| + C$

27. $\left\{u = \ln x,\ du = \dfrac{dx}{x}\right\}$; $\displaystyle\int \dfrac{dx}{x\ln x} = \int \dfrac{du}{u} = \ln u + C = \ln|\ln x| + C$

29. $\left\{u = \ln x,\ du = \dfrac{dx}{x}\right\}$; $\displaystyle\int \dfrac{dx}{x(\ln x)^2} = \int \dfrac{du}{u^2} = -\dfrac{1}{u} + C = -\dfrac{1}{\ln x} + C$

31. $\left\{\begin{array}{l} u = \sin x + \cos x \\ du = (\cos x - \sin x)\,dx \end{array}\right\}$;

$$\int \dfrac{\sin x - \cos x}{\sin x + \cos x}\,dx = -\int \dfrac{1}{u}\,du = -\ln|u| + C = -\ln|\sin x + \cos x| + C$$

33. $\left\{u = 1 + x\sqrt{x}, \ du = \dfrac{3}{2}x^{1/2}\,dx\right\}; \quad \displaystyle\int \frac{\sqrt{x}}{1 + x\sqrt{x}}\,dx = \frac{2}{3}\int \frac{du}{u} = \frac{2}{3}\ln|u| + C$

$$= \frac{2}{3}\ln\left|1 + x\sqrt{x}\,\right| + C$$

35. $\displaystyle\int (1 + \sec x)^2\,dx = \int \left(1 + 2\sec x + \sec^2 x\right)dx = x + 2\ln|\sec x + \tan x| + \tan x + C$

37. $\displaystyle\int_1^e \frac{dx}{x} = [\ln x]_1^e = \ln e - \ln 1 = 1 - 0 = 1$

39. $\displaystyle\int_e^{e^2} \frac{dx}{x} = [\ln x]_e^{e^2} = \ln e^2 - \ln e = 2 - 1 = 1$

41. $\displaystyle\int_4^5 \frac{x}{x^2 - 1}\,dx = \left[\frac{1}{2}\ln|x^2 - 1|\right]_4^5 = \frac{1}{2}(\ln 24 - \ln 15) = \frac{1}{2}\ln\frac{8}{5}$

43. $\left\{\begin{array}{l} u = 1 + \sin x \\ du = \cos x\,dx \end{array}\middle| \begin{array}{l} x = \pi/6 \implies u = 3/2 \\ x = \pi/2 \implies u = 2 \end{array}\right\}; \quad \displaystyle\int_{\pi/6}^{\pi/2} \frac{\cos x}{1 + \sin x}\,dx = \int_{3/2}^2 \frac{du}{u} = [\ln u]_{3/2}^2 = \ln\frac{4}{3}$

45. $\displaystyle\int_{\pi/4}^{\pi/2} \cot x\,dx = [\ln|\sin x|]_{\pi/4}^{\pi/2} = \ln 1 - \ln\frac{\sqrt{2}}{2} = \ln\sqrt{2} = \frac{1}{2}\ln 2$

47. The integrand $\dfrac{1}{x - 2}$ is not defined at $x = 2$.

49. $\ln|g(x)| = 2\ln(x^2 + 1) + 5\ln|x - 1| + 3\ln x$

$$\frac{g'(x)}{g(x)} = 2\left(\frac{2x}{x^2 + 1}\right) + \frac{5}{x - 1} + \frac{3}{x}$$

$$g'(x) = (x^2 + 1)^2 (x - 1)^5 x^3 \left(\frac{4x}{x^2 + 1} + \frac{5}{x - 1} + \frac{3}{x}\right); \quad g'(1) = 0$$

51. $\ln|g(x)| = 4\ln|x| + \ln|x - 1| - \ln|x + 2| - \ln(x^2 + 1)$

$$\frac{g'(x)}{g(x)} = \frac{4}{x} + \frac{1}{x - 1} - \frac{1}{x + 2} - \frac{2x}{x^2 + 1}$$

$$g'(x) = \frac{x^4(x - 1)}{(x + 2)(x^2 + 1)}\left(\frac{4}{x} + \frac{1}{x - 1} - \frac{1}{x + 2} - \frac{2x}{x^2 + 1}\right); \quad g'(0) = 0$$

53.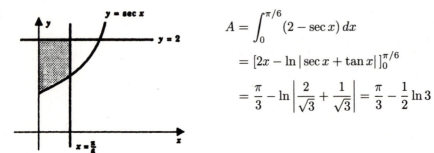

$$A = \int_0^{\pi/6} (2 - \sec x)\,dx$$

$$= [2x - \ln|\sec x + \tan x|]_0^{\pi/6}$$

$$= \frac{\pi}{3} - \ln\left|\frac{2}{\sqrt{3}} + \frac{1}{\sqrt{3}}\right| = \frac{\pi}{3} - \frac{1}{2}\ln 3$$

55.

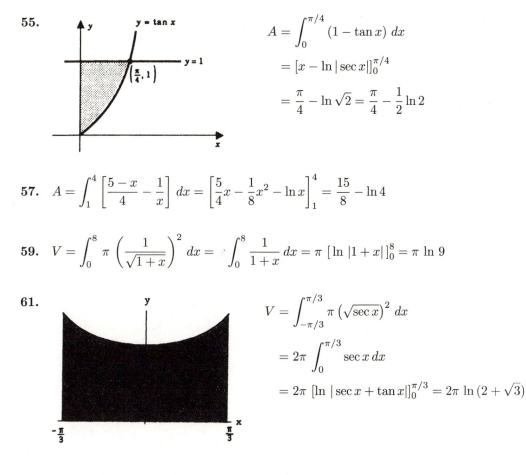

$$A = \int_0^{\pi/4} (1 - \tan x)\, dx$$
$$= [x - \ln|\sec x|]_0^{\pi/4}$$
$$= \frac{\pi}{4} - \ln\sqrt{2} = \frac{\pi}{4} - \frac{1}{2}\ln 2$$

57. $\quad A = \int_1^4 \left[\frac{5-x}{4} - \frac{1}{x}\right] dx = \left[\frac{5}{4}x - \frac{1}{8}x^2 - \ln x\right]_1^4 = \frac{15}{8} - \ln 4$

59. $\quad V = \int_0^8 \pi\left(\frac{1}{\sqrt{1+x}}\right)^2 dx = \int_0^8 \frac{1}{1+x}\, dx = \pi\left[\ln|1+x|\right]_0^8 = \pi\ln 9$

61.

$$V = \int_{-\pi/3}^{\pi/3} \pi\left(\sqrt{\sec x}\right)^2 dx$$
$$= 2\pi \int_0^{\pi/3} \sec x\, dx$$
$$= 2\pi\left[\ln|\sec x + \tan x|\right]_0^{\pi/3} = 2\pi\ln\left(2 + \sqrt{3}\right)$$

63. $\quad v(t) = \int a(t)\, dt = \int -(t+1)^{-2}\, dt = \frac{1}{t+1} + C.$

Since $v(0) = 1$, we get $1 = 1 + C$ so that $C = 0$. Then

$$s = \int_0^4 |v(t)|\, dt = \int_0^4 \frac{dt}{t+1} = \left[\ln(t+1)\right]_0^4 = \ln 5.$$

The particle traveled $\ln 5$ ft.

65. $\quad \dfrac{d}{dx}(\ln x) = \dfrac{1}{x}$

$\dfrac{d^2}{dx^2}(\ln x) = -\dfrac{1}{x^2}$

$\dfrac{d^3}{dx^3}(\ln x) = \dfrac{2}{x^3}$

$\dfrac{d^4}{dx^4}(\ln x) = -\dfrac{2\cdot 3}{x^4}$

$\vdots$

$\dfrac{d^n}{dx^n}(\ln x) = (-1)^{n-1}\dfrac{(n-1)!}{x^n}$

67. $\int \csc x \, dx = \int \dfrac{\csc x (\csc x - \cot x)}{\csc x - \cot x} \, dx = \int \dfrac{\csc^2 - \csc x \cot x}{\csc x - \cot x} \, dx$

$\left\{ \begin{array}{l} u = \csc x - \cot x \\ du = (-\csc x \cot x + \csc^2 x) \, dx \end{array} \right\} ; \quad \int \csc x \, dx = \int \dfrac{du}{u} = \ln |u| + C$

$$= \ln |\csc x - \cot x| + C$$

69. $f(x) = \ln (4 - x), \quad x < 4$

$f'(x) = \dfrac{1}{x - 4}$

$f''(x) = \dfrac{-1}{(x - 4)^2}$

(i)   domain $(-\infty, 4)$

(ii)   decreases throughout

(iii)   no extreme values

(iv)   concave down throughout: no pts of inflection

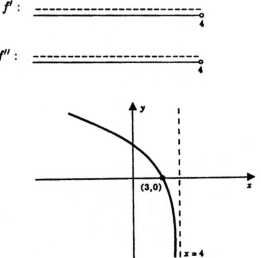

71. $f(x) = x^2 \ln x, \quad x > 0$

$f'(x) = 2x \ln x + x$

$f''(x) = 2 \ln x + 3$

(i) domain $(0, \infty)$

(ii) decreases on $(0, 1/\sqrt{e}]$,   increases on $[1/\sqrt{e}, \infty)$

(iii) $f(1/\sqrt{e}) = -1/2e$ local and absolute min

(iv) concave down on $(0, 1/e^{3/2})$,

concave up on $(1/e^{3/2}, \infty)$;

pt of inflection at $(1/e^{3/2}, -3/2e^3)$

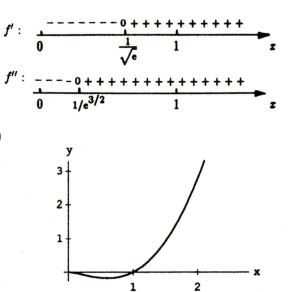

73. $f(x) = \ln\left[\dfrac{x}{1+x^2}\right], \ x > 0$

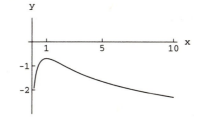

$f'(x) = \dfrac{1-x^2}{x+x^3}$

$f''(x) = \dfrac{x^4 - 4x^2 - 1}{(x+x^3)^2}$

(i) domain $(0, \infty)$

(ii) increases on $(0, 1]$,   decreases on $[1, \infty)$

(iii) $f(1) = $ local and absolute max

(iv) concave down on $(0, 2.0582)$,

   concave up on $(2.0582, \infty)$;

   pt of inflection at $(2.0582, -0.9338)$ (approx.)

75. Average slope $= \dfrac{1}{b-a}\displaystyle\int_a^b \dfrac{1}{x}\,dx = \dfrac{1}{(b-a)}\ln\dfrac{b}{a}$

77.

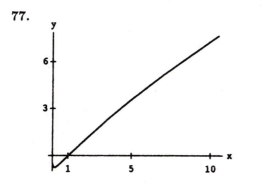

x-intercept: 1; abs min at $x = 1/e^2$;
abs max at $x = 10$

79.

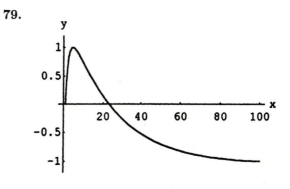

x-intercepts: $1, 23.1407$; abs min at $x = 100$;
abs max at $x = 4.8105$

81. (a) $v(t) - v(0) = \displaystyle\int_0^t a(u)\,du, \quad 0 \le t \le 3$

   $v(t) = \displaystyle\int_0^t \left[4 - 2(u+1) + \dfrac{3}{u+1}\right]\,du + 2$

   $= \left[2u - u^2 + 3\ln|u+1|\right]_0^t = 2 + 2t - t^2 + 3\ln(t+1)$

(b)

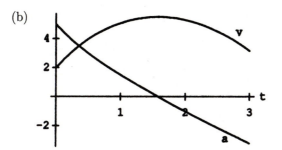

(c) max velocity at $t = 1.5811$; min velocity at $t = 0$

83. (b) x-coordinates of the points of intersection: $x = 1, \ 3.3028$

(c) $A \cong 2.34042$

85. (a) $f(x) = \dfrac{\ln x}{x^2}, \qquad f'(x) = \dfrac{1 - 2\ln x}{x^3}, \qquad f''(x) = \dfrac{-5 + 6\ln x}{x^4}$

(b) $f(1) = 0, \qquad f'(e^{1/2}) = 0, \qquad f''(e^{5/6}) = 0$

(c) $f(x) > 0$ on $(1, \infty) \qquad f'(x) > 0$ on $(0, e^{1/2}) \qquad f''(x) > 0$ on $(e^{5/6}, \infty)$

$\qquad f(x) < 0$ on $(0, 1) \qquad f'(x) < 0$ on $(e^{1/2}, \infty) \qquad f''(x) > 0$ on $(0, e^{5/6})$

(d) $f(e^{1/2})$ local and absolute maximum

SECTION 7.4

1. $\dfrac{dy}{dx} = e^{-2x}\dfrac{d}{dx}(-2x) = -2e^{-2x}$
$\qquad$ **3.** $\dfrac{dy}{dx} = e^{x^2-1}\dfrac{d}{dx}(x^2 - 1) = 2xe^{x^2-1}$

5. $\dfrac{dy}{dx} = e^x\dfrac{d}{dx}(\ln x) + \ln x\dfrac{d}{dx}(e^x) = e^x\left(\dfrac{1}{x} + \ln x\right)$

7. $\dfrac{dy}{dx} = x^{-1}\dfrac{d}{dx}(e^{-x}) + e^{-x}\dfrac{d}{dx}(x^{-1}) = -x^{-1}e^{-x} - e^{-x}x^{-2} = -(x^{-1} + x^{-2})e^{-x}$

9. $\dfrac{dy}{dx} = \dfrac{1}{2}(e^x - e^{-x})$

11. $\dfrac{dy}{dx} = e^{\sqrt{x}}\dfrac{d}{dx}\left(\ln\sqrt{x}\,\right) + \ln\sqrt{x}\,\dfrac{d}{dx}(e^{\sqrt{x}}) = e^{\sqrt{x}}\left(\dfrac{1}{\sqrt{x}} \cdot \dfrac{1}{2\sqrt{x}}\right) + \ln\sqrt{x}\,\dfrac{e^{\sqrt{x}}}{2\sqrt{x}} = \dfrac{1}{2}e^{\sqrt{x}}\left(\dfrac{1}{x} + \dfrac{\ln\sqrt{x}}{\sqrt{x}}\right)$

13. $\dfrac{dy}{dx} = 2(e^{x^2} + 1)\dfrac{d}{dx}(e^{x^2} + 1) = 2(e^{x^2} + 1)e^{x^2}\dfrac{d}{dx}(x^2) = 4xe^{x^2}(e^{x^2} + 1)$

15. $\dfrac{dy}{dx} = (x^2 - 2x + 2)\dfrac{d}{dx}(e^x) + e^x(2x - 2) = x^2e^x$

17. $\dfrac{dy}{dx} = \dfrac{(e^x + 1)\,e^x - (e^x - 1)\,e^x}{(e^x + 1)^2} = \dfrac{2e^x}{(e^x + 1)^2}$

19. $y = e^{4\ln x} = (e^{\ln x})^4 = x^4$ so $\dfrac{dy}{dx} = 4x^3.$
$\qquad$ **21.** $f'(x) = \cos(e^{2x})\,e^{2x} \cdot 2 = 2e^{2x}\cos(e^{2x})$

23. $f'(x) = e^{-2x}(-\sin x) + e^{-2x}(-2)\cos x = -e^{-2x}(2\cos x + \sin x)$

25. $\displaystyle\int e^{2x}\,dx = \dfrac{1}{2}e^{2x} + C$
$\qquad$ **27.** $\displaystyle\int e^{kx}\,dx = \dfrac{1}{k}e^{kx} + C$

29. $\{u = x^2, \quad du = 2x\,dx\};\quad \displaystyle\int xe^{x^2}\,dx = \dfrac{1}{2}\int e^u\,du = \dfrac{1}{2}e^u + C = \dfrac{1}{2}e^{x^2} + C$

31. $\left\{u = \dfrac{1}{x}, \quad du = -\dfrac{1}{x^2}\,dx\right\};\quad \displaystyle\int\dfrac{e^{1/x}}{x^2}\,dx = -\int e^u\,du = -e^u + C = -e^{1/x} + C$

33. $\int \ln e^x \, dx = \int x \, dx = \frac{1}{2}x^2 + C$

35. $\int \frac{4}{\sqrt{e^x}} \, dx = \int 4e^{-x/2} \, dx = -8e^{-x/2} + C$

37. $\left\{ \begin{array}{l} u = e^x + 1 \\ du = e^x \, dx \end{array} \right\};\quad \int \frac{e^x}{\sqrt{e^x + 1}} \, dx = \int \frac{du}{\sqrt{u}} = \int u^{-1/2} \, du = 2u^{1/2} + C = 2\sqrt{e^x + 1} + C$

39. $\left\{ \begin{array}{l} u = 2e^{2x} + 3 \\ du = 4e^{2x} \, dx \end{array} \right\};\quad \int \frac{e^{2x}}{2e^{2x} + 3} \, dx = \frac{1}{4} \int \frac{du}{u} = \frac{1}{4} \ln u + C = \frac{1}{4} \ln(2e^{2x} + 3) + C$

41. $\{ u = \sin x, \quad du = \cos x \, dx \};\quad \int \cos x \, e^{\sin x} \, dx = \int e^u \, du = e^u + C = e^{\sin x} + C$

43. $\int_0^1 e^x \, dx = [\, e^x \,]_0^1 = e - 1$

45. $\int_0^{\ln \pi} e^{-6x} \, dx = \left[-\frac{1}{6}e^{-6x} \right]_0^{\ln \pi} = -\frac{1}{6}e^{-6\ln \pi} + \frac{1}{6}e^0 = \frac{1}{6}\left(1 - \pi^{-6}\right)$

47. $\int_0^1 \frac{e^x + 1}{e^x} \, dx = \int_0^1 (1 + e^{-x}) \, dx = [x - e^{-x}]_0^1 = \left(1 - e^{-1}\right) - (0 - 1) = 2 - \frac{1}{e}$

49. $\int_0^{\ln 2} \frac{e^x}{e^x + 1} \, dx = [\ln(e^x + 1)]_0^{\ln 2} = \ln(e^{\ln 2} + 1) - \ln(e^0 + 1) = \ln 3 - \ln 2 = \ln \frac{3}{2}$

51. $\int_0^1 x(e^{x^2} + 2) \, dx = \int_0^1 (xe^{x^2} + 2x) \, dx = \left[\frac{1}{2}e^{x^2} + x^2 \right]_0^1 = (\tfrac{1}{2}e + 1) - (\tfrac{1}{2} + 0) = \tfrac{1}{2}(e + 1)$

53. (a) $f(x) = e^{ax}, \; f'(x) = ae^{ax}, \; f''(x) = a^2 e^{ax}, \; \ldots, \; f^{(n)}(x) = a^n e^{ax}$

　　(b) $f(x) = e^{-ax}, \; f'(x) = -ae^{-ax}, \; f''(x) = a^2 e^{-ax}, \; \ldots, \; f^{(n)}(x) = (-1)^n a^n e^{-ax}$

55. $A = 2xe^{-x^2}$

$A' = 2x(-2xe^{-x^2}) + 2e^{-x^2} = 2e^{-x^2}(1 - 2x^2) = 0 \quad \Longrightarrow \quad x = \pm\sqrt{\frac{1}{2}} \text{ and } y = \frac{1}{\sqrt{e}}.$

Put the vertices at $\left(\pm\dfrac{1}{\sqrt{2}}, \dfrac{1}{\sqrt{e}} \right).$

57. $f(x) = e^{-x^2}$

$f'(x) = -2xe^{-x^2}$

$f''(x) = (4x^2 - 2)e^{-x^2}$

(a)　symmetric with respect to the y-axis
　　　$f(-x) = f(x)$

(b)　increases on $(-\infty, 0]$,　decreases on $[0, \infty)$

(c)　$f(0) = 1$ local and absolute max

(d)　concave up on $\left(-\infty, -1/\sqrt{2} \right) \cup \left(1/\sqrt{2}, \infty \right)$,　concave down on $\left(-1/\sqrt{2}, 1/\sqrt{2} \right)$

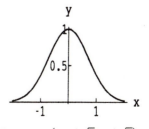

59. (a) $V = \int_0^1 2\pi x e^{-x^2}\, dx = \pi\left[-e^{-x^2}\right]_0^1 = \pi\left[1 - \frac{1}{e}\right]$

(b) $V = \int_0^1 \pi\left[e^{-x^2}\right]^2 dx = \pi\int_0^1 e^{-2x^2}\, dx$

61.

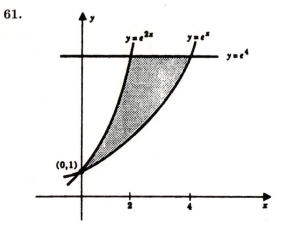

$A = \int_0^2 (e^{2x} - e^x)\, dx + \int_2^4 (e^4 - e^x)\, dx$

$= \left[\tfrac{1}{2}e^{2x} - e^x\right]_0^2 + \left[e^4 x - e^x\right]_2^4$

$= \left(\tfrac{1}{2}e^4 - e^2 - \tfrac{1}{2} + 1\right) + \left(4e^4 - e^4 - 2e^4 + e^2\right)$

$= \tfrac{1}{2}(3e^4 + 1)$

63.

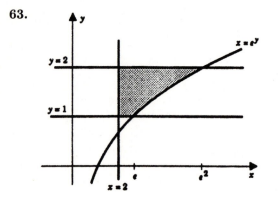

$A = \int_1^2 (e^y - 2)\, dy$

$= \left[e^y - 2y\right]_1^2 = e^2 - e - 2$

65. $f(x) = e^{(1/x)^2}$

$f'(x) = \dfrac{-2}{x^3}\, e^{(1/x)^2}$

$f''(x) = \dfrac{6x^2 + 4}{x^6}\, e^{(1/x)^2}$

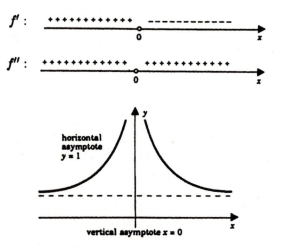

(i) domain $(-\infty, 0) \cup (0, \infty)$

(ii) increases on $(-\infty, 0)$,   decreases on $(0, \infty)$

(iii) no extreme values

(iv) concave up on $(-\infty, 0)$ and on $(0, \infty)$

67. $f(x) = x^2 \ln x$

$f'(x) = 2x \ln x + x$

$f''(x) = 3 + 2 \ln x$

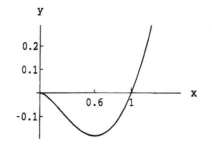

(i) domain $(0, \infty)$

(ii) decreases on $(0, e^{-1/2})$,   increases on $(e^{-1/2}, \infty)$

(iii) $f(e^{-1/2}) = -1/2e$ is a local and absolute min

(iv) concave down on $(0, e^{-3/2})$ and concave up

on $(e^{-3/2}, \infty)$; $(e^{-3/2}, -3/2e^3)$ is a point of inflection

69. $\displaystyle\int_0^{x_n} e^x \, dx = [e^x]_0^{x_n} = e^{x_n} - 1; \quad e^{x_n} - 1 = n \quad \Longrightarrow \quad x_n = \ln(n+1).$

71.   (a)   For $y = e^{ax}$ we have $dy/dx = ae^{ax}$. Therefore the line tangent to the curve $y = e^{ax}$

at an arbitrary point (x_0, e^{ax_0}) has equation

$$y - e^{ax_0} = ae^{ax_0}(x - x_0).$$

The line passes through the origin iff $e^{ax_0} = (ae^{ax_0})x_0$ iff $x_0 = 1/a$. The point of tangency

is $(1/a, e)$. This is point B. By symmetry, point A is $(-1/a, e)$.

(b)   The tangent line at B has equation $y = aex$.   By symmetry

$$A_{\mathrm{I}} = 2 \int_0^{1/a} (e^{ax} - aex) \, dx = 2 \left[\frac{1}{a}e^{ax} - \frac{1}{2}aex^2 \right]_0^{1/a} = \frac{1}{a}(e - 2).$$

(c)   The normal at B has equation

$$y - e = -\frac{1}{ae}\left(x - \frac{1}{a}\right).$$

This can be written

$$y = -\frac{1}{ae}x + \frac{a^2 e^2 + 1}{a^2 e}.$$

Therefore

$$A_{\mathrm{II}} = 2 \int_0^{1/a} \left(-\frac{1}{ae}x + \frac{a^2 e^2 + 1}{a^2 e} - e^{ax} \right) dx = \frac{1 + 2a^2 e}{a^3 e}.$$

73. For $x > (n+1)!$

$$e^x > 1 + x + \cdots + \frac{x^{n+1}}{(n+1)!} > \frac{x^{n+1}}{(n+1)!} = x^n \left[\frac{x}{(n+1)!} \right] > x^n.$$

75. (a)

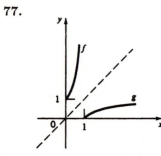

(b) $x_1 \cong -1.9646, x_2 \cong 1.0580$

(c) $A \cong \displaystyle\int_{-1.9646}^{1.0580} \left[4 - x^2 - e^x \right] \, dx = \left[4x - \frac{1}{3}x^3 - e^x \right]_{-1.9646}^{1.0580} \cong 6.4240$

77.

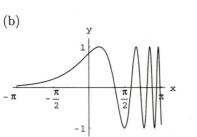

$$f\left(g(x) \right) = e^{\left(\sqrt{\ln x} \right)^2} = e^{\ln x} = x$$

79. (a) $f(x) = \sin(e^x); \quad f(x) = 0 \implies e^x = n\pi \implies x = \ln n\pi, \; n = 1, 2, \cdots.$

(b)

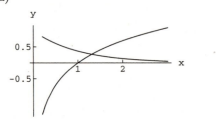

81. (a)

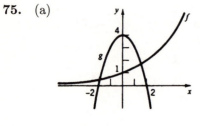

(b) $x \cong 1.3098$

(c) $f'(1.3098) \cong -0.26987; \; g'(1.3098) \cong 0.76348$

(d) the tangent lines are not perpendicular

83.   (a) $\displaystyle\int \frac{1}{1-e^x}\,dx = x - \ln|e^x - 1| + C$ (c) $\displaystyle\int \frac{e^{\tan x}}{\cos^2 x}\,dx = e^{\tan x} + C$

(b) $\displaystyle\int e^{-x}\left(\frac{1-e^x}{e^x}\right)^4 dx = -\frac{1}{5}\,e^{-5x} + e^{-4x} - 2\,e^{-3x} + 2\,e^{-2x} - e^{-x} + C$

PROJECT 7.4

Step 1. $\displaystyle \ln\left(1+\frac{1}{n}\right) = \int_1^{1+\frac{1}{n}} \frac{dt}{t} \le \int_1^{1+\frac{1}{n}} 1\,dt = \frac{1}{n}$

(since $\frac{1}{t} \le 1$ throughout the interval of integration) $= 1$.

$\displaystyle \ln\left(1+\frac{1}{n}\right) = \int_1^{1+\frac{1}{n}} \frac{dt}{t} \ge \int_1^{1+\frac{1}{n}} \frac{dt}{1+\frac{1}{n}} = \frac{1}{1+\frac{1}{n}}\,\frac{1}{n} = \frac{1}{n+1}.$

(since $\frac{1}{t} \ge \dfrac{1}{1+\frac{1}{n}}$ throughout the interval of integration)

Step 2. From Step 1, we get

$1 + \dfrac{1}{n} \le e^{1/n} \implies \left(1+\dfrac{1}{n}\right)^n \le e$ and $e^{1/n+1} \le 1 + \dfrac{1}{n} \implies e \le \left(1+\dfrac{1}{n}\right)^{n+1}$

Combining these two inequalities, we have

$$\left(1+\frac{1}{n}\right)^n \le e \le \left(1+\frac{1}{n}\right)^{n+1}$$

SECTION 7.5

1. $\log_2 64 = \log_2\left(2^6\right) = 6$

3. $\log_{64}\left(1/2\right) = \dfrac{\ln\left(1/2\right)}{\ln 64} = \dfrac{-\ln 2}{6\ln 2} = -\dfrac{1}{6}$

5. $\log_5 1 = \log_5\left(5^0\right) = 0$

7. $\log_5\left(125\right) = \log_5\left(5^3\right) = 3$

9. $\log_p xy = \dfrac{\ln xy}{\ln p} = \dfrac{\ln x + \ln y}{\ln p} = \dfrac{\ln x}{\ln p} + \dfrac{\ln y}{\ln p} = \log_p x + \log_p y$

11. $\log_p x^y = \dfrac{\ln x^y}{\ln p} = y\dfrac{\ln x}{\ln p} = y\log_p x$

13. $10^x = e^x \implies \left(e^{\ln 10^x}\right) = e^x \implies e^{x\ln 10} = e^x$

$\implies x\ln 10 = x \implies x(\ln 10 - 1) = 0 \implies$ Thus, $x = 0$.

15. $\log_x 10 = \log_4 100 \implies \dfrac{\ln 10}{\ln x} = \dfrac{\ln 100}{\ln 4} \implies \dfrac{\ln 10}{\ln x} = \dfrac{2\ln 10}{2\ln 2}$

$\implies \ln x = \ln 2$ Thus, $x = 2$.

17. The logarithm function is increasing. Thus,

$$e^{t_1} < a < e^{t_2} \implies t_1 = \ln e^{t_1} < \ln a < \ln e^{t_2} = t_2.$$

19. $f'(x) = 3^{2x}(\ln 3)(2) = 2(\ln 3)3^{2x}$

21. $f'(x) = 2^{5x}(\ln 2)(5)3^{\ln x} + 2^{5x}3^{\ln x}(\ln 3)\dfrac{1}{x} = 2^{5x}3^{\ln x}\left(5\ln 2 + \dfrac{\ln 3}{x}\right)$

23. $g'(x) = \frac{1}{2}\left(\log_3 x\right)^{-1/2}\left(\dfrac{1}{\ln 3}\right)\dfrac{1}{x} = \dfrac{1}{2(\ln 3)x\sqrt{\log_3 x}}$

25. $f'(x) = \sec^2\left(\log_5 x\right)(\ln 5)\dfrac{1}{x} = \dfrac{\sec^2\left(\log_5 x\right)}{x\ln 5}$

27. $F'(x) = -\sin\left(2^x + 2^{-x}\right)\left[2^x \ln 2 - 2^{-x}\ln 2\right] = \ln 2\left(2^{-x} - 2^x\right)\sin\left(2^x + 2^{-x}\right)$

29. $\displaystyle\int 3^x\,dx = \dfrac{3^x}{\ln 3} + C$

31. $\displaystyle\int (x^3 + 3^{-x})\,dx = \dfrac{1}{4}x^4 - \dfrac{3^{-x}}{\ln 3} + C$

33. $\displaystyle\int \dfrac{dx}{x\ln 5} = \dfrac{1}{\ln 5}\displaystyle\int \dfrac{dx}{x} = \dfrac{\ln|x|}{\ln 5} + C = \log_5|x| + C$

35.
$$\int \dfrac{\log_2 x^3}{x}\,dx = \dfrac{1}{\ln 2}\int \dfrac{\ln x^3}{x}\,dx = \dfrac{3}{\ln 2}\int \dfrac{\ln x}{x}\,dx$$

$$= \dfrac{3}{\ln 2}\left[\dfrac{1}{2}(\ln x)^2\right] + C = \dfrac{3}{\ln 4}(\ln x)^2 + C$$

37. $f'(x) = \dfrac{1}{x\ln 3}$ so $f'(e) = \dfrac{1}{e\ln 3}$

39. $f'(x) = \dfrac{1}{x\ln x}$ so $f'(e) = \dfrac{1}{e\ln e} = \dfrac{1}{e}$

41.
$$f(x) = p^x$$
$$\ln f(x) = x\ln p$$
$$\dfrac{f'(x)}{f(x)} = \ln p$$
$$f'(x) = f(x)\ln p$$
$$f'(x) = p^x\ln p$$

43.
$$y = (x+1)^x$$
$$\ln y = x\ln(x+1)$$
$$\dfrac{1}{y}\dfrac{dy}{dx} = \dfrac{x}{x+1} + \ln(x+1)$$
$$\dfrac{dy}{dx} = (x+1)^x\left[\dfrac{x}{x+1} + \ln(x+1)\right]$$

45. $y = (\ln x)^{\ln x}$

$\ln y = \ln x \left[\ln (\ln x)\right]$

$\dfrac{1}{y}\dfrac{dy}{dx} = \ln x \left[\dfrac{1}{x \ln x}\right] + \dfrac{1}{x} \left[\ln (\ln x)\right]$

$\dfrac{dy}{dx} = (\ln x)^{\ln x} \left[\dfrac{1 + \ln (\ln x)}{x}\right]$

47. $y = x^{\sin x}$

$\ln y = (\sin x)(\ln x)$

$\dfrac{1}{y}\dfrac{dy}{dx} = (\cos x)(\ln x) + \sin x \left(\dfrac{1}{x}\right)$

$\dfrac{dy}{dx} = x^{\sin x} \left[(\cos x)(\ln x) + \dfrac{\sin x}{x}\right]$

49. $y = (\sin x)^{\cos x}$

$\ln y = (\cos x)(\ln[\sin x])$

$\dfrac{1}{y}\dfrac{dy}{dx} = (-\sin x)(\ln[\sin x]) + (\cos x)\left(\dfrac{1}{\sin x}\right)(\cos x)$

$\dfrac{dy}{dx} = (\sin x)^{\cos x}\left[\dfrac{\cos^2 x}{\sin x} - (\sin x)(\ln[\sin x])\right]$

51. $y = x^{2^x}$

$\ln y = 2^x \ln x$

$\dfrac{1}{y}\dfrac{dy}{dx} = 2^x \ln 2 \ln x + 2^x\left(\dfrac{1}{x}\right)$

$\dfrac{dy}{dx} = x^{2^x}\left[2^x \ln 2 \ln x + \dfrac{2^x}{x}\right]$

53. From the definition of the derivative, the derivative of $f(x) = \ln x$ at $x = 1$ is

$$f'(1) = \lim_{h \to 0} \frac{\ln (1 + h) - \ln 1}{h}.$$

Since $f'(1) = 1$, we have

$$\frac{\ln (1 + h) - \ln 1}{h} = \frac{1}{h}\ln (1 + h) = \ln (1 + h)^{1/h} \to 1 \text{ as } h \to 0.$$

Set $x = 1/h$. Then $h \to 0 \implies x \to \infty$ and

$$\ln \left(1 + \frac{1}{x}\right)^x \to 1 \text{ as } x \to \infty$$

Therefore

$$\left(1 + \frac{1}{x}\right)^x = e^{\ln (1 + 1/x)^x} \to e^1 = e \text{ as } x \to \infty.$$

55.

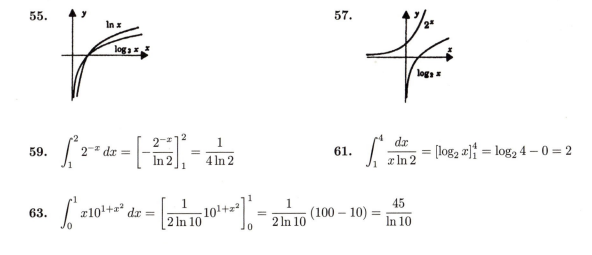

57.

59. $\displaystyle\int_1^2 2^{-x}\,dx = \left[-\dfrac{2^{-x}}{\ln 2}\right]_1^2 = \dfrac{1}{4\ln 2}$

61. $\displaystyle\int_1^4 \dfrac{dx}{x \ln 2} = \left[\log_2 x\right]_1^4 = \log_2 4 - 0 = 2$

63. $\displaystyle\int_0^1 x10^{1+x^2}\,dx = \left[\dfrac{1}{2\ln 10}10^{1+x^2}\right]_0^1 = \dfrac{1}{2\ln 10}(100 - 10) = \dfrac{45}{\ln 10}$

65. $\int_0^1 (2^x + x^2)\, dx = \left[\dfrac{2^x}{\ln 2} + \dfrac{x^3}{3}\right]_0^1 = \dfrac{1}{3} + \dfrac{1}{\ln 2}$

67. approx   16.99999; $5^{\ln 17/\ln 5} = \left(e^{\ln 5}\right)^{\ln 17/\ln 5} = e^{\ln 17} = 17$

69. (b) the x-coordinates of the points of intersection are: $x_1 \cong -1.198$, $x_2 = 3$ and $x_3 \cong 3.408$

(c) for the interval $[-1.198, 3]$, $A \cong 5.5376$; for the interval $[3, 3.408]$, $A \cong 0.01373$

SECTION 7.6

1. We begin with $A(t) = A_0 e^{rt}$

and take $A_0 = \$500$ and $t = 10$. The interest earned is given by

$$A(10) - A_0 = 500\left(e^{10r} - 1\right).$$

Thus, (a) $500\left(e^{0.6} - 1\right) \cong \$411.06$ (b) $500\left(e^{0.8} - 1\right) \cong \$612.77$

(c) $500\,(e - 1) \cong \$859.14$.

3. In general

$$A(t) = A_0 e^{rt}.$$

We set

$$3A_0 = A_0 e^{20r}$$

and solve for r:

$$3 = e^{20r}, \quad \ln 3 = 20r, \quad r = \dfrac{\ln 3}{20} \cong 5\tfrac{1}{2}\%.$$

5. $P(t) = \tfrac{9}{2}\, e^{\frac{t}{20}\ln(4/3)} = \tfrac{9}{2}\, e^{\ln(4/3)^{t/20}} = \tfrac{9}{2}\left(\tfrac{4}{3}\right)^{t/20}$.

7. (a) $P(t) = 10{,}000 e^{t\ln 2} = 10{,}000(2)^t$

(b) $P(26) = 10{,}000(2)^{26}$, $P(52) = 10{,}000(2)^{52}$

9. (a) $P(10) = P(0)e^{0.035(10)t} = P(0)e^{0.35t}$.   Thus it increases by $e^{0.35}$.

(b) $2P(0) = P(0)e^{15k} \implies k = \dfrac{\ln 2}{15}$.

11. Using the data from Ex. 10, the growth constant $k \cong 0.0121$. Therefore

$P(20) \cong 249\, e^{20k} \cong 249\, e^{0.242} \cong 317.1$ million.

$P(11) \cong 249\, e^{11k} \cong 249\, e^{0.1331} \cong 284.4$ million.

13. $4.5e^{0.0143t} = 30 \implies 0.0143t = \ln\dfrac{30}{4.5} \implies t \simeq 115.7$ years.

Thus maximum population will be reached in 2095.

15.
$$V'(t) = kV(t)$$
$$V'(t) - ktV(t) = 0$$
$$e^{-kt}V'(t) - ke^{-kt}V(t) = 0$$
$$\frac{d}{dt}\left[e^{-kt}V(t)\right] = 0$$
$$e^{-kt}V(t) = C$$
$$V(t) = Ce^{kt}.$$

Since $V(0) = C = 200,$ $V(t) = 200\,e^{kt}.$

Since $V(5) = 160,$ $200\,e^{5k} = 160,$ $e^{5k} = \frac{4}{5},$ $e^k = \left(\frac{4}{5}\right)^{1/5}$

and therefore $V(t) = 200\left(\frac{4}{5}\right)^{t/5}$ liters.

17. Take two years ago as time $t = 0$. In general

$(*)$ $A(t) = A_0 e^{kt}.$

We are given that

$$A_0 = 5 \quad \text{and} \quad A(2) = 4.$$

Thus,

$$4 = 5e^{2k} \quad \text{so that} \quad \tfrac{4}{5} = e^{2k} \quad \text{or} \quad e^k = \left(\tfrac{4}{5}\right)^{1/2}.$$

We can write
$$A(t) = 5\left(\tfrac{4}{5}\right)^{t/2}$$

and compute $A(5)$ as follows:
$$A(5) = 5\left(\tfrac{4}{5}\right)^{5/2} = 5e^{\frac{5}{2}\ln(4/5)} \cong 5e^{-0.56} \cong 2.86.$$

About 2.86 gm will remain 3 years from now.

19. A fundamental property of radioactive decay is that the percentage of substance that decays during any year is constant:

$$100\left[\frac{A(t) - A(t+1)}{A(t)}\right] = 100\left[\frac{A_0 e^{kt} - A_0 e^{k(t+1)}}{A_0 e^{kt}}\right] = 100(1 - e^k)$$

If the half-life is n years, then

$$\tfrac{1}{2}A_0 = A_0 e^{kn} \quad \text{so that} \quad e^k = \left(\tfrac{1}{2}\right)^{1/n}.$$

Thus, $100\left[1 - \left(\tfrac{1}{2}\right)^{1/n}\right]\%$ of the material decays during any one year.

21. (a) $A(1620) = A_0 e^{1620k} = \dfrac{1}{2}A_0 \implies k = \dfrac{\ln\dfrac{1}{2}}{1620} \simeq -0.00043.$

Thus $A(500) = A_0 e^{500k} = 0.807A_0.$ Hence 80.7% will remain.

(b) $0.25A_0 = A_0 e^{kt} \implies t = 3240$ years.

23. (a) $x_1(t) = 10^6 t, \quad x_2(t) = e^t - 1$

   (b) $\dfrac{d}{dt}[x_1(t) - x_2(t)] = \dfrac{d}{dt}[10^6 t - (e^t - 1)] = 10^6 - e^t$

   This derivative is zero at $t = 6 \ln 10 \cong 13.8$. After that the derivative is negative.

   (c) $x_2(15) < e^{15} = (e^3)^5 \cong 20^5 = 2^5(10^5) = 3.2(10^6) < 15(10^6) = x_1(15)$

   $x_2(18) = e^{18} - 1 = (e^3)^6 - 1 \cong 20^6 - 1 = 64(10^6) - 1 > 18(10^6) = x_1(18)$

   $x_2(18) - x_1(18) \cong 64(10^6) - 1 - 18(10^6) \cong 46(10^6)$

   (d) If by time t_1 EXP has passed LIN, then $t_1 > 6 \ln 10$. For all $t \geq t_1$ the speed of EXP is greater than the speed of LIN:

   $$\text{for} \quad t \geq t_1 > 6 \ln 10, \quad v_2(t) = e^t > 10^6 = v_1(t).$$

25. Let $p(h)$ denote the pressure at altitude h. The equation $\dfrac{dp}{dh} = kp$ gives

   (∗) $$p(h) = p_0 e^{kh}$$

   where p_0 is the pressure at altitude zero (sea level).
   Since $p_0 = 15$ and $p(10000) = 10$,

   $$10 = 15e^{10000k}, \quad \tfrac{2}{3} = e^{10000k}, \quad \tfrac{1}{10000} \ln \tfrac{2}{3} = k.$$

   Thus, (∗) can be written

   $$p(h) = 15 \left(\tfrac{2}{3}\right)^{h/10000}.$$

   (a) $p(5000) = 15 \left(\tfrac{2}{3}\right)^{1/2} \cong 12.25 \text{ lb/in.}^2$.

   (b) $p(15000) = 15 \left(\tfrac{2}{3}\right)^{3/2} \cong 8.16 \text{ lb/in.}^2$.

27. From Exercise 26, we have $6000 = 10{,}000e^{-8r}$. Thus

   $$e^{-8r} = \frac{6000}{10{,}000} = \frac{3}{5} \Rightarrow -8r = \ln(3/5) \quad \text{and} \quad r \cong 0.064 \text{ or } r = 6.4\%$$

29. The future value of $\$25{,}000$ at an interest rate r, t years from now is given by $Q(t) = 25{,}000\,e^{rt}$.

   Thus

   (a) For $r = 0.05$: $P(3) = 25{,}000\,e^{(0.05)3} \cong \$29{,}045.86$.

   (b) For $r = 0.08$: $P(3) = 25{,}000\,e^{(0.08)3} \cong \$31{,}781.23$.

   (c) For $r = 0.12$: $P(3) = 25{,}000\,e^{(0.12)3} \cong \$35{,}833.24$.

31. By Exercise 30

$$(*) \qquad\qquad\qquad v(t) = Ce^{-kt}, \quad t \text{ in seconds.}$$

We use the initial conditions

$$v(0) = C = 4 \text{ mph } = \tfrac{1}{900} \text{ mi/sec} \qquad \text{and} \qquad v(60) = 2 = \tfrac{1}{1800} \text{ mi/sec}$$

to determine e^{-k}:

$$\tfrac{1}{1800} = \tfrac{1}{900}e^{-60k}, \qquad e^{60k} = 2, \qquad e^k = 2^{1/60}.$$

Thus, $(*)$ can be written

$$v(t) = \tfrac{1}{900}\, 2^{-t/60}.$$

The distance traveled by the boat is

$$s = \int_0^{60} \frac{1}{900} 2^{-t/60}\, dt = \frac{1}{900}\left[\frac{-60}{\ln 2} 2^{-t/60}\right]_0^{60} = \frac{1}{30\ln 2} \text{ mi} = \frac{176}{\ln 2} \text{ ft} \quad (\text{about } 254 \text{ ft}).$$

33. Let $A(t)$ denote the amount of ^{14}C remaining t years after the organism dies. Then $A(t) = A(0)e^{kt}$ for some constant k. Since the half-life of ^{14}C is 5700 years, we have

$$\frac{1}{2} = e^{5700k} \;\Rightarrow\; k = -\frac{\ln 2}{5700} \cong 0.000122 \text{ and } A(t) = A(0)e^{-0.000122t}$$

If 25% of the original amount of ^{14}C remains after t years, then

$$0.25A(0) = A(0)e^{-0.000122t} \;\Rightarrow\; t = \frac{\ln 0.25}{-0.000122} \cong 11,400 \text{ (years)}$$

35.

$$f'(t) = tf(t)$$

$$f'(t) - tf(t) = 0$$

$$e^{-t^2/2}f'(t) - te^{-t^2/2}f(t) = 0$$

$$\frac{d}{dt}[e^{-t^2/2}f(t)] = 0$$

$$e^{-t^2/2}f(t) = C$$

$$f(t) = Ce^{t^2/2}$$

37.

$$f'(t) = \cos t\, f(t)$$

$$f'(t) - \cos t\, f(t) = 0$$

$$e^{-\sin t}f'(t) - \cos t\, e^{-\sin t}f(t) = 0$$

$$\frac{d}{dt}\left[e^{-\sin t}f(t)\right] = 0$$

$$e^{-\sin t}f(t) = C$$

$$f(t) = Ce^{\sin t}$$

SECTION 7.7

1. (a) 0 (b) $-\pi/3$ **3.** (a) $2\pi/3$ (b) $3\pi/4$ **5.** (a) $1/2$ (b) $\pi/4$

7. (a) does not exist   **9.** (a) $\sqrt{3}/2$ (b) $-7/25$

    (b) does not exist

11. $\dfrac{dy}{dx} = \dfrac{1}{1+(x+1)^2} = \dfrac{1}{x^2+2x+2}$

13. $f'(x) = \dfrac{1}{|2x^2|\sqrt{(2x^2)^2-1}}\dfrac{d}{dx}(2x^2) = \dfrac{2}{x\sqrt{4x^4-1}}$

15. $f'(x) = \arcsin 2x + x\dfrac{1}{\sqrt{1-(2x)^2}}\dfrac{d}{dx}(2x) = \arcsin 2x + \dfrac{2x}{\sqrt{1-4x^2}}$

17. $\dfrac{du}{dx} = 2\,(\arcsin x)\dfrac{d}{dx}(\arcsin x) = \dfrac{2\arcsin x}{\sqrt{1-x^2}}$

19. $\dfrac{dy}{dx} = \dfrac{x\left(\dfrac{1}{1+x^2}\right) - (1)\arctan x}{x^2} = \dfrac{x-(1+x^2)\arctan x}{x^2\,(1+x^2)}$

21. $f'(x) = \dfrac{1}{2}(\arctan 2x)^{-1/2}\dfrac{d}{dx}(\arctan 2x) = \dfrac{1}{2}(\arctan 2x)^{-1/2}\dfrac{2}{1+(2x)^2} = \dfrac{1}{(1+4x^2)\sqrt{\arctan 2x}}$

23. $\dfrac{dy}{dx} = \dfrac{1}{1+(\ln x)^2}\dfrac{d}{dx}(\ln x) = \dfrac{1}{x[1+(\ln x)^2]}$

25. $\dfrac{d\theta}{dr} = \dfrac{1}{\sqrt{1-(\sqrt{1-r^2})^2}}\dfrac{d}{dr}(\sqrt{1-r^2}) = \dfrac{1}{\sqrt{r^2}}\cdot\dfrac{-r}{\sqrt{1-r^2}} = -\dfrac{r}{|r|\sqrt{1-r^2}}$

27. $g'(x) = 2x\,\text{arcsec}\left(\dfrac{1}{x}\right) + x^2\cdot\dfrac{1}{\left|\dfrac{1}{x}\right|\sqrt{\dfrac{1}{x^2}-1}}\cdot\left(-\dfrac{1}{x^2}\right) = 2x\,\sec^{-1}\left(\dfrac{1}{x}\right) - \dfrac{x^2}{\sqrt{1-x^2}}$

29. $\dfrac{dy}{dx} = \cos\,[\text{arcsec}\,(\ln x)]\cdot\dfrac{1}{|\ln x|\sqrt{(\ln x)^2-1}}\cdot\dfrac{1}{x} = \dfrac{\cos\,[\text{arcsec}\,(\ln x)]}{x\,|\ln x|\sqrt{(\ln x)^2-1}}$

31. $f'(x) = \dfrac{-x}{\sqrt{c^2-x^2}} + \dfrac{c}{\sqrt{1-(x/c)^2}}\cdot\left(\dfrac{1}{c}\right) = \dfrac{c-x}{\sqrt{c^2-x^2}} = \sqrt{\dfrac{c-x}{c+x}}$

33. (a) $\sin\,(\arcsin x) = x$ (b) $\cos\,(\arcsin x) = \sqrt{1-x^2}$ (c) $\tan\,(\arcsin x) = \dfrac{x}{\sqrt{1-x^2}}$

      (d) $\cot\,(\arcsin x) = \dfrac{\sqrt{1-x^2}}{x}$ (e) $\sec\,(\arcsin x) = \dfrac{1}{\sqrt{1-x^2}}$ (f) $\csc\,(\arcsin x) = \dfrac{1}{x}$

35. $\left\{\begin{array}{l} au = x+b \\ a\,du = dx \end{array}\right\};\quad \displaystyle\int\dfrac{dx}{\sqrt{a^2-(x+b)^2}} = \int\dfrac{a\,du}{\sqrt{a^2-a^2u^2}} = \int\dfrac{du}{\sqrt{1-u^2}}$

$$= \arcsin u + C = \sin^{-1}\left(\dfrac{x+b}{a}\right) + C$$

37. $\left\{\begin{array}{l} au = x+b \\ a\,du = dx \end{array}\right\};\quad \displaystyle\int\dfrac{dx}{(x+b)\sqrt{(x+b)^2-a^2}} = \int\dfrac{a\,du}{au\sqrt{a^2u^2-a^2}} = \dfrac{1}{a}\int\dfrac{du}{u\sqrt{u^2-1}}$

$$= \dfrac{1}{a}\,\text{arcsec}\,|u| + C = \dfrac{1}{a}\,\text{arcsec}\left(\dfrac{|x+b|}{a}\right) + C$$

39. $\displaystyle\int_0^1\dfrac{dx}{1+x^2} = [\arctan x]_0^1 = \dfrac{\pi}{4}$

41. $\displaystyle\int_0^{1/\sqrt{2}}\dfrac{dx}{\sqrt{1-x^2}} = [\arcsin x]_0^{1/\sqrt{2}} = \dfrac{\pi}{4}$

43. $\displaystyle\int_0^5 \frac{dx}{25+x^2} = \left[\frac{1}{5}\arctan\frac{x}{5}\right]_0^5 = \frac{\pi}{20}$

45. $\left\{\begin{array}{l} 3u = 2x \\ 3\,du = 2\,dx \end{array}\middle| \begin{array}{l} x=0 \implies u=0 \\ x=3/2 \implies u=1 \end{array}\right\}$; $\displaystyle\int_0^{3/2} \frac{dx}{9+4x^2} = \frac{1}{6}\int_0^1 \frac{du}{1+u^2} = \frac{1}{6}[\arctan u]_0^1 = \frac{\pi}{24}$

47. $\left\{\begin{array}{l} u = 4x \\ du = 4\,dx \end{array}\middle| \begin{array}{l} x=3/2 \implies u=6 \\ x=3 \implies u=12 \end{array}\right\}$;

$\displaystyle\int_{3/2}^3 \frac{dx}{x\sqrt{16x^2-9}} = \int_6^{12} \frac{du/4}{(u/4)\sqrt{u^2-9}} = \frac{1}{3}\left[\operatorname{arcsec}\left(\frac{|u|}{3}\right)\right]_6^{12} = \frac{1}{3}\operatorname{arcsec}4 - \frac{\pi}{9}$

49. $\displaystyle\int_{-3}^{-2} \frac{dx}{\sqrt{4-(x+3)^2}} = \left[\arcsin\left(\frac{x+3}{2}\right)\right]_{-3}^{-2} = \frac{\pi}{6}$

51. $\left\{\begin{array}{l} u = e^x \\ du = e^x\,dx \end{array}\middle| \begin{array}{l} x=0 \implies u=1 \\ x=\ln 2 \implies u=2 \end{array}\right\}$;

$\displaystyle\int_0^{\ln 2} \frac{e^x}{1+e^{2x}}\,dx = \int_1^2 \frac{du}{1+u^2} = [\arctan u]_1^2 = \arctan 2 - \frac{\pi}{4} \cong 0.322$

53. $\left\{\begin{array}{l} u = x^2 \\ du = 2x\,dx \end{array}\right\}$; $\displaystyle\int \frac{x}{\sqrt{1-x^4}}\,dx = \frac{1}{2}\int \frac{du}{\sqrt{1-u^2}} = \frac{1}{2}\arcsin u + C = \frac{1}{2}\arcsin x^2 + C$

55. $\left\{\begin{array}{l} u = x^2 \\ du = 2x\,dx \end{array}\right\}$; $\displaystyle\int \frac{x}{1+x^4}\,dx = \frac{1}{2}\int \frac{du}{1+u^2} = \frac{1}{2}\arctan u + C = \frac{1}{2}\arctan x^2 + C$

57. $\left\{\begin{array}{l} u = \tan x \\ du = \sec^2 x\,dx \end{array}\right\}$; $\displaystyle\int \frac{\sec^2 x}{9+\tan^2 x}\,dx = \int \frac{du}{9+u^2} = \frac{1}{3}\arctan\left(\frac{u}{3}\right) + C = \frac{1}{3}\arctan\left(\frac{\tan x}{3}\right) + C$

59. $\left\{\begin{array}{l} u = \arcsin x \\ du = \dfrac{1}{\sqrt{1-x^2}}\,dx \end{array}\right\}$; $\displaystyle\int \frac{\arcsin x}{\sqrt{1-x^2}}\,dx = \int u\,du = \frac{1}{2}u^2 + C = \frac{1}{2}(\arcsin x)^2 + C$

61. $\left\{\begin{array}{l} u = \ln x \\ du = \dfrac{1}{x}\,dx \end{array}\right\}$; $\displaystyle\int \frac{dx}{x\sqrt{1-(\ln x)^2}} = \int \frac{du}{\sqrt{1-u^2}} = \arcsin u + C = \arcsin(\ln x) + C$

63. $\displaystyle A = \int_{-1}^1 \frac{1}{\sqrt{4-x^2}}\,dx = 2\int_0^1 \frac{1}{\sqrt{4-x^2}}\,dx$

$\displaystyle = 2\left[\arcsin\left(\frac{x}{2}\right)\right]_0^1 = \frac{\pi}{3}$

65. $\dfrac{8}{x^2+4} = \dfrac{1}{4}x^2 \quad\Rightarrow\quad x = \pm 2$

$$A = \int_{-2}^{2}\left(\frac{8}{x^2+4} - \frac{1}{4}x^2\right)dx = 2\int_{0}^{2}\left(\frac{8}{x^2+4} - \frac{1}{4}x^2\right)dx$$

$$= 2\left[8\cdot\frac{1}{2}\arctan\left(\frac{x}{2}\right) - \frac{1}{12}x^3\right]_{0}^{2} = 2\pi - \frac{4}{3}$$

67. $V = \displaystyle\int_{0}^{2} 2\pi x\,\frac{1}{\sqrt{4+x^2}}\,dx = \pi\int_{0}^{2}\frac{2x}{\sqrt{4+x^2}}\,dx = 2\pi\left[\sqrt{4+x^2}\right]_{0}^{2} = 4\pi\left(\sqrt{2}-1\right)$

69.   Let x be the distance between the motorist and the point on the road where the line determined by the sign intersects the road. Then, from the given figure,

$$\theta = \arctan\left(\frac{s+k}{x}\right) - \arctan\frac{s}{x}, \quad 0 < x < \infty$$

and

$$\frac{d\theta}{dx} = \frac{1}{1 + \dfrac{(s+k)^2}{x^2}}\left(-\frac{s+k}{x^2}\right) - \frac{1}{1 + \dfrac{s^2}{x^2}}\left(-\frac{s}{x^2}\right)$$

$$= \frac{-(s+k)}{x^2+(s+k)^2} + \frac{s}{x^2+s^2} = \frac{s^2k + sk^2 - kx^2}{[x^2+(s+k)^2]\,[x^2+s^2]}$$

Setting $d\theta/dx = 0$ we get $x = \sqrt{s^2+sk}$. Since θ is essentially 0 when x is close to 0 and when x is "large," we can conclude that θ is a maximum when $x = \sqrt{s^2+sk}$.

71.   (b) $\displaystyle\int_{-a}^{a}\sqrt{a^2-x^2}\,dx = \left[\frac{x}{2}\sqrt{a^2-x^2} + \frac{a^2}{2}\arcsin\left(\frac{x}{a}\right)\right]_{-a}^{a} = \frac{\pi a^2}{2}$

The graph of $f(x) = \sqrt{a^2-x^2}$ on the interval $[-a, a]$ is the upper half of the circle of radius a centered at the origin. Thus, the integral gives the area of the semi-circle: $A = \frac{1}{2}\pi a^2$.

73.   Set $y = \operatorname{arccot} x$. Then $\cot y = x$ and, by the hint, $\tan\left(\frac{1}{2}\pi - y\right) = x$. Therefore

$$\frac{1}{2}\pi - y = \arctan x, \qquad \arctan x + y = \frac{1}{2}\pi, \qquad \arctan x + \operatorname{arccot} x = \frac{1}{2}\pi.$$

75.   The integrand is undefined for $x \geq 1$.

77. $I = \displaystyle\int_{0}^{0.5}\frac{1}{\sqrt{1-x^2}}\,dx \cong \frac{1}{10}\left[f(0.05) + f(0.15) + f(0.25) + f(0.35) + f(0.45)\right] \cong 0.523$;

and $\sin(0.523) \cong 0.499$. Explanation: $I = \arcsin(0.5)$ and $\sin\left[\arcsin(0.5)\right] = 0.5$.

PROJECT 7.7

1. (a)

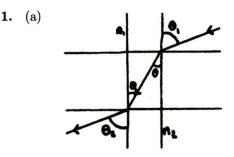

$$n_1 \sin \theta_1 = n \sin \theta = n_2 \sin \theta_2$$

SECTION 7.8

1. $\dfrac{dy}{dx} = \cosh x^2 \dfrac{d}{dx}\left(x^2\right) = 2x \cosh x^2$

3. $\dfrac{dy}{dx} = \dfrac{1}{2}\left(\cosh ax\right)^{-1/2}\left(a \sinh ax\right) = \dfrac{a \sinh ax}{2\sqrt{\cosh ax}}$

5. $\dfrac{dy}{dx} = \dfrac{(\cosh x - 1)(\cosh x) - \sinh x\,(\sinh x)}{(\cosh x - 1)^2} = \dfrac{1}{1 - \cosh x}$

7. $\dfrac{dy}{dx} = ab \cosh bx - ab \sinh ax = ab\,(\cosh bx - \sinh ax)$

9. $\dfrac{dy}{dx} = \dfrac{1}{\sinh ax}\,(a \cosh ax) = a \coth ax$

11. $\dfrac{dy}{dx} = \cosh\left(e^{2x}\right)e^{2x}(2) = 2e^{2x}\cosh\left(e^{2x}\right)$

13. $\dfrac{dy}{dx} = -e^{-x}\cosh 2x + 2e^{-x}\sinh 2x$

15. $\dfrac{dy}{dx} = \dfrac{1}{\cosh x}\,(\sinh x) = \tanh x$

17. $\ln y = x \ln \sinh x$; $\quad \dfrac{1}{y}\dfrac{dy}{dx} = \ln \sinh x + x\,\dfrac{\cosh x}{\sinh x}\quad$ and $\quad \dfrac{dy}{dx} = (\sinh x)^x\left[\ln \sinh x + x \coth x\right]$

19. $\cosh^2 t - \sinh^2 t = \left(\dfrac{e^t + e^{-t}}{2}\right)^2 - \left(\dfrac{e^t - e^{-t}}{2}\right)^2$

$$= \dfrac{1}{4}\left\{\left(e^{2t} + 2 + e^{-2t}\right) - \left(e^{2t} - 2 + e^{-2t}\right)\right\} = \dfrac{4}{4} = 1$$

21. $\cosh t \cosh s + \sinh t \sinh s = \left(\dfrac{e^t + e^{-t}}{2}\right)\left(\dfrac{e^s + e^{-s}}{2}\right) + \left(\dfrac{e^t - e^{-t}}{2}\right)\left(\dfrac{e^s - e^{-s}}{2}\right)$

$$= \dfrac{1}{4}\left\{2e^{t+s} + 2e^{-(t+s)}\right\} = \dfrac{e^{t+s} + e^{-(t+s)}}{2} = \cosh(t + s)$$

23. Set $s = t$ in $\cosh(t + s) = \cosh t \cosh s + \sinh t \sinh s$ to get $\cosh(2t) = \cosh^2 t + \sinh^2 t$.

Then use Exercise 19 to obtain the other two identities.

25. $\sinh(-t) = \dfrac{e^{(-t)} - e^{-(-t)}}{2} = -\dfrac{e^t - e^{-t}}{2} = -\sinh t$

27.

$$y = -5\cosh x + 4\sinh x = -\frac{5}{2}\left(e^x + e^{-x}\right) + \frac{4}{2}\left(e^x - e^{-x}\right) = -\frac{1}{2}e^x - \frac{9}{2}e^{-x}$$

$$\frac{dy}{dx} = -\frac{1}{2}e^x + \frac{9}{2}e^{-x} = \frac{e^{-x}}{2}\left(9 - e^{2x}\right); \quad \frac{dy}{dx} = 0 \implies e^x = 3 \text{ or } x = \ln 3$$

$$\frac{d^2y}{dx^2} = -\frac{1}{2}e^x - \frac{9}{2}e^{-x} < 0 \quad \text{for all} \quad x \quad \text{so abs max occurs at} \quad x = \ln 3.$$

$$\text{The abs max is } y = -\frac{1}{2}e^{\ln 3} - \frac{9}{2}e^{-\ln 3} = -\frac{1}{2}(3) - \frac{9}{2}\left(\frac{1}{3}\right) = -3.$$

29.

$$[\cosh x + \sinh x]^n = \left[\frac{e^x + e^{-x}}{2} + \frac{e^x - e^{-x}}{2}\right]^n$$

$$= [e^x]^n = e^{nx} = \frac{e^{nx} + e^{-nx}}{2} + \frac{e^{nx} - e^{-nx}}{2} = \cosh nx + \sinh nx$$

31.

$$y = A\cosh cx + B\sinh cx; \qquad\qquad y(0) = 2 \implies 2 = A.$$

$$y' = Ac\sinh cx + Bc\cosh cx; \qquad\quad y'(0) = 1 \implies 1 = Bc.$$

$$y'' = Ac^2\cosh cx + Bc^2\sinh cx = c^2 y; \quad y'' - 9y = 0 \implies \left(c^2 - 9\right)y = 0.$$

Thus, $c = 3$, $B = \frac{1}{3}$, and $A = 2$.

33. $\dfrac{1}{a}\sinh ax + C$ **35.** $\dfrac{1}{3a}\sinh^3 ax + C$ **37.** $\dfrac{1}{a}\ln\left(\cosh ax\right) + C$ **39.** $-\dfrac{1}{a\cosh ax} + C$

41. From the identity $\cosh 2t = 2\cosh^2 t - 1$ (Exercise 23), we get $\cosh^2 t = \frac{1}{2}\left(1 + \cosh 2t\right)$. Thus,

$$\int \cosh^2 x\, dx = \frac{1}{2}\int\left(1 + \cosh 2x\right)dx$$

$$= \frac{1}{2}\left(x + \frac{1}{2}\sinh 2x\right) + C$$

$$= \frac{1}{2}\left(x + \sinh x\,\cosh x\right) + C$$

43. $\left\{\begin{array}{l} u = \sqrt{x} \\ du = dx/2\sqrt{x} \end{array}\right\}; \quad \displaystyle\int \frac{\sinh\sqrt{x}}{\sqrt{x}}\, dx = 2\int \sinh u\, du = 2\cosh u + C = 2\cosh\sqrt{x} + C$

45. $A.V. = \dfrac{1}{1 - (-1)}\displaystyle\int_{-1}^{1}\cosh x\, dx = \frac{1}{2}\left[\sinh x\right]_{-1}^{1} = \dfrac{e^2 - 1}{2e} \cong 1.175$

47. $A = \displaystyle\int_{0}^{\ln 10}\sinh x\, dx = \left[\cosh x\right]_{0}^{\ln 10} = \dfrac{e^{\ln 10} + e^{-\ln 10}}{2} - 1 = \dfrac{81}{20}$

49. $V = \int_0^1 \pi \left(\cosh^2 x - \sinh^2 x\right) dx = \int_0^1 \pi \, dx = \pi$

51. $V = \int_{-\ln 5}^{\ln 5} \pi [\cosh 2x]^2 \, dx = 2\pi \int_0^{\ln 5} \cosh^2 2x \, dx$

$$= \tfrac{1}{2}\pi \int_0^{\ln 5} \left(e^{4x} + 2 + e^{-4x}\right) dx$$

$$= \tfrac{1}{2}\pi \left[\tfrac{1}{4}e^{4x} + 2x + \tfrac{1}{4}e^{-4x}\right]_0^{\ln 5} = \pi \left[\tfrac{1}{4}\sinh(4\ln 5) + \ln 5\right]$$

53. (a) $(0.69315, 1.25)$ (b) $A \cong 0.38629$

SECTION 7.9

1. $\dfrac{dy}{dx} = 2 \tanh x \, \text{sech}^2 x$

3. $\dfrac{dy}{dx} = \dfrac{1}{\tanh x} \text{sech}^2 x = \text{sech}\, x \, \text{csch}\, x$

5. $\dfrac{dy}{dx} = \cosh\left(\arctan e^{2x}\right) \dfrac{d}{dx}\left(\arctan e^{2x}\right) = \dfrac{2e^{2x}\cosh\left(\arctan e^{2x}\right)}{1 + e^{4x}}$

7. $\dfrac{dy}{dx} = -\text{csch}^2\left(\sqrt{x^2 + 1}\,\right) \dfrac{d}{dx}\left(\sqrt{x^2 + 1}\,\right) = -\dfrac{x}{\sqrt{x^2 + 1}} \text{csch}^2\left(\sqrt{x^2 + 1}\,\right)$

9. $\dfrac{dy}{dx} = \dfrac{(1 + \cosh x)\left(-\text{sech}\, x \tanh x\right) - \text{sech}\, x\, (\sinh x)}{(1 + \cosh x)^2}$

$$= \dfrac{-\text{sech}\, x\, (\tanh x + \cosh x \tanh x + \sinh x)}{(1 + \cosh x)^2} = \dfrac{-\text{sech}\, x\, (\tanh x + 2\sinh x)}{(1 + \cosh x)^2}$$

11. $\dfrac{d}{dx}(\coth x) = \dfrac{d}{dx}\left[\dfrac{\cosh x}{\sinh x}\right] = \dfrac{\sinh x\, (\sinh x) - \cosh x\, (\cosh x)}{\sinh^2 x}$

$$= -\dfrac{\cosh^2 x - \sinh^2 x}{\sinh^2 x} = \dfrac{-1}{\sinh^2 x} = -\text{csch}^2 x$$

13. $\dfrac{d}{dx}(\text{csch}\, x) = \dfrac{d}{dx}\left[\dfrac{1}{\sinh x}\right] = -\dfrac{\cosh x}{\sinh^2 x} = -\text{csch}\, x \coth x$

15. (a) By the hint $\text{sech}^2 x_0 = \dfrac{9}{25}$. Take $\text{sech}\, x_0 = \dfrac{3}{5}$ since $\text{sech}\, x = \dfrac{1}{\cosh x} > 0$ for all x.

   (b) $\cosh x_0 = \dfrac{1}{\text{sech}\, x_0} = \dfrac{5}{3}$ (c) $\sinh x_0 = \cosh x_0 \tanh x_0 = \left(\dfrac{5}{3}\right)\left(\dfrac{4}{5}\right) = \dfrac{4}{3}$

   (d) $\coth x_0 = \dfrac{\cosh x_0}{\sinh x_0} = \dfrac{5/3}{4/3} = \dfrac{5}{4}$ (e) $\text{csch}\, x_0 = \dfrac{1}{\sinh x_0} = \dfrac{3}{4}$

17. If $x \leq 0$, the result is obvious. Suppose then that $x > 0$. Since $x^2 \geq 1$, we have $x \geq 1$. Consequently

$$x - 1 = \sqrt{x-1}\,\sqrt{x-1} \leq \sqrt{x-1}\,\sqrt{x+1} = \sqrt{x^2 - 1}$$

and therefore

$$x - \sqrt{x^2 - 1} \leq 1.$$

19. By Theorem 7.9.2,

$$\frac{d}{dx}\left(\sinh^{-1}x\right) = \frac{d}{dx}\left[\ln\left(x + \sqrt{x^2+1}\,\right)\right] = \frac{1}{x + \sqrt{x^2+1}}\left(1 + \frac{x}{\sqrt{x^2+1}}\right) = \frac{1}{\sqrt{x^2+1}}.$$

21. By Theorem 7.9.2

$$\frac{d}{dx}\left(\operatorname{arctan}x\right) = \frac{d}{dx}\left[\frac{1}{2}\ln\left(\frac{1+x}{1-x}\right)\right] = \frac{1}{2}\frac{1}{\left(\dfrac{1+x}{1-x}\right)}\left(\frac{(1-x)(1) - (1+x)(-1)}{(1-x)^2}\right)$$

$$= \frac{1}{\left(\dfrac{1+x}{1-x}\right)(1-x)^2} = \frac{1}{1-x^2}.$$

23. Let $y = \operatorname{csch}^{-1}x$. Then $\operatorname{csch} y = x$ and $\sinh y = \dfrac{1}{x}$.

$$\sinh y = \frac{1}{x}$$

$$\cosh y\,\frac{dy}{dx} = -\frac{1}{x^2}$$

$$\frac{dy}{dx} = -\frac{1}{x^2\cosh y} = -\frac{1}{x^2\sqrt{1+(1/x)^2}} = -\frac{1}{|x|\sqrt{1+x^2}}$$

25.

(a) $\dfrac{dy}{dx} = -\operatorname{sech}x\,\tanh x = -\dfrac{\sinh x}{\cosh^2 x}$

$\dfrac{dy}{dx} = 0$ at $x = 0$;

$\dfrac{dy}{dx} > 0$ if $x < 0$; $\dfrac{dy}{dx} < 0$ if $x > 0$

f is increasing on $(-\infty, 0]$ and decreasing on $[0, \infty)$; $f(0) = 1$ is the absolute maximum of f.

(b) $\dfrac{d^2y}{dx^2} = -\dfrac{\cosh^2 x - 2\sinh^2 x}{\cosh^3 x} = \dfrac{\sinh^2 x - 1}{\cosh^3 x}$

$\dfrac{d^2y}{dx^2} = 0 \implies \sinh x = \pm 1$

$\sinh x = 1 \implies \dfrac{e^x - e^{-x}}{2} = 1 \implies e^{2x} - 2e^x - 1 = 0 \implies x = \ln\left(1 + \sqrt{2}\right) \cong 0.881$;

$\sinh x = -1 \implies \dfrac{e^x - e^{-x}}{2} = -1 \implies e^{2x} + 2e^x - 1 = 0 \implies x = -\ln\left(1 + \sqrt{2}\right) = -0.881$

(c) The graph of f is concave up on $(-\infty, -0.881) \cup (0.881, \infty)$ and concave down on $(-0.881, 0.881)$; points of inflection at $x = \pm 0.881$

27. $y = \sinh x; \quad \dfrac{dy}{dx} = \cosh x; \quad \dfrac{d^2y}{dx^2} = \sinh x.$

$\dfrac{d^2y}{dx^2} = 0 \;\Rightarrow\; \sinh x = 0 \;\Rightarrow\; x = 0.$

$y = \sinh^{-1} x = \ln\left(x + \sqrt{x^2 + 1}\right); \quad \dfrac{dy}{dx} = \dfrac{1}{\sqrt{x^2+1}}; \quad \dfrac{d^2y}{dx^2} = -\dfrac{x}{(x^2+1)^{3/2}}.$

$\dfrac{d^2y}{dx^2} = 0 \;\Rightarrow\; -\dfrac{x}{(x^2+1)^{3/2}} = 0 \;\Rightarrow\; x = 0.$

It is easy to verify that $(0,0)$ is a point of inflection for both graphs.

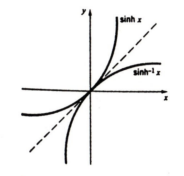

29. (a) $\tan\phi = \sinh x$

  (b) $\sinh x = \tan\phi$

$\phi = \arctan(\sinh x)$

$\dfrac{d\phi}{dx} = \dfrac{\cosh x}{1 + \sinh^2 x}$

$= \dfrac{\cosh x}{\cosh^2 x} = \dfrac{1}{\cosh x} = \operatorname{sech} x$

$x = \sinh^{-1}(\tan\phi)$

$= \ln\left(\tan\phi + \sqrt{\tan^2\phi + 1}\,\right)$

$= \ln(\tan\phi + \sec\phi)$

$= \ln(\sec\phi + \tan\phi)$

  (c) $x = \ln(\sec\phi + \tan\phi)$

$\dfrac{dx}{d\phi} = \dfrac{\sec\phi\tan\phi + \sec^2\phi}{\tan\phi + \sec\phi} = \sec\phi$

31. $\displaystyle\int \tanh x\,dx = \int \dfrac{\sinh x}{\cosh x}\,dx$

$\left\{\begin{array}{l} u = \cosh x \\ du = \sinh x\,dx \end{array}\right\}; \quad \displaystyle\int \dfrac{\sinh x}{\cosh x}\,dx = \int \dfrac{1}{u}\,du = \ln|u| + C = \ln\cosh x + C$

33. $\displaystyle\int \operatorname{sech} x\,dx = \int \dfrac{1}{\cosh x}\,dx = \int \dfrac{2}{e^x + e^{-x}}\,dx = \int \dfrac{2e^x}{e^{2x}+1}\,dx$

$\left\{\begin{array}{l} u = e^x \\ du = e^x\,dx \end{array}\right\}; \quad \displaystyle\int \dfrac{2e^x}{e^{2x}+1}\,dx = 2\int \dfrac{1}{u^2+1}\,du = 2\arctan u + C = 2\arctan\left(e^x\right) + C$

35. $\begin{cases} u = \operatorname{sech} x \\ du = -\operatorname{sech} s \tanh x \, dx \end{cases}$; $\displaystyle \int \operatorname{sech}^3 x \tanh x \, dx = -\int u^2 \, du = -\frac{1}{3} u^3 + C$

$$= -\tfrac{1}{3} \operatorname{sech}^3 x + C$$

37. $\begin{cases} u = \ln(\cosh x) \\ du = \tanh x \, dx \end{cases}$; $\displaystyle \int \tanh x \ln(\cosh x) \, dx = \int u \, du = \frac{1}{2} u^2 + C = \frac{1}{2} \left[\ln(\cosh x)\right]^2 + C$

39. $\begin{cases} u = 1 + \tanh x \\ du = \operatorname{sech}^2 x \, dx \end{cases}$; $\displaystyle \int \frac{\operatorname{sech}^2 x}{1 + \tanh x} \, dx = \int \frac{1}{u} \, du = \ln|u| + C = \ln|1 + \tanh x| + C$

41. $\begin{cases} x = a \sinh u \\ dx = a \cosh u \, du \end{cases}$; $\displaystyle \int \frac{dx}{\sqrt{a^2 + x^2}} \, dx = \int \frac{a \cosh u}{\sqrt{a^2 + a^2 \sinh^2 u}} \, du$

$$= \int du = u + C = \sinh^{-1}\left(\frac{x}{a}\right) + C$$

43. Suppose $|x| < a$.

$\begin{cases} x = a \tanh u \\ dx = a \operatorname{sech}^2 u \, du \end{cases}$; $\displaystyle \int \frac{dx}{a^2 - x^2} \, dx = \int \frac{a \operatorname{sech}^2 u}{a^2 - a^2 \tanh^2 u} \, du$

$$= \frac{1}{a} \int du = \frac{u}{a} + C = \frac{1}{a} \tanh^{-1}\left(\frac{x}{a}\right) + C$$

The other case is done in the same way.

REVIEW EXERCISES

1. $f'(x) = \frac{1}{3} x^{-2/3} > 0$ except at $x = 0$; f is increasing, it is one-to-one; $f^{-1}(x) = (x - 2)^3$.

3. Suppose $f(x_1) = f(x_2)$. Then

$$\frac{x_1 + 1}{x_1 - 1} = \frac{x_2 + 1}{x_2 - 1}$$
$$x_1 x_2 + x_2 - x_1 - 1 = x_1 x_2 + x_1 - x_2 - 1$$
$$2x_2 = 2x_1$$
$$x_1 = x_2$$

Thus f is one-to-one. $f^{-1}(x) = \dfrac{x + 1}{x - 1}$.

5. $f'(x) = -\dfrac{1}{x^2} e^{\frac{1}{x}} < 0$ except at $x = 0$; f is decreasing, it is one-to-one; $f^{-1}(x) = \frac{1}{\ln x}$.

7. f is not one-to-one. Reason: $f'(x) = 1 + \ln x \Longrightarrow f$ decreases on $(0, 1/e]$ and increases on $[1/e, \infty)$. There exist horizontal lines that intersect the graph in two points.

9. $f'(x) = \dfrac{-e^x}{(1 + e^x)^2} < 0$ for all x; f is one to one and has an inverse function.

Since $f(0) = \frac{1}{2}$, $(f^{-1})'\left(\frac{1}{2}\right) = \dfrac{1}{f'(0)} = -4$

11. $f'(x) = \sqrt{4 + x^2} > 0 \implies f$ has an inverse.

Since $f(0) = 0$, $(f^{-1})'(0) = \dfrac{1}{f'(0)} = \dfrac{1}{2}$

13. $f'(x) = 3\left(\ln x^2\right)^2 \dfrac{1}{x^2}(2x) = \dfrac{6\left(\ln x^2\right)^2}{x} = \dfrac{24\left(\ln x\right)^2}{x}$

15. $g'(x) = \dfrac{\left(1 + e^{2x}\right)e^x - e^x\left(2e^{2x}\right)}{\left(1 + e^{2x}\right)^2} = \dfrac{e^x(1 - e^{2x})}{(1 + e^{2x})^2}$

17. $\dfrac{dy}{dx} = \dfrac{1}{x^3 + 3^x}\left(3x^2 + 3^x \ln 3\right) = \dfrac{3x^2 + 3^x \ln 3}{x^3 + 3^x}$

19. $\ln f(x) = \dfrac{1}{x}\ln \cosh x = \dfrac{\ln \cosh x}{x}$

$$\dfrac{f'(x)}{f(x)} = \dfrac{x\dfrac{\sinh x}{\cosh x} - \ln \cosh x}{x^2} = \dfrac{\sinh x}{x\cosh x} - \dfrac{\ln \cosh x}{x^2}.$$

Therefore $f'(x) = (\cosh x)^{1/x}\left[\dfrac{\sinh x}{x\cosh x} - \dfrac{\ln \cosh x}{x^2}\right]$

21. $f(x) = \dfrac{1}{\ln 3}\ln\dfrac{1 + x}{1 - x}$; $f'(x) = \dfrac{1}{\ln 3}\left(\dfrac{1 - x}{1 + x}\right)\left(\dfrac{2}{(1 - x)^2}\right) = \dfrac{2}{\ln 3(1 - x^2)}$

23. Substitution: Set $u = e^x$, $du = e^x\, dx$.

$$\int \dfrac{e^x}{\sqrt{1 - e^2x}}\, dx = \int \dfrac{1}{\sqrt{1 - u^2}}\, dx = \arcsin u + C = \arcsin e^x + C$$

25. Substitution: let $u = \sin x$, $du = \cos x\, dx$

$$\int \dfrac{\cos x}{4 + \sin^2 x}\, dx = \int \dfrac{1}{4 + u^2}\, du = \dfrac{1}{2}\arctan\left(\dfrac{u}{2}\right) + C = \dfrac{1}{2}\arctan\left(\dfrac{\sin x}{2}\right) + C$$

27. Substitution: let $u = \sqrt{x}$, $du = \dfrac{1}{2}\dfrac{1}{\sqrt{x}}\, dx$

$$\int \dfrac{\sec\sqrt{x}}{\sqrt{x}}\, dx = 2\int \sec u\, du = 2\ln|\sec u + \tan u| + C = 2\ln|\sec\sqrt{x} + \tan\sqrt{x}| + C$$

29. Substitution: let $u = \ln x$, $du = \dfrac{1}{x}\, dx$

$$\int \dfrac{5^{\ln x}}{x}\, dx = \dfrac{5^{\ln x}}{\ln 5} + C$$

31. Substitution: $u = x^{4/3} + 1$, $du = \frac{4}{3}x^{1/3}\, dx$, $u(1) = 2$, $u(8) = 17$

$$\int_1^8 \dfrac{x^{1/3}}{x^{4/3} + 1}\, dx = \dfrac{3}{4}\int_2^{17} \dfrac{1}{u}\, du = \dfrac{3}{4}[\ln u]_2^{17} = \dfrac{3}{4}\left(\ln 17 - \ln 2\right) = \dfrac{3}{4}\ln\left(\dfrac{17}{2}\right)$$

33. Substitution: $u = 2^x$, $du = 2^x \ln 2\, dx$

$$\int 2^x \sinh 2^x\, dx = \frac{1}{\ln 2} \int \sinh u\, du = \frac{1}{\ln 2} \cosh u + C = \frac{1}{\ln 2} \cosh 2^x + C$$

35. $\displaystyle \int_2^5 \frac{1}{x^2 - 4x + 13}\, dx = \int_2^5 \frac{1}{(x-2)^2 + 3^2}\, dx = \frac{1}{3}\left[\arctan \frac{x-2}{3}\right]_2^5 = \frac{\pi}{12}$

37. $\displaystyle \int_0^2 \operatorname{sech}^2\left(\frac{x}{2}\right) dx = 2\left[\tanh \frac{x}{2}\right]_0^2 = 2\tanh 1 = 2\frac{e^2 - 1}{e^2 + 1}$

39. $\displaystyle A = \int_0^1 \frac{x}{1 + x^2}\, dx = \frac{1}{2}\left[\ln(x^2 + 1)\right]_0^1 = \frac{\ln 2}{2}$

41. $\displaystyle A = \int_0^{1/2} \frac{1}{\sqrt{1 - x^2}}\, dx = [\arcsin x]_0^{1/2} = \frac{\pi}{6}$

43. Fix $x > 0$. Then, by the mean-value theorem,

$$\ln(1 + x) - \ln 1 = \frac{1}{1 + c}(x - 0) = \frac{x}{1 + c} \quad \text{for some } c \in (0, x)$$

Since $\quad \dfrac{x}{1 + x} < \dfrac{x}{1 + c} < x, \quad$ it follows that

$$\frac{x}{1 + x} < \ln(1 + x) < x$$

Now fix $x \in (-1, 0)$. Then,

$$\ln 1 - \ln(1 + x) = \frac{1}{1 + c}(0 - x) = \frac{-x}{1 + c} \quad \text{for some } c \in (x, 0)$$

Since $\quad -x < \dfrac{-x}{1 + c} < \dfrac{-x}{1 + x}, \quad$ it follows that

$$\frac{x}{1 + x} < \ln(1 + x) < x$$

Thus $\quad \dfrac{x}{1 + x} < \ln(1 + x) < x \quad$ for all $x > -1$ (the result is obvious if $x = 0$).

(b) The result follows from part (a) and the pinching theorem (Theorem 2.5.1)

45. Assume $a > 0$: $\displaystyle A = \int_a^{2a} \frac{a^2}{x}\, dx = \left[a^2 \ln x\right]_a^{2a} = a^2(\ln 2a - \ln a) = a^2 \ln 2$

47. (a) $\displaystyle V = \int_0^{\sqrt 3} \pi \frac{1}{1 + x^2}\, dx = \pi[\arctan x]_0^{\sqrt 3} = \frac{\pi^2}{3}$

    (b) $\displaystyle V = \int_0^{\sqrt 3} 2\pi x(1 + x^2)^{-1/2}\, dx = 2\pi\left[\sqrt{1 + x^2}\right]_0^{\sqrt 3} = 2\pi$

49. $f(x) = \dfrac{\ln x}{x}$, domain: $(0, \infty)$

$f'(x) = \dfrac{1 - \ln x}{x^2}$

critical pt. $x = e$

$f''(x) = \dfrac{2 \ln x - 3}{x^3}$

$f'(x) > 0$ on $(0, e)$,

$f'(x) < 0$ on (e, ∞)

$f''(x) > 0$ on $\left(e^{3/2}, \infty\right)$

$f''(x) < 0$ on $\left(0, e^{3/2}\right)$

51. $\displaystyle\int_0^1 \dfrac{b}{\sqrt{1 - b^2 x^2}}\, dx$: substitution: $u = bx$, $du = b\, dx$, $u(0) = 0$, $u(1) = b$

$\displaystyle\int_0^1 \dfrac{b}{\sqrt{1 - b^2 x^2}}\, dx = \int_0^b \dfrac{1}{\sqrt{1 - u^2}}\, du = [\arcsin u]_0^b = \arcsin b.$

$\displaystyle\int_0^a \dfrac{1}{\sqrt{1 - x^2}}\, dx = \arcsin a; \Longrightarrow a = b$

53. Let 6 a.m. correspond to $t = 0$; measure time in hours.

$$P(t) = 20e^{kt}; \quad P(2) = 40 = 20e^{2k} \quad \Longrightarrow \quad k = \dfrac{\ln 2}{2} \quad \Longrightarrow \quad P(t) = 20e^{\frac{t}{2}\ln 2} = 20\,(2)^{t/2}$$

(a) $P(6) = 20(2)^3 = 20(8) = 160$ grams

(b) $20(2)^{t/2} = 200 \quad \Longrightarrow \quad \dfrac{t}{2}\ln 2 = \ln 10 \quad \Longrightarrow \quad t = \dfrac{2\ln 10}{\ln 2} \cong 6.64$ hours

55. (a) $A(t) = 100e^{kt}$. From the relation $k = \dfrac{-\ln 2}{T}$, where T is the half-life, $A(t) = 100e^{-\frac{t}{140}\ln 2}$

(b) $100e^{-\frac{t}{140}\ln 2} = 75 \quad \Longrightarrow \quad \dfrac{-t}{140}\ln 2 = \ln 0.75 \quad \Longrightarrow \quad t = \dfrac{-140\ln 0.75}{\ln 2} \cong 58.11$ days

57. $P(t) = P_0 e^{kt}$. Since the doubling time is 2 years, $k = \frac{1}{2}\ln 2$ and $P(t) = P_0 e^{\frac{t}{2}\ln 2} = P_0(2)^{t/2}$.

(a) $40,000 = P(4) = P_0(2)^2 \quad \Longrightarrow \quad 4P_0 = 25,000 \Longrightarrow P_0 = 6250$.

(b) $P(t) = 6250(2)^{t/2} = 40,000 \quad \Longrightarrow \quad (2)^{t/2} = 6.4 \quad \Longrightarrow \quad t = \dfrac{2\ln 6.4}{\ln 2} \cong 5.36$; it will take

approximately 5.36 years for the population to reach 40,000.

CHAPTER 8

SECTION 8.1

1. $\displaystyle\int e^{2-x}dx = -e^{2-x} + C$

3. $\displaystyle\int_0^1 \sin \pi x\, dx = \left[-\frac{1}{\pi}\cos \pi x\right]_0^1 = \frac{2}{\pi}$

5. $\displaystyle\int \sec^2(1-x)\,dx = -\tan(1-x) + C$

7. $\displaystyle\int_{\pi/6}^{\pi/3} \cot x\, dx = \left[\ln(\sin x)\right]_{\pi/6}^{\pi/3} = \ln\frac{\sqrt{3}}{2} - \ln\frac{1}{2} = \frac{1}{2}\ln 3$

9. $\left\{\begin{array}{l} u = 1 - x^2 \\ du = -2x\, dx \end{array}\right\};\quad \displaystyle\int \frac{x\, dx}{\sqrt{1-x^2}} = -\frac{1}{2}\int u^{-1/2}\, du = -u^{1/2} + C = -\sqrt{1-x^2} + C$

11. $\displaystyle\int_{-\pi/4}^{\pi/4} \frac{\sin x}{\cos^2 x}\, dx = \int_{-\pi/4}^{\pi/4} \sec x \tan x\, dx = \left[\sec x\right]_{-\pi/4}^{\pi/4} = 0$

13. $\left\{\begin{array}{l|ll} u = 1/x & x = 1 & \Rightarrow \quad u = 1 \\ du = dx/x^2 & x = 2 & \Rightarrow \quad u = 1/2 \end{array}\right\};$

$\displaystyle\int_1^2 \frac{e^{1/x}}{x^2}\, dx = \int_1^{1/2} -e^u\, du = \left[-e^u\right]_1^{1/2} = e - \sqrt{e}$

15. $\displaystyle\int_0^c \frac{dx}{x^2 + c^2} = \left[\frac{1}{c}\arctan\left(\frac{x}{c}\right)\right]_0^c = \frac{\pi}{4c}$

17. $\left\{\begin{array}{l} u = 3\tan\theta + 1 \\ du = 3\sec^2\theta\, d\theta \end{array}\right\};$

$\displaystyle\int \frac{\sec^2\theta}{\sqrt{3\tan\theta+1}}\, d\theta = \frac{1}{3}\int u^{-1/2}\, du = \frac{2}{3}u^{1/2} + C = \frac{2}{3}\sqrt{3\tan\theta+1} + C$

19. $\displaystyle\int \frac{e^x}{ae^x - b}\, dx = \frac{1}{a}\ln|ae^x - b| + C$

21. $\left\{\begin{array}{l} u = x + 1 \\ du = dx \end{array}\right\};$

$\displaystyle\int \frac{x}{(x+1)^2 + 4}\, dx = \int \frac{u-1}{u^2+4}\, du = \int \frac{u}{u^2+4}\, du - \int \frac{du}{u^2+4}$

$= \frac{1}{2}\ln|u^2 + 4| - \frac{1}{2}\arctan\frac{u}{2} + C$

$= \frac{1}{2}\ln|(x+1)^2 + 4| - \frac{1}{2}\arctan\left(\frac{x+1}{2}\right) + C$

23. $\left\{\begin{array}{l} u = x^2 \\ du = 2x\,dx \end{array}\right\};$ $\quad \int \dfrac{x}{\sqrt{1-x^4}}\,dx = \dfrac{1}{2}\int \dfrac{du}{\sqrt{1-u^2}} = \dfrac{1}{2}\arcsin u + C = \tfrac{1}{2}\arcsin\left(x^2\right) + C$

25. $\left\{\begin{array}{l} u = x + 3 \\ du = dx \end{array}\right\};\int \dfrac{dx}{x^2+6x+10} = \int \dfrac{dx}{(x+3)^2+1} = \int \dfrac{du}{u^2+1} = \arctan u + C = \arctan\left(x+3\right) + C$

27. $\displaystyle\int x \sin x^2\,dx = -\dfrac{1}{2}\cos x^2 + C$

29. $\displaystyle\int \tan^2 x\,dx = \int (\sec^2 x - 1)\,dx = \tan x - x + C$

31. $\left\{\begin{array}{l l} u = \ln x & x = 1 \quad \Rightarrow \quad u = 0 \\ du = dx/x & x = e \quad \Rightarrow \quad u = 1 \end{array}\right\};$

$\displaystyle\int_1^e \dfrac{\ln x^3}{x}\,dx = \int_1^e \dfrac{3\ln x}{x}\,dx = 3\int_0^1 u\,du = 3\left[\dfrac{u^2}{2}\right]_0^1 = \dfrac{3}{2}$

33. $\left\{\begin{array}{l} u = \arcsin x \\ du = \dfrac{dx}{\sqrt{1-x^2}} \end{array}\right\};$ $\quad \int \dfrac{\arcsin x}{\sqrt{1-x^2}}\,dx = \int u\,du = \dfrac{1}{2}u^2 + C = \dfrac{1}{2}\left(\arcsin x\right)^2 + C$

35. $\left\{\begin{array}{l} u = \ln x \\ du = dx/x \end{array}\right\};$ $\quad \int \dfrac{1}{x\ln x}\,dx = \int \dfrac{1}{u}\,du = \ln|u| + C = \ln|\ln x| + C$

37. $\left\{\begin{array}{l l} u = \cos x & x = 0 \quad \Rightarrow \quad u = 1 \\ du = -\sin x\,dx & x = \pi/4 \quad \Rightarrow \quad u = \sqrt{2}/2 \end{array}\right\};$

$\displaystyle\int_0^{\pi/4} \dfrac{1+\sin x}{\cos^2 x}\,dx = \int_0^{\pi/4} \sec^2 x\,dx + \int_0^{\pi/4} \dfrac{\sin x}{\cos^2 x}\,dx$

$\displaystyle= \Big[\tan x\Big]_0^{\pi/4} - \int_1^{\sqrt{2}/2} \dfrac{du}{u^2} = 1 + \left[\dfrac{1}{u}\right]_1^{\sqrt{2}/2} = \sqrt{2}$

39. (formula 99) $\displaystyle\int \sqrt{x^2-4}\,dx = \dfrac{x}{2}\sqrt{x^2-4} - 2\ln\left|x + \sqrt{x^2-4}\right| + C$

41. (formula 18) $\displaystyle\int \cos^3 2t\,dt = \dfrac{1}{2}[\sin 2t - \dfrac{1}{3}\sin^3 2t] + C$

43. (formula 108) $\displaystyle\int \dfrac{1}{x(2x+3)}\,dx = \dfrac{1}{3}\ln\left|\dfrac{x}{2x+3}\right| + C$

45. (formula 81) $\displaystyle\int \dfrac{\sqrt{x^2+9}}{x^2}\,dx = -\dfrac{\sqrt{x^2+9}}{x} + \ln\left|x + \sqrt{x^2+9}\right| + C$

47. (formula 11) $\displaystyle\int x^3 \ln x\,dx = x^4\left(\dfrac{\ln x}{4} - \dfrac{1}{16}\right) + C$

49.
$$\int_0^\pi \sqrt{1 + \cos x}\, dx = \int_0^\pi \sqrt{2\cos^2\left(\frac{x}{2}\right)}\, dx$$

$$= \sqrt{2}\int_0^\pi \cos\left(\frac{x}{2}\right)\, dx \qquad \left[\cos\left(\frac{x}{2}\right) \geq 0 \text{ on } [0, \pi]\right]$$

$$= 2\sqrt{2}\left[\sin\left(\frac{x}{2}\right)\right]_0^\pi = 2\sqrt{2}$$

51. (a) $\displaystyle\int_0^\pi \sin^2 nx\, dx = \int_0^\pi \left[\frac{1}{2} - \frac{\cos 2nx}{2}\right]\, dx = \left[\frac{x}{2} - \frac{\sin 2nx}{4n}\right]_0^\pi = \frac{\pi}{2}$

(b) $\displaystyle\int_0^\pi \sin nx \cos nx\, dx = \frac{1}{2}\int_0^\pi \sin 2nx\, dx = -\left[\frac{\cos 2nx}{4n}\right]_0^\pi = 0$

(c) $\displaystyle\int_0^{\pi/n} \sin nx \cos nx\, dx = \frac{1}{2}\int_0^{\pi/n} \sin 2nx\, dx = -\left[\frac{\cos 2nx}{4n}\right]_0^{\pi/n} = 0$

53. (a) $\displaystyle\int \tan^3 x\, dx = \int \tan^2 x \tan x\, dx = \int \left(\sec^2 x - 1\right)\tan x\, dx$

$$= \int \sec^2 x \tan x\, dx - \int \tan x\, dx$$

$$= \int u\, du - \int \tan x\, dx \quad (u = \tan x, \ du = \sec^2 x\, dx)$$

$$= \frac{1}{2}u^2 - \ln|\sec x| + C = \frac{1}{2}\tan^2 x - \ln|\sec x| + C$$

(b) $\displaystyle\int \tan^5 x\, dx = \int \tan^3 x \tan^2 x\, dx = \int \tan^3 x \left(\sec^2 x - 1\right)\, dx$

$$= \int \tan^3 x \sec^2 x\, dx - \int \tan^3 x\, dx$$

$$= \int u^3\, du - \int \tan^3 x\, dx \quad (u = \tan x \ du = \sec^2 x\, dx)$$

$$= \frac{1}{4}u^4 - \frac{1}{2}\tan^2 x + \ln|\sec x| + C$$

$$= \frac{1}{4}\tan^4 x - \frac{1}{2}\tan^2 x + \ln|\sec x| + C$$

(c) $\displaystyle\int \tan^7 x\, dx = \int \tan^5 x \sec^2 x\, dx - \int \tan^5 x\, dx$

$$= \frac{1}{6}\tan^6 x - \frac{1}{4}\tan^4 x + \frac{1}{2}\tan^2 x - \ln|\sec x| + C$$

(d) $\displaystyle\int \tan^{2k+1} x\, dx = \int \tan^{2k-1} x \tan^2 x\, dx = \int \tan^{2k-1} x \sec^2 x\, dx - \int \tan^{2k-1} x\, dx$

$$= \frac{1}{2k}\tan^{2k} x - \int \tan^{2k-1} x\, dx$$

55. (a)

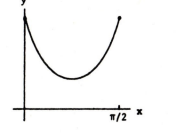

(b) $\sin x + \cos x = \sqrt{2}\left[\sin x \cos(\pi/4) + \cos x \sin(\pi/4)\right]$

$$= \sqrt{2}\,\sin\left(x + \frac{\pi}{4}\right);$$

$$A = \sqrt{2}, \quad B = \pi/4$$

(c) $\text{Area} = \displaystyle\int_0^{\pi/2} \frac{1}{\sin x + \cos x}\,dx = \frac{1}{\sqrt{2}}\int_0^{\pi/2} \frac{1}{\sin\left(x + \frac{\pi}{4}\right)}\,dx$

$$\begin{Bmatrix} u = x + \pi/4 & x = 0 & \Rightarrow & u = \pi/4 \\ du = dx & x = \pi/2 & \Rightarrow & u = 3\pi/4 \end{Bmatrix};$$

$$\frac{1}{\sqrt{2}}\int_0^{\pi/2} \frac{1}{\sin\left(x + \frac{\pi}{4}\right)}\,dx = \frac{\sqrt{2}}{2}\int_{\pi/4}^{3\pi/4} \frac{1}{\sin u}\,du$$

$$= \frac{\sqrt{2}}{2}\int_{\pi/4}^{3\pi/4} \csc u\,du$$

$$= \frac{\sqrt{2}}{2}\left[\ln\left|\csc u - \cot u\right|\right]_{\pi/4}^{3\pi/4}$$

$$= \frac{\sqrt{2}}{2}\ln\left[\frac{\sqrt{2}+1}{\sqrt{2}-1}\right]$$

57. (a)

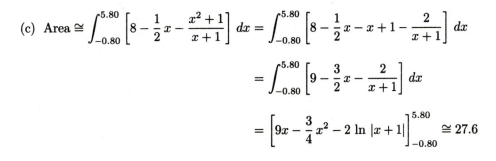

(b) $x_1 \cong -0.80, \quad x_2 \cong 5.80$

(c) $\text{Area} \cong \displaystyle\int_{-0.80}^{5.80}\left[8 - \frac{1}{2}x - \frac{x^2+1}{x+1}\right]dx = \int_{-0.80}^{5.80}\left[8 - \frac{1}{2}x - x + 1 - \frac{2}{x+1}\right]dx$

$$= \int_{-0.80}^{5.80}\left[9 - \frac{3}{2}x - \frac{2}{x+1}\right]dx$$

$$= \left[9x - \frac{3}{4}x^2 - 2\ln|x+1|\right]_{-0.80}^{5.80} \cong 27.6$$

SECTION 8.2

1.
$$\begin{array}{ll} u = x, & dv = e^{-x}\,dx \\ du = dx, & v = -e^{-x} \end{array}$$
$$\int xe^{-x}\,dx = -xe^{-x} - \int -e^{-x}\,dx = -xe^{-x} - e^{-x} + C$$

3.
$$\left\{ \begin{array}{l} t = -x^3 \\ dt = -3x^2\,dx \end{array} \right\};\quad \int x^2 e^{-x^3}\,dx = -\frac{1}{3}\int e^t\,dt = -\frac{1}{3}e^t + C = -\frac{1}{3}e^{-x^3} + C$$

5.
$$\int x^2 e^{-x}\,dx = -x^2 e^{-x} - \int -2xe^{-x}\,dx = -x^2 e^{-x} + 2\int xe^{-x}\,dx$$

$$\begin{array}{ll} u = x^2, & dv = e^{-x}\,dx \\ du = 2x\,dx, & v = -e^{-x} \end{array}$$
$$= -x^2 e^{-x} + 2\left[-xe^{-x} - \int -e^{-x}\,dx \right]$$

$$\begin{array}{ll} u = x, & dv = e^{-x}\,dx \\ du = dx, & v = -e^{-x} \end{array}$$
$$= -x^2 e^{-x} + 2\left(-xe^{-x} - e^{-x} \right) + C$$

$$= -e^{-x}\left(x^2 + 2x + 2 \right) + C$$

$$\int_0^1 x^2 e^{-x}\,dx = \left[-e^{-x}\left(x^2 + 2x + 2 \right) \right]_0^1 = 2 - 5e^{-1}$$

7.
$$\int x^2 \left(1 - x \right)^{-1/2}\,dx = -2x^2 \left(1 - x \right)^{1/2} + 4\int x\left(1 - x \right)^{1/2}\,dx$$

$$\begin{array}{ll} u = x^2, & dv = \left(1 - x \right)^{-1/2}\,dx \\ du = 2x\,dx, & v = -2\left(1 - x \right)^{1/2} \end{array}$$
$$= -2x^2 \left(1 - x \right)^{1/2} + 4\left[-\frac{2x}{3}\left(1 - x \right)^{3/2} + \int \frac{2}{3}\left(1 - x \right)^{3/2}\,dx \right]$$

$$\begin{array}{ll} u = x, & dv = \left(1 - x \right)^{1/2}\,dx \\ du = dx, & v = -\frac{2}{3}\left(1 - x \right)^{3/2} \end{array}$$
$$= -2x^2 \left(1 - x \right)^{1/2} - \frac{8x}{3}\left(1 - x \right)^{3/2} - \frac{16}{15}\left(1 - x \right)^{5/2} + C$$

Or, use the substitution $t = 1 - x$ (no integration by parts needed) to obtain:

$$-2\left(1 - x \right)^{1/2} + \tfrac{4}{3}\left(1 - x \right)^{3/2} - \tfrac{2}{5}\left(1 - x \right)^{5/2} + C.$$

9.
$$\int x\ln\sqrt{x}\,dx = \frac{1}{2}\int x\ln x\,dx = \frac{1}{2}\left[\frac{1}{2}x^2 \ln x - \frac{1}{2}\int x\,dx \right]$$

$$\begin{array}{ll} u = \ln x, & dv = x\,dx \\ du = \dfrac{dx}{x}, & v = \dfrac{1}{2}x^2 \end{array}$$
$$= \tfrac{1}{4}x^2 \ln x - \tfrac{1}{8}x^2 + C$$

$$\int_1^{e^2} x\ln\sqrt{x}\,dx = \left[\tfrac{1}{4}x^2 \ln x - \tfrac{1}{8}x^2 \right]_1^{e^2} = \tfrac{3}{8}e^4 + \tfrac{1}{8}$$

11.
$$\int \frac{\ln(x+1)}{\sqrt{x+1}}\, dx = 2\sqrt{x+1}\,\ln(x+1) - \int \frac{2\, dx}{\sqrt{x+1}}$$

$$\boxed{\begin{array}{ll} u = \ln(x+1), & dv = \dfrac{dx}{\sqrt{x+1}} \\[2mm] du = \dfrac{dx}{x+1}, & v = 2\sqrt{x+1} \end{array}} \qquad = 2\sqrt{x+1}\,\ln(x+1) - 4\sqrt{x+1} + C$$

13.
$$\int (\ln x)^2\, dx = x\,(\ln x)^2 - 2\int \ln x\, dx$$

$$\boxed{\begin{array}{ll} u = (\ln x)^2, & dv = dx \\[2mm] du = \dfrac{2\ln x}{x}\, dx, & v = x \end{array}}$$

$$= x\,(\ln x)^2 - 2\left[x\ln x - \int dx \right]$$

$$\boxed{\begin{array}{ll} u = \ln x, & dv = dx \\[2mm] du = \dfrac{dx}{x}, & v = x \end{array}}$$

$$= x\,(\ln x)^2 - 2x\ln x + 2x + C$$

15.
$$\int x^3\, 3^x\, dx = \frac{x^3\, 3^x}{\ln 3} - \frac{3}{\ln 3}\int x^2\, 3^x\, dx$$

$$\boxed{\begin{array}{ll} u = x^3, & dv = 3^x\, dx \\[2mm] du = 3x^2\, dx, & v = \dfrac{3^x}{\ln 3} \end{array}}$$

$$= \frac{x^3\, 3^x}{\ln 3} - \frac{3}{\ln 3}\left[\frac{x^2\, 3^x}{\ln 3} - \frac{2}{\ln 3}\int x\, 3^x\, dx \right]$$

$$\boxed{\begin{array}{ll} u = x^2, & dv = 3^x\, dx \\[2mm] du = 2x\, dx & v = \dfrac{3^x}{\ln 3} \end{array}}$$

$$= \frac{x^3\, 3^x}{\ln 3} - \frac{3x^2\, 3^x}{(\ln 3)^2} + \frac{6}{(\ln 3)^2}\int x\, 3^x\, dx$$

$$= \frac{x^3\, 3^x}{\ln 3} - \frac{3x^2\, 3^x}{(\ln 3)^2} + \frac{6}{(\ln 3)^2}\left[\frac{x3^x}{\ln 3} - \frac{1}{\ln 3}\int 3^x\, dx \right]$$

$$\boxed{\begin{array}{ll} u = x, & dv = 3^x\, dx \\[2mm] du = dx, & v = \dfrac{3^x}{\ln 3} \end{array}}$$

$$= 3^x\left[\frac{x^3}{\ln 3} - \frac{3x^2}{(\ln 3)^2} + \frac{6x}{(\ln 3)^3} - \frac{6}{(\ln 3)^4} \right] + C$$

17.
$$\int x\,(x+5)^{14}\, dx = \frac{x}{15}\,(x+5)^{15} - \frac{1}{15}\int (x+5)^{15}\, dx$$

$$\boxed{\begin{array}{ll} u = x, & dv = (x+5)^{14}\, dx \\[2mm] du = dx, & v = \tfrac{1}{15}(x+5)^{15} \end{array}} \qquad = \tfrac{1}{15}x\,(x+5)^{15} - \tfrac{1}{240}(x+5)^{16} + C$$

Or, use the substitution $t = x + 5$ (integration by parts not needed) to obtain:

$$\tfrac{1}{16}(x+5)^{16} - \tfrac{1}{3}(x+5)^{15} + C.$$

19.

$$\int x \cos \pi x \, dx = \frac{1}{\pi} x \sin x - \frac{1}{\pi} \int \sin \pi x \, dx$$

$$\boxed{\begin{array}{ll} u = x, & dv = \cos \pi x \, dx \\ du = dx, & v = \frac{1}{\pi} \sin \pi x \end{array}} \qquad = \frac{1}{\pi} x \sin \pi x + \frac{1}{\pi^2} \cos \pi x + C$$

$$\int_0^{1/2} x \cos \pi x \, dx = \left[\frac{1}{\pi} x \sin \pi x + \frac{1}{\pi^2} \cos \pi x \right]_0^{1/2} = \frac{1}{2\pi} - \frac{1}{\pi^2}$$

21.

$$\int x^2 (x+1)^9 \, dx = \frac{x^2}{10}(x+1)^{10} - \frac{1}{5} \int x(x+1)^{10} \, dx$$

$$\boxed{\begin{array}{ll} u = x^2, & dv = (x+1)^9 \, dx \\ du = 2x \, dx, & v = \frac{1}{10}(x+1)^{10} \end{array}} \qquad = \frac{x^2}{10}(x+1)^{10} - \frac{1}{5}\left[\frac{x}{11}(x+1)^{11} - \frac{1}{11}\int (x+1)^{11} \, dx\right]$$

$$\boxed{\begin{array}{ll} u = x, & dv = (x+1)^{10} \, dx \\ du = dx, & v = \frac{1}{11}(x+1)^{11} \end{array}} \qquad = \frac{x^2}{10}(x+1)^{10} - \frac{x}{55}(x+1)^{11} + \frac{1}{660}(x+1)^{12} + C$$

23.

$$\int e^x \sin x \, dx = -e^x \cos x + \int e^x \cos x \, dx$$

$$\boxed{\begin{array}{ll} u = e^x, & dv = \sin x \, dx \\ du = e^x dx, & v = -\cos x \end{array}}$$

$$= -e^x \cos x + e^x \sin x - \int e^x \sin x \, dx$$

$$\boxed{\begin{array}{ll} u = e^x, & dv = \cos x \, dx \\ du = e^x dx, & v = \sin x \end{array}}$$

Adding $\int e^x \sin x \, dx$ to both sides, we get

$$2 \int e^x \sin x \, dx = -e^x \cos x + e^x \sin x$$

so that

$$\int e^x \sin x \, dx = \frac{1}{2} e^x (\sin x - \cos x) + C.$$

25.
$$\int \ln\left(1+x^2\right) dx = x \ln\left(1+x^2\right) - 2 \int \frac{x^2}{1+x^2} dx$$

$$\boxed{\begin{array}{ll} u = \ln(1+x^2), & dv = dx \\ du = \dfrac{2x}{1+x^2}\, dx, & v = x \end{array}}$$
$$= x \ln\left(1+x^2\right) - 2 \int \frac{x^2+1-1}{1+x^2} dx$$

$$= x \ln\left(1+x^2\right) - 2 \int \left(1 - \frac{1}{1+x^2}\right) dx$$

$$= x \ln\left(1+x^2\right) - 2x + 2\arctan x + C$$

$$\int_0^1 \ln(1+x^2)\, dx = \left[x \ln(1+x^2) - 2x + 2\arctan x\right]_0^1 = \ln 2 - 2 + \frac{\pi}{2}$$

27.
$$\int x^n \ln x\, dx = \frac{x^{n+1} \ln x}{n+1} - \frac{1}{n+1} \int x^n\, dx$$

$$\boxed{\begin{array}{ll} u = \ln x, & dv = x^n\, dx \\ du = \dfrac{dx}{x}, & v = \dfrac{x^{n+1}}{n+1} \end{array}}$$
$$= \frac{x^{n+1} \ln x}{n+1} - \frac{x^{n+1}}{(n+1)^2} + C$$

29.
$$\left\{t = x^2, \quad dt = 2x\, dx\right\}; \quad \int x^3 \sin x^2\, dx = \frac{1}{2} \int t \sin t\, dt$$

$$\boxed{\begin{array}{ll} u = t, & dv = \sin t\, dt \\ du = dt, & v = -\cos t \end{array}}$$
$$= \frac{1}{2}\left[-t \cos t + \int \cos t\, dt\right]$$

$$= \tfrac{1}{2}\left(-t \cos t + \sin t\right) + C$$

$$= -\tfrac{1}{2}x^2 \cos x^2 + \tfrac{1}{2}\sin x^2 + C$$

31.
$$\left\{\begin{array}{l|l} u = 2x & x = 0 \implies u = 0 \\ du = 2\, dx & x = 1/4 \implies u = 1/2 \end{array}\right\};$$

$$\int_0^{1/4} \arcsin 2x\, dx = \frac{1}{2} \int_0^{1/2} \arcsin u\, du$$

$$= \frac{1}{2}\left[u \arcsin u + \sqrt{1-u^2}\right]_0^{1/2} \qquad [\text{by}(8.2.5)]$$

$$= \frac{1}{2}\left[\frac{\pi}{12} + \frac{\sqrt{3}}{2} - 1\right] = \frac{\pi}{24} + \frac{\sqrt{3}-2}{4}$$

33.
$$\left\{\begin{array}{l|l} u = x^2 & x = 0 \implies u = 0 \\ du = 2x\, dx & x = 1 \implies u = 1 \end{array}\right\};$$

$$\int_0^1 x \arctan\left(x\right)^2 dx = \frac{1}{2} \int_0^1 \arctan u\, du$$

$$= \frac{1}{2}\left[u \arctan u - \tfrac{1}{2}\left(1+u^2\right)\right]_0^1 \qquad [\text{by } 8.2.6]$$

$$= \frac{\pi}{8} - \frac{1}{4}\ln 2$$

35.
$$\int x^2 \cosh 2x \, dx = \tfrac{1}{2}x^2 \sinh 2x - \int x \sinh 2x \, dx$$

$u = x^2,$	$dv = \cosh 2x \, dx$
$du = 2x \, dx,$	$v = \tfrac{1}{2}\sinh 2x$

$$= \tfrac{1}{2}x^2 \sinh 2x - \tfrac{1}{2}x \cosh 2x + \tfrac{1}{2}\int \cosh 2x \, dx$$

$u = x,$	$dv = \sinh 2x \, dx$
$du = dx,$	$v = \tfrac{1}{2}\cosh 2x$

$$= \tfrac{1}{2}x^2 \sinh 2x - \tfrac{1}{2}x \cosh 2x + \tfrac{1}{4}\sinh 2x + C$$

37. Let $u = \ln x, \ \ du = \tfrac{1}{x} dx.$ Then

$$\int \frac{1}{x}\arcsin(\ln x)\,dx = \int \arcsin u \, du = u \arcsin u + \sqrt{(1-u^2)} + C$$

$$= (\ln x)\arcsin(\ln x) + \sqrt{1 - (\ln x)^2} + C$$

39.
$$\int \sin(\ln x)\,dx = x\sin(\ln x) - \int \cos(\ln x)\,dx$$

$u = \sin(\ln x),$	$dv = dx$
$du = \cos(\ln x)\dfrac{1}{x}\,dx,$	$v = x$

$$= x\sin(\ln x) - x\cos(\ln x) - \int \sin(\ln x)\,dx$$

$u = \cos(\ln x),$	$dv = dx$
$du = -\sin(\ln x)\dfrac{1}{x}\,dx,$	$v = x$

Adding $\int \sin(\ln x)\,dx$ to both sides, we get

$$2\int \sin(\ln x)\,dx = x\sin(\ln x) - x\cos(\ln x)$$

so that

$$\int \sin(\ln x)\,dx = \tfrac{1}{2}\left[x\sin(\ln x) - x\cos(\ln x)\right] + C$$

41.

$u = \ln x$	$dv = dx$
$du = \dfrac{1}{x}\,dx$	$v = x$

$$\int \ln x \, dx = x\ln x - \int dx = x\ln x - x + C$$

43.

$u = \ln x$	$dv = x^k dx$
$du = \dfrac{1}{x}\,dx$	$v = \dfrac{1}{k+1}x^{k+1}$

$$\int x^k \ln x \, dx = \frac{1}{k+1}x^{k+1}\ln x - \int \frac{1}{k+1}x^k \, dx$$

$$= \frac{1}{k+1}x^{k+1}\ln x - \frac{x^{k+1}}{(k+1)^2} + C$$

45.

$$\boxed{\begin{array}{ll} u = e^{ax} & dv = \sin bx\, dx \\ du = ae^{ax}\, dx & v = -\dfrac{1}{b}\cos bx \end{array}} \qquad \int e^{ax}\sin bx\, dx = -\frac{1}{b}e^{ax}\cos bx + \frac{a}{b}\int e^{ax}\cos bx\, dx$$

$$\boxed{\begin{array}{ll} u = e^{ax} & dv = \cos bx\, dx \\ du = ae^{ax}\, dx & v = \dfrac{1}{b}\sin bx \end{array}} \qquad = -\frac{1}{b}e^{ax}\cos bx + \frac{a}{b^2}e^{ax}\sin bx - \frac{a^2}{b^2}\int e^{ax}\sin bx\, dx.$$

$$\implies \quad \int e^{ax}\sin bx\, dx = \frac{e^{ax}}{a^2+b^2}\left(a\sin bx - b\cos bx\right) + C$$

47. $\displaystyle\int_0^\pi x\sin x\, dx = \left[-x\cos x + \sin x\right]_0^\pi = \pi$

49. $A = \displaystyle\int_0^{1/2}\arcsin x\, dx = \left[x\arcsin x + \sqrt{1-x^2}\right]_0^{1/2} = \frac{\pi}{12} + \frac{\sqrt{3}-2}{2}$

51.   (a) $A = \displaystyle\int_1^e \ln x\, dx = [x\ln x - x]_1^e = 1$

    (b) $\bar{x}A = \displaystyle\int_1^e x\ln x\, dx = \left[\frac{1}{2}x^2\ln x - \frac{1}{4}x^2\right]_1^e = \frac{1}{4}\left(e^2+1\right), \quad \bar{x} = \frac{1}{4}\left(e^2+1\right)$

      $\bar{y}A = \displaystyle\int_1^e \frac{1}{2}\left(\ln x\right)^2 dx = \frac{1}{2}\left[x\left(\ln x\right)^2 - 2x\ln x + 2x\right]_1^e = \frac{1}{2}e - 1, \quad \bar{y} = \frac{1}{2}e - 1$

    (c) $V_x = 2\pi\bar{y}A = \pi\left(e-2\right), \quad V_y = 2\pi\bar{x}A = \frac{1}{2}\pi\left(e^2+1\right)$

53. $\bar{x} = \dfrac{1}{e-1}, \qquad \bar{y} = \dfrac{1}{4}\left(e+1\right)$ **55.** $\bar{x} = \frac{1}{2}\pi, \qquad \bar{y} = \frac{1}{8}\pi$

57.   (a) $M = \displaystyle\int_0^1 e^{kx}\, dx = \frac{1}{k}\left(e^k - 1\right)$

    (b) $x_M M = \displaystyle\int_0^1 xe^{kx}\, dx = \frac{(k-1)e^k + 1}{k^2}, \qquad x_M = \frac{(k-1)e^k + 1}{k\left(e^k - 1\right)}$

59. $V_y = \displaystyle\int_0^1 2\pi x\cos\frac{1}{2}\pi x\, dx = \left[4x\sin\frac{1}{2}\pi x + \frac{8}{\pi}\cos\frac{1}{2}\pi x\right]_0^1 = 4 - \frac{8}{\pi}$

61. $V_y = \displaystyle\int_0^1 2\pi x^2 e^x\, dx = 2\pi\left(e-2\right)$ (see Example 6)

63.

$$V_x = \int_0^1 \pi e^{2x}\, dx = \pi\left[\frac{1}{2}e^{2x}\right]_0^1 = \frac{1}{2}\pi\left(e^2 - 1\right)$$

$$\bar{x}V_x = \int_0^1 \pi x e^{2x}\, dx = \pi\left[\frac{1}{2}xe^{2x} - \frac{1}{4}e^{2x}\right]_0^1 = \frac{1}{4}\pi\left(e^2 + 1\right);$$

$$\bar{x} = \frac{e^2+1}{2\left(e^2-1\right)}$$

65.
$$A = \int_0^1 \cosh x \, dx = [\sinh x]_0^1 = \sinh 1 = \frac{e - e^{-1}}{2} = \frac{e^2 - 1}{2e}$$

$$\overline{x}A = \int_0^1 x \cosh x \, dx = [x \sinh x - \cosh x]_0^1 = \sinh 1 - \cosh 1 + 1 = \frac{2(e-1)}{2e}$$

$$\overline{y}A = \int_0^1 \frac{1}{2} \cosh^2 x \, dx = \frac{1}{4} [\sinh x \cosh x + x]_0^1 = \frac{1}{4}(\sinh 1 \cosh 1 + 1) = \frac{e^4 + 4e^2 - 1}{16e^2}$$

Therefore $\overline{x} = \dfrac{2}{e+1}$ and $\overline{y} = \dfrac{e^4 + 4e^2 - 1}{8e(e^2 - 1)}$.

67.
$$\boxed{\begin{array}{ll} u = x^2, & dv = e^{-x} \, dx \\ du = 2x \, dx, & v = -e^{-x} \end{array}} \qquad \int x^n e^{ax} \, dx = \frac{x^n e^{ax}}{a} - \frac{n}{a} \int x^{n-1} e^{ax} \, dx$$

69.
$$\int x^3 e^{2x} \, dx = \frac{1}{2} x^3 e^{2x} - \frac{3}{2} \int x^2 e^{2x} \, dx = \frac{1}{2} x^3 e^{2x} - \frac{3}{2} \left[\frac{1}{2} x^2 e^{2x} - \int x e^{2x} \, dx \right]$$

$$= \frac{1}{2} x^3 e^{2x} - \frac{3}{4} x^2 e^{2x} - \frac{3}{4} x e^{2x} - \frac{3}{4} \int e^{2x} \, dx$$

$$= \frac{1}{2} x^3 e^{2x} - \frac{3}{4} x^2 e^{2x} + \frac{3}{4} x e^{2x} - \frac{3}{8} e^{2x} + C$$

71.
$$\int (\ln x)^3 \, dx = x(\ln x)^3 - 3 \int (\ln x)^2 \, dx = x(\ln x)^3 - 3x(\ln x)^2 + 6 \int \ln x \, dx$$

$$= x(\ln x)^3 - 3x(\ln x)^2 + 6x \ln x - 6 \int dx$$

$$= x(\ln x)^3 - 3x(\ln x)^2 + 6x \ln x - 6x + C$$

73. (a) Differentiating, $x^3 e^x = Ax^3 e^x + 3Ax^2 e^x + 2Bx e^x + Bx^2 e^x + Ce^x + Cx e^x + De^x$
$\implies A = 1, \ B = -3, \ C = 6, \ \text{and} \ D = -6$

(b)
$$\int x^3 e^x \, dx = x^3 e^x - 3 \int x^2 e^x \, dx$$

$$= x^3 e^x - 3x^2 e^x + 6 \int x e^x \, dx$$

$$= x^3 e^x - 3x^2 e^x + 6x e^x - 6 \int e^x \, dx$$

$$= x^3 e^x - 3x^2 e^x + 6x e^x - 6e^x + C$$

75. (a) Set $\displaystyle \int (x^2 - 3x + 1) e^x \, dx = Ax^2 e^x + Bx e^x + Ce^x$ and differentiate both sides of the equation.

$$x^2 e^x - 3x e^x + e^x = Ax^2 e^x + 2Ax e^x + Bx e^x + Be^x + Ce^x$$

$$= Ax^2 e^x + (2A + B)x e^x + (B + C)e^x$$

Equating the coefficients, we find that $A = 1, \ B = -5, \ C = 6$

Thus $\displaystyle \int (x^2 - 3x + 1) e^x \, dx = x^2 e^x - 5x e^x + 6e^x + K$

(b) Set $\int \left(x^3 - 2x\right) e^x \, dx = Ax^3 e^x + Bx^2 e^x + Cxe^x + De^x$ and differentiate both sides of the equation.

$$x^3 e^x - 2xe^x = Ax^3 e^x + 3Ax^2 e^x + Bx^2 e^x + 2Bxe^x + Cxe^x + ce^x + De^x$$

$$= Ax^3 e^x + (3A + B)x^2 e^x + (2B + C)xe^x + (C + D)e^x$$

Equating the coefficients, we find that $A = 1, \ B = -3, \ C = 4, \ D = -4$

Thus $\int \left(x^3 - 2x\right) e^x \, dx = x^3 e^x - 3x^2 e^x + 4xe^x - 4e^x + K$

77. Let $u = f(x), \ dv = g''(x) \, dx$. Then $du = f'(x) \, dx, \ v = g'(x)$, and

$$\int_a^b f(x)g''(x) \, dx = [f(x)g'(x)]_a^b - \int_a^b f'(x)g'(x) \, dx = -\int_a^b f'(x)g'(x) \, dx \quad \text{since } f(a) = f(b) = 0$$

Now let $u = f'(x), \ dv = g'(x) \, dx$. Then $du = f''(x) \, dx, \ v = g(x)$, and

$$-\int_a^b f'(x)g'(x) \, dx = [-f'(x)g(x)]_a^b + \int f''(x)g(x) \, dx = \int g(x)f''(x) \, dx \quad \text{since } g(a) = g(b) = 0$$

Therefore, if f and g have continuous second derivatives, and if $f(a) = g(a) = f(b) = g(b) = 0$, then

$$\int_a^b f(x)g''(x) \, dx = \int_a^b g(x)f''(x) \, dx$$

79. (a) From Exercise 47, $A = \int_0^\pi x \sin x \, dx = \pi$

(b) $f(x) = x \sin x < 0$ on $(\pi, 2\pi)$. Therefore

$$A = -\int_\pi^{2\pi} x \sin x \, dx = \left[-x \cos x + \sin x \right]_\pi^{2\pi} = 3\pi$$

(c) $A = \int_{2\pi}^{3\pi} x \sin x \, dx = \left[-x \cos x + \sin x \right]_{2\pi}^{3\pi} = 5\pi$

(d) $A = (2n + 1)\pi, \ n = 1, 2, 3, \ldots$

81. (a) area of R: $\int_0^\pi (1 - \sin x) \, dx = \left[x + \cos x \right]_0^\pi = \pi - 2$

(b) $V = \int_0^\pi 2\pi x \left(1 - \sin x\right) \, dx = 2\pi \left[\frac{1}{2} x^2 + x \cos x - \sin x \right]_0^\pi = \pi^3 - 2\pi^2$

(c) The region is symmetric about the line $x = \frac{1}{2}\pi$. Therefore $\bar{x} = \frac{1}{2}\pi$.

$$\bar{y}A = \frac{1}{2} \int_0^\pi (1 - \sin x)^2 \, dx$$

$$= \frac{1}{2} \int_0^\pi \left(1 - 2\sin x + \sin^2 x\right) \, dx$$

$$= \frac{1}{2} \int_0^\pi \left(\frac{3}{2} - 2\sin x - \frac{1}{2}\cos 2x\right) \, dx$$

$$= \frac{1}{2} \left[\frac{3}{2}x + 2\cos x - \frac{1}{4}\sin 2x \right]_0^\pi = \frac{3}{4}\pi - 2$$

Therefore, $\bar{y} = \dfrac{\frac{3}{4}\pi - 2}{\pi - 2} \cong 0.31202$

PROJECT 8.2

1. $\displaystyle\int_0^{2\pi} \sin^2 nx\,dx = \int_0^{2\pi} \left(\frac{1}{2} - \frac{1}{2}\cos 2nx\right) dx = \left[\frac{1}{2}x - \frac{1}{4n}\sin 2nx\right]_0^{2\pi} = \pi.$

$\displaystyle\int_0^{2\pi} \cos^2 nx\,dx = \int_0^{2\pi} \left(\frac{1}{2} + \frac{1}{2}\cos 2nx\right) dx = \left[\frac{1}{2}x + \frac{1}{4n}\sin 2nx\right]_0^{2\pi} = \pi$

SECTION 8.3

1. $\displaystyle\int \sin^3 x\,dx = \int (1 - \cos^2 x)\sin x\,dx = \frac{1}{3}\cos^3 x - \cos x + C$

3. $\displaystyle\int_0^{\pi/6} \sin^2 3x\,dx = \int_0^{\pi/6} \frac{1 - \cos 6x}{2}\,dx = \left[\frac{1}{2}x - \frac{1}{12}\sin 6x\right]_0^{\pi/6} = \frac{\pi}{12}$

5.
$$\int \cos^4 x \sin^3 x\,dx = \int \cos^4 x\,(1 - \cos^2 x)\sin x\,dx$$
$$= \int (\cos^4 x - \cos^6 x)\sin x\,dx$$
$$= -\tfrac{1}{5}\cos^5 x + \tfrac{1}{7}\cos^7 x + C$$

7.
$$\int \sin^3 x \cos^3 x\,dx = \int \sin^3 x\,(1 - \sin^2 x)\cos x\,dx = \int (\sin^3 x - \sin^5 x)\cos x\,dx$$
$$= \tfrac{1}{4}\sin^4 x - \tfrac{1}{6}\sin^6 x + C$$

9. $\displaystyle\int \sec^2 \pi x\,dx = \frac{1}{\pi}\tan \pi x + C$

11.
$$\int \tan^3 x\,dx = \int (\sec^2 x - 1)\tan x\,dx$$
$$= \int \tan x \sec^2 x\,dx - \int \tan x\,dx$$
$$= \tfrac{1}{2}\tan^2 x + \ln|\cos x| + C$$

13.
$$\int \sin^4 x\,dx = \int \left(\frac{1 - \cos 2x}{2}\right)^2 dx$$
$$= \frac{1}{4}\int \left(1 - 2\cos 2x + \cos^2 2x\right) dx$$
$$= \frac{1}{4}\int \left(1 - 2\cos 2x + \frac{1 + \cos 4x}{2}\right) dx$$
$$= \int \left(\frac{3}{8} - \frac{1}{2}\cos 2x + \frac{1}{8}\cos 4x\right) dx$$
$$= \tfrac{3}{8}x - \tfrac{1}{4}\sin 2x + \tfrac{1}{32}\sin 4x + C$$
$$\int_0^{\pi} \sin^4 x\,dx = \left[\tfrac{3}{8}x - \tfrac{1}{4}\sin 2x + \tfrac{1}{32}\sin 4x\right]_0^{\pi} = \tfrac{3}{8}\pi$$

15.

$$\int \sin 2x \cos 3x \, dx = \int \frac{1}{2} \left[\sin \left(-x \right) + \sin 5x \right] dx$$

$$= \int \frac{1}{2} \left(-\sin x + \sin 5x \right) dx$$

$$= \tfrac{1}{2} \cos x - \tfrac{1}{10} \cos 5x + C$$

17. $\displaystyle \int \tan^2 x \sec^2 x \, dx = \frac{1}{3} \tan^3 x + C$

19.

$$\int \sin^2 x \sin 2x \, dx = \int \sin^2 x \left(2 \sin x \cos x \right) dx$$

$$= 2 \int \sin^3 x \cos x \, dx$$

$$= \tfrac{1}{2} \sin^4 x + C$$

21.

$$\int \sin^6 x \, dx = \int \left(\frac{1 - \cos 2x}{2} \right)^3 dx$$

$$= \frac{1}{8} \int \left(1 - 3 \cos 2x + 3 \cos^2 2x - \cos^3 2x \right) dx$$

$$= \frac{1}{8} \int \left[1 - 3 \cos 2x + 3 \left(\frac{1 + \cos 4x}{2} \right) - \cos 2x \left(1 - \sin^2 2x \right) \right] dx$$

$$= \frac{1}{8} \int \left(\frac{5}{2} - 4 \cos 2x + \frac{3}{2} \cos 4x + \sin^2 2x \cos 2x \right) dx$$

$$= \tfrac{5}{16} x - \tfrac{1}{4} \sin 2x + \tfrac{3}{64} \sin 4x + \tfrac{1}{48} \sin^3 2x + C$$

23. $\displaystyle \int_{\pi/6}^{\pi/2} \cot^2 x \, dx = \int_{\pi/6}^{\pi/2} \left(\csc^2 x - 1 \right) dx = \left[-\cot x - x \right]_{\pi/6}^{\pi/2} = \sqrt{3} - \frac{\pi}{3}$

25.

$$\int \cot^3 x \csc^3 x \, dx = \int \left(\csc^2 x - 1 \right) \csc^3 x \cot x \, dx$$

$$= \int \left(\csc^4 x - \csc^2 x \right) \csc x \cot x \, dx$$

$$= -\frac{1}{5} \csc^5 x + \frac{1}{3} \csc^3 x + C$$

27.

$$\int \sin 5x \sin 2x \, dx = \int \frac{1}{2} \left(\cos 3x - \cos 7x \right) dx$$

$$= \tfrac{1}{6} \sin 3x - \tfrac{1}{14} \sin 7x + C$$

29.

$$\int \sin^{5/2} x \cos^3 x \, dx = \int \sin^{5/2} x \left(1 - \sin^2 x \right) \cos x \, dx$$

$$= \int \sin^{5/2} x \cos x \, dx - \int \sin^{9/2} x \cos x \, dx$$

$$= \tfrac{2}{7} \sin^{7/2} x - \tfrac{2}{11} \sin^{11/2} x + C$$

31.
$$\int \tan^5 3x \, dx = \int \tan^3 3x \left(\sec^2 3x - 1 \right) dx$$

$$= \int \tan^3 3x \sec^2 3x \, dx - \int \tan^3 3x \, dx$$

$$= \int \tan^3 3x \sec^2 3x \, dx - \int \left(\tan 3x \sec^2 3x - \tan 3x \right) dx$$

$$= \tfrac{1}{12} \tan^4 3x - \tfrac{1}{6} \tan^2 3x + \tfrac{1}{3} \ln |\sec 3x| + C$$

33.
$$\int_{-1/6}^{1/3} \sin^4 3\pi x \cos^3 3\pi x \, dx = \int_{-1/6}^{1/3} \sin^4 3\pi x \cos^2 3\pi x \cos 3\pi x \, dx$$

$$= \int_{-1/6}^{1/3} \sin^4 3\pi x \left(1 - \sin^2 3\pi x \cos 3\pi x \right) dx$$

$$= \int_{-1}^{0} u^4 (1 - u^2) \frac{1}{3\pi} \, du \quad [u = \sin 3\pi x, \quad du = 3\pi \cos 3\pi x \, dx]$$

$$= \frac{1}{3\pi} \left[\tfrac{1}{5} u^5 - \tfrac{1}{7} u^7 \right]_{-1}^{0} = \frac{2}{105\pi}$$

35.
$$\int_0^{\pi/4} \cos 4x \sin 2x \, dx = \int_0^{\pi/4} \frac{1}{2} \left(\sin 6x - \sin 2x \right) dx$$

$$= \left[-\tfrac{1}{12} \cos 6x + \tfrac{1}{4} \cos 2x \right]_0^{\pi/4} = -\tfrac{1}{6}$$

37.
$$\int \tan^4 x \sec^4 x \, dx = \int \tan^4 x \left(\tan^2 x + 1 \right) \sec^2 x \, dx$$

$$= \int \left(\tan^6 x + \tan^4 x \right) \sec^2 x \, dx$$

$$= \frac{1}{7} \tan^7 x + \frac{1}{5} \tan^5 x + C$$

39. $\displaystyle \int \sin(x/2) \cos 2x \, dx = \int \tfrac{1}{2} \left(\sin \left(\tfrac{5}{2} x \right) - \sin \left(\tfrac{3}{2} x \right) \right) dx = \tfrac{1}{3} \cos \left(\tfrac{3}{2} x \right) - \tfrac{1}{5} \cos \left(\tfrac{5}{2} x \right) + C$

41. $\displaystyle \left\{ \begin{array}{l|l} u = \tan x & x = 0 \Rightarrow u = 0 \\ du = \sec^2 x \, dx & x = \pi/4 \Rightarrow u = 1 \end{array} \right\}; \quad \int_0^{\pi/4} \tan^3 x \sec^2 x \, dx = \int_0^1 u^3 \, du = \left[\tfrac{1}{4} u^4 \right]_0^1 = \tfrac{1}{4}$

43. $\displaystyle \int_0^{\pi/6} \tan^2 2x \, dx = \int_0^{\pi/6} \left(\sec^2 2x - 1 \right) dx = \left[\tfrac{1}{2} \tan 2x - x \right]_0^{\pi/6} = \frac{\sqrt{3}}{2} - \frac{\pi}{6}$

45. $\displaystyle A = \int_0^{\pi} \sin^2 x \, dx = \int_0^{\pi} \tfrac{1}{2} (1 - \cos 2x) \, dx = \tfrac{1}{2} \left[x - \tfrac{1}{2} \sin 2x \right]_0^{\pi} = \frac{\pi}{2}$

47.
$$V = \int_0^\pi \pi \left(\sin^2 x\right)^2 dx = \pi \int_0^\pi \sin^4 x \, dx = \pi \int_0^\pi \left(\tfrac{1}{2}(1 - \cos 2x\right)^2 dx$$

$$= \frac{\pi}{4} \int_0^\pi \left(1 - 2\cos 2x + \cos^2 2x\right) dx$$

$$= \frac{\pi}{4} \left[x - \sin 2x\right]_0^\pi + \frac{\pi}{8} \int_0^\pi \left(1 + \cos 4x\right) dx$$

$$= \frac{\pi^2}{4} + \frac{\pi}{8} \left[x + \tfrac{1}{4} \sin 4x\right]_0^\pi = \frac{3\pi^2}{8}$$

49. $V = \int_0^{\pi/4} \pi \left[1^2 - \tan^2 x\right] dx = \pi \int_0^{\pi/4} \left[2 - \sec^2 x\right] dx = \pi \left[2x - \tan x\right]_0^{\pi/4} = \frac{\pi^2}{2} - \pi$

51.
$$V = \int_0^{\pi/4} \pi \left[(\tan x + 1)^2 - 1^2\right] dx = \pi \int_0^{\pi/4} \left[\tan^2 x + 2\tan x\right] dx$$

$$= \pi \int_0^{\pi/4} \left(\sec^2 x + 2\tan x - 1\right) dx$$

$$= \pi \left[\tan x + 2\ln |\sec x| - x\right]_0^{\pi/4} = \pi \left[\ln 2 + 1 - \frac{\pi}{4}\right]$$

53. **(a)** $\int \sin^n x \, dx = \int \sin^{n-1} x \sin x \, dx;$ $\quad \left\{ \begin{array}{ll} u = \sin^{n-1} x & dv = \sin x \, dx \\ du = (n-1)\sin^{n-2} x \, dx & v = -\cos x \end{array} \right\}$

$$\int \sin^n x \, dx = -\sin^{n-1} x \cos x + (n-1) \int \sin^{n-2} x \cos^2 x \, dx$$

$$= -\sin^{n-1} x \cos x + (n-1) \int \sin^{n-2} x \left(1 - \sin^2 x\right) dx$$

Therefore,

$$n \int \sin^n x \, dx = -\sin^{n-1} x \cos x + (n-1) \int \sin^{n-2} x \, dx$$

and

$$\int \sin^n x \, dx = -\frac{1}{n} \sin^{n-1} x \cos x + \frac{n-1}{n} \int \sin^{n-2} x \, dx.$$

(b) $\int_0^{\pi/2} \sin^n x \, dx = \left[-\frac{1}{n} \sin^{n-1} x \cos x\right]_0^{\pi/2} + \frac{n-1}{n} \int_0^{\pi/2} \sin^{n-2} x \, dx = \frac{n-1}{n} \int_0^{\pi/2} \sin^{n-2} x \, dx$

(c) n even:

$$\int_0^{\pi/2} \sin^n x \, dx = \frac{n-1}{n} \int_0^{\pi/2} \sin^{n-2} x \, dx = \frac{n-1}{n} \frac{n-3}{n-2} \int_0^{\pi/2} \sin^{n-4} x \, dx$$

and so on,

$$\int_0^{\pi/2} \sin^n x \, dx = \frac{n-1}{n} \frac{n-3}{n-2} \cdots \frac{3}{4} \frac{1}{2} \int_0^{\pi/2} 1 \, dx = \frac{n-1}{n} \frac{n-3}{n-2} \cdots \frac{3}{4} \frac{1}{2} \frac{\pi}{2}.$$

n odd:

$$\int_0^{\pi/2} \sin^n x \, dx = \frac{n-1}{n} \frac{n-3}{n-2} \cdots \frac{4}{5} \frac{2}{3} \int_0^{\pi/2} \sin x \, dx$$

$$= \frac{n-1}{n} \frac{n-3}{n-2} \cdots \frac{4}{5} \frac{2}{3} \left[-\cos x\right]_0^{\pi/2} = \frac{n-1}{n} \frac{n-3}{n-2} \cdots \frac{4}{5} \frac{2}{3}.$$

55. (a) $\displaystyle\int_0^{\pi/2} \sin^7 x \, dx = \frac{6 \cdot 4 \cdot 2}{7 \cdot 5 \cdot 3} = \frac{16}{35}$

(b) $\displaystyle\int_0^{\pi/2} \cos^6 x \, dx = \left(\frac{5 \cdot 3 \cdot 1}{6 \cdot 4 \cdot 2}\right)\frac{\pi}{2} = \frac{5\pi}{32}$

57. (a) $V \cong 4.9348$

(b) $\displaystyle V = \int_0^{\sqrt{\pi}} 2\pi x \sin^2\left(x^2\right) dx = \pi \int_0^{\pi} \sin^2 u \; du = \tfrac{1}{2}\pi \int_0^{\pi} (1 - \cos 2u) \, du$

$u = x^2 \underline{\qquad\uparrow}$

$\qquad\qquad\qquad = \tfrac{1}{2}\pi \left[u - \tfrac{1}{2}\sin 2u\right]_0^{\pi}$

$\qquad\qquad\qquad = \tfrac{1}{2}\pi^2 \cong 4.9348$

SECTION 8.4

1. $\left\{\begin{array}{l} x = a\sin u \\ dx = a\cos u\, du \end{array}\right\};$ $\displaystyle\int \frac{dx}{\sqrt{a^2 - x^2}} = \int \frac{a\cos u\, du}{a\cos u}$

$\qquad\qquad\qquad\qquad = \int du = u + C = \arcsin\left(\frac{x}{a}\right) + C$

3. $\left\{\begin{array}{l} x = \sec u \\ dx = \sec u\tan u\, du \end{array}\right\};$ $\displaystyle\int \sqrt{x^2 - 1}\, dx = \int \tan^2 u\sec u\, du$

$\qquad\qquad\qquad\qquad = \int (\sec^3 u - \sec u)\, du$

$\qquad\qquad\qquad\qquad = \frac{1}{2}\sec u\tan u - \frac{1}{2}\ln|\sec u + \tan u| + C$

$\qquad\qquad\qquad\qquad\text{Example 8, Section 8.3}$

$\qquad\qquad\qquad\qquad = \frac{1}{2}x\sqrt{x^2 - 1} - \frac{1}{2}\ln|x + \sqrt{x^2 - 1}| + C$

5. $\left\{\begin{array}{l} x = 2\sin u \\ dx = 2\cos u\, du \end{array}\right\};$ $\displaystyle\int \frac{x^2}{\sqrt{4 - x^2}}\, dx = \int \frac{4\sin^2 u}{2\cos u}2\cos u\, du$

$\qquad\qquad\qquad\qquad = 2\int (1 - \cos 2u)\, du$

$\qquad\qquad\qquad\qquad = 2u - \sin 2u + C$

$\qquad\qquad\qquad\qquad = 2u - 2\sin u\cos u + C$

$\qquad\qquad\qquad\qquad = 2\arcsin\left(\frac{x}{2}\right) - \frac{1}{2}x\sqrt{4 - x^2} + C$

7. $\left\{\begin{array}{l} u = 1 - x^2 \\ du = -2x\, dx \end{array}\right\};$ $\displaystyle\int \frac{x}{(1 - x^2)^{3/2}}\, dx = -\frac{1}{2}\int \frac{du}{u^{3/2}} = u^{-1/2} + C = \frac{1}{\sqrt{1 - x^2}} + C$

9. $\left\{\begin{array}{l} x = \sin u \\ dx = \cos u\, du \end{array}\right\};$ $\displaystyle\int \frac{x^2}{(1 - x^2)^{3/2}}\, dx = \int \frac{\sin^2 u}{\cos^3 u}\cos u\, du = \int \tan^2 u\, du$

$\qquad\qquad\qquad\qquad = \int (\sec^2 u - 1)\, du = \tan u - u + C$

$\qquad\qquad\qquad\qquad = \frac{x}{\sqrt{1 - x^2}} - \arcsin x + C$

$\qquad\qquad \displaystyle\int_0^{1/2} \frac{x^2}{(1 - x^2)^{3/2}}\, dx = \left[\frac{x}{\sqrt{1 - x^2}} - \arcsin x\right]_0^{1/2} = \frac{2\sqrt{3} - \pi}{6}$

11. $\left\{\begin{array}{l} u = 4 - x^2 \\ du = -2x\,dx \end{array}\right\}$; $\displaystyle\int x\sqrt{4 - x^2}\,dx = -\frac{1}{2}\int u^{1/2}\,du = -\frac{1}{3}u^{3/2} + C$

$$= -\frac{1}{3}(4 - x^2)^{3/2} + C$$

13. $\left\{\begin{array}{ll} x = 5\sin u|\ x = 0 & \Longrightarrow\quad u = 0 \\ dx = 5\cos u\,du|\ x = 5 & \Longrightarrow\quad u = \pi/2 \end{array}\right\}$;

$$\int_0^5 x^2\sqrt{25 - x^2}\,dx = \int_0^{\pi/2}(5\sin u)^2(5\cos u)^2\,du$$

$$= 625\int_0^{\pi/2}\left(\sin^2 u - \sin^4 u\right)\,du$$

$$= 625\left[\frac{1}{2}\cdot\frac{\pi}{2} - \frac{3\cdot 1}{4\cdot 2}\cdot\frac{\pi}{2}\right] = \frac{625\pi}{16} \qquad \text{[see Exercise 53, Section 8.3]}$$

15. $\left\{\begin{array}{l} x = \sqrt{8}\tan u \\ dx = \sqrt{8}\sec^2 u\,du \end{array}\right\}$; $\displaystyle\int \frac{x^2}{(x^2 + 8)^{3/2}}\,dx = \int \frac{8\tan^2 u}{(8\sec^2 u)^{3/2}}\sqrt{8}\sec^2 u\,du$

$$= \int \frac{\tan^2 u}{\sec u}\,du = \int \frac{\sec^2 u - 1}{\sec u}\,du$$

$$= \int(\sec u - \cos u)\,du$$

$$= \ln|\sec u + \tan u| - \sin u + C$$

$$= \ln\left(\frac{\sqrt{x^2 + 8} + x}{\sqrt{8}}\right) - \frac{x}{\sqrt{x^2 + 8}} + C$$

$(\text{absorb} - \ln\sqrt{8}\text{ in }C)$

$$= \ln\left(\sqrt{x^2 + 8} + x\right) - \frac{x}{\sqrt{x^2 + 8}} + C$$

17. $\left\{\begin{array}{l} x = a\sin u \\ dx = a\cos u\,du \end{array}\right\}$; $\displaystyle\int \frac{dx}{x\sqrt{a^2 - x^2}} = \int \frac{a\cos u\,du}{a\sin u\,(a\cos u)} = \frac{1}{a}\int \csc u\,du$

$$= \frac{1}{a}\ln|\csc u - \cot u| + C$$

$$= \frac{1}{a}\ln\left|\frac{a - \sqrt{a^2 - x^2}}{x}\right| + C$$

19. $\left\{\begin{array}{l} x = 3\tan u \\ dx = 3\sec^2 u\,du \end{array}\right\}$; $\displaystyle\int \frac{x^3}{\sqrt{9 + x^2}}\,dx = \int \frac{27\tan^3 u}{3\sec u}\cdot 3\sec^2 u\,du$

$$= 27\int \tan^3 u\,\sec u\,du$$

$$= 27\int\left(\sec^2 u - 1\right)\sec u\,\tan u\,du$$

$$= 27\left[\tfrac{1}{3}\sec^3 u - \sec u\right] + C$$

$$= \tfrac{1}{3}\left(9 + x^2\right)^{3/2} - 9\left(9 + x^2\right)^{1/2} + C$$

$$\int_0^3 \frac{x^3}{\sqrt{9 + x^2}}\,dx = \left[\tfrac{1}{3}\left(9 + x^2\right)^{3/2} - 9\left(9 + x^2\right)^{1/2}\right]_0^3 = 18 - 9\sqrt{2}$$

21. $\left\{\begin{array}{l} x = a\tan u \\ dx = a\sec^2 u\,du \end{array}\right\};\quad \displaystyle\int \frac{dx}{x^2\sqrt{a^2+x^2}} = \int \frac{a\sec^2 u\,du}{a^2\tan^2 u\,(a\sec u)}$

$$= \frac{1}{a^2}\int \frac{\sec u}{\tan^2 u}\,du$$

$$= \frac{1}{a^2}\int \cot u\,\csc u\,du$$

$$= -\frac{1}{a^2}\cos u + C = -\frac{1}{a^2 x}\sqrt{a^2+x^2} + C$$

23. $\left\{\begin{array}{l} x = \sqrt{5}\sin u \\ dx = \sqrt{5}\cos u\,du \end{array}\right\};\quad \displaystyle\int \frac{dx}{(5-x^2)^{3/2}} = \int \frac{\sqrt{5}\cos u\,du}{(5\cos^2 u)^{3/2}}$

$$= \frac{1}{5}\int \sec^2 u\,du$$

$$= \frac{1}{5}\tan u + C = \frac{x}{5\sqrt{5-x^2}} + C$$

$$\int_0^1 \frac{dx}{(5-x^2)^{3/2}} = \left[\frac{x}{5\sqrt{5-x^2}}\right]_0^1 = \frac{1}{10}$$

25. $\left\{\begin{array}{l} x = a\sec u \\ dx = a\sec u\tan u\,du \end{array}\right\};\quad \displaystyle\int \frac{dx}{x^2\sqrt{x^2-a^2}} = \int \frac{a\sec u\tan u\,du}{a^2\sec^2 u\,(a\tan u)}$

$$= \frac{1}{a^2}\int \cos u\,du$$

$$= \frac{1}{a^2}\sin u + C$$

$$= \frac{1}{a^2 x}\sqrt{x^2-a^2} + C$$

27. $\left\{\begin{array}{l} e^x = 3\sec u \\ e^x dx = 3\sec u\tan u\,du \end{array}\right\};\quad \displaystyle\int \frac{dx}{e^x\sqrt{e^{2x}-9}} = \int \frac{\tan u\,du}{3\sec u\,(3\tan u)}$

$$= \frac{1}{9}\int \cos u\,du$$

$$= \tfrac{1}{9}\sin u + C$$

$$= \tfrac{1}{9}e^{-x}\sqrt{e^{2x}-9} + C$$

29. $(x^2-4x+4)^{\frac{3}{2}} = \begin{cases} (x-2)^3, & x > 2 \\ (2-x)^3, & x < 2 \end{cases}$

$$\int \frac{dx}{(x^2-4x+4)^{3/2}} = \begin{cases} -\dfrac{1}{2(x-2)^2} + C, & x > 2 \\[2mm] \dfrac{1}{2(2-x)^2} + C, & x < 2 \end{cases}$$

31. $\begin{cases} x - 3 = \sin u \\ dx = \cos u\, du \end{cases}$; $\displaystyle \int x\sqrt{6x - x^2 - 8}\, dx = \int x\sqrt{1 - (x-3)^2}\, dx$

$$= \int (3 + \sin u)(\cos u)\cos u\, du$$

$$= \int (3\cos^2 u + \cos^2 u \sin u)\, du$$

$$= \int \left[3\left(\frac{1 + \cos 2u}{2} \right) + \cos^2 u \sin u \right] du$$

$$= \frac{3u}{2} + \frac{3}{4}\sin 2u - \frac{1}{3}\cos^3 u + C$$

$$= \frac{3}{2}\arcsin(x - 3) + \frac{3}{2}(x - 3)\sqrt{6x - x^2 - 8} - \frac{1}{3}(6x - x^2 - 8)^{3/2} + C$$

33. $\begin{cases} x + 1 = 2\tan u \\ dx = 2\sec^2 u\, du \end{cases}$; $\displaystyle \int \frac{x}{(x^2 + 2x + 5)^2}\, dx = \int \frac{x}{[(x+1)^2 + 4]^2}\, dx$

$$= \int \frac{2\tan u - 1}{(4\sec^2 u)^2} 2\sec^2 u\, du$$

$$= \frac{1}{8} \int \frac{2\tan u - 1}{\sec^2 u}\, du$$

$$= \frac{1}{8} \int (2\sin u \cos u - \cos^2 u)\, du$$

$$= \frac{1}{8} \int \left(2\sin u \cos u - \frac{1 + \cos 2u}{2} \right) du$$

$$= \frac{1}{8} \left(\sin^2 u - \frac{u}{2} - \frac{\sin 2u}{4} \right) + C$$

$$= \frac{1}{8} \left[\left(\frac{x+1}{\sqrt{x^2 + 2x + 5}} \right)^2 - \frac{1}{2}\arctan\left(\frac{x+1}{2} \right) - \frac{1}{4}\overbrace{(2)\left(\frac{x+1}{\sqrt{x^2 + 2x + 5}} \right)\left(\frac{2}{\sqrt{x^2 + 2x + 5}} \right)}^{\sin 2u = 2\sin u \cos u} \right] + C$$

$$= \frac{x^2 + x}{8(x^2 + 2x + 5)} - \frac{1}{16}\arctan\left(\frac{x+1}{2} \right) + C$$

35. $\boxed{\begin{array}{ll} u = \sec^{-1} x & dv = dx \\ du = \dfrac{1}{x\sqrt{x^2 - 1}}\, dx & v = x \end{array}}$ $\displaystyle \int \sec^{-1} x\, dx = x\sec^{-1} x - \int \frac{1}{\sqrt{x^2 - 1}}\, dx$

$\boxed{x = \sec u \quad dx = \sec u \tan u\, du}$ $\displaystyle \qquad = x\sec^{-1} x - \int \sec u\, du$

$$= x\sec^{-1} x - \ln[\sec u + \tan u] + C$$

$$= x\sec^{-1} x - \ln\left[x + \sqrt{x^2 - 1} \right] + C$$

37.   Let $x = a \tan u$. Then $dx = a \sec^2 u \, du$, $\sqrt{x^2 + a^2} = a \sec u$, and

$$\int \frac{dx}{(x^2 + a^2)^n} = \int \frac{a \sec^2 u}{a^{2n} \sec^{2n} u} \, du = \frac{1}{a^{2n-1}} \int \cos^{2n-2} u \, du$$

39. $$\int \frac{1}{(x^2 + 1)^3} \, dx = \int \cos^4 u \, du$$

$$= \frac{1}{4} \cos^3 u \, \sin u + \frac{3}{8} \cos u \, \sin u + \frac{3}{8} u + C \quad \text{[by the reduction formula (8.3.2)]}$$

$$= \frac{3}{8} \arctan x + \frac{3x}{8 \, (x^2 + 1)} + \frac{x}{4 \, (x^2 + 1)^2} + C$$

41.

$$\boxed{\begin{array}{ll} u = \arcsin x & dv = x \, dx \\ du = \dfrac{1}{\sqrt{1 - x^2}} \, dx & v = \dfrac{x^2}{2} \end{array}}$$

$$\int x \arcsin x \, dx = \frac{x^2}{2} \arcsin x - \frac{1}{2} \int \frac{x^2}{\sqrt{1 - x^2}} \, dx$$

$$\boxed{x = \sin u \quad dx = \cos u \, du} \qquad = \frac{x^2}{2} \arcsin x - \frac{1}{2} \int \sin^2 u \, du$$

$$= \frac{x^2}{2} \arcsin x - \frac{1}{2} \left[\frac{1}{2} u - \frac{1}{4} \sin 2u \right] + C$$

$$= \frac{x^2}{2} \arcsin x - \frac{1}{4} \arcsin x + \frac{1}{8} \left(2x \sqrt{1 - x^2} \right) + C$$

$$= \arcsin x \left(\frac{x^2}{2} - \frac{1}{4} \right) + \frac{1}{4} x \sqrt{1 - x^2} + C$$

43.

$$V = \int_0^1 \pi \left(\frac{1}{1 + x^2} \right)^2 dx = \pi \int_0^1 \frac{1}{(1 + x^2)^2} \, dx$$

$$= \pi \int_0^{\pi/4} \cos^2 u \, du \qquad [x = \tan u, \ \text{see Ex. 45}]$$

$$= \frac{\pi}{2} \int_0^{\pi/4} (1 + \cos 2u) \, du$$

$$= \frac{\pi}{2} \left[u + \tfrac{1}{2} \sin 2u \right]_0^{\pi/4} = \frac{\pi^2}{8} + \frac{\pi}{4}$$

45.   We need only consider angles θ between 0 and π. Assume first that $0 \le \theta \le \frac{\pi}{2}$.

The area of the triangle is: $\frac{1}{2} r^2 \sin \theta \, \cos \theta$.

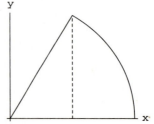

The area of the other region is given by:

$$\int_{r\cos\theta}^{r} \sqrt{r^2 - x^2}\, dx = \left[\frac{x}{2}\sqrt{r^2 - x^2} + \frac{r^2}{2}\arcsin\frac{x}{r}\right]_{r\cos\theta}^{r}$$

$$= \frac{\pi r^2}{4} - \frac{r^2}{2}\sin\theta\cos\theta - \frac{r^2}{2}\arcsin(\cos\theta)$$

$$= \frac{r^2\theta}{2} - \frac{r^2}{2}\sin\theta\cos\theta$$

Thus, the area of the sector is $A = \frac{1}{2}r^2\theta$. If $\dfrac{\pi}{2} < \theta \le \pi$, then

$$A = \tfrac{1}{2}\pi r^2 - \tfrac{1}{2}r^2(\pi - \theta) = \tfrac{1}{2}r^2\theta.$$

47. $A = 2\displaystyle\int_3^5 4\sqrt{\frac{x^2}{9} - 1}\,dx = 24\int_1^{\frac{5}{3}} \sqrt{u^2 - 1}\,du \quad$ (where $u = \frac{x}{3}$)

$$= 24\left[\frac{u}{2}\sqrt{u^2 - 1} - \frac{1}{2}\ln|u + \sqrt{u^2 - 1}|\right]_1^{\frac{5}{3}} = \frac{80}{3} - 12\ln 3.$$

49. $M = \displaystyle\int_0^a \frac{dx}{\sqrt{x^2 + a^2}} = \left[\ln\left(x + \sqrt{x^2 + a^2}\right)\right]_0^a = \ln\left(1 + \sqrt{2}\right)$

$$x_M M = \int_0^a \frac{x}{\sqrt{x^2 + a^2}}\,dx = \left[\sqrt{x^2 + a^2}\right]_0^a = (\sqrt{2} - 1)a \qquad x_M = \frac{(\sqrt{2} - 1)a}{\ln\left(1 + \sqrt{2}\right)}$$

51.

$$A = \int_a^{\sqrt{2}a} \sqrt{x^2 - a^2}\,dx = \left[\frac{1}{2}x\sqrt{x^2 - a^2} - \frac{1}{2}a^2\ln|x + \sqrt{x^2 - a^2}|\right]_a^{\sqrt{2}a}$$

$$= \tfrac{1}{2}a^2[\sqrt{2} - \ln(\sqrt{2} + 1)]$$

$$\bar{x}A = \int_a^{\sqrt{2}a} x\sqrt{x^2 - a^2}\,dx = \frac{1}{3}a^3, \qquad \bar{y}A = \int_a^{\sqrt{2}a}\left[\frac{1}{2}(x^2 - a^2)\right]dx = \frac{1}{6}a^3(2 - \sqrt{2})$$

$$\bar{x} = \frac{2a}{3[\sqrt{2} - \ln(\sqrt{2} + 1)]}, \qquad \bar{y} = \frac{(2 - \sqrt{2})a}{3[\sqrt{2} - \ln(\sqrt{2} + 1)]}$$

53. $V_y = 2\pi\overline{R}A = \dfrac{2}{3}\pi a^3 \qquad \bar{y}V_y = \displaystyle\int_a^{\sqrt{2}a}\pi x(x^2 - a^2)\,dx = \frac{1}{4}\pi a^4, \quad \bar{y} = \frac{3}{8}a$

55.   Let $x = a\sec u$. Then $dx = a\sec u\tan u\,du$ and $\sqrt{x^2 - a^2} = a\tan u$.

$$\int \frac{1}{\sqrt{x^2 - a^2}}\,dx = \int \sec u\,du = \ln|\sec u + \tan u| + C$$

$$= \ln\left|\frac{x}{a} + \frac{\sqrt{x^2 - a^2}}{a}\right| + C$$

$$= \ln|x + \sqrt{x^2 - a^2}| + K, \quad K = C - \ln a$$

57. (a)

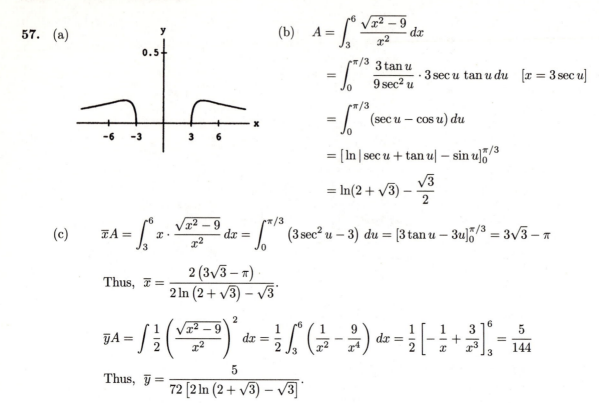

(b) $\quad A = \displaystyle\int_3^6 \frac{\sqrt{x^2-9}}{x^2}\,dx$

$$= \int_0^{\pi/3} \frac{3\tan u}{9\sec^2 u}\cdot 3\sec u\,\tan u\,du \quad [x = 3\sec u]$$

$$= \int_0^{\pi/3} (\sec u - \cos u)\,du$$

$$= \big[\ln|\sec u + \tan u| - \sin u\big]_0^{\pi/3}$$

$$= \ln(2+\sqrt{3}) - \frac{\sqrt{3}}{2}$$

(c) $\quad \overline{x}A = \displaystyle\int_3^6 x\cdot\frac{\sqrt{x^2-9}}{x^2}\,dx = \int_0^{\pi/3}\left(3\sec^2 u - 3\right)\,du = [3\tan u - 3u]_0^{\pi/3} = 3\sqrt{3} - \pi$

Thus, $\quad \overline{x} = \dfrac{2\left(3\sqrt{3}-\pi\right)}{2\ln\left(2+\sqrt{3}\right) - \sqrt{3}}.$

$$\overline{y}A = \int \frac{1}{2}\left(\frac{\sqrt{x^2-9}}{x^2}\right)^2 dx = \frac{1}{2}\int_3^6 \left(\frac{1}{x^2} - \frac{9}{x^4}\right)dx = \frac{1}{2}\left[-\frac{1}{x} + \frac{3}{x^3}\right]_3^6 = \frac{5}{144}$$

Thus, $\quad \overline{y} = \dfrac{5}{72\left[2\ln\left(2+\sqrt{3}\right) - \sqrt{3}\right]}.$

SECTION 8.5

1. $\quad \dfrac{1}{x^2 + 7x + 6} = \dfrac{1}{(x+1)(x+6)} = \dfrac{A}{x+1} + \dfrac{B}{x+6}$

$1 = A(x+6) + B(x+1)$

$x = -6: \quad 1 = -5B \quad \Longrightarrow \quad B = -1/5$

$x = -1: \quad 1 = 5A \quad \Longrightarrow \quad A = 1/5$

$\dfrac{1}{x^2+7x+6} = \dfrac{1/5}{x+1} - \dfrac{1/5}{x+6}$

3. $\quad \dfrac{x}{x^4-1} = \dfrac{x}{(x^2+1)(x+1)(x-1)} = \dfrac{Ax+B}{x^2+1} + \dfrac{C}{x+1} + \dfrac{D}{x-1}$

$x = (Ax+B)(x^2-1) + C(x-1)(x^2+1) + D(x+1)(x^2+1)$

$x = 1: \quad 1 = 4D \quad \Longrightarrow \quad D = 1/4$

$x = -1: \quad -1 = -4C \quad \Longrightarrow \quad C = 1/4$

$x = 0: \quad -B - C + D = 0 \Longrightarrow B = 0$

$x = 2: \quad 6A + 5C + 15D = 2 \Longrightarrow A = -1/2$

$\dfrac{x}{x^4-1} = \dfrac{1/4}{x-1} + \dfrac{1/4}{x+1} - \dfrac{x/2}{x^2+1}$

5. $\dfrac{x^2 - 3x - 1}{x^3 + x^2 - 2x} = \dfrac{x^2 - 3x - 1}{x(x+2)(x+1)} = \dfrac{A}{x} + \dfrac{B}{x+2} + \dfrac{C}{x-1}$

$x^2 - 3x - 1 = A(x+2)(x-1) + Bx(x-1) + Cx(x+2)$

$x = 0: \quad -1 = -2A \quad \Longrightarrow \quad A = 1/2$

$x = -2: \quad 9 = 6B \quad \Longrightarrow \quad B = 3/2$

$x = 1: \quad -3 = 3C \Longrightarrow C = -1$

$\dfrac{x^2 - 3x - 1}{x^3 + x^2 - 2x} = \dfrac{1/2}{x} + \dfrac{3/2}{x+2} - \dfrac{1}{x-1}$

7. $\dfrac{2x^2 + 1}{x^3 - 6x^2 + 11x - 6} = \dfrac{2x^2 + 1}{(x-1)(x-2)(x-3)} = \dfrac{A}{x-1} + \dfrac{B}{x-2} + \dfrac{C}{x-3}$

$2x^2 + 1 = A(x-2)(x-3) + B(x-1)(x-3) + C(x-1)(x-2)$

$x = 1: \quad 3 = 2A \quad \Longrightarrow \quad A = 3/2$

$x = 2: \quad 9 = -B \quad \Longrightarrow \quad B = -9$

$x = 3: \quad 19 = 2C \Longrightarrow C = 19/2$

$\dfrac{2x^2 + 1}{x^3 - 6x^2 + 11x - 6} = \dfrac{3/2}{x-1} - \dfrac{9}{x-2} + \dfrac{19/2}{x-3}$

9. $\dfrac{7}{(x-2)(x+5)} = \dfrac{A}{x-2} + \dfrac{B}{x+5}$

$7 = A(x+5) + B(x-2)$

$x = -5: \quad 7 = -7B \quad \Longrightarrow \quad B = -1$

$x = 2: \quad 7 = 7A \quad \Longrightarrow \quad A = 1$

$\displaystyle\int \dfrac{7}{(x-2)(x+5)}\, dx = \int\left(\dfrac{1}{x-2} - \dfrac{1}{x+5}\right) = \ln|x-2| - \ln|x+5| + C = \ln\left|\dfrac{x-2}{x+5}\right| + C$

11. We carry out the division until the numerator has degree smaller than the denominator:

$$\dfrac{2x^4 - 4x^3 + 4x^2 + 3}{x^3 - x^2} = 2x - 2 + \dfrac{2x^2 + 3}{x^2(x-1)}$$

$\dfrac{2x^2 + 3}{x^2(x-1)} = \dfrac{A}{x} + \dfrac{B}{x^2} + \dfrac{C}{x-1} \quad \Longrightarrow \quad 2x^2 + 3 = Ax(x-1) + B(x-1) + Cx^2$

$x = 0: \quad 3 = -B \quad \Longrightarrow \quad B = -3$

$x = 1: \quad 5 = C \quad \Longrightarrow \quad C = 5$

$x = -1: \quad 5 = 2A - 2B + C \quad \Longrightarrow \quad A = -3$

$\displaystyle\int\left(2x - 2 + \dfrac{2x^2 + 3}{x^2(x-1)}\right)dx = x^2 - 2x + \int\left(-\dfrac{3}{x} - \dfrac{3}{x^2} + \dfrac{5}{x-1}\right)dx$

$$= x^2 - 2x - 3\ln|x| + \dfrac{3}{x} + 5\ln|x-1| + C$$

13. We carry out the division until the numerator has degree smaller than the denominator:

$$\dfrac{x^5}{(x-2)^2} = \dfrac{x^5}{x^2 - 4x + 4} = x^3 + 4x^2 + 12x + 32 + \dfrac{80x - 128}{(x-2)^2}.$$

Then,

$$\frac{80x - 128}{(x-2)^2} = \frac{80x - 160 + 32}{(x-2)^2} = \frac{80}{x-2} + \frac{32}{(x-2)^2}$$

$$\int \frac{x^5}{(x-2)^2}\,dx = \int \left(x^3 + 4x^2 + 12x + 32 + \frac{80}{x-2} + \frac{32}{(x-2)^2} \right) dx$$

$$= \frac{1}{4}x^4 + \frac{4}{3}x^3 + 6x^2 + 32x + 80\ln|x-2| - \frac{32}{x-2} + C.$$

15. $\dfrac{x+3}{x^2 - 3x + 2} = \dfrac{A}{x-1} + \dfrac{B}{x-2}$

$$x + 3 = A(x-2) + B(x-1)$$

$x = 1:\quad 4 = -A \quad\Longrightarrow\quad A = -4$

$x = 2:\quad 5 = B \quad\Longrightarrow\quad B = 5$

$$\int \frac{x+3}{x^2 - 3x + 2}\,dx = \int \left(\frac{-4}{x-1} + \frac{5}{x-2} \right) dx = -4\ln|x-1| + 5\ln|x-2| + C$$

17. $\displaystyle\int \frac{dx}{(x-1)^3} = \int (x-1)^{-3}\,dx = -\frac{1}{2}(x-1)^{-2} + C = -\frac{1}{2(x-1)^2} + C$

19. $\dfrac{x^2}{(x-1)^2(x+1)} = \dfrac{A}{x-1} + \dfrac{B}{(x-1)^2} + \dfrac{C}{x+1}$

$$x^2 = A(x-1)(x+1) + B(x+1) + C(x-1)^2$$

$x = 1:\quad 1 = 2B \quad\Longrightarrow\quad B = 1/2$

$x = -1:\quad 1 = 4C \quad\Longrightarrow\quad C = 1/4$

$x = 0:\quad 0 = -A + B + C \quad\Longrightarrow\quad A = 3/4$

$$\int \frac{x^2}{(x-1)^2(x+1)}\,dx = \int \left(\frac{3/4}{x-1} + \frac{1/2}{(x-1)^2} + \frac{1/4}{x+1} \right) dx$$

$$= \frac{3}{4}\ln|x-1| - \frac{1}{2(x-1)} + \frac{1}{4}\ln|x+1| + C$$

21. $x^4 - 16 = (x^2 - 4)(x^2 + 4) = (x-2)(x+2)(x^2 + 4)$

$$\frac{1}{x^4 - 16} = \frac{A}{x-2} + \frac{B}{x+2} + \frac{Cx + D}{x^2 + 4}$$

$$1 = A(x+2)(x^2 + 4) + B(x-2)(x^2 + 4) + (Cx + D)(x^2 - 4)$$

$x = 2:\quad 1 = 32A \quad\Longrightarrow\quad A = 1/32$

$x = -2:\quad 1 = -32B \quad\Longrightarrow\quad B = -1/32$

$x = 0:\quad 1 = 8A - 8B - 4D \quad\Longrightarrow\quad D = -1/8$

$x = 1:\quad 1 = 15A - 5B - 3C - 3D \quad\Longrightarrow\quad C = 0$

$$\int \frac{dx}{x^4 - 16} = \int \left(\frac{1/32}{x-2} - \frac{1/32}{x+2} - \frac{1/8}{x^2+4} \right) dx$$

$$= \frac{1}{32} \ln |x-2| - \frac{1}{32} \ln |x+2| - \frac{1}{8} \left(\frac{1}{2} \arctan \frac{x}{2} \right) + C$$

$$= \frac{1}{32} \ln \left| \frac{x-2}{x+2} \right| - \frac{1}{16} \arctan \frac{x}{2} + C$$

23. $\quad \dfrac{x^3 + 4x^2 - 4x - 1}{(x^2+1)^2} = \dfrac{Ax+B}{x^2+1} + \dfrac{Cx+D}{(x^2+1)^2}$

$\qquad x^3 + 4x^2 - 4x - 1 = (Ax+B)(x^2+1) + (Cx+D)$

$x = 0: \qquad -1 = B + D \qquad\qquad\qquad \Longrightarrow \quad D = -B - 1$

$\qquad\qquad\qquad\qquad\qquad\qquad\qquad\qquad\qquad \Longrightarrow \quad B = 4, \ \ D = -5$

$x = 1: \qquad 0 = 2A + 2B + C + D \qquad \Longrightarrow \quad 6 = 4B + 2D$

$x = -1: \qquad 6 = -2A + 2B - C + D \qquad\quad 6 = -2A + 8 - C - 5$

$\qquad\qquad\qquad\qquad\qquad\qquad\qquad\qquad\qquad\qquad\qquad\qquad \Longrightarrow \quad \begin{array}{l} A = 1, \\ C = -5 \end{array}$

$x = 2: \qquad 15 = 10A + 5B + 2C + D \qquad 15 = 10A + 20 + 2C - 5$

$$\int \frac{x^3 + 4x^2 - 4x - 1}{(x^2+1)^2} \, dx = \int \left(\frac{x}{x^2+1} + \frac{4}{x^2+1} - \frac{5x}{(x^2+1)^2} - \frac{5}{(x^2+1)^2} \right) dx$$

$(*) \qquad\qquad = \dfrac{1}{2} \ln (x^2 + 1) + 4 \arctan x + \dfrac{5}{2(x^2+1)} - 5 \displaystyle\int \dfrac{dx}{(x^2+1)^2}$

For this last integral we set

$$\left\{ \begin{array}{l} x = \tan u \\ dx = \sec^2 u \, du \end{array} \right\}; \quad \int \frac{dx}{(x^2+1)^2} = \int \frac{\sec^2 u \, du}{(1 + \tan^2 u)^2} = \int \cos^2 u \, du$$

$$= \frac{1}{2} \int (1 + \cos 2u) \, du$$

$$= \tfrac{1}{2} \left(u + \tfrac{1}{2} \sin 2u \right) + C = \tfrac{1}{2} (u + \sin u \cos u) + C$$

$$= \frac{1}{2} \left(\arctan x + \frac{x}{1 + x^2} \right) + C.$$

Substituting this result in $(*)$ and rearranging the terms, we get

$$\int \frac{x^3 + 4x^2 - 4x + 1}{(x^2+1)^2} \, dx = \frac{1}{2} \ln (x^2 + 1) + \frac{3}{2} \arctan x + \frac{5(1-x)}{2(1+x^2)} + C.$$

25. $\quad \dfrac{1}{x^4 + 4} = \dfrac{Ax+B}{x^2 + 2x + 2} + \dfrac{Cx+D}{x^2 - 2x + 2} \qquad$ (using the hint)

$\qquad\qquad 1 = (Ax+B)(x^2 - 2x + 2) + (Cx+D)(x^2 + 2x + 2)$

$\begin{array}{ll} x = \ \ 0: & 1 = 2B + 2D \\ x = \ \ 1: & 1 = A + B + 5C + 5D \\ x = -1: & 1 = -5A + 5B - C + D \\ x = \ \ 2: & 1 = 4A + 2B + 20C + 10D \end{array} \Bigg\} \Longrightarrow \begin{array}{ll} A = & 1/8 \\ B = & 1/4 \\ C = & -1/8 \\ D = & 1/4 \end{array}$

$$\int \frac{dx}{x^4 + 4} = \frac{1}{8} \int \frac{x + 2}{x^2 + 2x + 2} \, dx - \frac{1}{8} \int \frac{x - 2}{x^2 - 2x + 2} \, dx$$

$$= \frac{1}{8} \int \frac{x + 1}{x^2 + 2x + 2} \, dx + \frac{1}{8} \int \frac{dx}{(x + 1)^2 + 1} - \frac{1}{8} \int \frac{x - 1}{x^2 - 2x + 2} \, dx + \frac{1}{8} \int \frac{dx}{(x - 1)^2 + 1}$$

$$= \frac{1}{16} \ln (x^2 + 2x + 2) + \frac{1}{8} \arctan (x + 1) - \frac{1}{16} \ln (x^2 - 2x + 2) + \frac{1}{8} \arctan (x - 1) + C$$

$$= \frac{1}{16} \ln \left(\frac{x^2 + 2x + 2}{x^2 - 2x + 2} \right) + \frac{1}{8} \arctan (x + 1) + \frac{1}{8} \arctan (x - 1) + C$$

27.
$$\frac{x - 3}{x^3 + x^2} = \frac{x - 3}{x^2 (x + 1)} = \frac{A}{x} + \frac{B}{x^2} + \frac{C}{x + 1}$$

$$x - 3 = Ax(x + 1) + B(x + 1) + Cx^2$$

$x = 0: \quad -3 = B$

$x = -1: \quad -4 = C$

$x = 1: \quad -2 = 2A + 2B + C \quad \Longrightarrow \quad A = 4$

$$\int \frac{x - 3}{x^3 + x^2} \, dx = 4 \int \frac{1}{x} \, dx - 3 \int \frac{1}{x^2} \, dx - 4 \int \frac{1}{x + 1} \, dx$$

$$= 4 \ln |x| + \frac{3}{x} - 4 \ln |x + 1| + C = \frac{3}{x} + 4 \ln \left| \frac{x}{x + 1} \right| + C$$

29.
$$\frac{x + 1}{x^3 + x^2 - 6x} = \frac{x + 1}{x(x - 2)(x + 3)} = \frac{A}{x} + \frac{B}{x - 2} + \frac{C}{x + 3}$$

$$x + 1 = A(x - 2)(x + 3) + Bx(x + 3) + Cx(x - 2)$$

$x = 0: \quad 1 = -6A \quad \Longrightarrow \quad A = -1/6$

$x = 2: \quad 3 = 10B \quad \Longrightarrow \quad B = 3/10$

$x = -3: \quad -2 = 15C \quad \Longrightarrow \quad C = -2/15$

$$\int \frac{x + 1}{x^3 + x^2 - 6x} \, dx = -\tfrac{1}{6} \int \tfrac{1}{x} \, dx + \tfrac{3}{10} \int \tfrac{1}{x - 2} \, dx - \tfrac{2}{15} \int \tfrac{1}{x + 3} \, dx$$

$$= -\tfrac{1}{6} \ln |x| + \tfrac{3}{10} \ln |x - 2| - \tfrac{2}{15} \ln |x + 3| + C$$

31.
$$\int_0^2 \frac{x}{x^2 + 5x + 6} \, dx = \int_0^2 \frac{x}{(x + 2)(x + 3)} \, dx = \int_0^2 \left(\frac{3}{x + 3} - \frac{2}{x + 2} \right) dx$$

$$= [3 \ln |x + 3| - 2 \ln |x + 2|]_0^2 = \ln \left(\frac{125}{108} \right)$$

33.
$$\int_1^3 \frac{x^2 - 4x + 3}{x^3 + 2x^2 + x} \, dx = \int_1^3 \frac{x^2 - 4x + 3}{x(x + 1)^2} \, dx = \int_1^3 \left(\frac{3}{x} - \frac{2}{x + 1} - \frac{8}{(x + 1)^2} \right) dx$$

$$= \left[3 \ln |x| - 2 \ln |x + 1| + \frac{8}{x + 1} \right]_1^3 = \ln \left(\frac{27}{4} \right) - 2$$

35. $\displaystyle\int \frac{\cos\theta}{\sin^2\theta - 2\sin\theta - 8}\, d\theta = \frac{1}{6}\int \frac{\cos\theta}{\sin\theta - 4}\, d\theta - \frac{1}{6}\int \frac{\cos\theta}{\sin\theta + 2}\, d\theta$

$$= \frac{1}{6}\ln|\sin\theta - 4| - \frac{1}{6}\ln|\sin\theta + 2| + C$$

$$= \frac{1}{6}\ln\left|\frac{\sin\theta - 4}{\sin\theta + 2}\right| + C$$

37. $\displaystyle\int \frac{1}{t[(\ln t)^2 - 4]}\, dt = \frac{1}{4}\int \frac{1}{t(\ln t - 2)}\, dt - \frac{1}{4}\int \frac{1}{t(\ln t + 2)}\, dt$

$$= \frac{1}{4}\ln|\ln t - 2| - \frac{1}{4}\ln|\ln t + 2| + C$$

$$= \frac{1}{4}\ln\left|\frac{\ln t - 2}{\ln t + 2}\right| + C$$

39. First we carry out the division: $\quad\displaystyle\frac{u}{a + bu} = \frac{1}{b} - \frac{a/b}{a + bu}.$

$$\int \frac{u}{a + bu}\, du = \int \left(\frac{1}{b} - \frac{a/b}{a + bu}\right) du$$

$$= \frac{1}{b}u - \frac{a}{b^2}\ln|a + bu| + K$$

$$= \frac{1}{b^2}\left(a + bu - a\ln|a + bu|\right) + C, \quad C = K - \frac{a}{b^2}$$

41. $\displaystyle\int \frac{1}{u^2(a + bu)}\, du = -\frac{b}{a^2}\int \frac{1}{u}\, du + \frac{1}{a}\int \frac{1}{u^2}\, du + \frac{b}{a^2}\int \frac{b}{a + bu}\, du$

$$= \frac{b}{a^2}\left(\ln|a + bu| - \ln|u|\right) - \frac{1}{au} + C$$

$$= \frac{b}{a^2}\ln\left|\frac{u}{a + bu}\right| - \frac{1}{au} + C$$

43. $\displaystyle\int \frac{1}{a^2 - u^2}\, du = \frac{1}{2a}\int \frac{1}{a + u}\, du + \frac{1}{2a}\int \frac{1}{a - u}\, du$

$$= \frac{1}{2a}\left(\ln\left|\frac{a + u}{a - u}\right|\right) + C$$

45. $\displaystyle\int \frac{u^2}{a^2 - u^2}\, du = \int \frac{a^2}{a^2 - u^2}\, du - \int du$

$$= a^2\left(\frac{1}{2a}\right)\left(\ln\left|\frac{a + u}{a - u}\right|\right) - u + C \quad \text{by Exercise 44}$$

$$= -u + \frac{a}{2}\left(\ln\left|\frac{a + u}{a - u}\right|\right) + C$$

47. $\displaystyle\int \frac{1}{(a + bu)(c + du)}\, du = -\frac{1}{ad - bc}\int \frac{b}{a + bu}\, du + \frac{1}{ad - bc}\int \frac{d}{c + du}\, du$

$$= \frac{1}{ad - bc}\left(\ln\left|\frac{c + du}{a + bu}\right|\right) + C$$

49. (a) $V = \int_0^{3/2} \pi \left(\dfrac{1}{\sqrt{4-x^2}}\right)^2 dx = \pi \int_0^{3/2} \dfrac{1}{4-x^2}\, dx = \dfrac{\pi}{4}\ln\left|\dfrac{2+x}{2-x}\right|_0^{3/2} = \tfrac{1}{4}\pi\ln 7$

(b) $V = \int_0^{3/2} 2\pi x \dfrac{1}{\sqrt{4-x^2}}\, dx = \pi \int_0^{3/2} \dfrac{2x}{\sqrt{4-x^2}}\, dx = -2\pi\left[\sqrt{4-x^2}\right]_0^{3/2} = \pi\left[4 - \sqrt{7}\right]$

51.

$$A = \int_0^1 \dfrac{dx}{x^2+1} = [\arctan x]_0^1 = \dfrac{1}{4}\pi$$

$$\overline{x}A = \int_0^1 \dfrac{x}{x^2+1}\, dx = \dfrac{1}{2}\left[\ln(x^2+1)\right]_0^1 = \dfrac{1}{2}\ln 2$$

$$\overline{y}A = \int_0^1 \dfrac{dx}{2(x^2+1)^2} = \dfrac{1}{2}\left[\arctan x + \dfrac{x}{x^2+1}\right]_0^1 = \dfrac{1}{8}(\pi+2)$$

$$\overline{x} = \dfrac{2\ln 2}{\pi}; \quad \overline{y} = \dfrac{\pi+2}{2\pi}$$

53. (a) $\dfrac{6x^4 + 11x^3 - 2x^2 - 5x - 2}{x^2(x+1)^3} = \dfrac{1}{x} - \dfrac{2}{x^2} + \dfrac{5}{x+1} - \dfrac{4}{(x+1)^3}$

(b) $-\dfrac{x^3 + 20x^2 + 4x + 93}{(x^2+4)(x^2-9)} = \dfrac{1}{x^2+4} + \dfrac{3}{x+3} - \dfrac{4}{x-3}$

(c) $\dfrac{x^2 + 7x + 12}{x(x^2 + 2x + 4)} = \dfrac{2x-1}{x^2+2x+4} - \dfrac{3}{x}$

55. $\dfrac{x^8 + 2x^7 + 7x^6 + 23x^5 + 10x^4 + 95x^3 - 19x^2 + 133x - 52}{x^6 + 2x^5 + 5x^4 + 16x^3 - 8x^2 + 32x - 48}$

$$= x^2 + 2 + \dfrac{2}{x-1} + \dfrac{1}{x+3} + \dfrac{x}{(x^2+4)^2} + \dfrac{3}{x^2+4}$$

57. (a)

(b) $A = \int_0^4 \dfrac{x}{x^2 + 5x + 6}\, dx$

$= \int_0^4 \dfrac{x}{(x+2)(x+3)}\, dx$

$= \int_0^4 \left(\dfrac{3}{x+3} - \dfrac{2}{x+2}\right) dx$

$= \left[3\ln|x+3| - 2\ln|x+2|\right]_0^4 = 3\ln 7 - 5\ln 3$

59. (a)

(b) $A = \int_{-2}^9 \dfrac{9-x}{(x+3)^2}\, dx$

$= \int_{-2}^9 \left(\dfrac{12}{(x+3)^2} - \dfrac{1}{x+3}\right) dx$

$= \left[-\ln|x+3| - \dfrac{12}{x+3}\right]_{-2}^9 = 11 - \ln 12$

SECTION 8.6

1. $\left\{ \begin{array}{l} x = u^2 \\ dx = 2u\,du \end{array} \right\}$; $\displaystyle\int \frac{dx}{1 - \sqrt{x}} = \int \frac{2u\,du}{1 - u} = 2 \int \left(\frac{1}{1 - u} - 1 \right) du$

$$= -2(\ln|1 - u| + u) + C = -2(\sqrt{x} + \ln|1 - \sqrt{x}|) + C$$

3. $\left\{ \begin{array}{l} u^2 = 1 + e^x \\ 2u\,du = e^x\,dx \end{array} \right\}$; $\displaystyle\int \sqrt{1 + e^x}\,dx = \int u \cdot \frac{2u\,du}{u^2 - 1} = 2 \int \left(1 + \frac{1}{u^2 - 1} \right) du$

$$= 2 \int \left(1 + \frac{1}{2}\left[\frac{1}{u - 1} - \frac{1}{u + 1} \right] \right) du$$

$$= 2u + \ln|u - 1| - \ln|u + 1| + C$$

$$= 2\sqrt{1 + e^x} + \ln\left[\frac{\sqrt{1 + e^x} - 1}{\sqrt{1 + e^x} + 1} \right] + C$$

$$= 2\sqrt{1 + e^x} + \ln\left[\frac{(\sqrt{1 + e^x} - 1)^2}{e^x} \right] + C$$

$$= 2\sqrt{1 + e^x} + 2\ln(\sqrt{1 + e^x} - 1) - x + C$$

5. (a) $\left\{ \begin{array}{l} u^2 = 1 + x \\ 2u\,du = dx \end{array} \right\}$; $\displaystyle\int x\sqrt{1 + x}\,dx = \int (u^2 - 1)(u)2u\,du$

$$= \int (2u^4 - 2u^2)\,du$$

$$= \tfrac{2}{5}u^5 - \tfrac{2}{3}u^3 + C$$

$$= \tfrac{2}{5}(x + 1)^{5/2} - \tfrac{2}{3}(x + 1)^{3/2} + C$$

(b) $\left\{ \begin{array}{l} u = 1 + x \\ du = dx \end{array} \right\}$; $\displaystyle\int x\sqrt{1 + x}\,dx = \int (u - 1)\sqrt{u}\,du = \int \left(u^{3/2} - u^{1/2} \right) du$

$$= \tfrac{2}{5}u^{5/2} - \tfrac{2}{3}u^{3/2} + C$$

$$= \tfrac{2}{5}(1 + x)^{5/2} - \tfrac{2}{3}(1 + x)^{3/2} + C$$

7. $\left\{ \begin{array}{l} u^2 = x - 1 \\ 2u\,du = dx \end{array} \right\}$; $\displaystyle\int (x + 2)\sqrt{x - 1}\,dx = \int (u + 3)(u)2u\,du$

$$= \int (2u^4 + 6u^2)\,du$$

$$= \tfrac{2}{5}u^5 + 2u^3 + C$$

$$= \tfrac{2}{5}(x - 1)^{5/2} + 2(x - 1)^{3/2} + C$$

9. $\left\{ \begin{array}{l} u^2 = 1 + x^2 \\ 2u\,du = 2x\,dx \end{array} \right\}$;

$$\int \frac{x^3}{(1+x^2)^3}\,dx = \int \frac{x^2}{(1+x^2)^3}\,x\,dx = \int \frac{u^2 - 1}{u^6}\,u\,du$$

$$= \int (u^{-3} - u^{-5})\,du = \frac{1}{2}u^{-2} + \frac{1}{4}u^{-4} + C$$

$$= \frac{1}{4(1+x^2)^2} - \frac{1}{2(1+x^2)} + C$$

$$= -\frac{1 + 2x^2}{4(1+x^2)^2} + C$$

11. $\left\{ \begin{array}{l} u^2 = x \\ 2u\,du = dx \end{array} \right\}$;

$$\int \frac{\sqrt{x}}{\sqrt{x} - 1}\,dx = \int \left(\frac{u}{u-1}\right) 2u\,du = 2\int \left(u + 1 + \frac{1}{u-1}\right) du$$

$$= u^2 + 2u + 2\ln|u - 1| + C$$

$$= x + 2\sqrt{x} + 2\ln|\sqrt{x} - 1| + C$$

13. $\left\{ \begin{array}{l} u^2 = x - 1 \\ 2u\,du = dx \end{array} \right\}$;

$$\int \frac{\sqrt{x-1} + 1}{\sqrt{x-1} - 1}\,dx = \int \frac{u+1}{u-1}\,2u\,du = \int \left(2u + 4 + \frac{4}{u-1}\right) du$$

$$= u^2 + 4u + 4\ln|u - 1| + C$$

$$= x - 1 + 4\sqrt{x-1} + 4\ln|\sqrt{x-1} - 1| + C$$

(absorb -1 in C)

$$= x + 4\sqrt{x-1} + 4\ln|\sqrt{x-1} - 1| + C$$

15. $\left\{ \begin{array}{l} u^2 = 1 + e^x \\ 2u\,du = e^x\,dx \end{array} \right\}$;

$$\int \frac{dx}{\sqrt{1+e^x}} = \int \left(\frac{1}{u}\right) \frac{2u\,du}{u^2 - 1} = \int \left[\frac{1}{u-1} - \frac{1}{u+1}\right] du$$

$$= \ln|u - 1| - \ln|u + 1| + C$$

$$= \ln \left[\frac{\sqrt{1+e^x} - 1}{\sqrt{1+e^x} + 1}\right] + C$$

$$= \ln \left[\frac{(\sqrt{1+e^x} - 1)^2}{e^x}\right] + C$$

$$= 2\ln(\sqrt{1+e^x} - 1) - x + C$$

17. $\left\{ \begin{array}{l} u^2 = x + 4 \\ 2u\,du = dx \end{array} \right\}$;

$$\int \frac{x}{\sqrt{x+4}}\,dx = \int \frac{u^2 - 4}{u}\,2u\,du = \int (2u^2 - 8)\,du$$

$$= \frac{2}{3}u^3 - 8u + C$$

$$= \frac{2}{3}(x+4)^{3/2} - 8(x+4)^{1/2} + C$$

$$= \frac{2}{3}(x - 8)\sqrt{x+4} + C$$

19. $\left\{ \begin{array}{l} u^2 = 4x + 1 \\ 2u\,du = 4\,dx \end{array} \right\};$ $\displaystyle\int 2x^2(4x+1)^{-5/2}\,dx = \int 2\left(\frac{u^2-1}{4}\right)^2 (u^{-5})\frac{u}{2}\,du$

$$= \frac{1}{16}\int (1 - 2u^{-2} + u^{-4})\,du = \frac{1}{16}u + \frac{1}{8}u^{-1} - \frac{1}{48}u^{-3} + C$$

$$= \tfrac{1}{16}(4x+1)^{1/2} + \tfrac{1}{8}(4x+1)^{-1/2} - \tfrac{1}{48}(4x+1)^{-3/2} + C$$

21. $\left\{ \begin{array}{l} u^2 = ax + b \\ 2u\,du = a\,dx \end{array} \right\};$ $\displaystyle\int \frac{x}{(ax+b)^{3/2}}\,dx = \int \frac{\dfrac{u^2-b}{a}}{u^3}\frac{2u}{a}\,du$

$$= \frac{2}{a^2}\int (1 - bu^{-2})\,du$$

$$= \frac{2}{a^2}(u + bu^{-1}) + C = \frac{2u^2 + 2b}{a^2 u} + C$$

$$= \frac{4b + 2ax}{a^2\sqrt{ax+b}} + C$$

23. $\left\{ \begin{array}{ll} u = \tan(x/2), & dx = \dfrac{2}{1+u^2}\,du \\[2mm] \sin x = \dfrac{2u}{1+u^2}, & \cos x = \dfrac{1-u^2}{1+u^2} \end{array} \right\};$

$$\int \frac{1}{1 + \cos x - \sin x}\,dx = \int \frac{1}{1 + \dfrac{1-u^2}{1+u^2} - \dfrac{2u}{1+u^2}} \cdot \frac{2}{1+u^2}\,du$$

$$= \int \frac{1}{1-u}\,du$$

$$= -\ln|1-u| + C = -\ln\left|1 - \tan\left(\frac{x}{2}\right)\right| + C$$

25. $\left\{ \begin{array}{ll} u = \tan(x/2), & dx = \dfrac{2}{1+u^2}\,du \\[2mm] \sin x = \dfrac{2u}{1+u^2} \end{array} \right\};$

$$\int \frac{1}{2 + \sin x}\,dx = \int \frac{1}{2 + \dfrac{2u}{1+u^2}} \cdot \frac{2}{1+u^2}\,du$$

$$= \int \frac{1}{u^2 + u + 1}\,du = \int \frac{1}{\left(u + \frac{1}{2}\right)^2 + \left(\frac{\sqrt{3}}{2}\right)^2}\,du$$

$$= \frac{2}{\sqrt{3}}\arctan\left(\frac{u + \frac{1}{2}}{\frac{\sqrt{3}}{2}}\right) + C$$

$$= \frac{2}{\sqrt{3}}\arctan\left[\frac{1}{\sqrt{3}}(2\tan(x/2) + 1)\right] + C$$

27.
$$\left\{ \begin{aligned} u &= \tan(x/2), \quad dx = \frac{2}{1+u^2}\,du \\ \sin x &= \frac{2u}{1+u^2}, \quad \tan x = \frac{2u}{1-u^2} \end{aligned} \right\};$$

$$\int \frac{1}{\sin x + \tan x}\,dx = \int \frac{1}{\dfrac{2u}{1+u^2} + \dfrac{2u}{1-u^2}} \cdot \frac{2}{1+u^2}\,du$$

$$= \int \frac{1-u^2}{2u}\,du = \frac{1}{2}\int \left(\frac{1}{u} - u\right)\,du$$

$$= \tfrac{1}{2}\left(\ln|u| - \tfrac{1}{2}u^2\right) + C = \tfrac{1}{2}\ln|\tan(x/2)| - \tfrac{1}{4}[\tan(x/2)]^2 + C$$

29.
$$\left\{ \begin{aligned} u &= \tan(x/2), \quad dx = \frac{2}{1+u^2}\,du \\ \sin x &= \frac{2u}{1+u^2}, \quad \cos x = \frac{1-u^2}{1+u^2} \end{aligned} \right\};$$

$$\int \frac{1-\cos x}{1+\sin x}\,dx = \int \frac{1 - \dfrac{1-u^2}{1+u^2}}{1 + \dfrac{2u}{1+u^2}} \cdot \frac{2}{1+u^2}\,du$$

$$= \int \frac{4u^2}{(1+u^2)(u+1)^2}\,du$$

$$= \int \left[\frac{2u}{1+u^2} - \frac{2}{u+1} + \frac{2}{(u+1)^2}\right]\,du$$

$$= \ln(u^2+1) - 2\ln|u+1| - \frac{2}{u+1} + C$$

$$= \ln\left[\frac{u^2+1}{(u+1)^2} - \frac{2}{u+1}\right] + C$$

$$= \ln\left[\frac{\tan^2(x/2)+1}{(\tan(x/2)+1)^2}\right] - \frac{2}{\tan(x/2)+1} + C = \ln\left|\frac{1}{1+\sin x}\right| - \frac{2}{\tan(x/2)+1} + C$$

31.
$$\left\{ \begin{aligned} u^2 &= x \\ 2u\,du &= dx \end{aligned} \right\}; \qquad \int \frac{x^{3/2}}{x+1}\,dx = \int \frac{u^3}{u^2+1}\,2u\,du$$

$$= \int \frac{2u^4}{u^2+1}\,du = \int \left[2u^2 - 2 + \frac{2}{u^2+1}\right]\,du$$

$$= \frac{2}{3}u^3 - 2u + 2\arctan u + C = \frac{2}{3}x^{3/2} - 2x^{1/2} + 2\arctan x^{1/2} + C$$

$$\int_0^4 \frac{x^{3/2}}{x+1}\,dx = \left[\frac{2}{3}x^{3/2} - 2x^{1/2} + 2\arctan x^{1/2}\right]_0^4 = \frac{4}{3} + 2\arctan 2$$

33.
$$\int_0^{\pi/2} \frac{\sin 2x}{2 + \cos x} \, dx = \int_0^{\pi/2} \frac{2 \sin x \cos x}{2 + \cos x} \, dx$$

$$= \int_0^1 \frac{2u}{2 + u} \, du \qquad [u = \cos x, \ \ du = -\sin x \, dx]$$

$$= \int_0^1 \left(2 - \frac{4}{2 + u} \right) du$$

$$= [2u - 4 \ln |2 + u|]_0^1 = 2 + 4 \ln \left(\tfrac{2}{3} \right)$$

35.
$$\left\{ \begin{array}{ll} u = \tan(x/2), & dx = \dfrac{2}{1 + u^2} \, du \\[2mm] \sin x = \dfrac{2u}{1 + u^2}, & \cos x = \dfrac{1 - u^2}{1 + u^2} \end{array} \right\} ;$$

$$\int \frac{1}{\sin x - \cos x - 1} \, dx = \int \frac{1}{\dfrac{2u}{1 + u^2} - \dfrac{1 - u^2}{1 + u^2} - 1} \cdot \frac{2}{1 + u^2} \, du$$

$$= \int \frac{1}{u - 1} \, du$$

$$= \ln |u - 1| + C = \ln |\tan(x/2) - 1| + C$$

$$\int_0^{\pi/3} \frac{1}{\sin x - \cos x - 1} \, dx = [\ln |\tan(x/2) - 1|]_0^{\pi/3} = \ln \left(\frac{\sqrt{3} - 1}{\sqrt{3}} \right)$$

37.
$$\left\{ \begin{array}{ll} u = \tan(x/2), & dx = \dfrac{2}{1 + u^2} \, du \\[2mm] \sin x = \dfrac{2u}{1 + u^2}, & \cos x = \dfrac{1 - u^2}{1 + u^2} \end{array} \right\} ;$$

$$\int \sec x \, dx = \int \frac{1}{\cos x} \, dx = \int \frac{1}{\dfrac{1 - u^2}{1 + u^2}} \cdot \frac{2}{1 + u^2} \, du$$

$$= 2 \int \frac{1}{1 - u^2} \, du = 2 \int \left[\frac{1/2}{1 - u} + \frac{1/2}{1 + u} \right] du$$

$$= \int \left[\frac{1}{1 - u} + \frac{1}{1 + u} \right] du = -\ln |1 - u| + \ln |1 + u| + C$$

$$= \ln \left| \frac{1 + \tan(x/2)}{1 - \tan(x/2)} \right| + C$$

39.
$$\int \csc x \, dx = \int \frac{\sin x}{\sin^2 x} \, dx = \int \frac{\sin x}{1 - \cos^2 x} \, dx$$

$$= -\int \frac{1}{1 - u^2} \, du \qquad [u = \cos x, \ \ du = -\sin x \, dx]$$

$$= \frac{1}{2} \int \left[\frac{1}{u - 1} - \frac{1}{u + 1} \right] du$$

$$= \frac{1}{2} [\ln |u - 1| - \ln |u + 1|] + C = \ln \sqrt{\frac{1 - \cos x}{1 + \cos x}} + C$$

41.
$$\left\{ \begin{array}{ll} u = \tanh(x/2), & dx = \dfrac{2}{1-u^2}\,du \\[2mm] \cosh x = \dfrac{1+u^2}{1-u^2}, & \operatorname{sech} x = \dfrac{1-u^2}{1+u^2} \end{array} \right\};$$

$$\int \operatorname{sech} x\,dx = \int \frac{1-u^2}{1+u^2}\cdot\frac{2}{1-u^2}\,du$$

$$= \int \frac{2}{1+u^2}\,du = 2\arctan u + C = 2\arctan\left(\tanh(x/2)\right) + C$$

43.
$$\left\{ \begin{array}{ll} u = \tanh(x/2), & dx = \dfrac{2}{1-u^2}\,du \\[2mm] \sinh x = \dfrac{2u}{1-u^2}, & \cosh x = \dfrac{1+u^2}{1-u^2} \end{array} \right\};$$

$$\int \frac{1}{\sinh x + \cosh x}\,dx = \int \frac{1}{\dfrac{2u}{1-u^2} + \dfrac{1+u^2}{1-u^2}}\cdot\frac{2}{1-u^2}\,du$$

$$= \int \frac{2}{(1+u)^2}\,du = \frac{-2}{u+1} + C = \frac{-2}{\tanh(x/2)+1} + C$$

SECTION 8.7

1. (a) $\quad L_{12} = \frac{12}{12}[0+1+4+9+16+25+36+49+64+81+100+121] = 506$

(b) $\quad R_{12} = \frac{12}{12}[1+4+9+16+25+36+49+64+81+100+121+144] = 650$

(c) $\quad M_6 = \frac{12}{6}[1+9+25+49+81+121] = 572$

(d) $\quad T_{12} = \frac{12}{24}[0+2(1+4+9+16+25+36+49+64+81+100+121)+144] = 578$

(e) $\quad S_6 = \frac{12}{36}[0+144+2(4+16+36+64+100)+4(1+9+25+49+81+121)] = 576$

$$\int_0^{12} x^2\,dx = \left[\frac{1}{3}x^3\right]_0^{12} = 576$$

3. (a) $\quad L_6 = \dfrac{3}{6}\left[\dfrac{1}{1+0} + \dfrac{1}{1+1/8} + \dfrac{1}{1+1} + \dfrac{1}{1+27/8} + \dfrac{1}{1+8} + \dfrac{1}{1+125/8}\right]$

$\qquad = \frac{1}{2}\left[1 + \frac{8}{9} + \frac{1}{2} + \frac{8}{35} + \frac{1}{9} + \frac{8}{133}\right] \cong 1.394$

(b) $\quad R_6 = \dfrac{3}{6}\left[\dfrac{1}{1+1/8} + \dfrac{1}{1+1} + \dfrac{1}{1+27/8} + \dfrac{1}{1+8} + \dfrac{1}{1+125/8} + \dfrac{1}{1+27}\right]$

$\qquad = \frac{1}{2}\left[\frac{8}{9} + \frac{1}{2} + \frac{8}{35} + \frac{1}{9} + \frac{8}{133} + \frac{1}{28}\right] \cong 0.9122$

(c) $\quad M_3 = \dfrac{3}{3}\left[\dfrac{1}{1+1/8} + \dfrac{1}{1+27/8} + \dfrac{1}{1+125/8}\right] = \dfrac{8}{9} + \dfrac{8}{35} + \dfrac{8}{133} \cong 1.1776$

(d) $\quad T_6 = \frac{3}{12}\left[1 + 2\left(\frac{8}{9} + \frac{1}{2} + \frac{8}{35} + \frac{1}{9} + \frac{8}{133}\right) + \frac{1}{28}\right] \cong 1.1533$

(e) $\quad S_3 = \frac{3}{18}\left\{1 + \frac{1}{28} + 2\left[\frac{1}{2} + \frac{1}{9}\right] + 4\left[\frac{8}{9} + \frac{8}{35} + \frac{8}{133}\right]\right\} \cong 1.1614$

5. (a) $\frac{1}{4}\pi \cong T_4 = \frac{1}{8}\left[1 + 2\left(\frac{1}{1+1/16} + \frac{1}{1+1/4} + \frac{1}{1+9/16}\right) + \frac{1}{1+1}\right]$

$= \frac{1}{8}\left[1 + 2\left(\frac{16}{17} + \frac{4}{5} + \frac{16}{25}\right) + \frac{1}{2}\right] \cong 0.7828$

$\pi \cong 4(0.7828) = 3.1312$

(b) $\frac{1}{4}\pi \cong S_4 = \frac{1}{24}\left[1 + \frac{1}{2} + 2\left(\frac{16}{17} + \frac{4}{5} + \frac{16}{25}\right) + 4\left(\frac{64}{65} + \frac{64}{73} + \frac{64}{89} + \frac{64}{113}\right)\right] \cong 0.7854$

$\pi \cong 4(0.7854) = 3.1416$

7. (a) $M_4 = \frac{2}{4}\left[\cos\left(\frac{-3}{4}\right)^2 + \cos\left(\frac{-1}{4}\right)^2 + \cos\left(\frac{1}{4}\right)^2 + \cos\left(\frac{3}{4}\right)^2\right] \cong 1.8440$

(b) $T_8 = \frac{2}{16}\left[\cos(-1)^2 + 2\cos\left(\frac{-3}{4}\right)^2 + 2\cos\left(\frac{-1}{2}\right)^2 + 2\cos\left(\frac{-1}{4}\right)^2 + \right.$

$\left. 2\cos(0)^2 + 2\cos\left(\frac{1}{4}\right)^2 + 2\cos\left(\frac{1}{2}\right)^2 + 2\cos\left(\frac{3}{4}\right)^2 + \cos(1)^2\right] \cong 1.7915$

(c) $S_4 = \frac{2}{24}\left\{\cos(-1)^2 + \cos(1)^2 + 2\left[\cos\left(\frac{-1}{2}\right)^2 + \cos(0)^2 + \cos\left(\frac{1}{2}\right)^2\right] + \right.$

$\left. 4\left[\cos\left(\frac{-3}{4}\right)^2 + \cos\left(\frac{-1}{4}\right)^2 + \cos\left(\frac{1}{4}\right)^2 + \cos\left(\frac{3}{4}\right)^2\right]\right\} \cong 1.8090$

9. (a) $T_{10} = \frac{2}{20}\left[e^{-0^2} + 2e^{-(1/5)^2} + 2e^{-(2/5)^2} + 2e^{-(3/5)^2} + 2e^{-(4/5)^2} + 2e^{-1^2} + 2e^{-(6/5)^2} \right.$

$\left. + 2e^{-(7/5)^2} + 2e^{-(8/5)^2} + 2e^{-(9/5)^2} + e^{-2^2}\right] \cong 0.8818$

(b) $S_5 = \frac{2}{30}\left\{e^{-0^2} + e^{-2^2} + 2\left[e^{-(2/5)^2} + e^{-(4/5)^2} + e^{-(6/5)^2} + e^{-(8/5)^2}\right]\right.$

$\left. + 4\left[e^{-(1/5)^2} + e^{-(3/5)^2} + e^{-1^2} + e^{-(7/5)^2} + e^{-(9/5)^2}\right]\right\} \cong 0.8821$

11. Such a curve passes through the three points

$$(a_1, b_1), \quad (a_2, b_2), \quad (a_3, b_3)$$

iff

$$b_1 = a_1{}^2 A + a_1 B + C, \quad b_2 = a_2{}^2 A + a_2 B + C, \quad b_3 = a_3{}^2 A + a_3 B + C,$$

which happens iff

$$A = \frac{b_1(a_2 - a_3) - b_2(a_1 - a_3) + b_3(a_1 - a_2)}{(a_1 - a_3)(a_1 - a_2)(a_2 - a_3)},$$

$$B = -\frac{b_1(a_2{}^2 - a_3{}^2) - b_2(a_1{}^2 - a_3{}^2) + b_3(a_1{}^2 - a_2{}^2)}{(a_1 - a_3)(a_1 - a_2)(a_2 - a_3)},$$

$$C = \frac{a_1{}^2(a_2 b_3 - a_3 b_2) - a_2{}^2(a_1 b_3 - a_3 b_1) + a_3{}^2(a_1 b_2 - a_2 b_1)}{(a_1 - a_3)(a_1 - a_2)(a_2 - a_3)}.$$

13. (a) $\left|\dfrac{(b-a)^3}{12n^2}f''(c)\right| = \dfrac{27}{12n^2}\dfrac{1}{4c^{3/2}} \le \dfrac{9}{16n^2} < 0.01 \implies n^2 > \left(\dfrac{15}{2}\right)^2 \implies n \ge 8$

(b) $\left|\dfrac{(b-a)^5}{2880n^4}f^{(4)}(c)\right| = \dfrac{243}{2880n^4}\dfrac{15}{16c^{7/2}} \le \dfrac{81}{1024n^4} < 0.01 \implies n \ge 2$

15. (a) $\left|\dfrac{(b-a)^3}{12n^2}f''(c)\right| = \dfrac{27}{12n^2}\dfrac{1}{4c^{3/2}} \le \dfrac{9}{16n^2} < 0.00001 \implies n > 75\sqrt{10} \implies n \ge 238$

(b) $\left|\dfrac{(b-a)^5}{2880n^4}f^{(4)}(c)\right| = \dfrac{243}{2880n^4}\dfrac{15}{16c^{7/2}} \le \dfrac{81}{1024n^4} < 0.00001 \implies n \ge 10$

17. (a) $\left|\dfrac{(b-a)^3}{12n^2}f''(c)\right| = \dfrac{\pi^3}{12n^2}\sin c \le \dfrac{\pi^3}{12n^2} < 0.001 \implies n > 5\pi\sqrt{\dfrac{10\pi}{3}} \implies n \ge 51$

(b) $\left|\dfrac{(b-a)^5}{2880n^4}f^{(4)}(c)\right| = \dfrac{\pi^5}{2880n^4}\sin c \le \dfrac{\pi^5}{2880n^4} < 0.001 \implies n \ge 4$

19. (a) $\left|\dfrac{(b-a)^3}{12n^2}f''(c)\right| = \dfrac{8}{12n^2}e^c \le \dfrac{8}{12n^2}e^3 < 0.01 \implies n > 10e\sqrt{\dfrac{2e}{3}} \implies n \ge 37$

(b) $\left|\dfrac{(b-a)^5}{2880n^4}f^{(4)}(c)\right| = \dfrac{32}{2880n^4}e^c \le \dfrac{1}{90n^4}e^3 < 0.01 \implies n \ge 3$

21. (a) $\left|\dfrac{(b-a)^3}{12n^2}f''(c)\right| = \left|\dfrac{8}{12n^2}2e^{-c^2}(2c^2-1)\right| \le \dfrac{8}{3n^2}e^{-3/2} < 0.0001$

$\implies n > 100\sqrt{\tfrac{8}{3}e^{-3/2}} \implies n \ge 78$

(b) $\left|\dfrac{(b-a)^5}{2880n^4}f^{(4)}(c)\right| = \left|\dfrac{32}{2880n^4}4e^{-c^2}\left(4c^4-12c^2+3\right)\right| \le \dfrac{32}{2880n^4}12 < 0.0001$

$\implies n > 10\left[\dfrac{32\cdot 12}{2880}\right]^{1/4} \implies n \ge 7$

23. $f^{(4)}(x) = 0$ for all x;   therefore by (8.7.3) the theoretical error is zero

25. (a) $\left|T_2 - \displaystyle\int_0^1 x^2\,dx\right| = \dfrac{3}{8} - \dfrac{1}{3} = \dfrac{1}{24} = E_2^T$

(b) $\left|S_1 - \displaystyle\int_0^1 x^4\,dx\right| = \dfrac{5}{24} - \dfrac{1}{5} = \dfrac{1}{120} = E_1^S$

27. (a) Let f be twice differentiable on $[a,b]$ with $f(x) > 0$ and $f''(x) > 0$, and let $P = \{x_0, x_1, x_2, \ldots, x_n\}$ be a regular partition of $[a,b]$. Figure A shows a typical subinterval with the approximating trapezoid ABCD. Since the area under the curve is less than the area of the trapezoid, we can conclude that

$$\int_a^b f(x)\,dx \le T_n.$$

Figure A

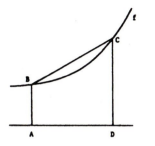

Figure B

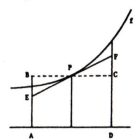

Now consider Figure B. Since the triangles EBP and PFC are congruent, the area of the rectangle ABCD equals the area of the trapezoid AEFD, and since the area under the curve is greater than the area of AEFD it follows that

$$M_n \leq \int_a^b f(x)\, dx.$$

(b) A similar argument in the case $f(x) > 0$, $f''(x) < 0$ gives

$$T_n \leq \int_a^b f(x)\, dx \leq M_n.$$

29. (a) $\displaystyle\int_0^{10} (x + \cos x)\, dx \cong 49.4578$

(b) $\displaystyle\int_{-4}^{7} \left(x^5 - 5x^4 + x^3 - 3x^2 - x + 4\right)\, dx \cong 1280.56$

31. $\displaystyle \left|E_{20}^S\right| \leq \frac{4^5}{2880(20)^4} \max\left|f^{(4)}(t)\right| \leq \frac{1024}{2880(20)^4}\,(0.1806) \leq 4.01 \times 10^{-7}$

REVIEW EXERCISES

1. Substitution: let $u = \sin x$, $du = \cos x\, dx$

$$\int \frac{\cos x}{4 + \sin^2 x}\, dx = \int \frac{1}{4 + u^2}\, du = \frac{1}{2}\arctan\left(\frac{u}{2}\right) + C = \frac{1}{2}\arctan\left(\frac{\sin x}{2}\right) + C$$

3. Integration by parts: $\begin{Bmatrix} u = x & dv = \sinh x\, dx \\ du = dx & v = \cosh x \end{Bmatrix}$

$$\int 2x \sinh x\, dx = 2\int x \sinh x\, dx = 2\left[x \cosh x - \int \cosh x\, dx\right] = 2x\cosh x - 2\sinh x + C$$

5. Partial fractions: $\dfrac{x-3}{x^2(x+1)} = \dfrac{4}{x} - \dfrac{3}{x^2} - \dfrac{4}{x+1}$

$$\int \frac{x-3}{x^2(x+1)}\,dx = \int \left(\frac{4}{x} - \frac{3}{x^2} - \frac{4}{x+1}\right) dx = 4\ln|x| + \frac{3}{x} - 4\ln|x+1| + C$$

7. $\displaystyle\int \sin 2x \cos x\,dx = \int (2\sin x \cos x)\cos x\,dx = 2\int \cos^2 x \sin x\,dx$

Substitution: $u = \cos x,\ du = -\sin x\,dx$

$$2\int \cos^2 x \sin x\,dx = -2\int u^2\,du = -\tfrac{2}{3}u^3 + C = -\tfrac{2}{3}\cos^3 x + C$$

9. $\displaystyle\int \ln\sqrt{x+1}\,dx = \int \ln(x+1)^{1/2}\,dx = \tfrac{1}{2}\int \ln(x+1)\,dx$

Integration by parts: $\left\{\begin{array}{ll} u = \ln(x+1) & dv = dx \\ du = \dfrac{1}{x+1}\,dx & v = x+1 \end{array}\right\}$

$$\frac{1}{2}\int \ln(x+1)\,dx = \frac{1}{2}\left[(x+1)\ln(x+1) - \int dx\right] = \frac{1}{2}\left[(x+1)\ln(x+1) - x\right] + C$$

$$\int_0^3 \ln\sqrt{x+1}\,dx = \frac{1}{2}\left[(x+1)\ln(x+1) - x\right]\Big|_0^3 = 4\ln 2 - \frac{3}{2}$$

Note: This answer can also be written: $\quad \tfrac{1}{2}\left[(x+1)\ln(x+1) - (x+1)\right] + C$;

set $w = x+1,\ dw = dx$, and integrate $\displaystyle\int \ln w\,dw$ by parts.

11. $\displaystyle\int \frac{\sin^3 x}{\cos x}\,dx = \int \frac{(1 - \cos^2 x)\sin x}{\cos x}\,dx = \int \tan x\,dx - \int \cos x \sin x\,dx = \ln|\sec x| - \frac{1}{2}\sin^2 x + C$

13. $\displaystyle\int_0^1 e^{-x}\cosh x\,dx = \int_0^1 e^{-x}\left[\frac{e^x + e^{-x}}{2}\right]dx = \frac{1}{2}\int_0^1 \left[1 + e^{-2x}\right]dx$

$$= \frac{1}{2}\left[x - \frac{e^{-2x}}{2}\right]_0^1 = \frac{1}{2}\left[1 - \frac{e^{-2}}{2} + \frac{1}{2}\right] = \frac{3}{4} - \frac{e^{-2}}{2}$$

15. $\displaystyle\int \frac{dx}{e^x - 4e^{-x}} = \int \frac{e^x}{e^{2x} - 4}\,dx;\quad$ substitution: $u = e^x,\ du = e^x\,dx$

$$\int \frac{dx}{e^x - 4e^{-x}} = \int \frac{1}{u^2 - 4}\,du = \int \left(\frac{1/4}{u-2} - \frac{1/4}{u+2}\right)du = \frac{1}{4}\left[\ln|u-2| - \ln|u+2|\right] + C$$

$$= \frac{1}{4}\left[\ln|e^x - 2| - \ln|e^x + 2|\right] + C = \frac{1}{4}\ln\left|\frac{e^x - 2}{e^x + 2}\right| + C$$

t

17. Integration by parts: $\left\{\begin{array}{ll} u = x & dv = 2^x\,dx \\ du = dx & v = 2^x/\ln 2 \end{array}\right\}$

$$\int x2^x\,dx = \frac{1}{\ln 2}x2^x - \frac{1}{\ln 2}\int 2^x\,dx = \frac{1}{\ln 2}\left(x2^x - \frac{1}{\ln 2}2^x\right) + C$$

19. Trigonometric substitution: set $x = a \sin u, \, dx = a \cos u \, du$

$$\int \frac{\sqrt{a^2 - x^2}}{x^2} \, dx = \int \frac{a \cos u}{a^2 \sin^2 u} \, a \cos u \, du = \int \frac{\cos^2 u}{\sin^2 u} \, du = \int \frac{1 - \sin^2 u}{\sin^2 u} \, du$$

$$= \int \csc^2 u \, du - \int du = -\cot u - u + c$$

$$= -\frac{\sqrt{a^2 - x^2}}{x} - \arcsin\left(\frac{x}{a}\right) + C$$

21. $\displaystyle\int x^3 e^{x^2} \, dx = \int x^2 \, x e^{x^2} \, dx$; integration by parts: $\left\{ \begin{array}{ll} u = x^2 & dv = x e^{x^2} \, dx \\ du = 2x \, dx & v = \frac{1}{2} e^{x^2} \end{array} \right\}$

$$\int x^3 e^{x^2} \, dx = \frac{1}{2} \left[x^2 e^{x^2} - \int 2x e^{x^2} \, dx^2 \right] = \frac{1}{2} \left[x^2 e^{x^2} - e^{x^2} \right] + C$$

23. $\displaystyle\int \frac{\sin^5 x}{\cos^7 x} \, dx = \int \tan^5 \sec^2 \, dx = \frac{1}{6} \tan^6 x + C$ (substitution: $u = \tan x$)

25.
$$\int_{\pi/6}^{\pi/3} \frac{\sin x}{\sin 2x} \, dx = \int_{\pi/6}^{\pi/3} \frac{\sin x}{2 \sin x \cos x} \, dx = \frac{1}{2} \int_{\pi/6}^{\pi/3} \sec x \, dx$$

$$= \frac{1}{2} \left[\ln|\sec x + \tan x| \right]_{\pi/6}^{\pi/3} = \frac{1}{2} \ln(2 + \sqrt{3}) - \frac{1}{4} \ln 3$$

27. $\displaystyle\int \frac{x^2 + x}{\sqrt{1 - x^2}} \, dx = \int \frac{x^2}{\sqrt{1 - x^2}} \, dx + \int \frac{x}{\sqrt{1 - x^2}} \, dx$

$$\int \frac{x}{\sqrt{1 - x^2}} \, dx = -\sqrt{1 - x^2} + C_1, \quad \text{substitution: set } u = 1 - x^2, \, du = -2 \, x dx$$

For the first integral, use the trigonometric substitution: $x = \sin u, \; dx = \cos u \, du$

$$\int \frac{x^2}{\sqrt{1 - x^2}} \, dx = \int \frac{\sin^2 u}{\cos u} \cos u \, du = \int \sin^2 u \, du = \int \frac{1 - \cos 2u}{2} \, du$$

$$= \tfrac{1}{2} u - \tfrac{1}{4} \sin 2u + C_2 = \tfrac{1}{2} u - \tfrac{1}{2} \sin u \cos u + C_2$$

$$= \frac{1}{2} \arcsin x - \frac{1}{2} x \sqrt{1 - x^2} + C_2$$

Therefore,
$$\int \frac{x^2 + x}{\sqrt{1 - x^2}} \, dx = \tfrac{1}{2} \arcsin x - \tfrac{1}{2} x \sqrt{1 - x^2} - \sqrt{1 - x^2} + C$$

29. Since $\dfrac{\cos^4 x}{\sin^2 x} = \dfrac{(1 - \sin^2 x)^2}{\sin^2 x} = \dfrac{1 - 2 \sin^2 x + \sin^4 x}{\sin^2 x} = \csc^x -2 + \sin x$

$$\int \frac{\cos^4 x}{\sin^2 x} \, dx = \int (\csc^x -2 + \sin x) dx = -\cot x - \frac{3}{2} x - \frac{1}{4} \sin 2x + C$$

31. $\displaystyle\int (\sin 2x + \cos 2x)^2 dx = \int (1 + 2 \sin 2x \cos 2x) \, dx = x + \int \sin 4x \, dx = x - \frac{1}{4} \cos 4x + C$

33. $\dfrac{5x+1}{(x+2)(x^2-2x+1)} = \dfrac{-1}{x+2} + \dfrac{1}{x-1} + \dfrac{2}{(x-1)^2}$

$$\int \frac{5x+1}{(x+2)(x^2-2x+1)}\,dx = \int \left\{ \frac{-1}{x+2} + \frac{1}{x-1} + \frac{2}{(x-1)^2} \right\} dx$$

$$= -\ln|x+2| + \ln|x-1| - \frac{2}{x-1} + C$$

35. $\displaystyle\int \frac{1}{\sqrt{x+1}-\sqrt{x}}\,dx = \int \frac{\sqrt{x+1}+\sqrt{x}}{(\sqrt{x+1}-\sqrt{x})(\sqrt{x+1}+\sqrt{x})}\,dx$

$$= \int (\sqrt{x+1}+\sqrt{x})\,dx = \frac{2}{3}(x+1)^{3/2} + \frac{2}{3}x^{3/2} + C$$

37. integration by parts: $\begin{cases} u = x^2 & dv = \cos 2x\,dx \\ du = 2x\,dx & v = \frac{1}{2}\sin 2x \end{cases}$

$$\int x^2 \cos 2x\,dx = \frac{1}{2}x^2 \sin 2x - \int x \sin 2x\,dx$$

integration by parts again: $\begin{cases} u = x & dv = \sin 2x\,dx \\ du = dx & v = -\frac{1}{2}\cos 2x \end{cases}$

$$\int x^2 \cos 2x\,dx = \frac{1}{2}x^2 \sin 2x - \left[-\frac{1}{2}x\cos 2x + \frac{1}{2}\int \cos 2x\,dx \right] = \frac{1}{2}x^2 \sin 2x + \frac{1}{2}x\cos 2x - \frac{1}{4}\sin 2x + C$$

39. $\dfrac{1-\sin 2x}{1+\sin 2x} = \dfrac{1-\sin 2x}{1+\sin 2x}\dfrac{1-\sin 2x}{1-\sin 2x} = \dfrac{1-2\sin 2x+\sin^2 2x}{1-\sin^2 2x} = \dfrac{1-2\sin 2x+\sin^2 2x}{\cos^2 2x}$

$$= \sec^2 2x - \frac{2\sin 2x}{\cos^2 2x} + \tan^2 2x = 2\sec^2 2x - 1 - \frac{2\sin 2x}{\cos^2 2x}$$

Therefore

$$\int \frac{1-\sin 2x}{1+\sin 2x}\,dx = \tan 2x - x - \frac{1}{\cos 2x} + C = \tan 2x - x - \sec 2x + C$$

41. (a) integration by parts: $\begin{cases} u = x^n & dv = \cos ax\,dx \\ du = nx^{n-1}\,dx & v = \frac{1}{a}\sin ax \end{cases}$

$$\int x^n \cos ax\,dx = \frac{x^n \sin ax}{a} - \frac{n}{a}\int x^{n-1}\sin ax\,dx$$

(b) integration by parts: $\begin{cases} u = x^n & dv = \sin ax\,dx \\ du = nx^{n-1}\,dx & v = -\frac{1}{a}\cos ax \end{cases}$

$$\int x^n \sin ax\,dx = -\frac{x^n \cos ax}{a} + \frac{n}{a}\int x^{n-1}\cos ax\,dx$$

43. (a) integration by parts: $\begin{cases} u = (\ln x)^n & dv = x^m\,dx \\ du = n(\ln x)^{n-1}\frac{1}{x}\,dx & v = \frac{1}{m+1}x^{m+1} \end{cases}$

$$\int x^m (\ln x)^n\,dx = \frac{x^{m+1}(\ln x)^n}{m+1} - \frac{n}{m+1}\int x^m (\ln x)^{n-1}\,dx$$

(b) From (a),

$$\int x^4 (\ln x)^3 dx = \frac{x^5 (\ln x)^3}{5} - \frac{3}{5} \int x^4 (\ln x)^2 dx$$

$$= \frac{x^5 (\ln x)^3}{5} - \frac{3}{5} \left\{ \frac{x^5 (\ln x)^2}{5} - \frac{2}{5} \int x^4 \ln x \, dx \right\}$$

$$= \frac{1}{5} x^5 (\ln x)^3 - \frac{3}{25} x^5 (\ln x)^2 + \frac{6}{25} \left\{ \frac{x^5 \ln x}{5} - \frac{1}{5} \int x^4 dx \right\}$$

$$= \frac{1}{5} x^5 (\ln x)^3 - \frac{3}{25} x^5 (\ln x)^2 + \frac{6}{125} x^5 \ln x - \frac{6}{125} x^5 dx + C$$

45.

$$A = \int_0^{\frac{1}{2}} (1 - x^2)^{-1/2} \, dx = \left[\arcsin x \right]_0^{1/2} = \frac{\pi}{6}$$

$$\bar{x} A = \int_0^{1/2} x(1 - x^2)^{-1/2} \, dx = -\left[\sqrt{1 - x^2} \right]_0^{1/2} = 1 - \frac{\sqrt{3}}{2} \implies \bar{x} = \frac{6}{\pi} \left(1 - \frac{\sqrt{3}}{2} \right)$$

$$\bar{y} A = \int_0^{1/2} \frac{1}{2} (1 - x^2)^{-1} \, dx = \frac{1}{4} \int_0^{1/2} \left[\frac{1}{1 - x} + \frac{1}{1 + x} \right] dx = \frac{1}{4} \left[\ln \left(\frac{1 + x}{1 - x} \right) \right]_0^{1/2} = \frac{1}{4} \ln 3$$

Therefore, $\bar{y} = \dfrac{3 \ln 3}{2\pi}$

47. Use Pappus's theorem and Problem 45.

(a) about the x-axis: $V = 2\pi \left(\frac{3 \ln 3}{2\pi} \right) \frac{\pi}{6} = \frac{\pi \ln 3}{2}$

(b) about the y-axis: $V = 2\pi \left[\frac{6}{\pi} \left(1 - \frac{\sqrt{3}}{2} \right) \right] \frac{\pi}{6} = 2\pi \left(1 - \frac{\sqrt{3}}{2} \right)$

49. Let $f = \sqrt{x^3 + x}$.

(a) $\displaystyle\int_0^2 \sqrt{x^3 + x} \, dx = \frac{2}{4} \left[f \left(\frac{0 + 1/2}{2} \right) + f \left(\frac{1/2 + 1}{2} \right) + f \left(\frac{1 + 3/2}{2} \right) + f \left(\frac{3/2 + 2}{2} \right) \right]$

$$= \frac{1}{2} [f(1/4) + f(3/4) + f(5/4) + f(7/4)] \approx 3.0270$$

(b) $\displaystyle\int_0^2 \sqrt{x^3 + x} \, dx$

$$= \frac{2}{16} [f(0) + 2f(1/4) + 2f(1/2) + 2f(3/4) + 2f(1) + 2f(5/4) + 2f(3/2) + 2f(7/4) + f(2)]$$

$$= \frac{1}{8} [2f(1/4) + 2f(1/2) + 2f(3/4) + 2f(1) + 2f(5/4) + 2f(3/2) + 2f(7/4) + f(2)] \approx 3.0120$$

(c) $\displaystyle\int_0^2 \sqrt{x^3 + x} \, dx$

$$= \frac{2}{24} \{ f(0) + 2[f(1/2) + 2f(1) + f(3/2)] + 4[f(1/4) + f(3/4) + f(5/4) + f(7/4)] + f(2) \}$$

$$= \frac{1}{12} \{ 2[f(1/2) + 2f(1) + f(3/2)] + 4[f(1/4) + f(3/4) + f(5/4) + f(7/4)] + f(2) \} \approx 3.0170$$

51.   (a) $|E_n^T| \leq \dfrac{(b-a)^3}{12n^2} M.$

With $a = 0$, $b = 2$ and

$$|f''(x)| = \frac{9}{4}(1 + 3x)^{-3/2} \leq \frac{9}{4} = M \quad \text{on } [0, 2],$$

we have

$$|E_n^T| \leq \frac{2^3}{12n^2} \frac{9}{4} = \frac{3}{2n^2}.$$

Solving $\dfrac{3}{2n^2} \leq 0.0001$ for n gives $n \geq 123$.

   (b) $|E_n^S| \leq \dfrac{(b-a)^5}{2880n^4} M.$

With $a = 0, b = 2$ and

$$|f^{(4)}(x)| = \frac{1215}{16}(1 + 3x)^{-7/2} \leq \frac{1215}{16} = M,$$

we have

$$|E_n^S| \leq= \frac{2^5}{2880n^4} \frac{1215}{16} = \frac{27}{32n^4}.$$

Solving $\dfrac{27}{32n^4} \leq 0.0001$ for n gives $n \geq 11$.

CHAPTER 9

SECTION 9.1

1. $y_1'(x) = \frac{1}{2} e^{x/2};$ $\quad 2y_1' - y_1 = 2\left(\frac{1}{2}\right) e^{x/2} - e^{x/2} = 0;$ $\quad y_1$ is a solution.

$y_2'(x) = 2x + e^{x/2};$ $\quad 2y_2' - y_2 = 2\left(2x + e^{x/2}\right) - \left(x^2 + 2e^{x/2}\right) = 4x - x^2 \neq 0;$

y_2 is not a solution.

3. $y_1'(x) = \dfrac{-e^x}{(e^x + 1)^2};$ $\quad y_1' + y_1 = \dfrac{-e^x}{(e^x + 1)^2} + \dfrac{1}{e^x + 1} = \dfrac{1}{(e^x + 1)^2} = y_1^2;$ $\quad y_1$ is a solution.

$y_2'(x) = \dfrac{-Ce^x}{(Ce^x + 1)^2};$ $\quad y_2' + y_2 = \dfrac{-Ce^x}{(Ce^x + 1)^2} + \dfrac{1}{Ce^x + 1} = \dfrac{1}{(Ce^x + 1)^2} = y_2^2;$

y_2 is a solution.

5. $y_1'(x) = 2e^{2x},$ $\quad y_1'' = 4e^{2x};$ $\quad y_1'' - 4y_1 = 4e^{2x} - 4e^{2x} = 0;$ $\quad y_1$ is a solution.

$y_2'(x) = 2C \cosh 2x,$ $\quad y_2'' = 4C \sinh 2x;$ $\quad y_2'' - 4y_2 = 4C \sinh 2x - 4C \sinh 2x = 0;$

y_2 is a solution.

7. $y' - 2y = 1;$ $\quad H(x) = \displaystyle\int (-2)\, dx = -2x,$ integrating factor: $\; e^{-2x}$

$$e^{-2x}y' - 2e^{-2x}y = e^{-2x}$$
$$\frac{d}{dx}\left[e^{-2x}y\right] = e^{-2x}$$
$$e^{-2x}y = -\frac{1}{2}e^{-2x} + C$$
$$y = -\frac{1}{2} + Ce^{2x}$$

9. $y' + \dfrac{5}{2}y = 1;$ $\quad H(x) = \displaystyle\int \left(\frac{5}{2}\right) dx = \frac{5}{2}x,$ integrating factor: $\; e^{5x/2}$

$$e^{5x/2}y' + \frac{5}{2}e^{5x/2}y = e^{5x/2}$$
$$\frac{d}{dx}\left[e^{5x/2}y\right] = e^{5x/2}$$
$$e^{5x/2}y = \frac{2}{5}e^{5x/2} + C$$
$$y = \frac{2}{5} + Ce^{-5x/2}$$

11. $y' - 2y = 1 - 2x;$ $\quad H(x) = \displaystyle\int (-2)\, dx = -2x,$ integrating factor: $\; e^{-2x}$

$$e^{-2x}y' - 2e^{-2x}y = e^{-2x} - 2xe^{-2x}$$
$$\frac{d}{dx}\left[e^{-2x}y\right] = e^{-2x} - 2xe^{-2x}$$
$$e^{-2x}y = -\frac{1}{2}e^{-2x} + x\, e^{-2x} + \frac{1}{2}e^{-2x} + C = x\, e^{-2x} + C$$
$$y = x + Ce^{2x}$$

13. $y' - \dfrac{4}{x} y = -2n;$ $\qquad H(x) = \displaystyle\int \left(-\dfrac{4}{x}\right) dx = -4 \ln x = \ln x^{-4},$ $\quad$ integrating factor: $\quad e^{\ln x^{-4}} = x^{-4}$

$$x^{-4} y' - \dfrac{4}{x} x^{-4} y = -2nx^{-4}$$

$$\dfrac{d}{dx}\left[x^{-4} y\right] = -2nx^{-4}$$

$$x^{-4} y = \dfrac{2}{3} nx^{-3} + C$$

$$y = \dfrac{2}{3} nx + Cx^4$$

15. $y' - e^x y = 0;$ $\qquad H(x) = \displaystyle\int -e^x \, dx = -e^x,$ $\quad$ integrating factor: $\quad e^{-e^x}$

$$e^{-e^x} y' - e^x e^{-e^x} y = 0$$

$$\dfrac{d}{dx}\left[e^{-e^x} y\right] = 0$$

$$e^{-e^x} y = C$$

$$y = Ce^{e^x}$$

17. $y' + \dfrac{1}{1 + e^x} y = \dfrac{1}{1 + e^x};$ $\qquad H(x) = \displaystyle\int \dfrac{1}{1 + e^x} dx = \ln \dfrac{e^x}{1 + e^x},$

integrating factor: $\quad e^{H(x)} = \dfrac{e^x}{1 + e^x}$

$$\dfrac{e^x}{1 + e^x} y' + \dfrac{1}{1 + e^x} \cdot \dfrac{e^x}{1 + e^x} y = \dfrac{1}{1 + e^x} \cdot \dfrac{e^x}{1 + e^x}$$

$$\dfrac{d}{dx}\left[\dfrac{e^x}{1 + e^x} y\right] = \dfrac{e^x}{(1 + e^x)^2}$$

$$\dfrac{e^x}{1 + e^x} y = -\dfrac{1}{1 + e^x} + C$$

$$y = -e^{-x} + C\left(1 + e^{-x}\right)$$

This solution can also be written: $\quad y = 1 + K\left(e^{-x} + 1\right),$ $\quad$ where K is an arbitrary constant.

19. $y' + 2xy = xe^{-x^2};$ $\qquad H(x) = \displaystyle\int 2x \, dx = x^2,$ $\quad$ integrating factor: $\quad e^{x^2}$

$$e^{x^2} y' + 2xe^{x^2} y = x$$

$$\dfrac{d}{dx}\left[e^{x^2} y\right] = x$$

$$e^{x^2} y = \dfrac{1}{2} x^2 + C$$

$$y = e^{-x^2} \left(\tfrac{1}{2} x^2 + C\right)$$

21. $y' + \dfrac{2}{x+1}\, y = 0;$ $H(x) = \displaystyle\int \dfrac{2}{x+1}\, dx = 2\ln(x+1) = \ln(x+1)^2,$

integrating factor: $e^{\ln(x+1)^2} = (x+1)^2$

$$(x+1)^2\, y' + 2(x+1)\, y = 0$$
$$\dfrac{d}{dx}\left[(x+1)^2\, y\right] = 0$$
$$(x+1)^2\, y = C$$
$$y = \dfrac{C}{(x+1)^2}$$

23. $y' + y = x;$ $H(x) = \displaystyle\int 1\, dx = x,$ integrating factor : e^x

$$e^x\, y' + e^x\, y = xe^x$$
$$\dfrac{d}{dx}\left[e^x\, y\right] = xe^x$$
$$e^x\, y = xe^x - e^x + C$$
$$y = (x-1) + Ce^{-x}$$

$y(0) = -1 + C = 1$ $\implies$ $C = 2.$ Therefore, $y = 2e^{-x} + x - 1$ is the solution which satisfies the initial condition.

25. $y' + y = \dfrac{1}{1+e^x};$ $H(x) = \displaystyle\int 1\, dx = x,$ integrating factor : e^x

$$e^x\, y' + e^x\, y = \dfrac{e^x}{1+e^x}$$
$$\dfrac{d}{dx}\left[e^x\, y\right] = \dfrac{e^x}{1+e^x}$$
$$e^x\, y = \ln\left(1+e^x\right) + C$$
$$y = e^{-x}\left[\ln\left(1+e^x\right) + C\right]$$

$y(0) = \ln 2 + C = e$ $\implies$ $C = e - \ln 2.$ Therefore, $y = e^{-x}\left[\ln\left(1+e^x\right) + e - \ln 2\right]$ is the solution which satisfies the initial condition.

27. $y' - \dfrac{2}{x}\, y = x^2 e^x;$ $H(x) = \displaystyle\int\left(-\dfrac{2}{x}\right)\, dx = -2\ln x = \ln x^{-2},$

integrating factor: $e^{\ln x^{-2}} = x^{-2}$

$$x^{-2}\, y' - 2x^{-3}\, y = e^x$$
$$\dfrac{d}{dx}\left[x^{-2}\, y\right] = e^x$$
$$x^{-2}\, y = e^x + C$$
$$y = x^2\left(e^x + C\right)$$

$y(1) = e + C = 0$ $\implies$ $C = -e.$ Therefore, $y = x^2\left(e^x - e\right)$ is the solution which satisfies the initial condition.

29. Set $z = y' - y$. Then $z' = y'' - y'$.

$$y' - y = y'' - y' \implies z = z' \implies z = C_1 e^x$$

Now,

$$z = y' - y = C_1 e^x \implies e^{-x}y' - e^{-x}y = C_1 \implies (e^{-x}y)' = C_1$$
$$\implies e^{-x}y = C_1 x + C_2 \implies y = C_1 x e^x + C_2 e^x$$

31. (a) Let y_1 and y_2 be solutions of $y' + p(x)y = 0$, and let $u = y_1 + y_2$. Then

$$u' + pu = (y_1 + y_2)' + p(y_1 + y_2)$$
$$= y_1' + y_2' + py_1 + py_2$$
$$= y_1' + py_1 + y_2' + py_2 = 0 + 0 = 0$$

Therefore u is a solution.

(b) Let $u = Cy$ where y is a solution of $y' + p(x)y = 0$. Then

$$u' + pu = (Cy)' + p(Cy) = Cy' + Cpy = C(y' + py) = C \times 0 = 0$$

Therefore u is a solution.

33. Let $y(x) = e^{-H(x)} \int_a^x q(t)\, e^{H(t)}\, dt$.

Note first that $y(a) = e^{-H(a)} \int_a^a q(t)\, e^{H(t)}\, dt = 0$ so y satisfies the initial condition.

Now,

$$y' + p(x)y = \left[e^{-H(x)} \int_a^x q(t)\, e^{H(t)}\, dt \right]' + p(x)\, e^{-H(x)} \int_a^x q(t)\, e^{H(t)}\, dt$$

$$= e^{-H(x)}q(x)\, e^{H(x)} + e^{-H(x)}[-p(x)] \int_a^x q(t)\, e^{H(t)}\, dt + p(x)\, e^{-H(x)} \int_a^x q(t)\, e^{H(t)}\, dt$$

$$= q(x)$$

Thus, $y(x) = e^{-H(x)} \int_a^x q(t)\, e^{H(t)}\, dt$ is the solution of the initial value problem.

35. According to Newton's Law of Cooling, the temperature T at any time t is given by

$$T(t) = 32 + [72 - 32]e^{-kt}$$

We can determine k by applying the condition $T(1/2) = 50°$:

$$50 = 32 + 40\, e^{-k/2}$$

$$e^{-k/2} = \frac{18}{40} = \frac{9}{20}$$

$$-\tfrac{1}{2}k = \ln(9/20)$$

$$k = -2\ln(9/20) \cong 1.5970$$

Therefore, $T(t) \cong 32 + 40\, e^{-1.5970t}$.

Now, $T(1) \cong 32 + 40\, e^{-1.5970} \cong 40.100$; the temperature after 1 minute is (approx.) $40.10°$.

To find how long it will take for the temperature to reach 35°, we solve

$$32 + 40\,e^{-1.5970t} = 35$$

for t:

$$32 + 40\,e^{-1.5970t} = 35$$

$$40\,e^{-1.5970t} = 3$$

$$-1.5970t = \ln(3/40)$$

$$t = \frac{\ln(3/40)}{-1.5970} \cong 1.62$$

It will take approximately 1.62 minutes for the thermometer to read 35°.

37. (a) The solution of the initial value problem $v' = 32 - kv,\ (k > 0)\ v(0) = 0$ is:

$$v(t) = \frac{32}{k}\left(1 - e^{-kt}\right).$$

(b) At each time t, $\quad 1 - e^{-kt} < 1$. Therefore

$$v(t) = \frac{32}{k}\left(1 - e^{-kt}\right) < \frac{32}{k} \quad \text{and} \quad \lim_{t\to\infty} v(t) = \frac{32}{k}$$

(c)

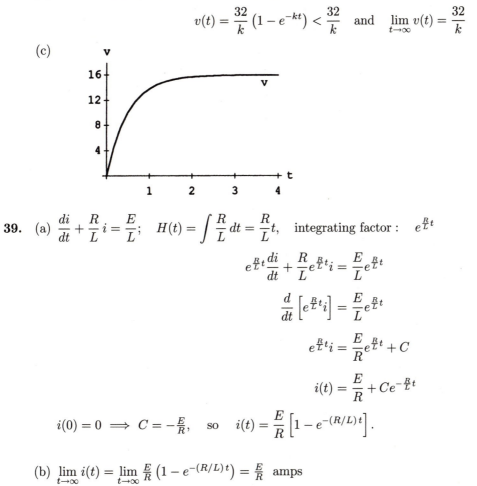

39. (a) $\dfrac{di}{dt} + \dfrac{R}{L}\,i = \dfrac{E}{L};\quad H(t) = \displaystyle\int \dfrac{R}{L}\,dt = \dfrac{R}{L}t,\quad$ integrating factor : $\quad e^{\frac{R}{L}t}$

$$e^{\frac{R}{L}t}\frac{di}{dt} + \frac{R}{L}e^{\frac{R}{L}t}i = \frac{E}{L}e^{\frac{R}{L}t}$$

$$\frac{d}{dt}\left[e^{\frac{R}{L}t}i\right] = \frac{E}{L}e^{\frac{R}{L}t}$$

$$e^{\frac{R}{L}t}i = \frac{E}{R}e^{\frac{R}{L}t} + C$$

$$i(t) = \frac{E}{R} + Ce^{-\frac{R}{L}t}$$

$$i(0) = 0 \implies C = -\frac{E}{R}, \quad \text{so} \quad i(t) = \frac{E}{R}\left[1 - e^{-(R/L)t}\right].$$

(b) $\displaystyle\lim_{t\to\infty} i(t) = \lim_{t\to\infty} \frac{E}{R}\left(1 - e^{-(R/L)t}\right) = \frac{E}{R}$ amps

(c) $i(t) = 0.9\frac{E}{R} \implies e^{-(R/L)t} = \frac{1}{10} \implies -\frac{R}{L}t = -\ln 10 \implies t = \frac{L}{R}\ln 10$ seconds.

41. (a) $V'(t) = kV(t) \implies V(t) = V_0 e^{kt}$

Loses 20% in 5 minutes, so $V(5) = V_0 e^{5k} = 0.8V_0 \implies k = \frac{1}{5}\ln 0.8$

$\implies V(t) = V_0 e^{\frac{1}{5}(\ln 0.8)t} = V_0 \left(e^{\ln 0.8}\right)^{t/5} = V_0 (0.8)^{t/5} = V_0 \left(\frac{4}{5}\right)^{t/5}$.

Since $V_0 = 200$ liters, we get $V(t) = 200\left(\frac{4}{5}\right)^{t/5}$

(b)

$$V'(t) = ktV(t)$$
$$V'(t) - ktV(t) = 0$$
$$e^{-kt^2/2}V'(t) - kte^{-kt^2/2}V(t) = 0$$
$$\frac{d}{dt}\left[e^{-kt^2/2}V(t)\right] = 0$$
$$e^{-kt^2/2}V(t) = C$$
$$V(t) = Ce^{kt^2/2}.$$

$V(0) = C = 200 \implies V(t) = 200e^{kt^2/2}.$

$V(5) = 160 \implies 200e^{k(25/2)} = 160, \quad e^{k(25/2)} = \frac{4}{5}, \quad e^k = \left(\frac{4}{5}\right)^{2/25}.$

Therefore $V(t) = 200\left(\frac{4}{5}\right)^{t^2/25}$ liters.

43. (a) $\dfrac{dP}{dt} = k(M - P)$

(b) $\dfrac{dP}{dt} + kP = kM;$ $H(t) = \displaystyle\int k\, dt = kt,$ integrating factor : e^{kt}

$$e^{kt}\frac{dP}{dt} + ke^{kt}P = kM\, e^{kt}$$
$$\frac{d}{dt}\left[e^{kt}P\right] = kM\, e^{kt}$$
$$e^{kt}P = M\, e^{kt} + C$$
$$P = M + Ce^{-kt}$$

$P(0) = M + C = 0 \implies C = -M$ and $P(t) = M\left(1 - e^{-kt}\right)$

$P(10) = M\left(1 - e^{-10k}\right) = 0.3M \implies k \cong 0.0357$ and $P(t) = M\left(1 - e^{-0.0357t}\right)$

(c) $P(t) = M\left(1 - e^{-0.0357t}\right) = 0.9M \implies e^{-0.0357t} = 0.1 \implies t \cong 65$

Therefore, it will take approximately 65 days for 90 % of the population to be aware of the product.

45. (a)

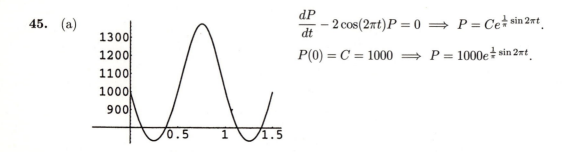

$\dfrac{dP}{dt} - 2\cos(2\pi t)P = 0 \implies P = Ce^{\frac{1}{\pi}\sin 2\pi t}.$

$P(0) = C = 1000 \implies P = 1000e^{\frac{1}{\pi}\sin 2\pi t}.$

(b)

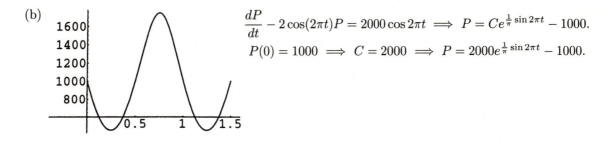

$$\frac{dP}{dt} - 2\cos(2\pi t)P = 2000\cos 2\pi t \implies P = Ce^{\frac{1}{\pi}\sin 2\pi t} - 1000.$$

$$P(0) = 1000 \implies C = 2000 \implies P = 2000e^{\frac{1}{\pi}\sin 2\pi t} - 1000.$$

SECTION 9.2

1.
$$y' = y\sin(2x + 3)$$

$$\frac{1}{y}\,dy = \sin(2x + 3)\,dx$$

$$\int \frac{1}{y}\,dy = \int \sin(2x + 3)\,dx$$

$$\ln|y| = -\frac{1}{2}\cos(2x + 3) + C$$

This solution can also be written: $y = Ce^{-(1/2)\cos(2x+3)}$.

3.
$$y' = (xy)^3$$

$$\frac{1}{y^3}\,dy = x^3\,dx, \qquad y \neq 0$$

$$\int \frac{1}{y^3}\,dy = \int x^3\,dx$$

$$-\frac{1}{2}y^{-2} = \frac{1}{4}x^4 + C$$

This solution can also be written: $x^4 + \dfrac{2}{y^2} = C, \quad \text{or} \quad y^2 = \dfrac{2}{C - x^4};$

5.
$$y' = \frac{\sin(1/x)}{x^2 y \cos y}$$

$$y\cos y\,dy = \frac{1}{x^2}\sin(1/x)\,dx$$

$$\int y\cos y\,dy = \int \frac{1}{x^2}\sin(1/x)\,dx$$

$$y\sin y + \cos y = \cos(1/x) + C$$

7.
$$y' = x\,e^{y+x}$$

$$e^{-y}\,dy = x\,e^x\,dx$$

$$\int e^{-y}\,dy = \int x\,e^x\,dx$$

$$-e^{-y} = x\,e^x - e^x + C$$

$$e^{-y} = e^x - x\,e^x + C$$

This solution can also be written: $y = -\ln(e^x - xe^x + C)$.

9.
$$(y \ln x)y' = \frac{(y+1)^2}{x}$$

$$\frac{y}{(y+1)^2} \, dy = \frac{1}{x \ln x} \, dx$$

$$\int \frac{y}{(y+1)^2} \, dy = \int \frac{1}{x \ln x} \, dx$$

$$\ln|y+1| + \frac{1}{y+1} = \ln|\ln x| + C$$

11.
$$(y \ln x)y' = \frac{y^2+1}{x}$$

$$\frac{y}{y^2+1} \, dy = \frac{1}{x \ln x} \, dx$$

$$\int \frac{y}{y^2+1} \, dy = \int \frac{1}{x \ln x} \, dx$$

$$\tfrac{1}{2} \ln(y^2+1) = \ln|\ln x| + K = \ln|C \ln x| \quad (K = \ln|C|)$$

$$\ln(y^2+1) = 2 \ln|C \ln x| = \ln(C \ln x)^2$$

$$y^2 = C(\ln x)^2 - 1$$

13.
$$y' = x\sqrt{\frac{1-y^2}{1-x^2}}, \qquad y(0) = 0$$

$$\frac{1}{\sqrt{1-y^2}} \, dy = \frac{x}{\sqrt{1-x^2}} \, dx$$

$$\int \frac{1}{\sqrt{1-y^2}} \, dy = \int \frac{x}{\sqrt{1-x^2}} \, dx$$

$$\sin^{-1} y = -\sqrt{1-x^2} + C$$

$$y(0) = 0 \quad \Longrightarrow \quad \arcsin 0 = -1 + C \quad \Longrightarrow \quad C = 1$$

Thus, $\quad \arcsin y = 1 - \sqrt{1-x^2}$.

15.
$$y' = \frac{x^2 y - y}{y+1}, \qquad y(3) = 1$$

$$\frac{y+1}{y} \, dy = (x^2 - 1) \, dx, \qquad y \neq 0$$

$$\int \frac{y+1}{y} \, dy = \int (x^2 - 1) \, dx$$

$$y + \ln|y| = \frac{1}{3} x^3 - x + C$$

$$y(3) = 1 \quad \Longrightarrow \quad 1 + \ln 1 = \frac{1}{3}(3)^3 - 3 + C \quad \Longrightarrow \quad C = -5.$$

Thus, $\quad y + \ln|y| = \tfrac{1}{3} x^3 - x - 5$.

17. $(xy^2 + y^2 + x + 1)\,dx + (y - 1)\,dy = 0, \qquad y(2) = 0$

$$(x + 1)(y^2 + 1)\,dx + (y - 1)\,dy = 0$$

$$(x + 1)\,dx + \frac{y - 1}{y^2 + 1}\,dy = 0$$

$$\int (x + 1)\,dx + \int \frac{y - 1}{y^2 + 1}\,dy = C$$

$$\frac{x^2}{2} + x + \frac{1}{2}\ln(y^2 + 1) - \tan^{-1} y = C$$

$y(2) = 0 \implies C = 4.$ Thus, $\frac{1}{2}x^2 + x + \frac{1}{2}\ln(y^2 + 1) - \tan^{-1} y = 4$

19. $y' = 6\,e^{2x - y}, \qquad y(0) = 0$

$$y' = 6\,e^{2x - y}$$

$$e^y\,dy = 6\,e^{2x}\,dx$$

$$e^y = 3\,e^{2x} + C$$

$y(0) = 0 \implies 1 = 3 + C \implies C = -2$

Thus, $e^y = 3\,e^{2x} - 2 \implies y = \ln\left[3\,e^{2x} - 2\right]$

21. We assume that $C = 0$ at time $t = 0$.   (a) Let $A_0 = B_0$. Then

$$\frac{dC}{dt} = k(A_0 - C)^2 \quad \text{and} \quad \frac{dC}{(A_0 - C)^2} = k\,dt.$$

Integrating, we get

$$\int \frac{1}{(A_0 - C)^2}\,dC = \int k\,dt$$

$$\frac{1}{A_0 - C} = kt + M \qquad M \text{ a constant.}$$

Since $C(0) = 0$, $M = \dfrac{1}{A_0}$ and

$$\frac{1}{A_0 - C} = kt + \frac{1}{A_0}.$$

Solving this equation for C gives

$$C(t) = \frac{kA_0^2 t}{1 + kA_0 t}.$$

(b) Suppose that $A_0 \neq B_0$. Then

$$\frac{dC}{dt} = k(A_0 - C)(B_0 - C) \quad \text{and} \quad \frac{dC}{(A_0 - C)(B_0 - C)} = k\,dt.$$

Integrating, we get

$$\int \frac{1}{(A_0 - C)(B_0 - C)} \, dC = \int k \, dt$$

$$\frac{1}{B_0 - A_0} \int \left(\frac{1}{A_0 - C} - \frac{1}{B_0 - C} \right) dC = \int k \, dt$$

$$\frac{1}{B_0 - A_0} \left[-\ln(A_0 - C) + \ln(B_0 - C) \right] = kt + M$$

$$\frac{1}{B_0 - A_0} \ln \left(\frac{B_0 - C}{A_0 - C} \right) = kt + M \qquad M \text{ an arbitrary constant}$$

Since $C(0) = 0$, $M = \frac{1}{B_0 - A_0} \ln \left(\frac{B_0}{A_0} \right)$ and

$$\frac{1}{B_0 - A_0} \ln \left(\frac{B_0 - C}{A_0 - C} \right) = kt + \frac{\ln(B_0/A_0)}{B_0 - A_0}.$$

Solving this equation for C, gives

$$C(t) = \frac{A_0 B_0 \left(e^{kA_0 t} - e^{kB_0 t} \right)}{A_0 e^{kA_0 t} - B_0 e^{kB_0 t}}.$$

23. (a)

$$m \frac{dv}{dt} = -\alpha v - \beta v^2$$

$$\frac{dv}{v(\alpha + \beta v)} = -\frac{1}{m} \, dt$$

$$\int \frac{1}{v(\alpha + \beta v)} \, dv = -\int \frac{1}{m} \, dt$$

$$\frac{1}{\alpha} \int \frac{1}{v} \, dv - \frac{\beta}{\alpha} \int \frac{1}{\alpha + \beta v} \, dv = -\int \frac{1}{m} \, dt$$

$$\frac{1}{\alpha} \ln v - \frac{1}{\alpha} \ln(\alpha + \beta v) = -\frac{1}{m} t + M, \quad M \text{ a constant}$$

$$\ln \left(\frac{v}{\alpha + \beta v} \right) = -\frac{\alpha}{m} t + \alpha M$$

$$\frac{v}{\alpha + \beta v} = K e^{-\alpha t/m} \quad \left[K = e^{\alpha M} \right]$$

Solving this equation for v we get $\quad v(t) = \dfrac{\alpha K}{e^{\alpha t/m} - \beta K} = \dfrac{\alpha}{C e^{\alpha t/m} - \beta} \quad [C = 1/K].$

(b) Setting $v(0) = v_0$, we get   (c) $\lim\limits_{t \to \infty} v(t) = 0$

$$C = \frac{\alpha + \beta v_0}{v_0} \quad \text{and}$$

$$v(t) = \frac{\alpha v_0}{(\alpha + \beta v_0)e^{\alpha t/m} - \beta v_0}$$

25. (a) Let $P = P(t)$ denote the number of people who have the disease at time t. Then, substituting into (9.2.4) with $M = 25,000$ and $R = 100$, we get

$$P(t) = \frac{25,000(100)}{100 + (249,00)e^{-25,000kt}} = \frac{25,000}{1 + 249e^{-25,000kt}}.$$

$$P(10)\frac{25,000}{1+249e^{-25,000(10k)}} = 400 \quad \Longrightarrow \quad -25,000k \cong -0.1398.$$

Therefore, $P(t) = \dfrac{25,000}{1+249e^{-0.1398t}}$.

(b) $\dfrac{25,000}{1+249e^{-0.1398\,t}} = 12,500 \quad \Longrightarrow \quad t \cong 40;$ (c)

It will take 40 days for half the

population to have the disease.

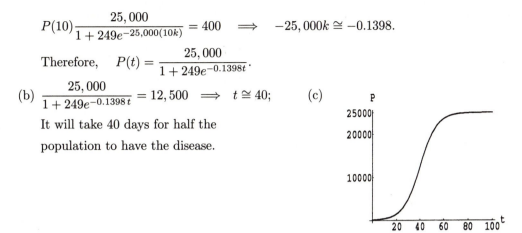

27.   Assume that the package is dropped from rest.

(a) Let $v = v(t)$ be the velocity at time t, $0 \le t \le 10$. Then

$$100\frac{dv}{dt} = 100g - 2v \quad \text{or} \quad \frac{dv}{dt} + \frac{1}{50}v = g \quad (g = 9.8 \text{ m/sec}^2)$$

This is a linear differential equation; $e^{t/50}$ is an integrating factor.

$$e^{t/50}\frac{dv}{dt} + \frac{1}{50}e^{t/50}v = g\,e^{t/50}$$

$$\frac{d}{dt}\left[e^{t/50}v\right] = g\,e^{t/50}$$

$$e^{t/50}v = 50g\,e^{t/50} + C$$

$$v = 50g + Ce^{-t/50}$$

Now, $v(0) = 0 \quad \Longrightarrow \quad C = -50g$ and $v(t) = 50g\left(1 - e^{-t/50}\right)$.

At the instant the parachute opens, $v(10) = 50g\left(1 - e^{-1/5}\right) \cong 50g(0.1813) \cong 88.82$ m/sec.

(b) Now let $v = v(t)$ denote the velocity of the package t seconds after the parachute opens. Then

$$100\frac{dv}{dt} = 100g - 4v^2 \quad \text{or} \quad \frac{dv}{dt} = g - \frac{1}{25}v^2$$

This is a separable differential equation:

$$\frac{dv}{dt} = g - \frac{1}{25}v^2 \quad \text{set } u = v/5, \; du = (1/5)dv$$

$$\frac{du}{g - u^2} = \frac{1}{5}\,dt$$

$$\frac{1}{2\sqrt{g}}\ln\left|\frac{u+\sqrt{g}}{u-\sqrt{g}}\right| = \frac{t}{5} + K$$

$$\ln\left|\frac{u+\sqrt{g}}{u-\sqrt{g}}\right| = \frac{2\sqrt{g}}{5}t + M$$

$$\frac{u+\sqrt{g}}{u-\sqrt{g}} = Ce^{2\sqrt{g}t/5} \cong Ce^{1.25\,t}$$

$$u = \sqrt{g}\,\frac{Ce^{1.25\,t}+1}{Ce^{1.25t}-1}$$

$$v = 5\sqrt{g}\,\frac{Ce^{1.25\,t}+1}{Ce^{1.25t}-1}$$

Now, $v(0) = 88.82 \implies 5\sqrt{g}\,\dfrac{C+1}{C-1} = 88.82 \implies C \cong 1.43.$

Therefore, $\quad v(t) = 5\sqrt{g}\,\dfrac{1.43e^{1.25\,t}+1}{1.43e^{1.25t}-1} = \dfrac{15.65\left(1+0.70e^{-1.25\,t}\right)}{1-0.70e^{-1.25\,t}}$

(c) From part (b), $\quad \lim\limits_{t\to\infty} v(t) = 15.65$ m/sec.

PROJECT 9.2

1. (a) $2x + 3y = C \implies 2 + 3y' = 0 \implies y' = -\dfrac{2}{3}$

The orthogonal trajectories are the solutions of:

$$y' = \frac{3}{2}.$$

$y' = \frac{3}{2} \implies y = \frac{3}{2}x + C$

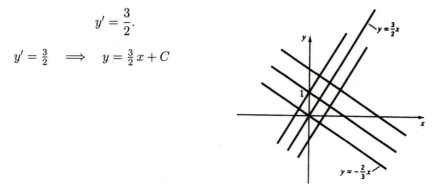

(b) Curves: $y = Cx, \quad y' = C = \dfrac{y}{x}$

orthogonal trajectories: $y' = -\dfrac{x}{y}$

$\displaystyle\int y\,dy + \int x\,dx = K_1; \quad x^2 + y^2 = K \;\; (= 2K_1)$

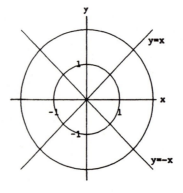

(c) $xy = C \implies y + xy' = 0 \implies y' = -\dfrac{y}{x}$

The orthogonal trajectories are the solutions of:

$$y' = \frac{x}{y}.$$

$$y' = \frac{x}{y}$$

$$\int x \, dx = \int y \, dy$$

$$\tfrac{1}{2} x^2 = \tfrac{1}{2} y^2 + C$$

or $x^2 - y^2 = C$

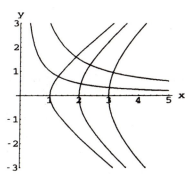

(d) $y = Cx^3, \quad y' = 3Cx^2 = \dfrac{3y}{x}$

orthogonal trajectories: $y' = -\dfrac{x}{3y}$

$$\int 3y \, dy + \int x \, dx = K_1; \quad 3y^2 + x^2 = K \;\; (= 2K_1)$$

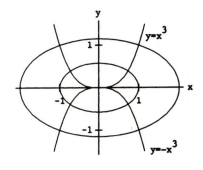

(e) $y = C\,e^x \implies y' = C\,e^x = y$

The orthogonal trajectories are the solutions of:

$$y' = -\frac{1}{y}.$$

$$y' = -\frac{1}{y}$$

$$\int y \, dy = -\int dx$$

$$\tfrac{1}{2} y^2 = -x + K$$

or $y^2 = -2x + C$

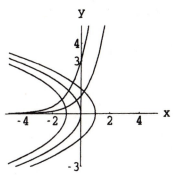

(f) $x = Cy^4, \quad 1 = 4Cy^3 y'; \quad y' = \dfrac{y}{4x}$

orthogonal trajectories: $\dfrac{dy}{dx} = -\dfrac{4x}{y}$

$$\int y \, dy + \int 4x \, dx = K_1; \quad y^2 + 4x^2 = K \;\; (= 2K_1)$$

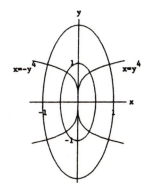

3. (a) A differential equation for the given family is:

$$y^2 = 2xyy' + y^2(y')^2$$

A differential equation for the family of orthogonal trajectories is found by replacing y' by $-1/y'$.

The result is:

$$y^2 = -\frac{2xy}{y'} + \frac{y^2}{(y')^2} \qquad \text{which simplifies to} \qquad y^2 = 2xyy' + y^2(y')^2$$

Thus, the given family is self-orthogonal.

(b) $\dfrac{x^2}{C^2} + \dfrac{y^2}{C^2 - 4} = 1 \implies \dfrac{2x}{C^2} + \dfrac{2yy'}{C^2 - 4} = 0 \implies C^2 = \dfrac{4x}{x + yy'}$

A differential equation for the given family is:

$$x^2 + xyy' - \frac{xy}{y'} - y^2 = 4$$

A differential equation for the family of orthogonal trajectories is found by replacing y' by $-1/y'$.

The result is:

$$x^2 - xy\frac{1}{y'} + xyy' - y^2 = 4$$

Thus, the given family is self-orthogonal.

SECTION 9.3

1. The characteristic equation is:

$$r^2 + 2r - 8 = 0 \qquad \text{or} \qquad (r + 4)(r - 2) = 0.$$

The roots are: $r = -4,\ 2$. The general solution is:

$$y = C_1 e^{-4x} + C_2 e^{2x}.$$

3. The characteristic equation is:

$$r^2 + 8r + 16 = 0 \qquad \text{or} \qquad (r + 4)^2 = 0.$$

There is only one root: $r = -4$. By Theorem 9.3.6 II, the general solution is:

$$y = C_1 e^{-4x} + C_2 x e^{-4x}.$$

5. The characteristic equation is: $r^2 + 2r + 5 = 0$.

The roots are complex: $r = -1 \pm 2i$. By Theorem 9.3.6 III, the general solution is:

$$y = e^{-x}\left(C_1 \cos 2x + C_2 \sin 2x\right).$$

7.   The characteristic equation is:

$$2r^2 + 5r - 3 = 0 \qquad \text{or} \qquad (2r-1)(r+3) = 0.$$

The roots are: $r = \frac{1}{2}, -3.$ The general solution is:

$$y = C_1 e^{x/2} + C_2 e^{-3x}.$$

9.   The characteristic equation is:

$$r^2 + 12 = 0.$$

The roots are complex: $r = \pm 2\sqrt{3}\,i.$ The general solution is:

$$y = C_1 \cos 2\sqrt{3}\,x + C_2 \sin 2\sqrt{3}\,x.$$

11.   The characteristic equation is:

$$5r^2 + \tfrac{11}{4}r - \tfrac{3}{4} = 0 \qquad \text{or} \qquad 20r^2 + 11r - 3 = (5r-1)(4r+3) = 0.$$

The roots are: $r = \frac{1}{5}, -\frac{3}{4}.$ The general solution is:

$$y = C_1 e^{x/5} + C_2 e^{-3x/4}.$$

13.   The characteristic equation is:

$$r^2 + 9 = 0.$$

The roots are complex: $r = \pm 3i.$ The general solution is:

$$y = C_1 \cos 3x + C_2 \sin 3x.$$

15.   The characteristic equation is:

$$2r^2 + 2r + 1 = 0.$$

The roots are complex: $r = -\frac{1}{2} \pm \frac{1}{2}i.$ The general solution is:

$$y = e^{-x/2}\left[C_1 \cos(x/2) + C_2 \sin(x/2)\right].$$

17.   The characteristic equation is:

$$8r^2 + 2r - 1 = 0 \qquad \text{or} \qquad (4r-1)(2r+1) = 0.$$

The roots are: $r = \frac{1}{4}, -\frac{1}{2}.$ The general solution is:

$$y = C_1 e^{x/4} + C_2 e^{-x/2}.$$

19. The characteristic equation is:

$$r^2 - 5r + 6 = 0 \quad \text{or} \quad (r-3)(r-2) = 0.$$

The roots are: $r = 3, 2$. The general solution and its derivative are:

$$y = C_1 e^{3x} + C_2 e^{2x}, \quad y' = 3C_1 e^{3x} + 2C_2 e^{2x}.$$

The conditions: $y(0) = 1$, $y'(0) = 1$ require that

$$C_1 + C_2 = 1 \quad \text{and} \quad 3C_1 + 2C_2 = 1.$$

Solving these equations simultaneously gives $C_1 = -1$, $C_2 = 2$.

The solution of the initial value problem is: $y = 2e^{2x} - e^{3x}$.

21. The characteristic equation is:

$$r^2 + \tfrac{1}{4} = 0.$$

The roots are: $r = \pm \tfrac{1}{2} i$. The general solution and its derivative are:

$$y = C_1 \cos(x/2) + C_2 \sin(x/2) \quad y' = -\tfrac{1}{2} C_1 \sin(x/2) + \tfrac{1}{2} C_2 \cos(x/2).$$

The conditions: $y(\pi) = 1$, $y'(\pi) = -1$ require that

$$C_2 = 1 \quad \text{and} \quad C_1 = 2.$$

The solution of the initial value problem is: $y = 2\cos(x/2) + \sin(x/2)$.

23. The characteristic equation is:

$$r^2 + 4r + 4 = 0 \quad \text{or} \quad (r+2)^2 = 0.$$

There is only one root: $r = -2$. The general solution and its derivative are:

$$y = C_1 e^{-2x} + C_2 x e^{-2x} \quad y' = -2C_1 e^{-2x} + C_2 e^{-2x} - 2C_2 x e^{-2x}.$$

The conditions: $y(-1) = 2$, $y'(-1) = 1$ require that

$$C_1 e^2 - C_2 e^2 = 2 \quad \text{and} \quad -2C_1 e^2 + 3C_2 e^2 = 1.$$

Solving these equations simultaneously gives $C_1 = 7e^{-2}$, $C_2 = 5e^{-2}$.

The solution of the initial value problem is: $y = 7e^{-2}e^{-2x} + 5e^{-2}xe^{-2x} = 7e^{-2(x+1)} + 5xe^{-2(x+1)}$.

25. The characteristic equation is:

$$r^2 - r - 2 = 0 \quad \text{or} \quad (r-2)(r+1) = 0.$$

The roots are: $r = 2, -1$. The general solution and its derivative are:

$$y = C_1 e^{2x} + C_2 e^{-x} \quad y' = 2C_1 e^{2x} - C_2 e^{-x}.$$

(a) $\quad y(0) = 1 \implies C_1 + C_2 = 1 \implies C_2 = 1 - C_1$.

Thus, the solutions that satisfy $y(0) = 1$ are: $\quad y = Ce^{2x} + (1 - C)e^{-x}$.

(b) $\quad y'(0) = 1 \implies 2C_1 - C_2 = 1 \implies C_2 = 2C_1 - 1$.

Thus, the solutions that satisfy $y'(0) = 1$ are: $\quad y = Ce^{2x} + (2C - 1)e^{-x}$.

(c) To satisfy both conditions, we must have $\quad 2C - 1 = 1 - C \implies C = \frac{2}{3}$.

The solution that satisfies $y(0) = 1$, $y'(0) = 1$ is:

$$y = \frac{2}{3}e^{2x} + \frac{1}{3}e^{-x}.$$

27. $\quad \alpha = \dfrac{r_1 + r_2}{2}, \qquad \beta = \dfrac{r_1 - r_2}{2};$

$y = k_1 e^{r_1 x} + k_2 e^{r_2 x} = e^{\alpha x}\left(C_1 \cosh \beta x + C_2 \sinh \beta x\right), \quad \text{where} \quad k_1 = \dfrac{C_1 + C_2}{2}, \quad k_2 = \dfrac{C_1 - C_2}{2}.$

29. (a) Let $y_1 = e^{\alpha x}, \quad y_2 = xe^{\alpha x}$. Then

$$W(x) = y_1 y_2' - y_2 y_1' = e^{\alpha x}\left[e^{\alpha x} + \alpha x e^{\alpha x}\right] - xe^{\alpha x}\left[\alpha e^{\alpha x}\right] = e^{2\alpha x} \neq 0$$

(b) Let $y_1 = e^{\alpha x}\cos\beta x, \quad y_2 = e^{\alpha x}\sin\beta x, \quad \beta \neq 0$. Then

$W(x) = y_1 y_2' - y_2 y_1'$

$\quad = e^{\alpha x}\cos\beta x\left[\alpha e^{\alpha x}\sin\beta x + \beta e^{\alpha x}\cos\beta x\right] - e^{\alpha x}\sin\beta x\left[\alpha e^{\alpha x}\cos\beta x - \beta e^{\alpha x}\sin\beta x\right]$

$\quad = \beta e^{2\alpha x} \neq 0$

31. (a) The solutions $y_1 = e^{2x}, \quad y_2 = e^{-4x}$ imply that the roots of the characteristic equation are $r_1 = 2, \quad r_2 = -4$. Therefore, the characteristic equation is:
$$(r - 2)(r + 4) = r^2 + 2r - 8 = 0$$
and the differential equation is: $\quad y'' + 2y' - 8y = 0$.

(b) The solutions $y_1 = 3e^{-x}, \quad y_2 = 4e^{5x}$ imply that the roots of the characteristic equation are $r_1 = -1, \quad r_2 = 5$. Therefore, the characteristic equation is
$$(r + 1)(r - 5) = r^2 - 4r - 5 = 0$$
and the differential equation is: $\quad y'' - 4y' - 5y = 0$.

(c) The solutions $y_1 = 2e^{3x}, \quad y_2 = xe^{3x}$ imply that 3 is the only root of the characteristic equation. Therefore, the characteristic equation is
$$(r - 3)^2 = r^2 - 6r + 9 = 0$$
and the differential equation is: $\quad y'' - 6y' + 9y = 0$.

33. (a) Let $y = e^{\alpha x}u$. Then

$$y' = \alpha e^{\alpha x}u + e^{\alpha x}u' \quad \text{and} \quad y'' = \alpha^2 e^{\alpha x}u + 2\alpha e^{\alpha x}u' + e^{\alpha x}u''$$

Now,

$$y'' - 2\alpha y + \alpha^2 y = \left(\alpha^2 e^{\alpha x}u + 2\alpha e^{\alpha x}u' + e^{\alpha x}u''\right) - 2\alpha \left(\alpha e^{\alpha x}u + e^{\alpha x}u'\right) + \alpha^2 e^{\alpha x}u$$

$$= e^{\alpha x}u''$$

Therefore, $\quad y'' - 2\alpha y + \alpha^2 y = 0 \quad \Longrightarrow \quad e^{\alpha x}u'' \quad \Longrightarrow \quad u'' = 0.$

(b) $y'' - 2\alpha y' + \left(\alpha^2 + \beta^2\right)y = y'' - 2\alpha y' + \alpha^2 y + \beta^2 y.$

From part (a) $y = e^{\alpha x}u \quad \Longrightarrow \quad y'' - 2\alpha y' + \alpha^2 y = e^{\alpha x}u''$. Therefore,

$$y'' - 2\alpha y' + \left(\alpha^2 + \beta^2\right)y = 0 \quad \Longrightarrow \quad e^{\alpha x}u'' + \beta^2 e^{\alpha x}u = 0 \quad \Longrightarrow \quad u'' + \beta^2 u = 0.$$

35. (a) If $a = 0$, $b > 0$, then the general solution of the differential equation is:

$$y = C_1 \cos \sqrt{b}\,x + C_2 \sin \sqrt{b}\,x = A \cos\left(\sqrt{b}\,x + \phi\right)$$

where A and ϕ are constants. Clearly $|y(x)| \le |A|$ for all x.

(b) If $a > 0$, $b = 0$, then the general solution of the differential equation is:

$$y = C_1 + C_2 e^{-ax} \quad \text{and} \quad \lim_{x \to \infty} y(x) = C_1.$$

The solution which satisfies the conditions: $y(0) = y_0$, $y'(0) = y_1$ is:

$$y = y_0 + \frac{y_1}{a} - \frac{y_1}{a}e^{-ax} \quad \text{and} \quad \lim_{x \to \infty} y(x) = y_0 + \frac{y_1}{a}; \qquad k = y_0 + \frac{y_1}{a}.$$

37. Let W be the Wronskian of y_1 and y_2. Then

$$W(a) = \begin{vmatrix} 0 & 0 \\ y_1'(a) & y_2'(a) \end{vmatrix} = 0$$

Therefore one of the solutions is a multiple of the other (see the Supplement to this Section).

39. From Exercise 38, the change of variable $z = \ln x$ transforms the equation

$$x^2 y'' - xy' - 8y = 0$$

into the differential equation with constant coefficients

$$\frac{d^2 y}{dz^2} - 2\frac{dy}{dz} - 8y = 0.$$

The characteristic equation is:

$$r^2 - 2r - 8 = 0 \quad \text{or} \quad (r - 4)(r + 2) = 0$$

The roots are: $r = 4$, $r = -2$, and the general solution (in terms of z) is:

$$y = C_1 e^{4z} + C_2 e^{-2z}.$$

Replacing z by $\ln x$ we get

$$y = C_1 e^{4\ln x} + C_2 e^{-2\ln x} = C_1 x^4 + C_2 x^{-2}.$$

41. From Exercise 38, the change of variable $z = \ln x$ transforms the equation

$$x^2 y'' - 3xy' + 4y = 0$$

into the differential equation with constant coefficients

$$\frac{d^2 y}{dz^2} - 4\frac{dy}{dz} + 4y = 0.$$

The characteristic equation is:

$$r^2 - 4r + 4 = 0 \quad \text{or} \quad (r-2)^2 = 0.$$

The only root is: $r = 2$, and the general solution (in terms of z) is:

$$y = C_1 e^{2z} + C_2 z e^{2z}.$$

Replacing z by $\ln x$ we get

$$y = C_1 e^{2\ln x} + C_2 \ln x \, e^{2\ln x} = C_1 x^2 + C_2 x^2 \ln x.$$

REVIEW EXERCISES

1. First calculate the integrating factor $e^{H(x)}$:

$$H(x) = \int 1 dx = x \quad \text{and} \quad e^{H(x)} = e^x$$

Multiplication by e^x gives

$$e^x y' + e^x y = 2e^{-x} \quad \text{which is} \quad \frac{d}{dx}(e^x y) = 2e^{-x}$$

Integrating this equation, we get

$$e^x y = -2e^{-x} + C$$

and

$$y = -2e^{-2x} + Ce^{-x}$$

3. The equation can be written

$$\cos^2 x dx - \frac{y}{y^2+1} dy = 0.$$

The equation is separable:

$$\int \cos^2 x dx - \int \frac{y}{y^2+1} dy = C \quad \text{and} \quad \frac{1}{4}\sin(2x) + \frac{1}{2}x - \frac{1}{2}\ln(y^2+1) = C$$

or $\quad \sin(2x) + 2x - 2\ln(y^2+1) = C.$

5. The equation can be written

$$y' + \frac{3}{x}y = \frac{\sin 2x}{x^2}.$$

Calculate the integrating factor $e^{H(x)}$:

$$H(x) = \int \frac{3}{x}dx = \ln x^3 \quad \text{and} \quad e^{H(x)} = x^3.$$

Multiplying by x^3 gives

$$x^3 y' + 3x^2 y = x \sin 2x \quad \text{which is} \quad \frac{d}{dx}(x^3 y) = x \sin 2x.$$

Integrating this equation, we get

$$x^3 y = \int x \sin 2x dx + C = -\frac{1}{2}x \cos 2x + \frac{1}{4}\sin 2x + C.$$

and

$$y = -\frac{1}{2x^2}\cos 2x + \frac{1}{4x^3}\sin 2x + \frac{C}{x^3}.$$

7. The equation can be written

$$1 + x^2 - \frac{1}{1+y^2}y' = 0.$$

The equation is separable:

$$\int (1+x^2)dx - \int \frac{1}{1+y^2}dy = C \quad \text{and} \quad x + \frac{x^3}{3} - \arctan y = C$$

or $\arctan y = x + \frac{x^3}{3} + C.$

9. The equation can be written

$$y' + \frac{2}{x}y = x^2.$$

Calculate the integrating factor $e^{H(x)}$:

$$H(x) = \int \frac{2}{x}dx = \ln x^2 \quad \text{and} \quad e^{H(x)} = x^2.$$

Multiplication by x^2 gives

$$x^2 y' + 2xy = x^4 \quad \text{which is} \quad \frac{d}{dx}(x^2 y) = x^4.$$

Integrating this equation, we get

$$x^2 y = \frac{1}{5}x^5 + C \quad \text{and} \quad y = \frac{1}{5}x^3 + \frac{C}{x^2}.$$

11. The equation can be written

$$y' + \frac{1}{x}y = \frac{2}{x^2} + 1$$

The integrating factor is

$$e^{H(x)} = e^{\ln x} = x.$$

Multiplication by x gives

$$xy' + y = \frac{2}{x} + x \quad \text{which is} \quad \frac{d}{dx}(xy) = \frac{2}{x} + x.$$

Integrating this equation, we get

$$xy = \ln x^2 + \frac{1}{2}x^2 + C \quad \text{and} \quad y = \frac{1}{x}\left(\ln x^2 + \frac{1}{2}x^2 + C\right).$$

Applying the initial condition $y(1) = 2$, we have

$$\ln 1 + \frac{1}{2} + C = 2 \quad \text{and} \quad C = \frac{3}{2}.$$

Therefore

$$y = \frac{1}{x}\left(\ln x^2 + \frac{1}{2}x^2 + \frac{3}{2}\right).$$

13. The equation can be written

$$e^{2x} + \frac{1}{2y-1}y' = 0.$$

The equation is separable:

$$\int e^{2x}\,dx + \int \frac{1}{2y-1}\,dy = C \quad \text{and} \quad \frac{1}{2}e^{2x} + \frac{1}{2}\ln|2y-1| = C.$$

Solving for y, we get

$$y = \frac{1}{2} + Ce^{-e^{2x}}.$$

To find the solution that satisfies $y(0) = \frac{1}{2} + \frac{1}{e}$, we set $x = 0$, $y = \frac{1}{2} + \frac{1}{e}$ and solve for C.

We have $C = 1$. Therefore $y = \frac{1}{2} + e^{-e^{2x}}$.

15. The characteristic equation is

$$r^2 - 2r + 2 = 0.$$

The roots are: $r = 1 \pm i$.

The general solution is

$$y = C_1 e^x \cos x + C_2 e^x \sin x$$

17. The characteristic equation is

$$r^2 - r - 2 = 0.$$

The roots are: $r = 2, -1$.

The general solution is

$$y = C_1 e^{2x} + C_2 e^{-x}$$

19. The characteristic equation is

$$r^2 - 6r + 9 = 0.$$

The roots are: $r = 3$ with multiplicity 2.

The general solution is

$$y = C_1 e^{3x} + C_2 x e^{3x}$$

21. The characteristic equation is

$$r^2 + 4r + 13 = 0$$

The roots are: $r = -2 \pm 3i$.

The general solution is

$$y = e^{-2x}(C_1 \cos 3x + C_2 \sin 3x)$$

23. The characteristic equation is

$$r^2 - r = 0.$$

The roots are: $r = 0, 1$.

The general solution is

$$y = C_1 + C_2 e^x.$$

Applying the initial conditions $y(0) = 1$ and $y'(0) = 0$, we have

$$C_1 + C_2 = 1, \quad C_2 = 0 \implies C_1 = 1.$$

The solution of the initial-value problem is: $y = 1$.

25. The characteristic equation is

$$r^2 - 6r + 13 = 0.$$

The roots are: $r = 3 \pm 2i$.

The general solution is

$$y = e^{3x}(C_1 \cos 2x + C_2 \sin 2x).$$

Applying the initial conditions $y(0) = 2, \quad y'(0) = 2$, we have

$$C_1 = 2, \quad 3C_1 + 2C_2 = 2 \quad \Longrightarrow \quad C_1 = 2, \; C_2 = -2.$$

The solution of the initial-value problem is: $y = e^{3x}(2 \cos 2x - 2 \sin 2x).$

27. Curves: $y = Ce^{2x}; \quad y' = 2Ce^{2x} \quad \Longrightarrow \quad y' = 2y$

Orthogonal trajectories: $y' = -\dfrac{1}{2y}; \quad \displaystyle\int 2y \, dy = -\int dx; \quad y^2 = C - x.$

29. Substituting $y = x^r$ into the equation, we get

$$r(r-1)x^r + 4rx^r + 2x^r = 0 \quad \text{or} \quad x^r(r^2 + 3r + 2) = 0 \quad \Longrightarrow \quad r^2 + 3r + 2 = (r+2)(r+1) = 0.$$

The solutions are: $r = -1, -2$.

31. Let $y(t)$ be the value of the business (measured in millions) at time t. Then $y(t)$ satisfies

$$\frac{dy}{dt} = ky^2$$

The general solution of this equation is

$$y(t) = \frac{1}{-kt + C}$$

Applying the given conditions, $y(0) = 1$ and $y(1) = 1.5$, to find C and k, we get $C = 1$ and $k = 1/3$.

1 year from now the business will be worth:

$$y(2) = \frac{1}{-\frac{1}{3} + 1} = 3 \, \text{million}.$$

1.5 years from now the business will be worth:

$$y(2.5) = \frac{1}{-\frac{1}{3}\left(\frac{5}{2}\right) + 1} = 6 \, \text{million}.$$

2 years from now the business will be worth:

$$y(3) = \frac{1}{-\frac{1}{3}(3) + 1} = \infty.$$

Obviously the business cannot continue to grow at a rate proportional to its value squared.

33. (a) The general solution of the differential equation is

$$y = \frac{a}{b} + Ce^{-bt}.$$

Applying the initial condition $y(0) = 0$, we get $C = -\frac{a}{b}$, and

$$y = \frac{a}{b}(1 - e^{-bt})$$

(b) $\lim\limits_{t \to \infty} y(t) = \frac{a}{b}$

(c) Setting $y = 0.9\frac{a}{b}$, we have

$$0.9\frac{a}{b} = \frac{a}{b}(1 - e^{-bt}).$$

The solution to this equation is $t = \dfrac{\ln 10}{b}$ hours.

35. Let $T(t)$ be the temperature of the object at time t. It follows from Newton's law of cooling that

$$T(t) = \tau + Ce^{-kt}.$$

By the conditions given in the problem, we have

$$\tau = 70, \quad T(10) = 20, \quad T(20) = 35.$$

Applying these conditions, we get

$$C = -\frac{500}{7} \quad \text{and} \quad k = \frac{1}{10}\ln(10/7).$$

(a) The temperature of the object at time t is

$$T = 70 - \frac{500}{7}e^{-(t/10)\ln(10/7)} = 70 - \frac{500}{7}\left(\frac{7}{10}\right)^{t/10}$$

(b) $T(0) = 70 - \dfrac{500}{7} = -\dfrac{10}{7}$

37. (a) Let T be the length of time needed to empty the tank. Then

$$80 - (8 - 4)T = 0 \quad \text{and} \quad T = 20 \text{ minutes}.$$

(b) Let $S(t)$ be the amount of salt in the tank at the time t. Then

$$\frac{dS}{dt} = 1 \times 4 - \frac{S}{80 - 4t} \times 8 = 4 - \frac{2S}{20 - t}, \quad S(0) = \frac{1}{8} \times 80 = 10$$

The solution to this initial-value problem is

$$S(t) = 4(20 - t) - \frac{7}{40}(20 - t)^2 \tag{1}$$

(c) Let t_0 be the time that the tank contains exactly 40 gallons. Then

$$80 - 4t_0 = 40 \quad \text{and} \quad t_0 = 10.$$

Substituting $t = 10$ into (1), we get $S(10) = 22.5$ pounds.

39. Let $P(t)$ be the number of people that have heard the rumor at time t. Then $P(t)$ satisfies:

$$\frac{dP}{dt} = kP(20,000 - P)$$

The general solution of this equation is

$$P(t) = \frac{20,000}{1 + Ce^{-20,000kt}}.$$

Now, $P(0) = \frac{20,000}{1 + C} = 500 \implies C = 39;$

$$P(10) = \frac{20,000}{1 + 39e^{-20,000(10k)}} = 1200 \implies 20,000k \cong -.0912.$$

Therefore, $P(t) = \frac{20,000}{1 + 39e^{-0.09120t}}.$

(a) $P(20) = \frac{20,000}{1 + 39e^{-0.0912(20)}} \cong 2742$

(b) The rumor will be spreading fastest when the number of people who have heard it is equal to the number of people who have not heard it:

$$\frac{20,000}{1 + 39e^{-0.0912t}} = 10,000.$$

The solution of this equation is: $t \cong 40$ days.

CHAPTER 10

SECTION 10.1

1. $y = \frac{1}{2}x^2$

vertex $(0,0)$

focus $(0, \frac{1}{2})$

axis $x = 0$

directrix $y = -\frac{1}{2}$

3. $y = \frac{1}{2}(x-1)^2$

vertex $(1,0)$

focus $(1, \frac{1}{2})$

axis $x = 1$

directrix $y = -\frac{1}{2}$

5. $y + 2 = \frac{1}{4}(x-2)^2$

vertex $(2, -2)$

focus $(2, -1)$

axis $x = 2$

directrix $y = -3$

7. $y = x^2 - 4x$

vertex $(2, -4)$

focus $(2, -\frac{15}{4})$

axis $x = 2$

directrix $y = -\frac{17}{4}$

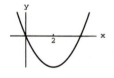

9. $\dfrac{x^2}{9} + \dfrac{y^2}{4} = 1$

center $(0,0)$

foci $(\pm\sqrt{5}, 0)$

length of major axis 6

length of minor axis 4

11. $\dfrac{x^2}{4} + \dfrac{y^2}{6} = 1$

center $(0,0)$

foci $(0, \pm\sqrt{2})$

length of major axis $2\sqrt{6}$

length of minor axis 4

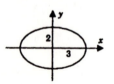

13. $\dfrac{x^2}{9} + \dfrac{(y-1)^2}{4} = 1$

center $(0, 1)$

foci $(\pm\sqrt{5}, 1)$

length of major axis 6

length of minor axis 4

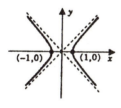

15. $\dfrac{(x-1)^2}{16} + \dfrac{y^2}{64} = 1$

center $(1, 0)$

foci $(1, \pm 4\sqrt{3})$

length of major axis 16

length of minor axis 8

17. $x^2 - y^2 = 1$

center $(0, 0)$

transverse axis 2

vertices $(\pm 1, 0)$

foci $(\pm\sqrt{2}, 0)$

asymptotes $y = \pm x$

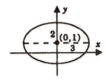

19. $\dfrac{x^2}{9} - \dfrac{y^2}{16} = 1$

center $(0, 0)$

transverse axis 6

vertices $(\pm 3, 0)$

foci $(\pm 5, 0)$

asymptotes $y = \pm\frac{4}{3}x$

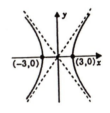

21. $\dfrac{y^2}{16} - \dfrac{x^2}{9} = 1$

center $(0, 0)$

transverse axis 8

vertices $(0, \pm 4)$

foci $(0, \pm 5)$

asymptotes $y = \pm\frac{4}{3}x$

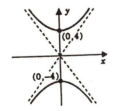

23. $\dfrac{(x-1)^2}{9} - \dfrac{(y-3)^2}{16} = 1$

center $(1, 3)$

transverse axis 6

vertices $(4, 3)$ and $(-2, 3)$

foci $(6, 3)$ and $(-4, 3)$

asymptotes $y - 3 = \pm\frac{4}{3}(x - 1)$

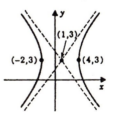

25. $\dfrac{(y-3)^2}{4} - \dfrac{(x-1)^2}{1} = 1$

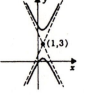

center $(1, 3)$

transverse axis 4

vertices $(1, 5)$ and $(1, 1)$

foci $(1, 3 \pm \sqrt{5})$

asymptotes $y - 3 = \pm 2(x - 1)$

27. We can choose the coordinate system so that the parabola has an equation of the form $y = \alpha x^2$, $\alpha > 0$. One of the points of intersection is then the origin and the other is of the form $(c, \alpha c^2)$. We will assume that $c > 0$.

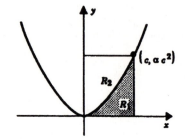

$$\text{area of } R_1 = \int_0^c \alpha x^2 \, dx = \frac{1}{3}\alpha c^3 = \frac{1}{3}A,$$

$$\text{area of } R_2 = A - \tfrac{1}{3}A = \tfrac{2}{3}A.$$

29. The equation of every such parabola takes the form

$$(x - x_0)^2 = 4c(y - y_0)$$

This equation can be written

$$y = \left(\frac{1}{4c}\right)x^2 - \left(\frac{x_0}{2c}\right)x + \left(y_0 + \frac{x_0^2}{4c}\right)$$

vertex $\left(-\dfrac{B}{2A}, \dfrac{4AC - B^2}{4A}\right)$, focus $\left(-\dfrac{B}{2A}, \dfrac{4AC - B^2 + 1}{4A}\right)$, directrix $y = \dfrac{4AC - B^2 - 1}{4A}$

31. By the hint, $xy = X^2 - Y^2 = 1$. In the XY-system $a = 1$, $b = 1$, $c = \sqrt{2}$. We have center $(0,0)$, vertices $(\pm 1, 0)$, foci $(\pm\sqrt{2}, 0)$ and asymptotes $Y = \pm X$. Using

$$x = X + Y \quad \text{and} \quad y = X - Y$$

to convert to the xy-system, we find center $(0, 0)$, vertices $(1, 1)$ and $(-1, -1)$, foci $(\sqrt{2}, \sqrt{2})$ and $(-\sqrt{2}, -\sqrt{2})$, asymptotes $y = 0$ and $x = 0$, transverse axis $2\sqrt{2}$.

33. $2\sqrt{\pi^2 a^4 - A^2} / \pi a$

35. In this case the length of the latus rectum is the width of the parabola at height $y = c$. With $y = c$, $4c^2 = x^2$, and $x = \pm 2c$. The length of the latus rectum is thus $4c$.

37. $A = \displaystyle\int_{-2c}^{2c} \left(c - \frac{x^2}{4c}\right) dx = 2\int_0^{2c} \left(c - \frac{x^2}{4c}\right) dx = 2\left[cx - \frac{x^3}{12c}\right]_0^{2c} = \frac{8}{3}c^2$

$\bar{x} = 0$ by symmetry

$$\bar{y}A = \int_{-2c}^{2c} \frac{1}{2}\left(c^2 - \frac{x^4}{16c^2}\right) dx = \int_0^{2c}\left(c^2 - \frac{x^4}{16c^2}\right) dx = \left[c^2 x - \frac{x^5}{80c^2}\right]_0^{2c} = \frac{8}{5}c^3$$

$$\bar{y} = \left(\frac{8}{5}c^3\right) / \left(\frac{8}{3}c^2\right) = \frac{3}{5}c$$

39. $\dfrac{kx}{p(0)} = \tan\theta = \dfrac{dy}{dx}, \quad y = \dfrac{k}{2\,p(0)}x^2 + C$

In our figure $C = y(0) = 0$. Thus the equation of the cable is $y = kx^2/2p(0)$, the equation of a parabola.

41. Start with any two parabolas γ_1, γ_2. By moving them we can see to it that they have equations of the following form:

$$\gamma_1: x^2 = 4c_1 y, \quad c_1 > 0; \qquad \gamma_2: x^2 = 4c_2 y, \quad c_2 > 0.$$

Now we change the scale for γ_2 so that the equation for γ_2 will look exactly like the equation for γ_1. Set $X = (c_1/c_2)\,x, \quad Y = (c_1/c_2)\,y.$ Then

$$x^2 = 4c_2 y \quad \Longrightarrow \quad (c_2/c_1)^2\, X^2 = 4c_2\,(c_2/c_1)\,Y \quad \Longrightarrow \quad X^2 = 4c_1 Y.$$

Now γ_2 has exactly the same equation as γ_1; only the scale, the units by which we measure distance, has changed.

43. $A = \dfrac{2b}{a} \displaystyle\int_a^{2a} \sqrt{x^2 - a^2}\, dx = \dfrac{2b}{a}\left[\dfrac{x}{2}\sqrt{x^2 - a^2} - \dfrac{a^2}{2}\ln\left(x + \sqrt{x^2 - a^2}\right)\right]_a^{2a}$

$$= [2\sqrt{3} - \ln(2 + \sqrt{3})]ab$$

45. $e = \dfrac{\sqrt{25 - 16}}{\sqrt{25}} = \dfrac{3}{5}$ **47.** $e = \dfrac{\sqrt{25 - 9}}{\sqrt{25}} = \dfrac{4}{5}$

49. E_1 is fatter than E_2,   more like a circle.

51. The ellipse tends to a line segment of length $2a$.

53. $a = 3, \quad c = ea = \frac{2}{3}\sqrt{2} \cdot 3 = 2\sqrt{2}, \quad b = \sqrt{9 - 8} = 1; \quad x^2/9 + y^2 = 1$

55. $e = \frac{5}{3}$ **57.** $e = \sqrt{2}$

59. The branches of H_1 open up less quickly than the branches of H_2.

61. The hyperbola tends to a pair of parallel lines separated by the transverse axis.

63. $P(x, y)$ is on the parabola with directrix $l: Ax + By + C = 0$ and focus $F(a, b)$ iff $d(P, l) = d(P, F)$ which happens iff $\dfrac{|Ax + By + C|}{\sqrt{A^2 + B^2}} = \sqrt{(x - a)^2 + (y - b)^2}.$

Squaring both sides of this equation and simplifying, we obtain

$$(Ay - Bx)^2 = (2aS + 2AC)x + (2bS + 2BC)y + c^2 - (a^2 + b^2)S$$

with $S = A^2 + B^2 \neq 0$.

SECTION 10.2

1–8.

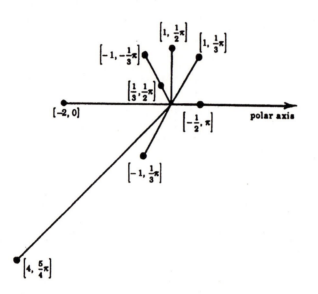

9. $x = 3\cos\frac{1}{2}\pi = 0$

$y = 3\sin\frac{1}{2}\pi = 3$

$(0, 3)$

11. $x = -\cos(-\pi) = 1$

$y = -\sin(-\pi) = 0$

$(1, 0)$

13. $x = -3\cos\left(-\frac{1}{3}\pi\right) = -\frac{3}{2}$

$y = -3\sin\left(-\frac{1}{3}\pi\right) = \frac{3}{2}\sqrt{3}$

$\left(-\frac{3}{2}, \frac{3}{2}\sqrt{3}\right)$

15. $x = 3\cos\left(-\frac{1}{2}\pi\right) = 0$

$y = 3\sin\left(-\frac{1}{2}\pi\right) = -3$

$(0, -3)$

17. $r^2 = 0^2 + 1^2, \quad r = \pm 1$

$r = 1: \quad \cos\theta = 0$ and $\sin\theta = 1$

$\theta = \frac{1}{2}\pi$ $\left\{\begin{array}{l}\end{array}\right.$ $\left[1, \frac{1}{2}\pi + 2n\pi\right], \quad \left[-1, \frac{3}{2}\pi + 2n\pi\right]$

19. $r^2 = (-3)^2 + 0^2 = 9, \quad r = \pm 3$

$r = 3: \quad \cos\theta = -1$ and $\sin\theta = 0$

$\theta = \pi$ $\left\{\begin{array}{l}\end{array}\right.$ $[3, \pi + 2n\pi], \quad [-3, 2n\pi]$

21. $r^2 = 2^2 + (-2)^2 = 8, \quad r = \pm 2\sqrt{2}$

$r = 2\sqrt{2}: \quad \cos\theta = \frac{1}{2}\sqrt{2}, \quad \sin\theta = -\frac{1}{2}\sqrt{2}$

$\theta = \frac{7}{4}\pi$ $\left\{\begin{array}{l}\end{array}\right.$ $\left[2\sqrt{2}, \frac{7}{4}\pi + 2n\pi\right], \quad \left[-2\sqrt{2}, \frac{3}{4}\pi + 2n\pi\right]$

23. $r^2 = \left(4\sqrt{3}\right)^2 + 4^2 = 64, \quad r \pm 8$

$r = 8: \quad \cos\theta = \frac{1}{2}\sqrt{3}, \quad \sin\theta = \frac{1}{2}$

$r = \frac{1}{6}\pi$ $\left\{\begin{array}{l}\end{array}\right.$ $\left[8, \frac{1}{6}\pi + 2n\pi\right], \quad \left[-8, \frac{7}{6}\pi + 2n\pi\right]$

25. $d^2 = (x_1 - x_2)^2 + (y_1 - y_2)^2 = (r_1 \cos\theta_1 - r_2\cos\theta_2)^2 + (r_1\sin\theta_1 - r_2\sin\theta_2)^2$

$$= r_1{}^2\cos^2\theta_1 - 2r_1r_2\cos\theta_1\cos\theta_2 + r_2{}^2\cos^2\theta_2$$

$$+ r_1{}^2\sin^2\theta_1 - 2r_1r_2\sin\theta_1\sin\theta_2 + r_2{}^2\sin^2\theta_2$$

$$= r_1{}^2 + r_2{}^2 - 2r_1r_2\left(\cos\theta_1\cos\theta_2 + \sin\theta_1\sin\theta_2\right)$$

$$= r_1{}^2 + r_2{}^2 - 2r_1r_2\cos\left(\theta_1 - \theta_2\right)$$

$d = \sqrt{r_1{}^2 + r_2{}^2 - 2r_1r_2\cos\left(\theta_1 - \theta_2\right)}$

27. (a) $\left[\frac{1}{2}, \frac{11}{6}\pi\right]$ (b) $\left[\frac{1}{2}, \frac{5}{6}\pi\right]$ (c) $\left[\frac{1}{2}, \frac{7}{6}\pi\right]$

29. (a) $\left[2, \frac{2}{3}\pi\right]$ (b) $\left[2, \frac{5}{3}\pi\right]$ (c) $\left[2, \frac{1}{3}\pi\right]$

31. about the x-axis?: $r = 2 + \cos(-\theta) \implies r = 2 + \cos\theta$,   yes.

    about the y-axis?: $r = 2 + \cos(\pi - \theta) \implies r = 2 - \cos\theta$,   no.

    about the origin?: $r = 2 + \cos(\pi + \theta) \implies r = 2 - \cos\theta$,   no.

33. about the x-axis?: $r\left(\sin(-\theta) + \cos(-\theta)\right) = 1 \implies r\left(-\sin\theta + \cos\theta\right) = 1$,   no.

    about the y-axis?: $r\left(\sin(\pi - \theta) + \cos(\pi - \theta)\right) = 1 \implies r\left(\sin\theta - \cos\theta\right) = 1$,   no.

    about the origin?: $r\left(\sin(\pi + \theta) + \cos(\pi + \theta)\right) = 1 \implies r\left(-\sin\theta - \cos\theta\right) = 1$,   no.

35. about the x-axis?: $r^2\sin(-2\theta) = 1 \implies -r^2\sin 2\theta = 1$,   no.

    about the y-axis?: $r^2\sin(2(\pi - \theta)) = 1 \implies -r^2\sin 2\theta = 1$,   no.

    about the origin?: $r^2\sin(2(\pi + \theta)) = 1 \implies r^2\sin 2\theta = 1$,   yes.

37. $x = 2$

     $r\cos\theta = 2$

39. $2xy = 1$

     $2(r\cos\theta)(r\sin\theta) = 1$

     $r^2\sin 2\theta = 1$

41. $x^2 + (y - 2)^2 = 4$

     $x^2 + y^2 - 4y = 0$

     $r^2 - 4r\sin\theta = 0$

         $r = 4\sin\theta$

43. $y = x$

     $r\sin\theta = r\cos\theta$

     $\tan\theta = 1$

     $\theta = \pi/4$

    [note: division by r okay

    since $[0, 0]$ is on the curve]

45. $x^2 + y^2 + x = \sqrt{x^2 + y^2}$

     $r^2 + r\cos\theta = r$

         $r = 1 - \cos\theta$

47. $(x^2 + y^2)^2 = 2xy$

     $r^4 = 2(r\cos\theta)(r\sin\theta)$

     $r^2 = \sin 2\theta$

49. The horizontal line $y = 4$

51. The line $y = \sqrt{3}x$

53.
$$r = 2\,(1 - \cos\theta)^{-1}$$
$$r - r\cos\theta = 2$$
$$\sqrt{x^2 + y^2} - x = 2$$
$$x^2 + y^2 = (x + 2)^2$$
$$y^2 = 4(x + 1)$$
a parabola

55.
$$r = 6\cos\theta$$
$$r^2 = 6r\cos\theta$$
$$x^2 + y^2 = 6x$$

57. The line $y = 2x$

59.
$$r = \frac{4}{2 - \cos\theta}$$
$$2r - r\cos\theta = 4$$
$$2\sqrt{x^2 + y^2} - x = 4$$
$$4(x^2 + y^2) = (x + 4)^2$$
$$3x^2 + 4y^2 - 8x = 16$$
an ellipse

61.
$$r = \frac{4}{1 - \cos\theta}$$
$$r - r\cos\theta = 4$$
$$\sqrt{x^2 + y^2} - x = 4$$
$$x^2 + y^2 = (x + 4)^2$$
$$y^2 = 8x + 16$$
a parabola

63.
$$r = 2a\,\sin\theta + 2b\,\cos\theta$$
$$r^2 = 2a\,r\,\sin\theta + 2b\,r\,\cos\theta$$
$$x^2 + y^2 = 2ay + 2bx$$
$$(x - b)^2 + (y - a)^2 = a^2 + b^2$$
center: (b, a);   radius: $\sqrt{a^2 + b^2}$

65. $\frac{1}{2}\,(r\cos\theta + d) = r$
$$r = \frac{d}{2 - \cos\theta}$$

SECTION 10.3

1.

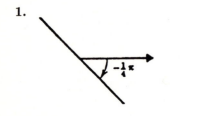

3.

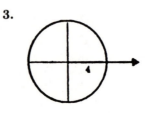

5.

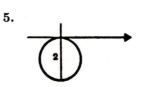

7.

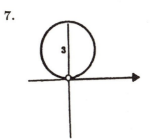

9.

11.

13.

15.

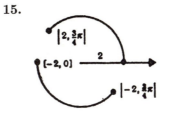

17.

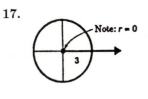

19.

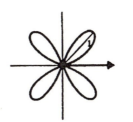

21.

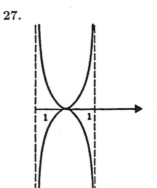

23.

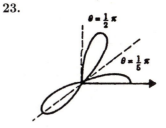

25.

27.

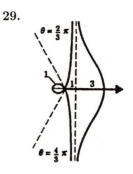

29.

31.

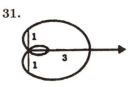

33. yes; $[1, \pi] = [-1, 0]$ and the pair $r = -1,\ \theta = 0$ satisfies the equation

35. yes;   the pair $r = \frac{1}{2},\ \theta = \frac{1}{2}\pi$ satisfies the equation

37. $[2, \pi] = [-2, 0]$. The coordinates $[-2, 0]$ satisfy the equation $r^2 = 4\cos\theta$, and the coordinates $[2, \pi]$ satisfy the equation $r = 3 + \cos\theta$.

39. $(0,0)$, $\left(-\frac{1}{2}, \frac{1}{2}\right)$

41. $(1,0)$, $(-1,0)$

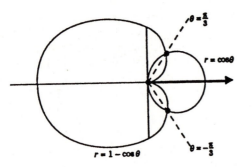

43. $(0,0)$, $\left(\frac{1}{4}, \frac{1}{4}\sqrt{3}\right)$, $\left(\frac{1}{4}, -\frac{1}{4}\sqrt{3}\right)$

45. $(0,0)$, $\left(\pm\frac{\sqrt{3}}{4}, \frac{3}{4}\right)$

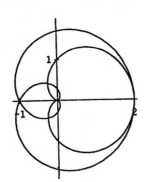

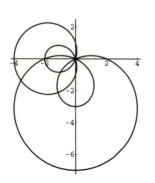

47. (a) The graph of $r = 1 + \cos\left(\theta - \frac{\pi}{3}\right)$ is the graph of $r = 1 + \cos\theta$ rotated counterclockwise $\pi/3$ radians; The graph of $r = 1 + \cos\left(\theta + \frac{\pi}{6}\right)$ is the graph of $r = 1 + \cos\theta$ rotated clockwise $\pi/6$ radians.

(b) The graph of $r = f(\theta - \alpha)$ is the graph of $r = f(\theta)$ rotated counterclockwise α radians.

49. (a)

(b) The curves intersect at:

the pole and $(2,0)$, $(-1,0)$, $(-0.25, \pm0.4330)$

51. (a)

(b) The curves intersect at the pole and at:

$r = 1 - 3\cos\theta$	$r = 2 - 5\sin\theta$
$[-2,0]$	$[2,\pi]$
$[3.800, 3.510]$	$[3.800, 3.510]$
$[2.412, 4.223]$	$[-2.412, 1.081]$
$[-1.267, 0.713]$	$[-1.267, 0.713]$

53. "Butterfly" curves. The graph for the case $k = 2$ is:

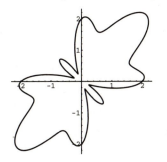

55. (a) $k = \frac{3}{2}$

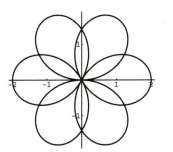

(b) $k = \frac{5}{2}$

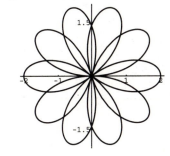

A petal curve with $2m$ petals.

PROJECT 10.3

1. $e = \dfrac{r}{d - r\cos\theta} \implies r = ed - er\cos\theta \implies r(1 + e\cos\theta) = ed \implies r = \dfrac{ed}{1 + e\cos\theta}$

3. (a) ellipse: $r = \dfrac{8}{4 + 3\cos\theta} = \dfrac{2}{1 + \frac{3}{4}\cos\theta}$.

Thus $e = \dfrac{3}{4}$ and $\dfrac{3}{4}d = 2 \implies d = \dfrac{8}{3}$.

Rectangular equation:

$$a = \frac{32}{7}, \quad c = \frac{24}{7}, \quad \text{so} \quad \frac{\left(x + \frac{24}{7}\right)^2}{\left(\frac{32}{7}\right)^2} + \frac{y^2}{\left(\frac{24}{7}\right)^2} = 1$$

(b) hyperbola: $r = \dfrac{6}{1 + 2\cos\theta}$

Thus $e = 2$ and $2d = 6 \implies d = 3$.

Rectangular equation:

$$a = 2, \quad c = 4, \quad \text{so} \quad \frac{(x - 4)^2}{4} - \frac{y^2}{12} = 1$$

(c) parabola: $r = \dfrac{6}{2 + 2\cos\theta} = \dfrac{3}{1 + \cos\theta}$

Thus $e = 1$ and $d = 3$.

Rectangular equation:

$$y^2 = -4\left(\frac{3}{2}\right)\left(x - \frac{3}{2}\right) = -6\left(x - \frac{3}{2}\right)$$

SECTION 10.4

1.

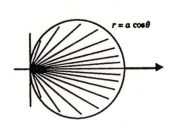

$$A = \int_{-\pi/2}^{\pi/2} \frac{1}{2} \left[a \cos \theta\right]^2 d\theta$$

$$= a^2 \int_0^{\pi/2} \frac{1 + \cos 2\theta}{2} \, d\theta$$

$$= a^2 \left[\frac{\theta}{2} + \frac{\sin 2\theta}{4}\right]_0^{\pi/2} = \frac{1}{4}\pi a^2$$

3.

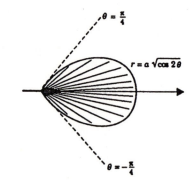

$$A = \int_{-\pi/4}^{\pi/4} \frac{1}{2} \left[a\sqrt{\cos 2\theta}\right]^2 d\theta$$

$$= a^2 \int_0^{\pi/4} \cos 2\theta \, d\theta$$

$$= a^2 \left[\frac{\sin 2\theta}{2}\right]_0^{\pi/4} = \frac{1}{2}a^2$$

5.

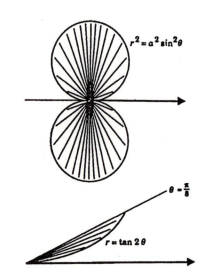

$$A = 2 \int_0^{\pi} \frac{1}{2} \left(a^2 \sin^2 \theta\right) d\theta$$

$$= a^2 \int_0^{\pi} \frac{1 - \cos 2\theta}{2} \, d\theta$$

$$= a^2 \left[\frac{\theta}{2} - \frac{\sin 2\theta}{4}\right]_0^{\pi} = \frac{1}{2}\pi a^2$$

7.

$$A = \int_0^{\pi/8} \frac{1}{2} \left[\tan 2\theta\right]^2 d\theta$$

$$= \frac{1}{2} \int_0^{\pi/8} \left(\sec^2 2\theta - 1\right) d\theta$$

$$= \frac{1}{2} \left[\frac{1}{2} \tan 2\theta - \theta\right]_0^{\pi/8} = \frac{1}{4} - \frac{\pi}{16}$$

9.

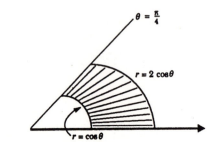

$$A = \int_0^{\pi/4} \frac{1}{2} \left([2 \cos \theta]^2 - [\cos \theta]^2\right) d\theta$$

$$= \frac{3}{2} \int_0^{\pi/4} \frac{1 + \cos 2\theta}{2} \, d\theta$$

$$= \frac{3}{2} \left[\frac{\theta}{2} + \frac{\sin 2\theta}{4}\right]_0^{\pi/4} = \frac{3}{16}\pi + \frac{3}{8}$$

11.

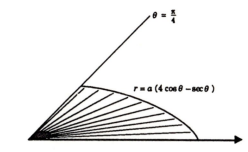

$$A = \int_0^{\pi/4} \frac{1}{2} \left[a \left(4\cos\theta - \sec\theta \right) \right]^2 d\theta$$

$$= \frac{a^2}{2} \int_0^{\pi/4} \left[16\cos^2\theta - 8 + \sec^2\theta \right] d\theta$$

$$= \frac{a^2}{2} \int_0^{\pi/4} \left[8\left(1 + \cos 2\theta\right) - 8 + \sec^2\theta \right] d\theta$$

$$= \frac{a^2}{2} \left[4\sin 2\theta + \tan\theta \right]_0^{\pi/4} = \frac{5}{2}a^2$$

13.

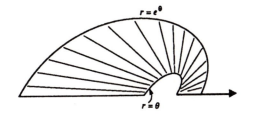

$$A = \int_0^{\pi} \frac{1}{2} \left(\left[e^\theta \right]^2 - \left[\theta \right]^2 \right) d\theta$$

$$= \frac{1}{2} \int_0^{\pi} \left(e^{2\theta} - \theta^2 \right) d\theta$$

$$= \tfrac{1}{2} \left[\tfrac{1}{2} e^{2\theta} - \tfrac{1}{3}\theta^3 \right]_0^{\pi} = \tfrac{1}{12} \left(3e^{2\pi} - 3 - 2\pi^3 \right)$$

15.

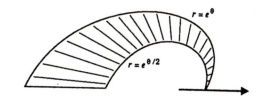

$$A = \int_0^{\pi} \frac{1}{2} \left(\left[e^\theta \right]^2 - \left[e^{\theta/2} \right]^2 \right) d\theta$$

$$= \frac{1}{2} \int_0^{\pi} \left(e^{2\theta} - e^\theta \right) d\theta$$

$$= \frac{1}{2} \left[\frac{1}{2} e^{2\theta} - e^\theta \right]_0^{\pi} = \frac{1}{4} \left(e^{2\pi} + 1 - 2e^\pi \right)$$

17.

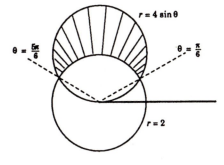

$$A = \int_{\pi/6}^{5\pi/6} \frac{1}{2} \left(\left[4\sin\theta \right]^2 - \left[2 \right]^2 \right) d\theta$$

19.

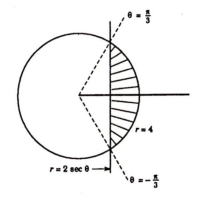

$$A = \int_{-\pi/3}^{\pi/3} \frac{1}{2} \left(\left[4 \right]^2 - \left[2\sec\theta \right]^2 \right) d\theta$$

21.

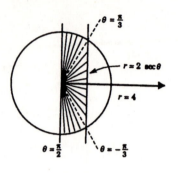

$$A = 2 \left\{ \int_0^{\pi/3} \frac{1}{2} (2 \sec \theta)^2 \, d\theta + \int_{\pi/3}^{\pi/2} \frac{1}{2} (4)^2 \, d\theta \right\}$$

23.

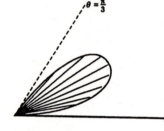

$$A = \int_0^{\pi/3} \frac{1}{2} (2 \sin 3\theta)^2 \, d\theta$$

25.

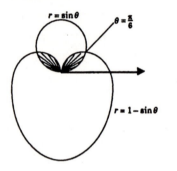

$$A = 2 \left\{ \int_0^{\pi/6} \frac{1}{2} (\sin \theta)^2 \, d\theta + \int_{\pi/6}^{\pi/2} \frac{1}{2} (1 - \sin \theta)^2 \, d\theta \right\}$$

27.

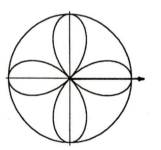

$$A = \pi - 8 \int_0^{\pi/4} \frac{1}{2} (\cos 2\theta)^2 \, d\theta$$

29.

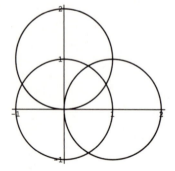

$$A = 2 \int_0^{\pi/6} \frac{1}{2} [2 \sin \theta]^2 \, d\theta + \frac{1}{12} \pi$$

$$= 4 \int_0^{\pi/6} \sin^2 \theta \, d\theta + \frac{1}{12} \pi$$

$$= 2 \left[\theta - \frac{1}{2} \sin 2\theta \right]_0^{\pi/6} + \frac{1}{12} \pi = \frac{5}{12} \pi - \frac{1}{2} \sqrt{3}$$

31. The area of one petal of the curve $r = a\cos 2n\theta$ is given by:

$$2\int_0^{\pi/4n} \tfrac{1}{2}(a\cos 2n\theta)^2\,d\theta = a^2\int_0^{\pi/4n} \cos^2 2n\theta\,d\theta$$

$$= a^2\int_0^{\pi/4n}\left(\frac{1}{2} + \frac{\cos 4n\theta}{2}\right)d\theta$$

$$= a^2\left[\frac{1}{2}\,\theta + \frac{\sin 4n\theta}{8n}\right]_0^{\pi/4n} = \frac{\pi a^2}{8n}$$

The total area enclosed by $r = a\cos 2n\theta$ is $\dfrac{\pi a^2}{2}$.

The area of one petal of the curve $r = a\sin 2n\theta$ is given by:

$$A = 2\int_0^{\pi/4n} \tfrac{1}{2}(a\sin 2n\theta)^2\,d\theta = a^2\int_0^{\pi/4n}\left(\frac{1}{2} + \frac{\cos 4n\theta}{2}\right)d\theta = \frac{\pi a^2}{8n}$$

and the total area enclosed by the curve is $\dfrac{\pi a^2}{2}$.

33. Let $P = \{\alpha = \theta_0, \theta_1, \theta_2, \ldots, \theta_n = \beta\}$ be a partition of the interval $[\alpha, \beta]$. Let θ_i^* be the midpoint of $[\theta_{i-1}, \theta_i]$ and let $r_i^* = f(\theta_i^*)$. The area of the ith "triangular" region is $\tfrac{1}{2}(r_i^*)\Delta\theta_i$, where $\Delta\theta_i = \theta_i - \theta_{i-1}$, and the rectangular coordinates of its centroid are(approximately) $\left(\tfrac{2}{3}r_i^*\cos\theta_i^*, \tfrac{2}{3}r_i^*\sin\theta_i^*\right)$.

The centroid $(\overline{x}_p, \overline{y}_p)$ of the union of the triangular regions satisfies the following equations

$$\overline{x}_p A_p = \frac{1}{3}(r_1^*)^3\cos\theta_1\Delta\theta_1 + \frac{1}{3}(r_2^*)^3\cos\theta_2\Delta\theta_2 + \cdots + \frac{1}{3}(r_n^*)^3\cos\theta_n\Delta\theta_n$$

$$\overline{y}_p A_p = \frac{1}{3}(r_1^*)^3\sin\theta_1\Delta\theta_1 + \frac{1}{3}(r_2^*)^3\sin\theta_2\Delta\theta_2 + \cdots + \frac{1}{3}(r_n^*)^3\sin\theta_n\Delta\theta_n$$

As $\|P\| \to 0$, the union of the triangular regions tends to the region Ω and the equations above tend to

$$\overline{x}\,A = \int_\alpha^\beta \frac{1}{3}r^3\cos\theta\,d\theta$$

$$\overline{y}\,A = \int_\alpha^\beta \frac{1}{3}r^3\sin\theta\,d\theta$$

The result follows from the fact that $A = \displaystyle\int_\alpha^\beta \frac{1}{2}r^2\cos\theta\,d\theta$.

35. Since the region enclosed by the cardioid $r = 1 + \cos\theta$ is symmetric with respect to the x-axis, $\overline{y} = 0$. To find $\overline{x}$:

$$A = \int_0^{2\pi} r^2\,d\theta = \int_0^{2\pi}(1 + \cos\theta)^2\,d\theta$$

$$= \int_0^{2\pi}(1 + 2\cos\theta + \cos^2\theta)\,d\theta$$

$$= \int_0^{2\pi}\left(\frac{3}{2} + 2\cos\theta\frac{1}{2}\cos 2\theta\right)d\theta$$

$$= \left[\frac{3}{2} + 2\sin\theta + \frac{1}{4}\sin 2\theta\right]_0^{2\pi} = 3\pi$$

and

$$\frac{2}{3} \int_0^{2\pi} r^3 \cos\theta \, d\theta = \frac{2}{3} \int_0^{2\pi} (1 + \cos\theta)^3 \cos\theta \, d\theta$$

$$= \frac{2}{3} \int_0^{2\pi} \left(\cos\theta + 3\cos^2\theta + 3\cos^3\theta + \cos^4\theta \right) d\theta$$

$$= \frac{2}{3} \int_0^{2\pi} \left(\frac{15}{8} + 4\cos\theta + 2\cos 2\theta + \frac{1}{8}\cos 4\theta - 3\sin^2\theta \cos\theta \right) d\theta$$

$$= \frac{2}{3} \left[\frac{15}{8}\theta + 4\sin\theta + \sin 2\theta + \frac{1}{32}\sin 4\theta - \sin^3\theta \right]_0^{2\pi} = \frac{5}{2}\pi$$

Thus $\overline{x} = \dfrac{5\pi/2}{3\pi} = \dfrac{5}{6}$.

37. $A = \displaystyle\int_0^{2\pi} \frac{1}{2}[2 + \cos\theta]^2 \, d\theta = \frac{1}{2}\int_0^{2\pi} (4 + 4\cos\theta + \cos^2\theta) \, d\theta = \frac{1}{2}\left[4\theta + 4\sin\theta + \frac{1}{2}\theta + \frac{1}{4}\sin 2\theta\right]_0^{2\pi} = \frac{9}{2}\pi$

39.

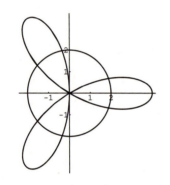

$$A = 6 \int_0^{\pi/9} \frac{1}{2}\left[(4\cos 3\theta)^2 - 4\right] d\theta$$

$$= 3 \int_0^{\pi/9} \left[16\cos^2 3\theta - 4\right] d\theta$$

$$= 3 \int_0^{\pi/9} (4 + 8\cos 6\theta) \, d\theta$$

$$= 3\left[4\theta + \frac{4}{3}\sin 6\theta\right]_0^{\pi/9} = \frac{4}{3}\pi + 2\sqrt{3}$$

41.   (a) $y^2 = x^2 \left(\dfrac{a - x}{a + x}\right)$

   (b)   Let $a = 2$

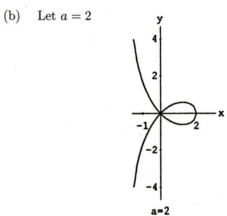

$$r^2 \sin^2\theta = r^2 \cos^2\theta \left(\frac{a - r\cos\theta}{a + r\cos\theta}\right)$$

$$\sin^2\theta(a + r\cos\theta) = \cos^2\theta(a - r\cos\theta)$$

$$r\cos\theta = a\cos 2\theta$$

$$r = a\cos 2\theta \sec\theta$$

   (c) $A = \displaystyle\int_{3\pi/4}^{5\pi/4} \frac{1}{2} a^2 \cos^2 2\theta \sec^2\theta \, d\theta$

$$= 2 \int_{3\pi/4}^{5\pi/4} \cos^2 2\theta \sec^2\theta \, d\theta \qquad (a = 2)$$

$$= 2 \int_{3\pi/4}^{5\pi/4} \frac{\left(2\cos^2\theta - 1\right)^2}{\cos^2\theta} \, d\theta$$

$$= 2 \int_{3\pi/4}^{5\pi/4} \left(4\cos^2\theta - 4 + \sec^2\theta\right) d\theta$$

a=2

$$= 2 \int_{3\pi/4}^{5\pi/4} \left(-2 + 2 \cos 2\theta + \sec^2 \theta \right) \, d\theta$$

$$= 2 \left[-2\theta + \sin 2\theta + \tan \theta \right]_{3\pi/4}^{5\pi/4} = 8 - 2\pi$$

SECTION 10.5

1. $\quad 4x = (y-1)^2$

3. $\quad y = 4x^2 + 1, \quad x \geq 0$

5. $\quad 9x^2 + 4y^2 = 36$

7. $\quad 1 + x^2 = y^2$

9. $\quad y = 2 - x^2, \quad -1 \leq x \leq 1$

11. $\quad 2y - 6 = x, \quad -4 \leq x \leq 4$

13. $\quad y = x - 1$

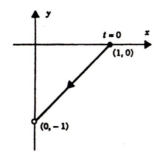

15. $\quad xy = 1$

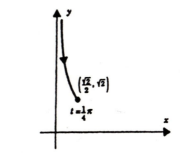

17. $\quad 2x + y = 11$

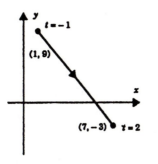

19. $\quad x = \sin \frac{1}{2}\pi y$

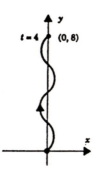

21. $\quad 1 + x^2 = y^2$

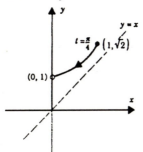

23. (a) $x(t) = -\sin 2\pi t, \quad y(t) = \cos 2\pi t$ (b) $x(t) = \sin 4\pi t, \quad y(t) = \cos 4\pi t$

(c) $x(t) = \cos \frac{1}{2}\pi t, \quad y(t) = \sin \frac{1}{2}\pi t$ (d) $x(t) = \cos \frac{3}{2}\pi t, \quad y(t) = -\sin \frac{3}{2}\pi t$

25. $x(t) = \tan \frac{1}{2}\pi t, \quad y(t) = 2$

27. $x(t) = 3 + 5t, \quad y(t) = 7 - 2t$ **29.** $x(t) = \sin^2 \pi t, \quad y(t) = -\cos \pi t$

31. $x(t) = (2-t)^2, \quad y(t) = (2-t)^3$

33. $\displaystyle \int_c^d y(t)x'(t)\,dt = \int_c^d f(x(t))x'(t)\,dt = \int_a^b f(x)\,dx = \text{area below } C$

35. $\displaystyle \int_c^d \pi\,[y(t)]^2\,x'(t)\,dt = \int_c^d \pi\,[f(x(t))]^2\,x'(t)\,dt = \int_a^b \pi\,[f(x)]^2\,dx = V_x$

$\displaystyle \int_c^d 2\pi x(t)y(t)x'(t)\,dt = \int_c^d 2\pi x(t)f(x(t))x'(t)\,dt = \int_a^b 2\pi x f(x)\,dx = V_y$

37.

$A = \displaystyle \int_0^{2\pi} x(t)\,y'(t)\,dt$

$= r^2 \displaystyle \int_0^{2\pi} (1 - \cos t)\,dt$

$= r^2\,[t - \sin t]_0^{2\pi} = 2\pi r^2$

39. (a) $V_x = 2\pi \overline{y} A = 2\pi \left(\frac{3}{4}r\right)\left(2\pi r^2\right) = 3\pi^2 r^3$

(b) $V_y = 2\pi \overline{x} A = 2\pi \left(\pi r\right) 2\pi r^2 = 4\pi^3 r^3$

41. $x(t) = -a\cos t, \quad y(t) = b\sin t \qquad t \in [0, \pi]$

43. (a) Equation for the ray: $y + 2x = 17, \quad x \geq 6.$

Equation for the circle: $(x-3)^2 + (y-1)^2 = 25.$

Simultaneous solution of these equations gives the points of intersection: $(6, 5)$ and $(8, 1)$.

(b) The particle on the ray is at $(6, 5)$ when $t = 0$.   However, when $t = 0$ the particle on the circle is at the point $(-2, 1)$. Thus, the intersection point $(6, 5)$ is not a collision point.

The particle on the ray is at $(8, 1)$ when $t = 1$. Since the particle on the circle is also at $(8, 1)$ when $t = 1$, the intersection point $(8, 1)$ is a collision point.

45. If $x(r) = x(s)$ and $r \neq s$, then

$$r^2 - 2r = s^2 - 2s$$

$$r^2 - s^2 = 2r - 2s$$

(1) $$r + s = 2.$$

If $y(r) = y(s)$ and $r \neq s$, then

$$r^3 - 3r^2 + 2r = s^3 - 3s^2 + 2s$$

$$\left(r^3 - s^3\right) - 3\left(r^2 - s^2\right) + 2\left(r - s\right) = 0$$

(2) $$\left(r^2 + rs + s^2\right) - 3\left(r + s\right) + 2 = 0.$$

Simultaneous solution of (1) and (2) gives $r = 0$ and $r = 2$. Since $(x(0), y(0)) = (0,0) = (x(2), y(2))$, the curve intersects itself at the origin.

47. Suppose that $\quad r, s \in [0,4] \quad$ and $\quad r \neq s$.

$$x(r) = x(s) \implies \quad \sin 2\pi r = \sin 2\pi s.$$

$$y(r) = y(s) \implies \quad 2r - r^2 = 2s - s^2 \quad \implies \quad 2(r - s) = r^2 - s^2 \quad \implies \quad 2 = r + s.$$

Now we solve the equations simultaneously:

$$\sin 2\pi r = \sin\left[2\pi\left(2 - r\right)\right] = -\sin 2\pi r$$

$$2 \sin 2\pi r = 0$$

$$\sin 2\pi r = 0.$$

Since $\quad r \in [0,4], \quad r = 0, \ \frac{1}{2}, \ 1, \ \frac{3}{2}, \ 2, \ \frac{5}{2}, \ 3, \ \frac{7}{2}, \ 4.$

Since $\quad s \in [0,4] \quad$ and $\quad r \neq s \quad$ and $\quad r + s = 2, \quad$ we are left with $\quad r = 0, \ \frac{1}{2}, \ \frac{3}{2}, \ 2.$ Note that
$(x(0), \ y(0)) = (0,0) = (x(2), \ y(2)) \quad$ and $\quad \left(x\left(\frac{1}{2}\right), \ y\left(\frac{1}{2}\right)\right) = \left(0, \ \frac{3}{4}\right) = \left(x\left(\frac{3}{2}\right), \ y\left(\frac{3}{2}\right)\right).$
The curve intersects itself at $(0,0)$ and $\left(0, \frac{3}{4}\right)$.

49. $x = 2t, \ y = 4t - t^2; \ 0 \le t \le 6:$ From the first equation, $t = \frac{1}{2}x$. Substituting this into the second equation, we get: $\quad y = 2x - \frac{1}{4}x^2, \quad$ a parabola. The limits on t imply that $0 \le x \le 12$. The particle moves along the parabola $y = 2x - \frac{1}{4}x^2$ from the point $(0,0)$ to the point $(12, -12)$.

51. $x = \cos(t^2 + t), \ y = \sin(t^2 + t) \implies x^2 = \cos^2(t^2 + t), \ y^2 = \sin^2(t^2 + t).$
Since $\cos^2(t^2 + t) + \sin^2(t^2 + t) = 1,$ we have $x^2 + y^2 = 1$ the unit circle.
The particle starts at the point $(1, 0)$ and moves around the unit circle in the counterclockwise direction.

53. $x(\theta) = \cos\theta(a - b\sin\theta), \quad y(\theta) = \sin\theta(a - b\sin\theta)$
(a) $a = 1, \ b = 2$ (b) $a = 2, \ b = 2$ (c) $a = 2, \ b = 1$

(d) The curves are limaçons; the curve has an inner loop if $a < b$ and no loop if $a > b$.

PROJECT 10.5

1. $x''(t) = 0 \implies x'(t) = C; \quad x'(0) = v_0 \cos\theta \implies x'(t) = v_0 \cos\theta.$

Integrating again, $\quad x(t) = (v_0 \cos\theta)t + x_0 \quad$ (since $x(0) = x_0$)

Similarly, since $y''(t) = -g \implies y'(t) = -gt + C; \quad y'(0) = v_0 \sin\theta \implies y'(t) = -gt + v_0 \sin\theta.$

Integrating again, $\quad y(t) = -\frac{1}{2}gt^2 + (v_0 \sin\theta)t + y_0 \quad$ (since $y(0) = y_0$)

3. (a) Set $x_0 = 0$, $y_0 = 0$, $g = 32$.

parametric equations: $\quad x = (v_0 \cos\theta)t, \quad y = -16t^2 + (v_0 \sin\theta)t$

rectangular equation: $\quad y = -\dfrac{16}{v_0{}^2}(\sec^2\theta)x^2 + (\tan\theta)x.$

(b) To find the range, set $y = 0$ and solve for x, $x \neq 0$:

$$-\frac{16}{v_0^2}\sec^2\theta\, x^2 + \tan\theta\, x = 0 \implies x = \frac{v_0^2}{16}\sin\theta\cos\theta$$

(c) $y(t) = 0$ (and $t \neq 0$) when $t = \dfrac{v_0}{16}\sin\theta$

(d) The range $\frac{1}{16}v_0{}^2 \sin\theta\cos\theta = \frac{1}{32}v_0{}^2 \sin 2\theta$ is maximal when $\theta = \frac{1}{4}\pi$ for then $\sin 2\theta = 1$.

(e) Set $x = \dfrac{1}{32}v_0{}^2 \sin 2\theta = b, \quad \theta = \dfrac{1}{2}\arcsin\left(\dfrac{32b}{v_0{}^2}\right).$

SECTION 10.6

1. $x'(1) = 1, \quad y'(1) = 3, \quad$ slope 3, $\quad$ point $(1, 0);$ $\quad$ tangent: $y = 3(x - 1)$

3. $x'(0) = 2, \quad y'(0) = 0, \quad$ slope 0, $\quad$ point $(0, 1);$ $\quad$ tangent: $y = 1$

5. $x'(1/2) = 1, \quad y'(1/2) = -3, \quad$ slope -3, $\quad$ point $\left(\frac{1}{4}, \frac{9}{4}\right);$ $\quad$ tangent: $y - \frac{9}{4} = -3\left(x - \frac{1}{4}\right)$

7. $x'\left(\dfrac{\pi}{4}\right) = -\dfrac{3}{4}\sqrt{2}, \quad y'\left(\dfrac{\pi}{4}\right) = \dfrac{3}{4}\sqrt{2}, \quad$ slope -1, $\quad$ point $\left(\dfrac{1}{4}\sqrt{2}, \dfrac{1}{4}\sqrt{2}\right);$

tangent: $y - \frac{1}{4}\sqrt{2} = -\left(x - \frac{1}{4}\sqrt{2}\right)$

9. $x(\theta) = \cos\theta\,(4 - 2\sin\theta), \quad y(\theta) = \sin\theta\,(4 - 2\sin\theta), \quad$ point $(4, 0)$

$x'(\theta) = -4\sin\theta - 2\left(\cos^2\theta - \sin^2\theta\right), \quad y'(\theta) = 4\cos\theta - 4\sin\theta\cos\theta$

$x'(0) = -2, \quad y'(0) = 4, \quad$ slope -2, $\quad$ tangent: $y = -2\,(x - 4)$

11. $x(\theta) = \dfrac{4\cos\theta}{5 - \cos\theta}, \quad y(\theta) = \dfrac{4\sin\theta}{5 - \cos\theta}, \quad$ point $\left(0, \dfrac{4}{5}\right)$

$x'(\theta) = \dfrac{-20\sin\theta}{(5 - \cos\theta)^2}, \quad y'(\theta) = \dfrac{4\,(5\cos\theta - 1)}{(5 - \cos\theta)^2}$

$x'\left(\dfrac{\pi}{2}\right) = -\dfrac{4}{5}, \quad y'\left(\dfrac{\pi}{2}\right) = -\dfrac{4}{25}, \quad$ slope $\dfrac{1}{5}$, $\quad$ tangent: $y - \dfrac{4}{5} = \dfrac{1}{5}x$

13. $x(\theta) = \dfrac{\cos\theta\,(\sin\theta - \cos\theta)}{\sin\theta + \cos\theta}$, $y(\theta) = \dfrac{\sin\theta\,(\sin\theta - \cos\theta)}{\sin\theta + \cos\theta}$,   point $(-1, 0)$

$x'(\theta) = \dfrac{\sin\theta\,\cos 2\theta + 2\cos\theta}{(\sin\theta + \cos\theta)^2}$, $y'(\theta) = \dfrac{2\sin\theta - \cos\theta\,\cos 2\theta}{(\sin\theta + \cos\theta)^2}$

$x'(0) = 2$, $y'(0) = -1$,   slope $-\tfrac{1}{2}$,   tangent $y = -\tfrac{1}{2}(x+1)$

15. $x(t) = t$, $y(t) = t^3$ **17.** $x(t) = t^{5/3}$, $y(t) = t$

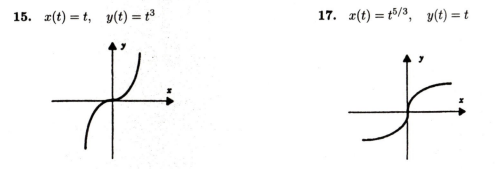

$x'(0) = 1$, $y'(0) = 0$,   slope 0 $x'(0) = 0$, $y'(0) = 1$,   slope undefined

tangent $y = 0$ tangent $x = 0$

19. $x'(t) = 3 - 3t^2$, $y'(t) = 1$

$x'(t) = 0 \implies t = \pm 1$; $y'(t) \neq 0$

(a)   none

(b)   at $(2, 2)$ and $(-2, 0)$

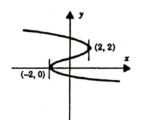

21. curve traced once completely with $t \in [0, 2\pi)$

$x'(t) = -4\cos t$, $y'(t) = -3\sin t$

$x'(t) = 0 \implies t = \dfrac{\pi}{2},\ \dfrac{3\pi}{2}$;

$y'(t) = 0 \implies t = 0, \pi$

(a)   at $(3, 7)$ and $(3, 1)$

(b)   at $(-1, 4)$ and $(7, 4)$

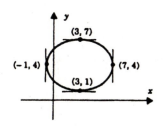

23. $x'(t) = 2t - 2$, $y'(t) = 3t^2 - 6t + 2$

$x'(t) = 0 \implies t = 1$

$y'(t) = 0 \implies t = 1 \pm \tfrac{1}{3}\sqrt{3}$

(a)   at $\left(-\tfrac{2}{3}, \pm\tfrac{2}{9}\sqrt{3}\right)$

(b)   at $(-1, 0)$

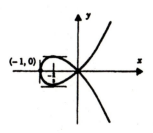

25. curve traced completely with $t \in [0, 2\pi)$

$x'(t) = -\sin t, \quad y'(t) = 2\cos 2t$

$x'(t) = 0 \implies t = 0, \pi$

$y'(t) = 0 \implies t = \dfrac{\pi}{4}, \dfrac{3\pi}{4}, \dfrac{5\pi}{4}, \dfrac{7\pi}{4}$

(a) at $\left(\pm\frac{1}{2}\sqrt{2}, \pm 1\right)$

(b) at $(\pm 1, 0)$

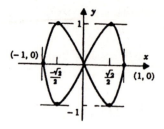

27. First, we find the values of t when the curve passes through $(2, 0)$.

$$y(t) = 0 \implies t^4 - 4t^2 = 0 \implies t = 0, \pm 2.$$

$$x(-2) = 2, \quad x(0) = 2, \quad x(2) = -2.$$

The curve passes through $(2, 0)$ at $t = -2$ and $t = 0$.

$$x'(t) = -1 - \frac{\pi}{2}\sin\frac{\pi t}{4}, \quad y'(t) = 4t^3 - 8t.$$

At $t = -2, \quad x'(-2) = \dfrac{\pi}{2} - 1, \quad y'(t) = -16, \quad$ tangent: $y = \dfrac{32}{2 - \pi}(x - 2)$.

At $t = 0, \quad x'(0) = -1, \quad y'(0) = 0, \quad$ tangent: $y = 0$.

29. The slope of $\overline{OP}$ is $\tan\theta_1$. The curve $r = f(\theta)$ can be parameterized by setting

$$x(\theta) = f(\theta)\cos\theta, \qquad y(\theta) = f(\theta)\sin\theta.$$

Differentiation gives

$$x'(\theta) = -f(\theta)\sin\theta + f'(\theta)\cos\theta, \qquad y'(\theta) = f(\theta)\cos\theta + f'(\theta)\sin\theta.$$

If $f'(\theta_1) = 0$, then

$$x'(\theta_1) = -f(\theta_1)\sin\theta_1, \qquad y'(\theta_1) = f(\theta_1)\cos\theta_1.$$

Since $f(\theta_1) \neq 0$, we have

$$m = \frac{y'(\theta_1)}{x'(\theta_1)} = -\cot\theta_1 = -\frac{1}{\text{slope of } \overline{OP}}.$$

31. $x'(t) = 3t^2, \quad y'(t) = 2t$

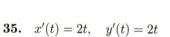

$x^2 = y^3$

33. $x'(t) = 5t^4, \quad y'(t) = 3t^2$

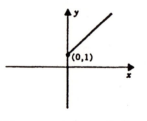

$x^3 = y^5$

35. $x'(t) = 2t, \quad y'(t) = 2t$

ray: $y = x + 1, \quad x \geq 0$

37. By (9.7.5), $\dfrac{d^2y}{dx^2} = \dfrac{(-\sin t)(-\sin t) - (\cos t)(-\cos t)}{(-\sin t)^3} = \dfrac{-1}{\sin^3 t}.$ At $t = \dfrac{\pi}{6}$, $\dfrac{d^2y}{dx^2} = -8.$

39. By (9.7.5), $\dfrac{d^2y}{dx^2} = \dfrac{(e^t)(e^{-t}) - (-e^{-t})(e^t)}{(e^t)^3} = 2e^{-3t}.$ At $t = 0$, $\dfrac{d^2y}{dx^2} = 2.$

41. $\dfrac{d^2y}{dx^2} = \cot^3 t$ **43.**   tangent line: $y - 2 = -\dfrac{16}{3}\left(x - \dfrac{1}{8}\right)$

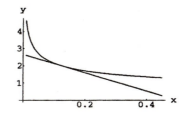

SECTION 10.7

1. $L = \displaystyle\int_0^1 \sqrt{1 + 2^2}\, dx = \sqrt{5}$

3. $L = \displaystyle\int_1^4 \sqrt{1 + \left[\dfrac{3}{2}\left(x - \dfrac{4}{9}\right)^{1/2}\right]^2}\, dx = \int_1^4 \dfrac{3}{2}\sqrt{x}\, dx = \left[x^{3/2}\right]_1^4 = 7$

5. $L = \displaystyle\int_0^3 \sqrt{1 + \left(\dfrac{1}{2}\sqrt{x} - \dfrac{1}{2\sqrt{x}}\right)^2}\, dx = \int_0^3 \left(\dfrac{1}{2}\sqrt{x} + \dfrac{1}{2\sqrt{x}}\right) dx = \left[\dfrac{1}{3}x^{3/2} + x^{1/2}\right]_0^3 = 2\sqrt{3}$

7. $L = \displaystyle\int_0^1 \sqrt{1 + \left[x\left(x^2 + 2\right)^{1/2}\right]^2}\, dx = \int_0^1 \left(x^2 + 1\right) dx = \left[\dfrac{1}{3}x^3 + x\right]_0^1 = \dfrac{4}{3}$

9. $L = \displaystyle\int_1^5 \sqrt{1 + \left[\dfrac{1}{2}\left(x - \dfrac{1}{x}\right)\right]^2}\, dx = \int_1^5 \dfrac{1}{2}\left(x + \dfrac{1}{x}\right) dx = \left[\dfrac{1}{2}\left(\dfrac{1}{2}x^2 + \ln x\right)\right]_1^5 = 6 + \dfrac{1}{2}\ln 5 \cong 6.80$

11. $L = \displaystyle\int_1^8 \sqrt{1 + \left[\dfrac{1}{2}\left(x^{1/3} - x^{-1/3}\right)\right]^2}\, dx = \int_1^8 \dfrac{1}{2}\left(x^{1/3} + x^{-1/3}\right) dx = \dfrac{1}{2}\left[\dfrac{3}{4}x^{4/3} + \dfrac{3}{2}x^{2/3}\right]_1^8 = \dfrac{63}{8}$

13. $L = \displaystyle\int_0^{\pi/4} \sqrt{1 + \tan^2 x}\, dx = \int_0^{\pi/4} \sec x\, dx = \left[\ln\left|\sec x + \tan x\right|\right]_0^{\pi/4} = \ln\left(1 + \sqrt{2}\right) \cong 0.88$

15. $L = \displaystyle\int_1^2 \sqrt{1 + \left(\sqrt{x^2 - 1}\right)^2}\, dx = \int_1^2 x\, dx = \left[\dfrac{1}{2}x^2\right]_1^2 = \dfrac{3}{2}$

17. $L = \displaystyle\int_0^1 \sqrt{1 + \left[\sqrt{3 - x^2}\right]^2}\, dx = \int_0^1 \sqrt{4 - x^2}\, dx = \int_0^{\pi/6} 4\cos^2 u\, du$

                   $(x = 2\sin u)$

    $= 2\displaystyle\int_0^{\pi/6} (1 + \cos 2u)\, du = 2\left[u + \dfrac{1}{2}\sin 2u\right]_0^{\pi/6} = \dfrac{1}{3}\pi + \dfrac{1}{2}\sqrt{3}$

19. $v(t) = \sqrt{(2t)^2 + 2^2} = 2\sqrt{t^2 + 1}$

initial speed $= v(0) = 2$,   terminal speed $= v(\sqrt{3}) = 4$

$$s = \int_0^{\sqrt{3}} 2\sqrt{t^2 + 1}\, dt = 2\int_0^{\pi/3} \sec^3 u\, du = 2\left[\frac{1}{2}\sec u \tan u + \frac{1}{2}\ln|\sec u + \tan u|\right]_0^{\pi/3}$$

$$\qquad\qquad (t = \tan u) \qquad\qquad\qquad\qquad\qquad \text{(by parts)}$$

$$= 2\sqrt{3} + \ln\left(2 + \sqrt{3}\right) \cong 4.78$$

21. $v(t) = \sqrt{(2t)^2 + (3t^2)^2}\, dt = t\left(4 + 9t^2\right)^{1/2}$

initial speed $= v(0) = 0$,   terminal speed $= v(1) = \sqrt{13}$

$$s = \int_0^1 t\left(4 + 9t^2\right)^{1/2} dt = \left[\frac{1}{27}\left(4 + 9t^2\right)^{3/2}\right]_0^1 = \frac{1}{27}\left(13\sqrt{13} - 8\right)$$

23. $v(t) = \sqrt{[e^t \cos t + e^t \sin t]^2 + [e^t \cos t - e^t \sin t]^2} = \sqrt{2}\, e^t$

initial speed $= v(0) = \sqrt{2}$,   terminal speed $= \sqrt{2}\, e^\pi$

$$s = \int_0^\pi \sqrt{2}\, e^t\, dt = \left[\sqrt{2}\, e^t\right]_0^\pi = \sqrt{2}\left(e^\pi - 1\right)$$

25. $L = \displaystyle\int_0^{2\pi} \sqrt{[x'(\theta)]^2 + [y'(\theta)]^2}\, d\theta = \int_0^{2\pi} \sqrt{a^2(1 - \cos\theta)^2 + a^2 \sin^2\theta}\, d\theta$

$$= a\int_0^{2\pi} \sqrt{2(1 - \cos\theta)}\, d\theta = 2a\int_0^{2\pi} \sin\frac{\theta}{2}\, d\theta = -4a\left[\cos\frac{\theta}{2}\right]_0^{2\pi} = 8a$$

27. (a) $L = \displaystyle\int_0^{2\pi} \sqrt{(-3a\sin\theta - 3a\sin 3\theta)^2 + (3a\cos\theta - 3a\cos 3\theta)^2}\, d\theta$

$$= 3a\int_0^{2\pi} \sqrt{\sin^2\theta + 2\sin\theta\sin 3\theta + \sin^2 3\theta + \cos^2\theta - 2\cos\theta\cos 3\theta + \cos^2 3\theta}\, d\theta$$

$$= 3a\int_0^{2\pi} \sqrt{2(1 - \cos 4\theta)}\, d\theta = 6a\int_0^{2\pi} |\sin 2\theta|\, d\theta$$

$$= 24a\int_0^{\pi/2} \sin 2\theta\, d\theta = -12a\left[\cos 2\theta\right]_0^{\pi/2} = 24a$$

(b) The result follows from the identities: $\cos 3\theta = 4\cos^3\theta - 3\cos\theta$; $\sin 3\theta = 3\sin\theta - 4\sin^3\theta$

29. $L = $ circumference of circle of radius $1 = 2\pi$

31. $L = \displaystyle\int_0^{4\pi} \sqrt{[e^\theta]^2 + [e^\theta]^2}\, d\theta = \int_0^{4\pi} \sqrt{2}\, e^\theta d\theta = \left[\sqrt{2}\, e^\theta\right]_0^{4\pi} = \sqrt{2}\left(e^{4\pi} - 1\right)$

33. $L = \displaystyle\int_0^{2\pi} \sqrt{[e^{2\theta}]^2 + [2e^{2\theta}]^2}\, d\theta = \int_0^{2\pi} \sqrt{5}\, e^{2\theta} d\theta = \left[\frac{1}{2}\sqrt{5}\, e^{2\theta}\right]_0^{2\pi} = \frac{1}{2}\sqrt{5}\left(e^{4\pi} - 1\right)$

35. $L = \int_0^{\pi/2} \sqrt{(1 - \cos\theta)^2 + \sin^2\theta}\, d\theta = \int_0^{\pi/2} \sqrt{2 - 2\cos\theta}\, d\theta$

$\qquad = \int_0^{\pi/2} \left(2\sin\frac{1}{2}\theta \right) d\theta = \left[-4\cos\frac{1}{2}\theta \right]_0^{\pi/2} = 4 - 2\sqrt{2}$

37. $s = \int_0^1 \sqrt{\left[\frac{1}{1+t^2} \right]^2 + \left[\frac{-t}{1+t^2} \right]^2}\, dt = \int_0^1 \frac{dt}{\sqrt{1+t^2}}$

$\qquad = \int_0^{\pi/4} \sec u\, du = [\ln|\sec u + \tan u|]_0^{\pi/4} = \ln\left(1 + \sqrt{2} \right)$

$\qquad\quad (t = \tan u)$

initial speed $= v(0) = 1,$ \qquad terminal speed $= v(1) = \frac{1}{2}\sqrt{2}$

39. $c = 1;$ \quad the curve $y = e^x$ is the curve $y = \ln x$ reflected in the line $y = x$

41. coordinates of the midpoint: \quad $\left(\frac{1}{2}, -\frac{7}{2} \right)$

43. $L = \int_a^b \sqrt{1 + \sinh^2 x}\, dx = \int_a^b \sqrt{\cosh^2 x}\, dx = \int_a^b \cosh x\, dx = A$

45.

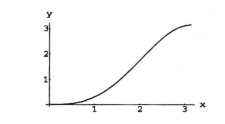

$f(x) = \sin x - x\cos x$

$f'(x) = x\sin x$

$L = \int_0^\pi \sqrt{1 + x^2\sin^2 x}\, dx \cong 4.6984$

47. $x = t^2, \quad y = t^3 - t$

$\qquad x' = 2t \quad y' = 3t^2 - 1$

$\qquad L = \int_{-1}^1 \sqrt{(2t)^2 + (3t^2 - 1)^2}\, dt \cong 2.7156$

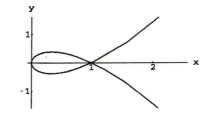

49.

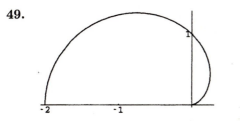

$r = 1 - \cos\theta, \quad r' = \sin\theta$

$L = \int_0^\pi \sqrt{(1 - \cos\theta)^2 + \sin^2\theta}\, d\theta = 4$

51. (a) $L = \int_0^{2\pi} \sqrt{a^2 \sin^2 t + b^2 \cos^2 t} \, dt = 4 \int_0^{\pi/2} \sqrt{a^2(1 - \cos^2 t) + b^2 \cos^2 t} \, dt$

$\qquad = 4a \int_0^{\pi/2} \sqrt{1 - e^2 \cos^2 t} \, dt, \quad \text{where} \quad e = \dfrac{\sqrt{a^2 - b^2}}{a}$

(b) $L = 4 \int_0^{\pi/2} \sqrt{25 - 9 \cos^2 t} \, dt \cong 28.3617$

53. $\sqrt{1 + [f(x)]^2} = \sqrt{1 + \tan^2 [\alpha(x)]} = |\sec [\alpha(x)]|$

SECTION 10.8

1. $L = $ length of the line segment $= 1$

$(\overline{x}, \overline{y}) = \left(\frac{1}{2}, 4\right)$ (the midpoint of the line segment)

$A_x = $ lateral surface area of cylinder of radius 4 and side $1 = 8\pi$.

3.

$L = \int_0^3 \sqrt{1 + \left(\dfrac{4}{3}\right)^2} \, dx = \left(\dfrac{5}{3}\right)^3 = 5$

$\overline{x}L = \int_0^3 x\sqrt{1 + \left(\dfrac{4}{3}\right)^2} \, dx = \dfrac{5}{3}\left[\dfrac{1}{2}x^2\right]_0^3 = \dfrac{15}{2}, \quad \overline{x} = \dfrac{3}{2}$

$\overline{y}L = \int_0^3 \dfrac{4}{3}x\sqrt{1 + \left(\dfrac{4}{3}\right)^2} \, dx = \left(\dfrac{4}{3}\right)\left(\dfrac{15}{2}\right) = 10, \quad \overline{y} = 2$

$A_x = 2\pi\overline{y}L = 2\pi(2)(5) = 20\pi$

5. $L = \int_0^2 \sqrt{(3)^2 + (4)^2} \, dt = (2)(5) = 10$

$\overline{x}L = \int_0^2 3t\sqrt{(3)^2 + (4)^2} \, dt = 15\left[\dfrac{1}{2}t^2\right]_0^2 = 30, \quad \overline{x} = 3$

$\overline{y}L = \int_0^2 4t\sqrt{(3)^2 + (4)^2} \, dt = 20\left[\dfrac{1}{2}t^2\right]_0^2 = 40, \quad \overline{y} = 4$

$A_x = 2\pi\overline{y}L = 2\pi(4)(10) = 80\pi$

7. $L = \int_0^{\pi/6} \sqrt{4\sin^2 t + 4\cos^2 t} \, dt = 2\left(\dfrac{\pi}{6}\right) = \dfrac{1}{3}\pi$

$\overline{x}L = \int_0^{\pi/6} 2\cos t\sqrt{4\sin^2 t + 4\cos^2 t} \, dt = 4\left[\sin t\right]_0^{\pi/6} = 2, \quad \overline{x} = \dfrac{6}{\pi}$

$\overline{y}L = \int_0^{\pi/6} 2\sin t\sqrt{4\sin^2 t + 4\cos^2 t} \, dt = 4\left[-\cos t\right]_0^{\pi/6} = 4 - 2\sqrt{3}, \quad \overline{y} = 6\left(2 - \sqrt{3}\right)/\pi$

$A_x = 2\pi\overline{y}L = 2\pi(6(2 - \sqrt{3})/\pi)\frac{1}{3}\pi = 4\pi(2 - \sqrt{3})$

9. $x(t) = a\cos t, \quad y = a\sin t; \quad t \in [\frac{1}{3}\pi, \frac{2}{3}\pi]$

$$L = \int_{\pi/3}^{2\pi/3} \sqrt{a^2\sin^2 t + a^2\cos^2 t}\, dt = \frac{1}{3}\pi a$$

by symmetry $\bar{x} = 0$

$$\bar{y}L = \int_{\pi/3}^{2\pi/3} a\sin t \sqrt{a^2\sin^2 t + a^2\cos^2 t}\, dt = a^2 \int_{\pi/3}^{2\pi/3} \sin t\, dt$$

$$= a^2\left[-\cos t\right]_{\pi/3}^{2\pi/3} = a^2, \quad \bar{y} = 3a/\pi$$

$$A_x = 2\pi\bar{y}L = 2\pi a^2$$

11. $A_x = \int_0^2 \frac{2}{3}\pi x^3 \sqrt{1 + x^4}\, dx = \frac{1}{9}\pi\left[(1 + x^4)^{3/2}\right]_0^2 = \frac{1}{9}\pi(17\sqrt{17} - 1) \cong 24.1179$

13. $A_x = \int_0^1 \frac{1}{2}\pi x^3 \sqrt{1 + \frac{9}{16}x^4}\, dx = \frac{4}{27}\pi\left[\left(1 + \frac{9}{16}x^4\right)^{3/2}\right]_0^1 = \frac{61}{432}\pi$

15. $A_x = \int_0^{\pi/2} 2\pi\cos x \sqrt{1 + \sin^2 x}\, dx = \int_0^1 2\pi\sqrt{1 + u^2}\, du$

$\underset{\underset{\displaystyle u = \sin x}{\big\uparrow}}{}$

$$= 2\pi\left[\tfrac{1}{2}u\sqrt{1 + u^2} + \tfrac{1}{2}\ln\left(u + \sqrt{1 + u^2}\right)\right]_0^1 = \pi\left[\sqrt{2} + \ln\left(1 + \sqrt{2}\right)\right]$$

$\underset{\underset{\displaystyle (8.5.1)}{\big\uparrow}}{}$

17. $A_x = \int_0^{\pi/2} 2\pi(e^\theta\sin\theta)\sqrt{[e^\theta\cos\theta - e^\theta\sin\theta]^2 + [e^\theta\sin\theta + e^\theta\cos\theta]^2}\, d\theta$

$$= 2\pi\sqrt{2}\int_0^{\pi/2} e^{2\theta}\sin\theta\, d\theta$$

$$= 2\pi\sqrt{2}\left[\tfrac{1}{5}\left(2e^{2\theta}\sin\theta - e^{2\theta}\cos\theta\right)\right]_0^{\pi/2} = \tfrac{2}{5}\sqrt{2}\,\pi\left(2e^\pi + 1\right)$$

(by parts twice)

19. (a) $A = \int_0^{2\pi} y(\theta)x'(\theta)\, d\theta \qquad [\text{see } (9.6.4)]$

$$= \int_0^{2\pi} a^2(1 - \cos\theta)^2\, d\theta$$

$$= a^2 \int_0^{2\pi} (1 - 2\cos\theta + \cos^2\theta)\, d\theta$$

$$= a^2 \int_0^{2\pi} \left(\frac{3}{2} - 2\cos\theta + \frac{1}{2}\cos 2\theta\right) d\theta$$

$$= a^2 \left[\frac{3}{2}\theta - 2\sin\theta + \frac{1}{4}\sin 2\theta\right]_0^{2\pi}$$

$$= 3\pi a^2$$

(b) $A = \int_0^{2\pi} 2\pi\, y(\theta) \sqrt{[x'(\theta)]^2 + [y'(\theta)]^2}\; d\theta$ (9.9.2)

$$= \int_0^{2\pi} 2\pi\, a(1 - \cos\theta)\sqrt{a^2(1 - \cos\theta)^2 + a^2 \sin^2\theta}\; d\theta$$

$$= 2\pi\, a^2 \int_0^{2\pi} (1 - \cos\theta)\sqrt{2 - 2\cos\theta}\; d\theta$$

$$= 4\pi\, a^2 \int_0^{2\pi} (1 - \cos\theta) \sin\frac{\theta}{2}\; d\theta$$

$$= 4\pi\, a^2 \int_0^{2\pi} \left(2\sin\frac{\theta}{2} - 2\cos^2\frac{\theta}{2}\sin\frac{\theta}{2} \right) d\theta$$

$$= 4\pi\, a^2 \left[-4\cos\frac{\theta}{2} \right]_0^{2\pi} + \frac{16\pi\, a^2}{3} \left[\cos^3(\theta/2) \right]_0^{2\pi} = \frac{64\pi\, a^2}{3}$$

21. $A = \frac{1}{2}\theta s_2{}^2 - \frac{1}{2}\theta s_1{}^2$

$$= \frac{1}{2}(\theta s_2 + \theta s_1)(s_2 - s_1)$$

$$= \frac{1}{2}(2\pi R + 2\pi r)s = \pi(R + r)s$$

23. (a) The centroids of the 3, 4, 5 sides are the midpoints $\left(\frac{3}{2}, 0\right)$, $(3, 2)$, $\left(\frac{3}{2}, 2\right)$.

(b) $\bar{x}(3 + 4 + 5) = \frac{3}{2}(3) + 3(4) + \frac{3}{2}(5)$, $12\bar{x} = 24$, $\bar{x} = 2$

$\bar{y}(3 + 4 + 5) = 0(3) + 2(4) + 2(5)$, $12\bar{y} = 18$, $\bar{y} = \dfrac{3}{2}$

(c) $A = \frac{1}{3}(3)(4) = 6$

$$\bar{x}A = \int_0^3 x\left(\frac{4}{3}x\right) dx = \int_0^3 \frac{4}{3}x^2 dx = \frac{4}{9}\left[x^3\right]_0^3 = 12,\quad \bar{x} = 2$$

$$\bar{y}A = \int_0^3 \frac{1}{2}\left(\frac{4}{3}x\right)^2 dx = \int_0^3 \frac{8}{9}x^2 dx = \frac{8}{27}\left[x^3\right]_0^3 = 8,\quad \bar{y} = \frac{4}{3}$$

(d) $\bar{x}(4 + 5) = 3(4) + \dfrac{3}{2}(5)$, $9\bar{x} = \dfrac{39}{2}$, $\bar{x} = \dfrac{13}{6}$

$\bar{y}(4 + 5) = 2(4) + 2(5)$, $9\bar{y} = 18$, $\bar{y} = 2$

(e) $A_x = 2\pi(2)(5) = 20\pi$

25. Set $y(t) = t$, $x(t) = \dfrac{1}{6a^2}\left(t^3 + \dfrac{3a^4}{t} \right)$ $t \in [a, 3a]$

$x'(t) = \dfrac{1}{6a^2}\left(3t^2 - \dfrac{3a^4}{t^2} \right)$, $y'(t) = 1$

$$A = \int_a^{3a} 2\pi t \sqrt{\frac{1}{36a^4}\left(3t^2 - \frac{3a^4}{t^2}\right)^2 + 1}\, dt = 2\pi \int_a^{3a} \frac{t}{6a^2}\left(3t^2 + \frac{3a^4}{t^2}\right) dt$$

$$= \frac{\pi}{a^2}\int_a^{3a}\left(t^3 + \frac{a^4}{t}\right) dt = \frac{\pi}{a^2}\left[\frac{t^4}{4} + a^4 \ln 4\right]_a^{3a} = \pi a^2(20 + \ln 3)$$

27. (a) No: $f'(x) = -x/\sqrt{r^2 - x^2}$ is not defined at $x = \pm r$. We can however begin with a smaller

interval $[-r+\epsilon, r-\epsilon]$, integrate, and take the limit as $\epsilon \to 0$.

(b) $A = \int_0^{2\pi} 2\pi r \sqrt{(\sin t)^2}\, dt = 2\pi r \int_0^{2\pi} |\sin t|\, dt = 8\pi r$

C is not simple. The curve [in this case the line segment that joins $(1, r)$ to $(-1, r)$] is traced

out twice.

27. The band can be obtained by revolving about the x-axis the graph of the function

$$f(x) = \sqrt{r^2 - x^2}, \qquad x \in [a, b].$$

A straightforward calculation shows that the surface area of the band is $2\pi r(b - a)$.

29. (a) Parameterize the upper half of the ellipse by

$$x(t) = a\cos t, \quad y(t) = b\sin t; \qquad t \in [0, \pi].$$

Here

$$\sqrt{[x'(t)]^2 + [y'(t)]^2} = \sqrt{a^2\sin^2 t + b^2\cos^2 t} = \sqrt{a^2 - (a^2 - b^2)\cos^2 t},$$

which, with $c = \sqrt{a^2 - b^2}$, can be written $\sqrt{a^2 - c^2\cos^2 t}$.   Therefore,

$$A = \int_0^\pi 2\pi b\sin t\sqrt{a^2 - c^2\cos^2 t}\, dt = 4\pi b\int_0^{\pi/2}\sin t\sqrt{a^2 - c^2\cos^2 t}\, dt.$$

Setting $u = c\cos t$, we have $du = -c\sin t$ and

$$A = -\frac{4\pi b}{c}\int_c^0 \sqrt{a^2 - u^2}\, du = \frac{4\pi b}{c}\left[\frac{u}{2}\sqrt{a^2 - u^2} + \frac{a^2}{2}\sin^{-1}\left(\frac{u}{a}\right)\right]_0^c$$

$$= 2\pi b^2 + \frac{2\pi a^2 b}{c}\sin^{-1}\left(\frac{c}{a}\right) = 2\pi b^2 + \frac{2\pi ab}{e}\sin^{-1} e$$

where e is the eccentricity of ellipse: $e = c/a$.

(b) Parameterize the right half of the ellipse by

$$x(t) = a\cos t, \quad y(t) = b\sin t; \quad t \in \left[-\frac{1}{2}\pi, \frac{1}{2}\pi\right].$$

Again $\sqrt{[x'(t)]^2 + [y'(t)]^2} = \sqrt{a^2 - c^2\cos^2 t}$ where $c = \sqrt{a^2 - b^2}$.

Therefore

$$A = \int_{-\pi/2}^{\pi/2} 2\pi a\cos t \sqrt{a^2 - c^2\cos^2 t}\, dt.$$

Set $u = c\sin t$. Then $du = c\cos t\, dt$ and

$$A = \frac{2\pi a}{c} \int_{-c}^{c} \sqrt{b^2 + u^2}\, du = \frac{2\pi a}{c} \left[\frac{u}{2}\sqrt{b^2 + u^2} + \frac{b^2}{2}\ln\left|u + \sqrt{b^2 + u^2}\right|\right]_{-c}^{c}$$

Routine calculation gives

$$A = 2\pi a^2 + \frac{\pi b^2}{e}\ln\left|\frac{1+e}{1-e}\right|.$$

31. Such a hemisphere can be obtained by revolving about the x-axis the curve

$$x(t) = r\cos t, \quad y(t) = r\sin t; \quad t \in \left[0, \tfrac{1}{2}\pi\right].$$

Therefore, $\quad \overline{x}A = \displaystyle\int_0^{\pi/2} 2\pi(r\cos t)(r\sin t)\sqrt{r^2\sin^2 t + r^2\cos^2 t}\, dt$

$$= \int_0^{\pi/2} 2\pi r^3 \sin t\cos t\, dt = \pi r^3 \left[\sin^2 t\right]_0^{\pi/2} = \pi r^3.$$

$$A = 2\pi r^2; \quad \overline{x} = \overline{x}A/A = \tfrac{1}{2}r.$$

The centroid lies on the midpoint of the axis of the hemisphere.

33. Such a surface can be obtained by revolving about the x-axis the graph of the function

$$f(x) = \left(\frac{R-r}{h}\right)x + r, \quad x \in [0, h].$$

Formula (9.9.8) gives

$$\overline{x}A = \int_0^h 2\pi x f(x)\sqrt{1 + [f'(x)]^2}\, dx$$

$$= \frac{2\pi}{h}\sqrt{h^2 + (R-h)^2} \int_0^h \left[\left(\frac{R-r}{h}\right)x^2 + rx\right] dx$$

$$= \frac{\pi}{3}\sqrt{h^2 + (R-r)^2}\,(2R+r)h$$

$$A = \pi(R+r)s = \pi(R+r)\sqrt{h^2 + (R-r)^2} \quad \text{and} \quad \overline{x} = \frac{\overline{x}A}{A} = \left(\frac{2R+r}{R+r}\right)\frac{h}{3}.$$

The centroid of the surface lies on the axis of the cone $\left(\dfrac{2R+r}{R+r}\right)\dfrac{h}{3}$ units from the base of radius r.

PROJECT 10.8

1. Referring to the figure we have

$$x(\theta) = \overline{OB} - \overline{AB} = R\theta - R\sin\theta = R(\theta - \sin\theta)$$

$$y(\theta) = \overline{BQ} - \overline{QC} = R - R\cos\theta = R(1 - \cos\theta).$$

3. (a) $\bar{x} = \pi R$ by symmetry

$$\bar{y}A = \int_0^{2\pi} \frac{1}{2}[y(\theta)]^2 x'(\theta)\,d\theta$$

$$= \int_0^{2\pi} \frac{1}{2}R^2(1 - \cos\theta)^2[R(1 - \cos\theta)]\,d\theta$$

$$= \frac{1}{2}R^3 \int_0^{2\pi} \left(1 - 3\cos\theta + 3\cos^2\theta - \cos^3\theta\right)d\theta = \frac{5}{2}\pi R^3$$

$$A = 3\pi R^2 \quad \text{(by Exercise 2b)} \qquad \bar{y} = \left(\tfrac{5}{2}\pi R^3\right)/\left(3\pi R^2\right) = \tfrac{5}{6}R$$

(b) $V_x = 2\pi\bar{y}A = 2\pi \left(\dfrac{5}{6}R\right)(3\pi R^2) = 5\pi^2 R^3$

(c) $V_y = 2\pi\bar{x}A = 2\pi(\pi R)(3\pi R^2) = 6\pi^3 R^3$

5. (a) $\dfrac{dy}{dx} = \dfrac{y'(\phi)}{x'(\phi)} = \dfrac{R\sin\phi}{R(1 + \cos\phi)} = \dfrac{2\sin\frac{1}{2}\phi\cos\frac{1}{2}\phi}{2\cos^2\frac{1}{2}\phi} = \tan\frac{1}{2}\phi; \quad \alpha = \frac{1}{2}\phi$

(b)
$$s = \int_0^{\phi} \sqrt{[x'(t)]^2 + [y'(t)]^2}\,dt = \int_0^{\phi} \sqrt{[R(1 + \cos t)]^2 + [R\sin t]^2}\,dt$$

$$= R\int_0^{\phi} \sqrt{2 + 2\cos t}\,dt = R\int_0^{\phi} 2\cos\frac{1}{2}t\,dt = 4R\left[\sin\frac{1}{2}t\right]_0^{\phi}$$

$$= 4R\sin\frac{1}{2}\phi = 4R\sin\alpha$$

Now note that the tangent at $(x(\phi), y(\phi))$ has slope

$$m = \frac{y'(\phi)}{x'(\phi)} = \frac{\sin\phi}{1 + \cos\phi} = \frac{\sin\phi(1 - \cos\phi)}{1 - \cos^2\phi} = \frac{1 - \cos\phi}{\sin\phi} = \frac{2\sin^2\frac{\phi}{2}}{2\sin\frac{\phi}{2}\cos\frac{\phi}{2}} = \tan\frac{\phi}{2}$$

So the inclination of the tangent is $\dfrac{\phi}{2} = \alpha$

7. $\dfrac{d^2s}{dt^2} = -g\sin\alpha = -g \cdot \dfrac{2R}{\sqrt{\pi^2 R^2 + 4R^2}} = -\dfrac{2g}{\sqrt{\pi^2 + 4}}$

Since $\dfrac{ds}{dt} = 0$ and $s = R\sqrt{\pi^2 + 4}$ when $t = 0$, integrating twice gives

$$s = -\frac{g}{\sqrt{\pi^2 + 4}}t^2 + R\sqrt{\pi^2 + 4}$$

$$s = 0 \text{ at } t^2 = \frac{1}{g}R(\pi^2 + 4), \quad \text{so } t = \sqrt{R(\pi^2 + 4)/g}$$

REVIEW EXERCISES

1. The equation can be written $x^2 = 4(y+1)$, a parabola.

vertex $(0,-1)$

focus $(0,0)$

axis y-axis

directrix $y = -2$

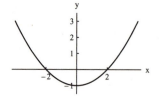

3. The equation can be written $\frac{(x+3)^2}{1} + \frac{y^2}{\frac{1}{4}} = 1$, an ellipse.

center $(-3,0)$

foci $\left(-3 \pm \frac{3\sqrt{3}}{2}, 0\right)$

length of the major axis: 2

length of the minor axis: 1

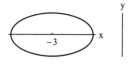

5. The equation can be written $\frac{(y+1)^2}{4} - \frac{(x-1)^2}{9} = 1$, a hyperbola.

center $(1,-1)$

foci $(1, -1 \pm \sqrt{13})$

vertices $(1,1)$, $(1,-3)$

asymptotes: $y = -1 \pm \frac{2}{3}(x-1)$

length of the transverse axis: 4

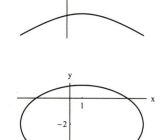

7. The equation can be written $\frac{(x-1)^2}{25} + \frac{(y+2)^2}{9} = 1$, an ellipse.

center $(1,-2)$

foci $(5,-2)$, $(-3,-2)$

length of the major axis: 10

length of the minor axis: 6

9. $x = -4\cos\frac{\pi}{3} = -2$, $y = -4\sin\frac{\pi}{3} = -2\sqrt{3}$; $(-2, -2\sqrt{3})$

11. $r = 4$, $\theta = 3\pi/2$; $[4, \frac{3}{2}\pi + 2n\pi]$, $n = 0, \pm 1, \pm 2, \cdots$; $[-4, \frac{\pi}{2} + 2n\pi]$, $n = 0, \pm 1, \pm 2, \cdots$

13. $r = \sqrt{16} = 4$, $\theta = -\frac{\pi}{6}$; $[4, -\frac{\pi}{6} + 2n\pi]$, $n = 0, \pm 1, \pm 2, \cdots$; $[-4, \frac{5}{6}\pi + 2n\pi]$, $n = 0, \pm 1, \pm 2, \cdots$

15. Set $y = r\sin\theta$ and $x = r\cos\theta$: $r\sin\theta = r^2\cos^2\theta \implies r = \sec\theta\tan\theta$.

17. Set $y = r\sin\theta$ and $x = r\cos\theta$:

$$r^2\cos^2\theta + r^2\sin^2\theta - 4r\cos\theta + 2r\sin\theta \implies r = 4\cos\theta - 2\sin\theta.$$

19. $x = 5$

21. Multiplication by r gives $r^2 = 3r\cos\theta + 4r\sin\theta \implies x^2 + y^2 - 3x - 4y = 0$.

23.

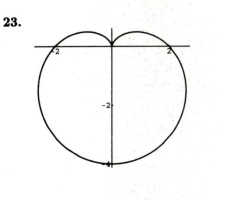

25.

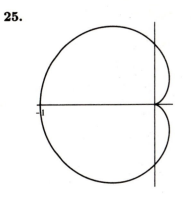

27.

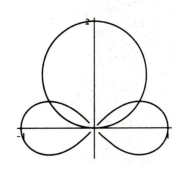

intersections: $[\frac{\sqrt{2}}{2}, \frac{1}{6}\pi], [\frac{\sqrt{2}}{2}, \frac{5}{6}\pi]$, the pole

29.

$$A = \int_0^{2\pi} \frac{1}{2}(2(1 - \cos\theta))^2 \, d\theta = 2 \int_0^{2\pi} (1 - 2\cos\theta + \cos^2\theta) \, d\theta$$

$$= 2 \int_0^{2\pi} \left(1 - 2\cos\theta + \frac{1 + \cos 2\theta}{2}\right) \theta$$

$$= 2 \left[\frac{3}{2}\theta - 2\sin\theta + \frac{1}{4}\sin 2\theta\right]_0^{2\pi} = 6\pi$$

31. One petal is formed when θ ranges from $-\pi/8$ to $\pi/8$. By symmetry, the area of one petal is

$$A = 2 \int_0^{\pi/8} \frac{1}{2}(4\cos^2 4\theta) \, d\theta = 4 \int_0^{\pi/8} \frac{1 + \cos 8\theta}{2} \, d\theta == 4 \left[\frac{1}{2}\theta + \frac{1}{16}\sin 8\theta\right]_0^{\pi/8} = \frac{\pi}{4}$$

33. $\int_0^{\frac{\pi}{4}} \frac{1}{2}(2\sin\theta)^2 d\theta + \int_{\frac{\pi}{4}}^{\frac{3}{4}\pi} \frac{1}{2}(\sin\theta + \cos\theta)^2 d\theta = \int_0^{\frac{\pi}{4}} (1 - \cos 2\theta) d\theta + \int_{\frac{\pi}{4}}^{\frac{3}{4}\pi} \frac{1}{2}(1 + \sin 2\theta) d\theta$

$$= \frac{\pi}{4} - \frac{1}{2} + \frac{\pi}{4} = \frac{\pi}{2} - \frac{1}{2}$$

35. $y = \cos 2t = 1 - 2\sin^2 \theta$

$y = 1 - 2x^2, \quad 0 \le x \le 1.$

37. $y = x^2 - 2x + 1 = (x-1)^2, \quad x \ge 1,$ graph: the right half of the parabola.

39. $x(t) = 1 + (5-1)t = 1 + 4t, \quad y(t) = 4 + (6-4)t = 4 + 2t, \quad t \in [0,1].$

41. $x(t) = 3\cos(-t + \frac{\pi}{2}) = 3\sin t, \quad y(t) = 2\sin(-t + \frac{\pi}{2}) = 2\cos t.$

43. $x(t) = 3e^t, \; x'(t) = 3e^t; \quad y(t) = 5e^{-t}, \; y'(t) = -5e^{-t}.$
At $t = 0: \; x(0) = 3, \; y(0) = 5, \; x'(0) = 3, \; y'(0) = -5;$ tangent line: $y - 5 = -\frac{5}{3}(x - 3)$ or
$5x + 3y = 30.$

45. (a) For a horizontal tangent, set $y'(t) = 0: \; 3t^2 + 3t - 6 = 0 \implies t = -2, \; 1.$
The curve has a horizontal tangent at: $(x(-2), y(-2)) = (0, 10)$ and $(x(1), y(1)) = (3, -7/2).$

(b) For a vertical tangent, set $x'(t) = 0$ (provided $y'(t) \ne 0$ simultaneously):
$x'(t) = 2t + 2 \implies t = -1.$
Since $y'(-1) \ne 0,$ the curve has a vertical tangent at $(x(-1), y(-1)) = (-1, 13/2).$

47. $x = 3t^2, \; x' = 6t, \; x'' = 6; \quad y = 4t^3, \; y' = 12t^2, \; y'' = 24t$

$$\frac{dy}{dx} = \frac{12t^2}{6t} = 2t; \quad \frac{d^2y}{dx^2} = \frac{6t(24t) - 12t^2(6)}{(6t)^3} = \frac{1}{3t}$$

49. $L = \displaystyle\int_0^{5/9} \sqrt{1 + (y'(x))^2}\, dx = \int_0^{5/9} \sqrt{1 + \frac{9}{4}x}\, dx = \frac{19}{27}.$

51. $L = \displaystyle\int_0^{\frac{\pi}{2}} \sqrt{\sin^2 t + 4\sin^2 t \cos^2 t}\, dt = \int_0^{\frac{\pi}{2}} 2\sin t \sqrt{\frac{1}{4} + \cos^2 t}\, dt$

$$= 2\int_0^1 \sqrt{\frac{1}{4} + u^2}\, du \quad \text{(substitution } u = \cos t\text{)}$$

$$= 2\left[\frac{u}{2}\sqrt{\frac{1}{4} + u^2} + \frac{1}{8}\ln\left(u + \sqrt{\frac{1}{4} + u^2}\right)\right]_0^1 = \frac{\sqrt{5}}{2} + \frac{1}{4}\ln(\sqrt{5} + 2)$$

53. By symmetry $L = \int_0^{2\pi} \sqrt{(1-\sin\theta)^2 + \cos^2\theta}\, d\theta = 2\int_0^{\pi} \sqrt{(1-\sin\theta)^2 + \cos^2\theta}\, d\theta.$

$$L = \int_0^{2\pi} \sqrt{(1-\sin\theta)^2 + \cos^2\theta}\, d\theta = 2\int_0^{\pi}\sqrt{2(1-\sin\theta)}\, d\theta$$

$$= 2\int_0^{\pi}\sqrt{4\cos^2(\theta/2)}\, d\theta = 4\int_0^{\pi}\cos(\theta/2)\, d\theta = 8\Big[\sin(\theta/2)\Big]_0^{\pi} = 8$$

55. $\displaystyle\int_0^{24} 2\pi(2\sqrt{x})\sqrt{1 + \frac{1}{x}}\, dx = 4\pi\int_0^{24}\sqrt{1+x}\, dx = \frac{8\pi}{3}\Big[(1+x)^{3/2}\Big]_0^{24} = \frac{992\pi}{3}$

57. $y = \dfrac{x^3}{6} + \dfrac{1}{2x}$

$$\int_1^3 2\pi\left(\frac{x^3}{6} + \frac{1}{2x}\right)\sqrt{1 + \left(\frac{x^2}{2} - \frac{1}{2x^2}\right)^2}\, dx = 2\pi\int_1^3\left(\frac{x^3}{6} + \frac{1}{2x}\right)\left(\frac{x^2}{2} + \frac{1}{2x^2}\right)dx$$

$$= 2\pi\int_1^3\left(\frac{x^5}{12} + \frac{x}{4} + \frac{x}{12} + \frac{1}{4x^3}\right)dx = \frac{2497\pi}{108}$$

59. Assume that $f'(x) > 0$. Then $(f^{-1})'(y) = \dfrac{1}{f'(x)}$, where $y = f(x)$, $dy = f'(x)\, dx$. The length of the graph of f^{-1} is given by:

$$\int_{f(a)}^{f(b)}\sqrt{1 + [(f^{-1})'(y)]^2}\, dy = \int_{f(a)}^{f(b)}\sqrt{1 + 1/[f'(x)]^2}\, dy$$

$$= \int_a^b \frac{1}{f'(x)}\sqrt{(f'(x))^2 + 1}\, f'(x)\, dx$$

$$= \int_a^b \sqrt{1 + (f'(x))^2}\, dx$$

61. If $\dfrac{x}{a} = \cos^3\theta$ and $\dfrac{y}{a} = \sin^3\theta$, then

$$\left(\frac{x}{a}\right)^{2/3} + \left(\frac{y}{a}\right)^{2/3} = \cos^2\theta + \sin^2\theta = 1 \quad\text{and}\quad x^{2/3} + y^{2/3} = a^{2/3}.$$

63. By symmetry, $\overline{x} = \overline{y}$.

$$\overline{x}\frac{L}{4} = \int_0^{\pi/2} a\cos^3\theta(3a\cos\theta\sin\theta)\, d\theta = -\frac{3a^2}{5}\Big[\cos^5\theta\Big]_0^{\frac{\pi}{2}} = \frac{3a^2}{5}$$

Therefore, $(\overline{x}, \overline{y}) = \left(\dfrac{2a}{5}, \dfrac{2a}{5}\right)$

CHAPTER 11

SECTION 11.1

1. lub = 2;   glb = 0

3. no lub;   glb = 0

5. lub = 2;   glb = −2

7. no lub;   glb = 2

9. lub = $2\frac{1}{2}$;   glb = 2

11. lub = 1;   glb = 0.9

13. lub = e;   glb = 0

15. lub = $\frac{1}{2}(-1 + \sqrt{5})$;   glb = $\frac{1}{2}(-1 - \sqrt{5})$

17. no lub;   no glb

19. no lub;   no glb

21. glb $S = 0$, $0 \le \left(\frac{1}{11}\right)^3 < 0 + 0.001$

23. glb $S = 0$, $0 \le \left(\frac{1}{10}\right)^{2n-1} < 0 + \left(\frac{1}{10}\right)^k$ $\left(n > \frac{k+1}{2}\right)$

25. Let $\epsilon > 0$. The condition $m \le s$ is satisfied by all numbers s in S. All we have to show therefore is that there is some number s in S such that

$$s < m + \epsilon.$$

Suppose on the contrary that there is no such number in S. We then have

$$m + \epsilon \le x \quad \text{for all} \quad x \in S.$$

This makes $m + \epsilon$ a lower bound for S. But this cannot be, for then $m + \epsilon$ is a lower bound for S that is *greater* than m, and by assumption, m is the *greatest* lower bound.

27. Let $c = \text{lub } S$.   Since $b \in S$, $b \le c$.   Since b is an upper bound for S, $c \le b$.   Thus, $b = c$.

29. (a) Suppose that K is an upper bound for S and k is a lower bound. Let t be any element of T. Then $t \in S$ which implies that $k \le t \le K$. Thus K is an upper bound for T and k is a lower bound, and T is bounded.

(b) Let $a = \text{glb } S$. Then $a \le t$ for all $t \in T$. Therefore, $a \le \text{glb } T$. Similarly, if $b = \text{lub } S$, then $t \le b$ for all $t \in T$,   so lub $T \le b$. It now follows that   glb $S \le \text{glb } T \le \text{lub } T \le \text{lub } S$.

31. Let c be a positive number and let $S = \{c, 2c, 3c, \ldots\}$. Choose any positive number M and consider the positive number M/c. Since the set of positive integers is not bounded above, there exists a positive integer k such that $k \ge M/c$. This implies that $kc \ge M$.   Since $kc \in S$, it follows that S is not bounded above.

33. (a) See Exercise 75 in Section 1.2.

(b) Suppose $x_0^2 > 2$. Choose a positive integer n such that

$$\frac{2x_0}{n} - \frac{1}{n^2} < x_0^2 - 2.$$

Then,

$$\frac{2x_0}{n} - \frac{1}{n^2} < x_0^2 - 2 \implies 2 < x_0^2 - \frac{2x_0}{n} + \frac{1}{n^2} = \left(x_0 - \tfrac{1}{n}\right)^2$$

(c) If $x_0^2 < 2$, then choose a positive integer n such that

$$\frac{2x_0}{n} + \frac{1}{n^2} < 2 - x_0^2.$$

Then

$$x_0^2 + \frac{2x_0}{n} + \frac{1}{n^2} < 2 \implies \left(x_0 + \tfrac{1}{n}\right)^2 < 2$$

35. (a) $n = 5 : 2.48832$; $n = 10 : 2.59374$; $n = 100 : 2.70481$; $n = 1000 : 2.71692$; $n = 10,000 : 2.71815$

(b) lub $= e$;   glb $= 2$

37. (a)

a_1	a_2	a_3	a_4	a_5
1.4142	1.6818	1.8340	1.9152	1.9571

a_6	a_7	a_8	a_9	a_{10}
1.9785	1.9892	1.9946	1.9973	1.9986

(b) Let S be the set of positive integers for which $a_n < 2$. Then $1 \in S$ since

$$a_1 = \sqrt{2} \cong 1.4142 < 2.$$

Assume that $k \in S$. Since $a_{k+1}^2 = 2a_k < 4$, it follows that $a_{k+1} < 2$. Thus $k + 1 \in S$ and S is the set of positive integers.

(c) 2 is the least upper bound.

(d) Let c be a positive number. Then c is the least upper bound of the set

$$S = \left\{ \sqrt{c}, \ \sqrt{c\sqrt{c}}, \ \sqrt{c\sqrt{c\sqrt{c}}}, \dots \right\}.$$

SECTION 11.2

1. $a_n = 2 + 3(n - 1) = 3n - 1, \quad n = 1, 2, 3, \dots$

3. $a_n = \dfrac{(-1)^{n-1}}{2n - 1}, \quad n = 1, 2, 3, \dots$

5. $a_n = \dfrac{n^2 + 1}{n}, \quad n = 1, 2, 3, \dots$

7. $a_n = \begin{cases} n & \text{if } n = 2k - 1 \\ 1/n, & \text{if } n = 2k, \end{cases}$ where $k = 1, 2, 3, \dots$

9. decreasing;   bounded below by 0 and above by 2

11. $\dfrac{n + (-1)^n}{n} = 1 + (-1)^n \dfrac{1}{n}$: not monotonic;   bounded below by 0 and above by $\dfrac{3}{2}$

13. decreasing; bounded below by 0 and above by 0.9

15. $\dfrac{n^2}{n+1} = n - 1 + \dfrac{1}{n+1}$: increasing;   bounded below by $\dfrac{1}{2}$ but not bounded above

17. $\dfrac{4n}{\sqrt{4n^2+1}} = \dfrac{2}{\sqrt{1 + 1/4n^2}}$ and $\dfrac{1}{4n^2}$ decreases to 0: increasing;

bounded below by $\frac{4}{5}\sqrt{5}$ and above by 2

19. increasing;   bounded below by $\frac{2}{51}$ but not bounded above

21. $\dfrac{2n}{n+1} = 2 - \dfrac{2}{n+1}$ increases toward 2: increasing;   bounded below by 0 and above by $\ln 2$.

23. decreasing;   bounded below by 1 and above by 4

25. increasing;   bounded below by $\sqrt{3}$ and above by 2

27. $(-1)^{2n+1}\sqrt{n} = -\sqrt{n}$: decreasing;   bounded above by -1 but not bounded below

29. $\dfrac{2^n - 1}{2^n} = 1 - \dfrac{1}{2^n}$: increasing;   bounded below by $\dfrac{1}{2}$ and above by 1

31. consider $\sin x$ as $x \to 0^+$: decreasing;   bounded below by 0 and above by 1

33. decreasing;   bounded below by 0 and above by $\frac{5}{6}$

35. $\dfrac{1}{n} - \dfrac{1}{n+1} = \dfrac{1}{n(n+1)}$: decreasing;   bounded below by 0 and above by $\dfrac{1}{2}$

37. Set $f(x) = \dfrac{\ln x}{x}$.   Then, $f'(x) = \dfrac{1 - \ln x}{x^2} < 0$ for $x > e$: decreasing;

bounded below by 0 and above by $\frac{1}{3}\ln 3$.

39. Set $a_n = \dfrac{3^n}{(n+1)^2}$.   Then, $\dfrac{a_{n+1}}{a_n} = 3\left(\dfrac{n+1}{n+2}\right)^2 > 1$: increasing;

bounded below by $\frac{3}{4}$ but not bounded above.

41. For $n \geq 5$

$$\dfrac{a_{n+1}}{a_n} = \dfrac{5^{n+1}}{(n+1)!} \cdot \dfrac{n!}{5^n} = \dfrac{5}{n+1} < 1 \quad \text{and thus} \quad a_{n+1} < a_n.$$

Sequence is not nonincreasing: $a_1 = 5 < \frac{25}{2} = a_2$.

43. boundedness: $0 < (c^n + d^n)^{1/n} < (2d^n)^{1/n} = 2^{1/n}d \le 2d$

monotonicity : $a_{n+1}^{n+1} = c^{n+1} + d^{n+1} = cc^n + dd^n$

$$< (c^n + d^n)^{1/n}c^n + (c^n + d^n)^{1/n}d^n$$
$$= (c^n + d^n)^{1+1/n}$$
$$= (c^n + d^n)^{(n+1)/n}$$
$$= a_n^{n+1}$$

Taking the $(n+1)$st root of each side we have $a_{n+1} < a_n$.   The sequence is decreasing.

45. $a_1 = 1,\quad a_2 = \frac{1}{2},\quad a_3 = \frac{1}{6},\quad a_4 = \frac{1}{24},\quad a_5 = \frac{1}{120},\quad a_6 = \frac{1}{720};\quad a_n = \frac{1}{n!}$

47. $a_1 = a_2 = a_3 = a_4 = a_5 = a_6 = 1;\quad a_n = 1$

49. $a_1 = 1,\quad a_2 = 3,\quad a_3 = 5,\quad a_4 = 7,\quad a_5 = 9,\quad a_6 = 11;\quad a_n = 2n - 1$

51. $a_1 = 1,\quad a_2 = 4,\quad a_3 = 9,\quad a_4 = 16,\quad a_5 = 25,\quad a_6 = 36;\quad a_n = n^2$

53. $a_1 = 1,\quad a_2 = 1,\quad a_3 = 2,\quad a_4 = 4,\quad a_5 = 8,\quad a_6 = 16;\quad a_n = 2^{n-2} \quad (n \ge 3)$

55. $a_1 = 1,\quad a_2 = 3,\quad a_3 = 5,\quad a_4 = 7,\quad a_5 = 9,\quad a_6 = 11;\quad a_n = 2n - 1$

57. First $a_1 = 2^1 - 1 = 1$.   Next suppose $a_k = 2^k - 1$ for some $k \ge 1$.   Then

$$a_{k+1} = 2a_k + 1 = 2\left(2^k - 1\right) + 1 = 2^{k+1} - 1.$$

59. First $a_1 = \dfrac{1}{2^0} = 1$.   Next suppose $a_k = \dfrac{k}{2^{k-1}}$ for some $k \ge 1$.   Then

$$a_{k+1} = \frac{k+1}{2k}a_k = \frac{k+1}{2k}\frac{k}{2^{k-1}} = \frac{k+1}{2^k}.$$

61. (a) If $r = 1$ then $S_n = n$ for $n = 1, 2, 3, \ldots$

(b)
$$S_n = 1 + r + r^2 + \cdots + r^{n-1}$$
$$rS_n = r + r^2 + \cdots + r^n$$
$$S_n - rS_n = 1 - r^n$$
$$S_n = \frac{1 - r^n}{1 - r}, \qquad r \ne 1.$$

63. (a) Let S_n denote the distance traveled between the nth and $(n+1)$st bounce. Then

$$S_1 = 75 + 75 = 150, \qquad S_2 = \tfrac{3}{4}(75) + \tfrac{3}{4}(75) = 150\left(\frac{3}{4}\right), \ldots, \qquad S_n = 150\left(\frac{3}{4}\right)^{n-1}.$$

(b) An object dropped from rest from a height h feet above the ground will hit the ground in $\frac{1}{4}\sqrt{h}$ seconds. Therefore it follows that the ball will be in the air

$$T_n = 2\left(\tfrac{1}{4}\right)\sqrt{\frac{S_n}{2}} = \frac{5\sqrt{3}}{2}\left(\frac{3}{4}\right)^{(n-1)/2} \qquad \text{seconds.}$$

65. $\{a_n\}$ is an increasing sequence; $a_n \to \frac{1}{2}$ as $n \to \infty$.

67. (a) Let S be the set of positive integers for which $a_{n+1} > a_n$. Since $a_2 = 1 + \sqrt{a_1} = 2 > 1$, $1 \in S$. Assume that $a_k = 1 + \sqrt{a_{k-1}} > a_{k-1}$. Then

$$a_{k+1} = 1 + \sqrt{a_k} > 1 + \sqrt{a_{k-1}} = a_k.$$

Thus, $k \in S$ implies $k + 1 \in S$. It now follows that $\{a_n\}$ is an increasing sequence.

(b) Since $\{a_n\}$ is an increasing sequence,

$$a_n = 1 + \sqrt{a_{n-1}} < 1 + \sqrt{a_n}, \quad \text{or} \quad a_n - \sqrt{a_n} - 1 < 0.$$

Rewriting the second inequality as

$$\left(\sqrt{a_n}\right)^2 - \sqrt{a_n} - 1 < 0$$

and solving for $\sqrt{a_n}$ it follows that $\sqrt{a_n} < \frac{1}{2}(1 + \sqrt{5})$. Hence, $a_n < \frac{1}{2}(3 + \sqrt{5})$ for all n.

(c) $a_2 = 2$, $a_3 \cong 2.4142$, $a_4 \cong 2.5538$, $a_5 \cong 2.5981, \ldots$, $a_9 \cong 2.6179$, $\ldots$, $a_{15} \cong 2.6180$;

(e) $\text{lub} \{a_n\} = \frac{1}{2}(3 + \sqrt{5}) \cong 2.6180$

SECTION 11.3

1. diverges

3. converges to 0

5. converges to 1: $\dfrac{n-1}{n} = 1 - \dfrac{1}{n} \to 1$

7. converges to 0: $\dfrac{n+1}{n^2} = \dfrac{1}{n} + \dfrac{1}{n^2} \to 0$

9. converges to 0: $0 < \dfrac{2^n}{4^n + 1} < \dfrac{2^n}{4^n} = \dfrac{1}{2^n} \to 0$ **11.** diverges

13. converges to 0

15. converges to 1: $\dfrac{n\pi}{4n+1} \to \dfrac{\pi}{4}$ so $\tan \dfrac{n\pi}{4n+1} \to \tan \dfrac{\pi}{4} = 1$

17. converges to $\dfrac{4}{9}$: $\dfrac{(2n+1)^2}{(3n-1)^2} = \dfrac{4 + 4/n + 1/n^2}{9 - 6/n + 1/n^2} \to \dfrac{4}{9}$

19. converges to $\dfrac{1}{2}\sqrt{2}$: $\dfrac{n^2}{\sqrt{2n^4 + 1}} = \dfrac{1}{\sqrt{2 + 1/n^4}} \to \dfrac{1}{\sqrt{2}}$

21. diverges: $\cos n\pi = (-1)^n$

23. converges to 1: $\dfrac{1}{\sqrt{n}} \to 0$ so $e^{1/\sqrt{n}} \to e^0 = 1$

25. converges to 0 : $\ln n - \ln (n+1) = \ln \left(\dfrac{n}{n+1} \right) \to \ln 1 = 0$

27. converges to $\dfrac{1}{2}$: $\dfrac{\sqrt{n+1}}{2\sqrt{n}} = \dfrac{1}{2}\sqrt{1 + \dfrac{1}{n}} \to \dfrac{1}{2}$

29. converges to e^2: $\left(1 + \dfrac{1}{n}\right)^{2n} = \left[\left(1 + \dfrac{1}{n}\right)^n\right]^2 \to e^2$

31. diverges;   since $2^n > n^3$ for $n \geq 10$, $\dfrac{2^n}{n^2} > \dfrac{n^3}{n^2} = n$

33. converges to 0: $\dfrac{|\sin n|}{\sqrt{n}} \leq \dfrac{1}{\sqrt{n}}$

35. converges to $1/2$: $\sqrt{n^2 + n} - n = \left(\sqrt{n^2 + n} - n\right) \dfrac{\sqrt{n^2 + n} + n}{\sqrt{n^2 + n} + n} = \dfrac{n}{\sqrt{n^2 + n} + n} \to \dfrac{1}{2}$

37. converges to $\pi/2$

39. converges to $-\pi/2$: $\dfrac{1 - n}{n} \to -1$; $\arcsin(-1) = \pi/2$.

41. (a) $\sqrt[n]{n} \to 1$ (b) $\dfrac{3^n}{n!} \to 0$

43. $b < \sqrt[n]{a^n + b^n} = b\sqrt[n]{(a/b)^n + 1} < b\sqrt[n]{2}$. Since $2^{1/n} \to 1$ as $n \to \infty$, it follows that $\sqrt[n]{a^n + b^n} \to b$ by the pinching theorem.

45. Set $\epsilon > 0$.   Since $a_n \to L$, there exists N_1 such that

$$\text{if} \quad n \geq N_1, \quad \text{then} \quad |a_n - L| < \epsilon/2.$$

Since $b_n \to M$, there exists N_2 such that

$$\text{if} \quad n \geq N_2, \quad \text{then} \quad |b_n - M| < \epsilon/2.$$

Now set $N = \max\{N_1, N_2\}$.   Then, for $n \geq N$,

$$|(a_n + b_n) - (L + M)| \leq |a_n - L| + |b_n - M| < \frac{\epsilon}{2} + \frac{\epsilon}{2} = \epsilon.$$

47. Since $\left(1 + \dfrac{1}{n}\right) \to 1$ and $\left(1 + \dfrac{1}{n}\right)^n \to e$,

$$\left(1 + \frac{1}{n}\right)^{n+1} = \left(1 + \frac{1}{n}\right)^n \left(1 + \frac{1}{n}\right) \to (e)(1) = e.$$

49. Suppose that $\{a_n\}$ is bounded and non-increasing. If L is the greatest lower bound of the range of this sequence, then $a_n \geq L$ for all n. Set $\epsilon > 0$. By Theorem 11.1.4 there exists a_k such that $a_k < L + \epsilon$. Since the sequence is non-increasing, $a_n \leq a_k$ for all $n \geq k$. Thus,

$$L \leq a_n < L + \epsilon \quad \text{or} \quad |a_n - L| < \epsilon \quad \text{for all} \quad n \geq k$$

and $a_n \to L$.

51. Let $\epsilon > 0$.   Choose k so that, for $n \geq k$,

$$L - \epsilon < a_n < L + \epsilon, \quad L - \epsilon < c_n < L + \epsilon \quad \text{and} \quad a_n \leq b_n \leq c_n.$$

For such n,

$$L - \epsilon < b_n < L + \epsilon.$$

53. Let $\epsilon > 0$. Since $a_n \to L$, there exists a positive integer N such that $L - \epsilon < a_n < L + \epsilon$ for all $n \geq N$. Now $a_n \leq M$ for all n, so $L - \epsilon < M$, or $L < M + \epsilon$. Since ϵ is arbitrary, $L \leq M$.

55. Assume $a_n \to 0$ as $n \to \infty$. Let $\epsilon > 0$. There exists a positive integer N such that $|a_n - 0| < \epsilon$ for all $n \geq N$. Since $||a_n| - 0| \leq |a_n - 0|$, it follows that $|a_n| \to 0$. Now assume that $|a_n| \to 0$. Since $-|a_n| \leq a_n \leq |a_n|$, $a_n \to 0$ by the pinching theorem.

57. By the continuity of f, $f(L) = f\left(\lim_{n\to\infty} a_n\right) = \lim_{n\to\infty} f(a_n) = \lim_{n\to\infty} a_{n+1} = L$.

59. Set $f(x) = x^{1/p}$. Since $\dfrac{1}{n} \to 0$ and f is continuous at 0, it follows by Theorem 11.3.12 that
$$\left(\frac{1}{n}\right)^{1/p} \to 0.$$

61. $a_n = e^{1-n} \to 0$

63. $a_n = \dfrac{1}{n!} \to 0$

65. $a_n = \dfrac{1}{2}[1 - (-1)^n]$ diverges

67. $L = 0, \quad n = 32$

69. $L = 0, \quad n = 4$

71. $L = 0, \quad n = 7$

73. $L = 0, \quad n = 65$

75. (a) $a_{n+1} = 1 + \sqrt{a_n}$ Suppose that $a_n \to L$ as $n \to \infty$. Then $a_{n+1} \to L$ as $n \to \infty$. Therefore $L = 1 + \sqrt{L}$ which, since $L > 1$, implies $L = \frac{1}{2}(3 + \sqrt{5})$.

    (b) $a_{n+1} = \sqrt{3a_n}$ Suppose that $a_n \to L$ as $n \to \infty$. Then $a_{n+1} \to L$ as $n \to \infty$. Therefore $L = \sqrt{3L}$ which, since $L > 1$, implies $L = 3$.

77. (a)

a_2	a_3	a_4	a_5	a_6	a_7	a_8	a_9	a_{10}
0.5403	0.8576	0.6543	0.7935	0.7014	0.7640	0.7221	0.7504	0.7314

    (b) L is a fixed point of $f(x) = \cos x$, that is, $\cos L = L$; $L \cong 0.739085$.

PROJECT 11.3

1. (a)

a_2	a_3	a_4	a_5	a_6	a_7	a_8
2.000000	1.750000	1.732143	1.732051	1.732051	1.732051	1.732051

    (b) $L = \dfrac{1}{2}\left(L + \dfrac{3}{L}\right)$ which implies $L^2 = 3$ or $L = \sqrt{3}$.

3. (a) $f(x) = x^3 - 8$,   so $x_n \to 2$ (b) $f(x) = \sin x - \frac{1}{2}$,   so $x_n \to \frac{\pi}{6}$

    (c) $f(x) = \ln x - 1$,   so $x_n \to e$

SECTION 11.4

1. converges to 1: $2^{2/n} = (2^{1/n})^2 \to 1^2 = 1$

3. converges to 0:   for $n > 3$, $0 < \left(\dfrac{2}{n}\right)^n < \left(\dfrac{2}{3}\right)^n \to 0$

5. converges to 0: $\dfrac{\ln(n+1)}{n} = \left[\dfrac{\ln(n+1)}{n+1}\right]\left(\dfrac{n+1}{n}\right) \to (0)(1) = 0$

7. converges to 0: $\dfrac{x^{100n}}{n!} = \dfrac{(x^{100})^n}{n!} \to 0$ 9.   converges to 1: $n^{\alpha/n} = (n^{1/n})^\alpha \to 1^\alpha = 1$

11. converges to 0: $\dfrac{3^{n+1}}{4^{n-1}} = 12\left(\dfrac{3^n}{4^n}\right) = 12\left(\dfrac{3}{4}\right)^n \to 12(0) = 0$

13. converges to 1: $(n+2)^{1/n} = e^{\frac{1}{n}\ln(n+2)}$ and, since

$$\frac{1}{n}\ln(n+2) = \left[\frac{\ln(n+2)}{n+2}\right]\left(\frac{n+2}{n}\right) \to (0)(1) = 0,$$

it follows that $(n+2)^{1/n} \to e^0 = 1.$

15. converges to 1: $\displaystyle\int_0^n e^{-x}\,dx = 1 - \dfrac{1}{e^n} \to 1$

17. converges to π: integral $= 2\displaystyle\int_0^n \dfrac{dx}{1+x^2} = 2\tan^{-1} n \to 2\left(\dfrac{\pi}{2}\right) = \pi$

19. converges to 1:   recall (11.4.6)   21.   converges to 0: $\dfrac{\ln(n^2)}{n} = 2\dfrac{\ln n}{n} \to 2(0) = 0$

23. diverges: Since $\displaystyle\lim_{x\to 0}\dfrac{\sin x}{x} = 1,$ $\dfrac{n}{\pi}\sin\dfrac{\pi}{n} = \dfrac{\sin(\pi/n)}{\pi/n} \to 1.$ Therefore,

$$n^2 \sin\frac{\pi}{n} = n\pi\left(\frac{n}{\pi}\sin\frac{\pi}{n}\right) \to n\pi.$$

25. converges to 0: $\dfrac{5^{n+1}}{4^{2n-1}} = 20\left(\dfrac{5}{16}\right)^n \to 0$

27. converges to e^{-1}: $\left(\dfrac{n+1}{n+2}\right)^n = \left(1 - \dfrac{1}{n+2}\right)^n = \dfrac{\left(1 + \dfrac{(-1)}{n+2}\right)^{n+2}}{\left(1 + \dfrac{(-1)}{n+2}\right)^2} \to \dfrac{e^{-1}}{1} = e^{-1}$

29. converges to 0: $0 < \displaystyle\int_n^{n+1} e^{-x^2}\,dx \le e^{-n^2}[(n+1) - n] = e^{-n^2} \to 0$

31. converges to 0: $\dfrac{n^n}{2^{n^2}} = \left(\dfrac{n}{2^n}\right)^n \to 0$ since $\dfrac{n}{2^n} \to 0$

33. converges to e^x: use (11.4.7)

35. converges to 0: $\left| \displaystyle\int_{-1/n}^{1/n} \sin x^2 \, dx \right| \leq \displaystyle\int_{-1/n}^{1/n} |\sin x^2| \, dx \leq \displaystyle\int_{-1/n}^{1/n} 1 \, dx = \dfrac{2}{n} \to 0$

37. converges: $\dfrac{\sin(6/n)}{\sin(3/n)} \to 2$

39. $\sqrt{n+1} - \sqrt{n} = \dfrac{\sqrt{n+1} - \sqrt{n}}{\sqrt{n+1} + \sqrt{n}} \left(\sqrt{n+1} + \sqrt{n} \right) = \dfrac{1}{\sqrt{n+1} + \sqrt{n}} \to 0$

41. (a) The length of each side of the polygon is $2r \sin(\pi/n)$. Therefore the perimeter, p_n, of the polygon is given by: $p_n = 2rn \sin(\pi/n)$.

  (b) $2rn \sin(\pi/n) \to 2\pi r$ as $n \to \infty$: The number $2rn \sin(\pi/n)$ is the perimeter of a regular polygon of n sides inscribed in a circle of radius r. As n tends to ∞, the perimeter of the polygon tends to the circumference of the circle.

43. By the hint, $\displaystyle\lim_{n\to\infty} \dfrac{1 + 2 + \ldots + n}{n^2} = \lim_{n\to\infty} \dfrac{n(n+1)}{2n^2} = \lim_{n\to\infty} \dfrac{1 + 1/n}{2} = \dfrac{1}{2}.$

45. By the hint, $\displaystyle\lim_{n\to\infty} \dfrac{1^3 + 2^3 + \ldots + n^3}{2n^4 + n - 1} = \lim_{n\to\infty} \dfrac{n^2(n+1)^2}{4(2n^4 + n - 1)} = \lim_{n\to\infty} \dfrac{1 + 2/n + 1/n^2}{8 + 4/n^3 - 4/n^4} = \dfrac{1}{8}.$

47. (a) $$m_{n+1} - m_n = \dfrac{1}{n+1}(a_1 + \cdots + a_n + a_{n+1}) - \dfrac{1}{n}(a_1 + \cdots + a_n)$$

$$= \dfrac{1}{n(n+1)} \left[na_{n+1} - \overbrace{(a_1 + \cdots + a_n)}^{n} \right]$$

$$> 0 \quad \text{since } \{a_n\} \text{ is increasing.}$$

  (b) We begin with the hint

$$m_n < \dfrac{|a_1 + \cdots + a_j|}{n} + \dfrac{\epsilon}{2} \left(\dfrac{n-j}{n} \right).$$

Since j is fixed,

$$\dfrac{|a_1 + \cdots + a_j|}{n} \to 0$$

and therefore for n sufficiently large

$$\dfrac{|a_1 + \cdots + a_j|}{n} < \dfrac{\epsilon}{2}.$$

Since

$$\dfrac{\epsilon}{2} \left(\dfrac{n-j}{n} \right) < \dfrac{\epsilon}{2},$$

we see that, for n sufficiently large, $|m_n| < \epsilon$. This shows that $m_n \to 0$.

49. (a) Let S be the set of positive integers n ($n \geq 2$) for which the inequalities hold. Since

$$\left(\sqrt{b} \right)^2 - 2\sqrt{ab} + \left(\sqrt{a} \right)^2 = \left(\sqrt{b} - \sqrt{a} \right)^2 > 0,$$

it follows that $\dfrac{a+b}{2} > \sqrt{ab}$ and so $a_1 > b_1$. Now,

$$a_2 = \frac{a_1 + b_1}{2} < a_1 \quad \text{and} \quad b_2 = \sqrt{a_1 b_1} > b_1.$$

Also, by the argument above,

$$a_2 = \frac{a_1 + b_1}{2} > \sqrt{a_1 b_1} = b_2,$$

and so $a_1 > a_2 > b_2 > b_1$. Thus $2 \in S$. Assume that $k \in S$. Then

$$a_{k+1} = \frac{a_k + b_k}{2} < \frac{a_k + a_k}{2} = a_k, \quad b_{k+1} = \sqrt{a_k b_k} > \sqrt{b_k^2} = b_k,$$

and

$$a_{k+1} = \frac{a_k + b_k}{2} > \sqrt{a_k b_k} = b_{k+1}.$$

Thus $k + 1 \in S$. Therefore, the inequalities hold for all $n \geq 2$.

(b) $\{a_n\}$ is a decreasing sequence which is bounded below.

   $\{b_n\}$ is an increasing sequence which is bounded above.

   Let $L_a = \lim\limits_{n\to\infty} a_n, \quad L_b = \lim\limits_{n\to\infty} b_n$. Then

$$a_n = \frac{a_{n-1} + b_{n-1}}{2} \quad \text{implies} \quad L_a = \frac{L_a + L_b}{2} \quad \text{and} \quad L_a = L_b.$$

51. The numerical work suggests $L \cong 1$. Justification: Set $f(x) = \sin x - x^2$. Note that $f(0) = 0$ and $f'(x) = \cos x - 2x > 0$ for x close to 0. Therefore $\sin x - x^2 > 0$ for x close to 0 and $\sin 1/n - 1/n^2 > 0$ for n large. Thus, for n large,

$$\frac{1}{n^2} < \sin \frac{1}{n} < \frac{1}{n} \quad (|\sin x| \leq |x| \quad \text{for all } x)$$

$$\left(\frac{1}{n^2}\right)^{1/n} < \left(\sin \frac{1}{n}\right)^{1/n} < \left(\frac{1}{n}\right)^{1/n}$$

$$\left(\frac{1}{n^{1/n}}\right)^2 < \left(\sin \frac{1}{n}\right)^{1/n} < \frac{1}{n^{1/n}}.$$

As $n \to \infty$ both bounds tend to 1 and therefore the middle term also tends to 1.

53. (a)

a_3	a_4	a_5	a_6	a_7	a_8	a_9	a_{10}
2	3	5	8	13	21	34	55

(b)

r_1	r_2	r_3	r_4	r_5	r_6
1	2	1.5	1.667	1.600	1.625

(c) Following the hint,

$$1 + \frac{1}{r_{n-1}} = 1 + \frac{1}{\dfrac{a_n}{a_{n-1}}} = 1 + \frac{a_{n-1}}{a_n} = \frac{a_n + a_{n-1}}{a_n} = \frac{a_{n+1}}{a_n} = r_n.$$

Now, if $r_n \to L$, then $r_{n-1} \to L$ and

$$1 + \frac{1}{L} = L \quad \text{which, since } L > 1, \text{ implies} \quad L = \frac{1 + \sqrt{5}}{2} \cong 1.618034.$$

SECTION 11.5

(We'll use ⋆ to indicate differentiation of numerator and denominator.)

1. $\lim\limits_{x\to 0^+} \dfrac{\sin x}{\sqrt{x}} \overset{\star}{=} \lim\limits_{x\to 0^+} 2\sqrt{x}\cos x = 0$

3. $\lim\limits_{x\to 0} \dfrac{e^x - 1}{\ln(1+x)} \overset{\star}{=} \lim\limits_{x\to 0}(1+x)e^x = 1$

5. $\lim\limits_{x\to \pi/2} \dfrac{\cos x}{\sin 2x} \overset{\star}{=} \lim\limits_{x\to \pi/2} \dfrac{-\sin x}{2\cos 2x} = \dfrac{1}{2}$

7. $\lim\limits_{x\to 0} \dfrac{2^x - 1}{x} \overset{\star}{=} \lim\limits_{x\to 0} 2^x \ln 2 = \ln 2$

9. $\lim\limits_{x\to 1} \dfrac{x^{1/2} - x^{1/4}}{x - 1} \overset{\star}{=} \lim\limits_{x\to 1}\left(\dfrac{1}{2}x^{-1/2} - \dfrac{1}{4}x^{-3/4}\right) = \dfrac{1}{4}$

11. $\lim\limits_{x\to 0} \dfrac{e^x - e^{-x}}{\sin x} \overset{\star}{=} \lim\limits_{x\to 0} \dfrac{e^x + e^{-x}}{\cos x} = 2$

13. $\lim\limits_{x\to 0} \dfrac{x + \sin \pi x}{x - \sin \pi x} \overset{\star}{=} \lim\limits_{x\to 0} \dfrac{1 + \pi \cos \pi x}{1 - \pi \cos \pi x} = \dfrac{1+\pi}{1-\pi}$

15. $\lim\limits_{x\to 0} \dfrac{e^x + e^{-x} - 2}{1 - \cos 2x} \overset{\star}{=} \lim\limits_{x\to 0} \dfrac{e^x - e^{-x}}{2\sin 2x} \overset{\star}{=} \lim\limits_{x\to 0} \dfrac{e^x + e^{-x}}{4\cos 2x} = \dfrac{1}{2}$

17. $\lim\limits_{x\to 0} \dfrac{\tan \pi x}{e^x - 1} \overset{\star}{=} \lim\limits_{x\to 0} \dfrac{\pi \sec^2 \pi x}{e^x} = \pi$

19. $\lim\limits_{x\to 0} \dfrac{1 + x - e^x}{x(e^x - 1)} \overset{\star}{=} \lim\limits_{x\to 0} \dfrac{1 - e^x}{xe^x + e^x - 1} \overset{\star}{=} \lim\limits_{x\to 0} \dfrac{-e^x}{xe^x + 2e^x} = -\dfrac{1}{2}$

21. $\lim\limits_{x\to 0} \dfrac{x - \tan x}{x - \sin x} \overset{\star}{=} \lim\limits_{x\to 0} \dfrac{1 - \sec^2 x}{1 - \cos x} \overset{\star}{=} \lim\limits_{x\to 0} \dfrac{-2\sec^2 x \tan x}{\sin x} = \lim\limits_{x\to 0} \dfrac{-2\sec^2 x}{\cos x} = -2$

23. $\lim\limits_{x\to 1^-} \dfrac{\sqrt{1 - x^2}}{\sqrt{1 - x^3}} = \lim\limits_{x\to 1^-} \sqrt{\dfrac{1 - x^2}{1 - x^3}} = \sqrt{\dfrac{2}{3}} = \dfrac{1}{3}\sqrt{6}$ since $\lim\limits_{x\to 1^-} \dfrac{1 - x^2}{1 - x^3} \overset{\star}{=} \lim\limits_{x\to 1^-} \dfrac{2x}{3x^2} = \dfrac{2}{3}$

25. $\lim\limits_{x\to \pi/2} \dfrac{\ln(\sin x)}{(\pi - 2x)^2} \overset{\star}{=} \lim\limits_{x\to \pi/2} \dfrac{-\cot x}{4(\pi - 2x)} \overset{\star}{=} \lim\limits_{x\to \pi/2} \dfrac{\csc^2 x}{-8} = -\dfrac{1}{8}$

27. $\lim\limits_{x\to 0} \dfrac{\cos x - \cos 3x}{\sin(x^2)} \overset{\star}{=} \lim\limits_{x\to 0} \dfrac{-\sin x + 3\sin 3x}{2x\cos(x^2)} \overset{\star}{=} \lim\limits_{x\to 0} \dfrac{-\cos x + 9\cos 3x}{2\cos(x^2) - 4x^2\sin(x^2)} = 4$

29.
$\lim\limits_{x\to \pi/4} \dfrac{\sec^2 x - 2\tan x}{1 + \cos 4x} \overset{\star}{=} \lim\limits_{x\to \pi/4} \dfrac{2\sec^2 x \tan x - 2\sec^2 x}{-4\sin 4x}$

$\overset{\star}{=} \lim\limits_{x\to \pi/4} \dfrac{2\sec^4 x + 4\sec^2 x \tan^2 x - 4\sec^2 x \tan x}{-16\cos 4x} = \dfrac{1}{2}$

31. $\lim\limits_{x\to 0} \dfrac{\tan^{-1} x}{\tan^{-1} 2x} \overset{\star}{=} \lim\limits_{x\to 0} \dfrac{1}{1 + x^2} \dfrac{1 + 4x^2}{2} = \dfrac{1}{2}$

33. 1 : $\lim\limits_{x\to \infty} \dfrac{\pi/2 - \tan^{-1} x}{1/x} \overset{\star}{=} \lim\limits_{x\to \infty} \dfrac{x^2}{1 + x^2} = 1$

35. 1 : $\lim\limits_{x\to \infty} \dfrac{1}{x[\ln(x+1) - \ln x]} = \lim\limits_{x\to \infty} \dfrac{1/x}{\ln(1 + 1/x)} = \lim\limits_{t\to 0^+} \dfrac{t}{\ln(1 + t)} \overset{\star}{=} \lim\limits_{t\to 0^+}(1 + t) = 1$

37. $\lim\limits_{x\to 0} \dfrac{x^3}{4^x - 1} = 0$

39. $\lim\limits_{x\to 0} \dfrac{x}{\frac{\pi}{2} - \arccos x} = 1$

41. $\lim\limits_{x\to 0} \dfrac{\tanh x}{x} = 1$

43. $\lim\limits_{x\to 0} (2 + x + \sin x) \neq 0, \quad \lim\limits_{x\to 0}(x^3 + x - \cos x) \neq 0$

45. The limit does not exist if $b \neq 1$. Therefore, $b = 1$.

$$\lim_{x\to 0} \frac{\cos ax - 1}{2x^2} \overset{\star}{=} \lim_{x\to 0} \frac{-a \sin ax}{4x} \overset{\star}{=} \lim_{x\to 0} \frac{-a^2 \cos ax}{4} = -\frac{a^2}{4}$$

Now, $-\dfrac{a^2}{4} = -4$ implies $a = \pm 4$.

47. Recall $\lim\limits_{x\to 0^+}(1 + x)^{1/x} = e$:

$$\lim_{x\to 0^+} \frac{(1+x)^{1/x} - e}{x} \overset{\star}{=} \lim_{x\to 0^+}\left[(1+x)^{1/x}\right]\left[\frac{x - (1+x)\ln(1+x)}{x^2 + x^3}\right]$$

$$= e \lim_{x\to 0^+} \frac{x - (1+x)\ln(1+x)}{x^2 + x^3}$$

$$\overset{\star}{=} e \lim_{x\to 0^+} \frac{-\ln(1+x)}{2x + 3x^2}$$

$$\overset{\star}{=} e \lim_{x\to 0^+} \frac{-1/(1+x)}{2 + 6x} = -\frac{e}{2}$$

49. $\lim\limits_{x\to 0} \dfrac{1}{x} \displaystyle\int_0^x f(t)\,dt \overset{\star}{=} \lim\limits_{x\to 0} \dfrac{f(x)}{1} = f(0)$

51. (a) $\lim\limits_{x\to 0} \dfrac{C(x)}{x} \overset{\star}{=} \lim\limits_{x\to 0} \dfrac{\cos^2 x}{1} = 1$

(b) $\lim\limits_{x\to 0} \dfrac{C(x) - x}{x^3} \overset{\star}{=} \lim\limits_{x\to 0} \dfrac{\cos^2 x - 1}{3x^2} \overset{\star}{=} \lim\limits_{x\to 0} \dfrac{-2\cos x \sin x}{6x} \overset{\star}{=} \lim\limits_{x\to 0} \dfrac{-2\cos^2 x + 2\sin^2 x}{6} = -\dfrac{1}{3}$

53. $A(b) = 2\displaystyle\int_0^{\sqrt{b}} (b - x^2)\,dx = 2\left[bx - \dfrac{x^3}{3}\right]_0^{\sqrt{b}} = \dfrac{4}{3}b\sqrt{b}$ and $T(b) = \dfrac{1}{2}\left(2\sqrt{b}\right)b = b\sqrt{b}$.

Thus, $\lim\limits_{b\to 0} \dfrac{T(b)}{A(b)} = \dfrac{b\sqrt{b}}{\frac{4}{3}b\sqrt{b}} = \dfrac{3}{4}$.

55. (a) $f(x) \to \infty$ as $x \to \pm\infty$

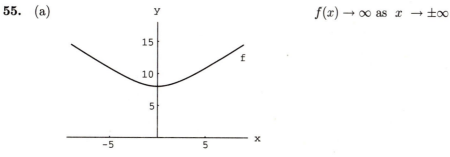

(b) $f(x) \to 10$ as $x \to 4$

Confirmation: $\lim\limits_{x\to 4} \dfrac{x^2 - 16}{\sqrt{x^2 + 9} - 5} \overset{\star}{=} \lim\limits_{x\to 4} \dfrac{2x}{x\,(x^2 + 9)^{-1/2}} = \lim\limits_{x\to 4} 2\sqrt{x^2 + 9} = 10$

57. (a) 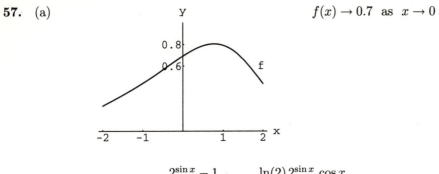 $f(x) \to 0.7$ as $x \to 0$

(b) Confirmation: $\displaystyle\lim_{x\to 0} \frac{2^{\sin x} - 1}{x} \overset{\star}{=} \lim_{x\to 0} \frac{\ln(2)\, 2^{\sin x} \cos x}{1} = \ln 2 \cong 0.6931$

SECTION 11.6

(We'll use $\star$ to indicate differentiation of numerator and denominator.)

1. $\displaystyle\lim_{x\to -\infty} \frac{x^2 + 1}{1 - x} \overset{\star}{=} \lim_{x\to -\infty} \frac{2x}{-1} = \infty$

3. $\displaystyle\lim_{x\to \infty} \frac{x^3}{1 - x^3} = \lim_{x\to \infty} \frac{1}{1/x^3 - 1} = -1$

5. $\displaystyle\lim_{x\to \infty} x^2 \sin \frac{1}{x} = \lim_{h\to 0^+} \left[\left(\frac{1}{h}\right)\left(\frac{\sin h}{h}\right)\right] = \infty$

7. $\displaystyle\lim_{x\to \frac{\pi}{2}^-} \frac{\tan 5x}{\tan x} = \lim_{x\to \frac{\pi}{2}^-} \left[\left(\frac{\sin 5x}{\sin x}\right)\left(\frac{\cos x}{\cos 5x}\right)\right] = \frac{1}{5}$ since

$$\lim_{x\to \frac{\pi}{2}^-} \frac{\sin 5x}{\sin x} = 1 \quad \text{and} \quad \lim_{x\to \frac{\pi}{2}^-} \frac{\cos x}{\cos 5x} \overset{\star}{=} \lim_{x\to \frac{\pi}{2}^-} \frac{\sin x}{5 \sin 5x} = \frac{1}{5}$$

9. $\displaystyle\lim_{x\to 0^+} x^{2x} = \lim_{x\to 0^+} (x^x)^2 = 1^2 = 1$ [see (11.6.4)]

11. $\displaystyle\lim_{x\to 0} x(\ln|x|)^2 = \lim_{x\to 0} \frac{(\ln|x|)^2}{1/x} \overset{\star}{=} \lim_{x\to 0} \frac{2\ln|x|}{-1/x} \overset{\star}{=} \lim_{x\to 0} \frac{2}{1/x} = \lim_{x\to 0} 2x = 0$

13. $\displaystyle\lim_{x\to \infty} \frac{1}{x} \int_0^x e^{t^2}\, dt \overset{\star}{=} \lim_{x\to \infty} \frac{e^{x^2}}{1} = \infty$

15.
$$\lim_{x\to 0} \left[\frac{1}{\sin^2 x} - \frac{1}{x^2}\right] = \lim_{x\to 0} \frac{x^2 - \sin^2 x}{x^2 \sin^2 x} \overset{\star}{=} \lim_{x\to 0} \frac{2x - 2\sin x \cos x}{2x^2 \sin x \cos x + 2x \sin^2 x}$$

$$= \lim_{x\to 0} \frac{2x - \sin 2x}{x^2 \sin 2x + 2x \sin^2 x} \overset{\star}{=} \lim_{x\to 0} \frac{2 - 2\cos 2x}{2x^2 \cos 2x + 4x \sin 2x + 2\sin^2 x}$$

$$\overset{\star}{=} \lim_{x\to 0} \frac{4\sin 2x}{-4x^2 \sin 2x + 12x \cos 2x + 6\sin 2x}$$

$$\overset{\star}{=} \lim_{x\to 0} \frac{8\cos 2x}{-8x^2 \cos 2x - 32x \sin 2x + 24\cos 2x} = \frac{1}{3}$$

17. $\displaystyle\lim_{x\to 1} x^{1/(x-1)} = e$ since $\displaystyle\lim_{x\to 1} \ln\left[x^{1/(x-1)}\right] = \lim_{x\to 1} \frac{\ln x}{x - 1} \overset{\star}{=} \lim_{x\to 1} \frac{1}{x} = 1$

19.

$$\lim_{x \to \infty} \left(\cos \frac{1}{x} \right)^x = 1 \quad \text{since} \quad \lim_{x \to \infty} \ln \left[\left(\cos \frac{1}{x} \right)^x \right] = \lim_{x \to \infty} \frac{\ln \left(\cos \frac{1}{x} \right)}{(1/x)}$$

$$\stackrel{\star}{=} \lim_{x \to \infty} \left(-\frac{\sin (1/x)}{\cos (1/x)} \right) = 0$$

21.

$$\lim_{x \to 0} \left[\frac{1}{\ln (1+x)} - \frac{1}{x} \right] = \lim_{x \to 0} \frac{x - \ln (1+x)}{x \ln (1+x)} \stackrel{\star}{=} \lim_{x \to 0} \frac{x}{x + (1+x) \ln (1+x)}$$

$$\stackrel{\star}{=} \lim_{x \to 0} \frac{1}{1 + 1 + \ln (1+x)} = \frac{1}{2}$$

23.

$$\lim_{x \to 0} \left[\frac{1}{x} - \cot x \right] = \lim_{x \to 0} \frac{\sin x - x \cos x}{x \sin x} \stackrel{\star}{=} \lim_{x \to 0} \frac{x \sin x}{\sin x + x \cos x}$$

$$\stackrel{\star}{=} \lim_{x \to 0} \frac{\sin x + x \cos x}{2 \cos x - x \sin x} = 0$$

25.

$$\lim_{x \to \infty} \left(\sqrt{x^2 + 2x} - x \right) = \lim_{x \to \infty} \left[\left(\sqrt{x^2 + 2x} - x \right) \left(\frac{\sqrt{x^2 + 2x} + x}{\sqrt{x^2 + 2x} + x} \right) \right]$$

$$= \lim_{x \to \infty} \frac{2x}{\sqrt{x^2 + 2x} + x} = \lim_{x \to \infty} \frac{2}{\sqrt{1 + 2/x} + 1} = 1$$

27. $\lim_{x \to \infty} \left(x^3 + 1 \right)^{1/\ln x} = e^3$ since

$$\lim_{x \to \infty} \ln \left[\left(x^3 + 1 \right)^{1/\ln x} \right] = \lim_{x \to \infty} \frac{\ln \left(x^3 + 1 \right)}{\ln x} \stackrel{\star}{=} \lim_{x \to \infty} \frac{\left(\frac{3x^2}{x^3 + 1} \right)}{1/x} = \lim_{x \to \infty} \frac{3}{1 + 1/x^3} = 3.$$

29. $\lim_{x \to \infty} (\cosh x)^{1/x} = e$ since

$$\lim_{x \to \infty} \ln \left[(\cosh x)^{1/x} \right] = \lim_{x \to \infty} \frac{\ln (\cosh x)}{x} \stackrel{\star}{=} \lim_{x \to \infty} \frac{\sinh x}{\cosh x} = 1.$$

31.

$$\lim_{x \to 0} \left(\frac{1}{\sin x} - \frac{1}{x} \right) = \lim_{x \to 0} \frac{x - \sin x}{x \sin x} \stackrel{\star}{=} \lim_{x \to 0} \frac{1 - \cos x}{\sin x + x \cos x}$$

$$\stackrel{\star}{=} \lim_{x \to 0} \frac{\sin x}{2 \cos x - x \sin x} = 0$$

33.

$$\lim_{x \to 1} \left(\frac{1}{\ln x} - \frac{x}{x-1} \right) = \lim_{x \to 1} \frac{x - 1 - x \ln x}{(x-1) \ln x} \stackrel{\star}{=} \lim_{x \to 1} \frac{-\ln x}{(x-1)(1/x) + \ln x}$$

$$= \lim_{x \to 1} \frac{-x \ln x}{x - 1 + x \ln x} \stackrel{\star}{=} \lim_{x \to 1} \frac{-\ln x - 1}{2 + \ln x} = -\frac{1}{2}$$

35. $0 : \quad \frac{1}{n} \ln \frac{1}{n} = -\frac{\ln n}{n} \to 0$ **37.** $1 : \quad \ln \left[(\ln n)^{1/n} \right] = \frac{1}{n} \ln (\ln n) \to 0$

39. $1 : \quad \ln \left[\left(n^2 + n \right)^{1/n} \right] = \frac{1}{n} \ln [n(n+1)] = \frac{\ln n}{n} + \frac{\ln (n+1)}{n} \to 0$

41. $0 : \quad 0 \le \frac{n^2 \ln n}{e^n} < \frac{n^3}{e^n}, \quad \lim_{x \to \infty} \frac{x^3}{e^x} = 0$

43. $\lim_{x \to 0} (\sin x)^x = 1$

45. $\lim_{x \to 0} \left(\dfrac{1}{\sin x} - \dfrac{1}{\tan x} \right) = 0$

47.

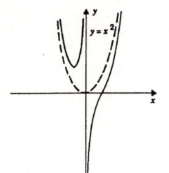

vertical asymptote y-axis

49.

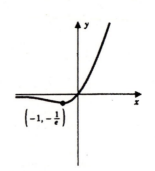

horizontal asymptote x-axis

51.

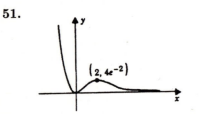

horizontal asymptote x-axis

53.
$$\dfrac{b}{a}\sqrt{x^2 - a^2} - \dfrac{b}{a}x =$$
$$\dfrac{\sqrt{x^2 - a^2} + x}{\sqrt{x^2 - a^2} + x} \left(\dfrac{b}{a} \right) \left(\sqrt{x^2 - a^2} - x \right) =$$
$$\dfrac{-ab}{\sqrt{x^2 - a^2} + x} \to 0 \quad \text{as} \quad x \to \infty$$

55. for instance, $\quad f(x) = x^2 + \dfrac{(x-1)(x-2)}{x^3}$

57. $\lim_{x \to 0^+} -\dfrac{2x}{\cos x} \neq \lim_{x \to 0^+} \dfrac{2}{-\sin x}$. L'Hospital's rule does not apply here since $\lim_{x \to 0^+} \cos x = 1$.

59. Let $y = \left(\dfrac{a^{1/x} + b^{1/x}}{2} \right)^x$. Then $\ln y = x \ln \left[\dfrac{a^{1/x} + b^{1/x}}{2} \right] = \dfrac{\ln \left(\dfrac{a^{1/x} + b^{1/x}}{2} \right)}{1/x}$.

Now,

$$\lim_{x \to \infty} \dfrac{\ln \left(\dfrac{a^{1/x} + b^{1/x}}{2} \right)}{1/x} \overset{\star}{=} \lim_{x \to \infty} \dfrac{2}{a^{1/x} + b^{1/x}} \cdot \dfrac{\dfrac{-a^{1/x} \ln a - b^{1/x} \ln b}{2x^2}}{-1/x^2}$$

$$= \lim_{x \to \infty} \dfrac{a^{1/x} \ln a + b^{1/x} \ln b}{a^{1/x} + b^{1/x}} = \tfrac{1}{2} \ln ab = \ln \sqrt{ab}$$

Thus, $\lim_{x \to \infty} \ln y = \ln \sqrt{ab} \implies \lim_{x \to \infty} y = \sqrt{ab}$.

61. (a) $A_b = 1 - (1 + b)e^{-b}$

(b) $\overline{x}_b = \dfrac{2 - \left(2 + 2b + b^2 \right) e^{-b}}{1 - (1 + b)e^{-b}}; \qquad \overline{y}_b = \dfrac{\tfrac{1}{4} - \tfrac{1}{4} \left(1 + 2b + 2b^2 \right) e^{-2b}}{2 \left[1 - (1 + b)e^{-b} \right]}$

(c) $\lim_{b \to \infty} A_b = 1; \quad \lim_{b \to \infty} \overline{x}_b = 2; \quad \lim_{b \to \infty} \overline{y}_b = \dfrac{1}{8}$

63. (a)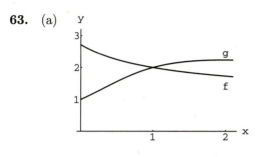

$$\lim_{x \to 0^+} \left(1 + x^2\right)^{1/x} = 1$$

(b) $\lim_{x \to 0^+} \left(1 + x^2\right)^{1/x} = 1$ since

$$\lim_{x \to 0^+} \ln\left[\left(1 + x^2\right)^{1/x}\right] = \lim_{x \to 0^+} \frac{\ln\left(1 + x^2\right)}{x} \overset{\star}{=} \lim_{x \to 0^+} \frac{2x}{\left(1 + x^2\right)} = 0$$

65. (a)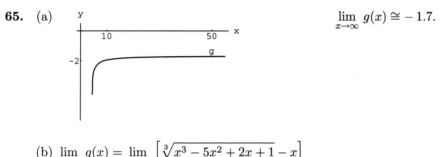

$$\lim_{x \to \infty} g(x) \cong -1.7.$$

(b) $\lim_{x \to \infty} g(x) = \lim_{x \to \infty} \left[\sqrt[3]{x^3 - 5x^2 + 2x + 1} - x\right]$

$$= \lim_{x \to \infty} \frac{-5x^2 + 2x + 1}{\left(\sqrt[3]{x^3 - 5x^2 + 2x + 1}\right)^2 + x\sqrt[3]{x^3 - 5x^2 + 2x + 1} + x^2}$$

$$= -\frac{5}{3} \cong -1.667$$

SECTION 11.7

1. 1: $\displaystyle\int_1^\infty \frac{dx}{x^2} = \lim_{b \to \infty} \int_1^b \frac{dx}{x^2} = \lim_{b \to \infty} \left[-\frac{1}{x}\right]_1^b = \lim_{b \to \infty} \left[1 - \frac{1}{b}\right] = 1$

3. $\dfrac{\pi}{4}$: $\displaystyle\int_0^\infty \frac{dx}{4 + x^2} = \lim_{b \to \infty} \int_0^b \frac{dx}{4 + x^2} = \lim_{b \to \infty} \left[\frac{1}{2}\tan^{-1}\frac{x}{2}\right]_0^b = \lim_{b \to \infty} \frac{1}{2}\tan^{-1}\left(\frac{b}{2}\right) = \frac{\pi}{4}$

5. diverges: $\displaystyle\int_0^\infty e^{px}\,dx = \lim_{b \to \infty} \int_0^b e^{px}\,dx = \lim_{b \to \infty} \left[\frac{1}{p}e^{px}\right]_0^b = \lim_{b \to \infty} \frac{1}{p}\left(e^{pb} - 1\right) = \infty$

7. 6: $\displaystyle\int_0^8 \frac{dx}{x^{2/3}} = \lim_{a \to 0^+} \int_a^8 x^{-2/3}\,dx = \lim_{a \to 0^+} \left[3x^{1/3}\right]_a^8 = \lim_{a \to 0^+} \left[6 - 3a^{1/3}\right] = 6$

9. $\dfrac{\pi}{2}$: $\displaystyle\int_0^1 \frac{dx}{\sqrt{1 - x^2}} = \lim_{b \to 1^-} \int_0^b \frac{dx}{\sqrt{1 - x^2}} = \lim_{b \to 1^-} \sin^{-1} b = \frac{\pi}{2}$

11. 2: $\displaystyle\int_0^2 \frac{x}{\sqrt{4 - x^2}}\,dx = \lim_{b \to 2^-} \int_0^b x\left(4 - x^2\right)^{-1/2}\,dx = \lim_{b \to 2^-} \left[-\left(4 - x^2\right)^{1/2}\right]_0^b$

$$= \lim_{b \to 2^-} \left(2 - \sqrt{4 - b^2}\right) = 2$$

13. diverges:

$$\int_e^\infty \frac{\ln x}{x}\, dx = \lim_{b\to\infty} \int_e^b \frac{\ln x}{x}\, dx = \lim_{b\to\infty} \left[\frac{1}{2}(\ln x)^2 \right]_e^b = \lim_{b\to\infty} \left[\frac{1}{2}(\ln b)^2 - \frac{1}{2} \right] = \infty$$

15. $-\dfrac{1}{4}$:

$$\int_0^1 x\ln x\, dx = \lim_{a\to 0^+} \int_a^1 x\ln x\, dx = \lim_{a\to 0^+} \left[\frac{1}{2}x^2\ln x - \frac{1}{4}x^2 \right]_a^1$$

$$\text{(by parts)}\underline{\qquad}\uparrow$$

$$= \lim_{a\to 0^+} \left[\frac{1}{4}a^2 - \frac{1}{2}a^2\ln a - \frac{1}{4} \right] = -\frac{1}{4}$$

Note: $\displaystyle \lim_{t\to 0^+} t^2\ln t = \lim_{t\to 0^+} \frac{\ln t}{1/t^2} \overset{\star}{=} \lim_{t\to 0^+} \frac{1/t}{-2/t^3} = -\frac{1}{2}\lim_{t\to 0^+} t^2 = 0.$

17. π:

$$\int_{-\infty}^\infty \frac{dx}{1+x^2} = \lim_{a\to -\infty} \int_a^0 \frac{dx}{1+x^2} + \lim_{b\to\infty} \int_0^b \frac{dx}{1+x^2}$$

$$= \lim_{a\to -\infty} \left[\tan^{-1} x \right]_a^0 + \lim_{b\to\infty} \left[\tan^{-1} x \right]_0^b = -\left(-\frac{\pi}{2} \right) + \frac{\pi}{2} = \pi$$

19. diverges: $\displaystyle \int_{-\infty}^\infty \frac{dx}{x^2} = \lim_{a\to -\infty} \int_a^{-1} \frac{dx}{x^2} + \lim_{b\to 0^-} \int_{-1}^b \frac{dx}{x^2} + \lim_{c\to 0^+} \int_c^1 \frac{dx}{x^2} + \lim_{d\to\infty} \int_1^d \frac{dx}{x^2};$

and, $\displaystyle \lim_{c\to 0^+} \int_c^1 \frac{dx}{x^2} = \lim_{c\to 0^+} \left[-\frac{1}{x} \right]_c^1 = \lim_{c\to 0^+} \left[\frac{1}{c} - 1 \right] = \infty$

21. $\ln 2$:

$$\int_1^\infty \frac{dx}{x(x+1)} = \lim_{b\to\infty} \int_1^b \left[\frac{1}{x} - \frac{1}{x+1} \right] dx$$

$$= \lim_{b\to\infty} \left[\ln\left(\frac{x}{x+1} \right) \right]_1^b = \lim_{b\to\infty} \left[\ln\left(\frac{b}{b+1} \right) - \ln\left(\frac{1}{2} \right) \right] = \ln 2$$

23. 4:

$$\int_3^5 \frac{x}{\sqrt{x^2-9}}\, dx = \lim_{a\to 3^-} \int_a^5 x\left(x^2 - 9 \right)^{-1/2} dx$$

$$= \lim_{a\to 3^-} \left[\left(x^2 - 9 \right)^{1/2} \right]_a^5 = \lim_{a\to 3^-} \left[4 - \left(a^2 - 9 \right)^{1/2} \right] = 4$$

25. $\displaystyle \int_{-3}^3 \frac{dx}{x(x+1)}$ diverges since $\displaystyle \int_0^3 \frac{dx}{x(x+1)}$ diverges:

$$\int_0^3 \frac{dx}{x(x+1)} = \lim_{a\to 0^+} \int_a^3 \left(\frac{1}{x} - \frac{1}{x+1} \right) dx = \lim_{a\to 0^+} \left[\ln|x| - \ln|x+1| \right]_a^3$$

$$= \lim_{a\to 0^+} \left[\ln 3 - \ln 4 - \ln a + \ln(a+1) \right] = \infty.$$

27. $\int_{-3}^{1} \dfrac{dx}{x^2 - 4}$ diverges since $\int_{-2}^{1} \dfrac{dx}{x^2 - 4}$ diverges:

$$\int_{-2}^{1} \frac{dx}{x^2 - 4} = \lim_{a \to -2+} \int_{a}^{1} \frac{1}{4} \left[\frac{1}{x - 2} - \frac{1}{x + 2} \right] dx$$

$$= \lim_{a \to -2+} \left[\frac{1}{4} \left(\ln |x - 2| - \ln |x + 2| \right) \right]_{a}^{1}$$

$$= \lim_{a \to -2+} \frac{1}{4} \left[-\ln 3 - \ln |a - 2| + \ln |a + 2| \right] = -\infty.$$

29. diverges: $\displaystyle\int_{0}^{\infty} \cosh x \, dx = \lim_{b \to \infty} \int_{0}^{b} \cosh x \, dx = \lim_{b \to \infty} [\sinh x]_{0}^{b} = \infty$

31. $\dfrac{1}{2}$: $\displaystyle\int_{0}^{\infty} e^{-x} \sin x \, dx = \lim_{b \to \infty} \int_{0}^{b} e^{-x} \sin x \, dx = \lim_{b \to \infty} -\frac{1}{2} \left[e^{-x} \cos x + e^{-x} \sin x \right]_{0}^{b}$

(by parts)

$$= \lim_{b \to \infty} \frac{1}{2} \left[1 - e^{-b} \cos b - e^{-b} \sin b \right] = \frac{1}{2}$$

33. $2e - 2$: $\displaystyle\int_{0}^{1} \frac{e^{\sqrt{x}}}{\sqrt{x}} \, dx = \lim_{a \to 0+} \int_{a}^{1} \frac{e^{\sqrt{x}}}{\sqrt{x}} \, dx = \lim_{a \to 0+} \left[2 e^{\sqrt{x}} \right]_{a}^{1} = 2(e - 1)$

35. (a) converges: $\displaystyle\int_{0}^{\infty} \frac{x}{(16 + x^2)^2} \, dx = \frac{1}{32}$ (b) converges: $\displaystyle\int_{0}^{\infty} \frac{x^2}{(16 + x^2)^2} \, dx = \frac{\pi}{16}$

(c) converges: $\displaystyle\int_{0}^{\infty} \frac{x}{16 + x^4} \, dx = \frac{\pi}{16}$ (d) diverges

37. $\displaystyle\int_{0}^{1} \sin^{-1} x \, dx = \left[x \sin^{-1} x \right]_{0}^{1} - \int_{0}^{1} \frac{x}{\sqrt{1 - x^2}} \, dx = \frac{\pi}{2} - \lim_{a \to 1^-} \int_{0}^{a} \frac{x}{\sqrt{1 - x^2}} \, dx$

(by parts)

Now, $\displaystyle\int_{0}^{a} \frac{x}{\sqrt{1 - x^2}} \, dx = -\frac{1}{2} \int_{1}^{1 - a^2} \frac{1}{\sqrt{u}} \, du = \left[-\sqrt{u} \right]_{1}^{1 - a^2} = 1 - \sqrt{1 - a^2}$

$u = 1 - x^2$

Thus, $\displaystyle\int_{0}^{1} \sin^{-1} x \, dx = \frac{\pi}{2} - \lim_{a \to 1^-} \left(1 - \sqrt{1 - a^2} \right) = \frac{\pi}{2} - 1.$

39. $$\int_{0}^{\infty} \frac{1}{\sqrt{x} (1 + x)} \, dx = \int_{0}^{1} \frac{1}{\sqrt{x} (1 + x)} \, dx + \int_{1}^{\infty} \frac{1}{\sqrt{x} (1 + x)} \, dx$$

$$= \lim_{a \to 0+} \int_{a}^{1} \frac{1}{\sqrt{x} (1 + x)} \, dx + \lim_{b \to \infty} \int_{1}^{b} \frac{1}{\sqrt{x} (1 + x)} \, dx$$

Now, $\displaystyle\int \frac{1}{\sqrt{x} (1 + x)} \, dx = \int \frac{2}{1 + u^2} \, du = 2 \arctan u + C = 2 \arctan \sqrt{x} + C.$

$u = \sqrt{x}$

Therefore, $\displaystyle\lim_{a\to 0^+}\int_a^1 \frac{1}{\sqrt{x}\,(1+x)}\,dx = \lim_{a\to 0^+}\left[2\arctan\sqrt{x}\right]_a^1 = \lim_{a\to 0^+} 2\left[\pi/4 - \arctan\sqrt{a}\right)] = \frac{\pi}{2}$

and $\displaystyle\lim_{b\to\infty}\int_1^b \frac{1}{\sqrt{x}\,(1+x)}\,dx = \lim_{b\to\infty}\left[2\arctan\sqrt{x}\right]_1^b = \lim_{b\to\infty} 2\left[\arctan\sqrt{b} - \pi/4\right] = \frac{\pi}{2}.$

Thus, $\displaystyle\int_0^\infty \frac{1}{\sqrt{x}\,(1+x)}\,dx = \pi.$

41. surface area $\displaystyle S = \int_1^\infty 2\pi\left(\frac{1}{x}\right)\sqrt{1+\frac{1}{x^4}}\,dx = 2\pi\int_1^\infty \frac{\sqrt{x^4+1}}{x^3}\,dx = \infty$ by comparison with $\displaystyle\int_1^\infty \frac{1}{x}\,dx$

43. (a)

(b) $\displaystyle A = \int_0^1 \frac{1}{\sqrt{x}}\,dx = \lim_{a\to 0^+}\int_a^1 \frac{1}{\sqrt{x}}\,dx$

$\displaystyle = \lim_{a\to 0^+}\left[2\sqrt{x}\right]_a^1 = 2$

(c) $\displaystyle V = \int_0^1 \pi\left(\frac{1}{\sqrt{x}}\right)^2 dx = \pi\int_0^1 \frac{1}{x}\,dx = \pi\lim_{a\to 0^+}\int_a^1 \frac{1}{x}\,dx = \pi\lim_{a\to 0^+}\left[\ln x\right]_a^1$ diverges

45. (a)

(b) $\displaystyle A = \int_0^\infty e^{-x}\,dx = 1$

(c) $\displaystyle V_x = \int_0^\infty \pi e^{-2x}\,dx = \pi/2$

(d) $\displaystyle V_y = \int_0^\infty 2\pi x e^{-x}\,dx = \lim_{b\to\infty}\int_0^b 2\pi x e^{-x}\,dx = \lim_{b\to\infty}\left[2\pi(-x-1)e^{-x}\right]_0^b$

(by parts)

$\displaystyle = 2\pi\left(1 - \lim_{b\to\infty}\frac{b+1}{e^b}\right) = 2\pi(1-0) = 2\pi$

(e) $A = \int_0^\infty 2\pi e^{-x} \sqrt{1 + e^{-2x}} \, dx = \lim_{b \to \infty} \int_0^b 2\pi e^{-x} \sqrt{1 + e^{-2x}} \, dx$

$\int_0^b 2\pi e^{-x} \sqrt{1 + e^{-2x}} \, dx = -2\pi \int_1^{e^{-b}} \sqrt{1 + u^2} \, du$

$u = e^{-x}$

$= -\pi \left[u\sqrt{1 + u^2} + \ln\left(u + \sqrt{1 + u^2}\right) \right]_1^{e^{-b}}$

$= \pi \left[\sqrt{2} + \ln\left(1 + \sqrt{2}\right) - e^{-b}\sqrt{1 + e^{-2b}} - \ln\left(e^{-b} + \sqrt{1 + e^{-2b}}\right) \right]$

Taking the limit of this last expression as $b \to \infty$, we have

$$A = \pi \left[\sqrt{2} + \ln\left(1 + \sqrt{2}\right) \right].$$

47. (a) The interval $[0, 1]$ causes no problem. For $x \geq 1$, $e^{-x^2} \leq e^{-x}$ and $\int_1^\infty e^{-x} \, dx$ is finite.

(b) $V_y = \int_0^\infty 2\pi x e^{-x^2} \, dx = \lim_{b \to \infty} \int_0^b 2\pi x e^{-x^2} \, dx = \lim_{b \to \infty} \pi \left[-e^{-x^2} \right]_0^b = \lim_{b \to \infty} \pi \left(1 - e^{-b^2}\right) = \pi$

49. (a)

(b) $A = \lim_{a \to 0^+} \int_a^1 x^{-1/4} \, dx = \lim_{a \to 0^+} \left[\frac{4}{3} x^{3/4} \right]_a^1 = \frac{4}{3}$

(c) $V_x = \lim_{a \to 0^+} \int_a^1 \pi x^{-1/2} \, dx = \lim_{a \to 0^+} \left[2\pi x^{1/2} \right]_a^1 = 2\pi$

(d) $V_y = \lim_{a \to 0^+} \int_a^1 2\pi x^{3/4} \, dx = \lim_{a \to 0^+} \left[\frac{8\pi}{7} x^{7/4} \right]_a^1 = \frac{8}{7}\pi$

51. converges by comparison with $\int_1^\infty \frac{dx}{x^{3/2}}$

53. diverges since for x large the integrand is greater than $\frac{1}{x}$ and $\int_1^\infty \frac{dx}{x}$ diverges

55. converges by comparison with $\int_1^\infty \frac{dx}{x^{3/2}}$

57. (a) $\lim_{b \to \infty} \int_0^b \frac{2x}{1 + x^2} \, dx = \lim_{b \to \infty} \left[\ln\left(1 + x^2\right) \right]_0^b = \infty$

Thus, the improper integral $\int_0^\infty \frac{2x}{1 + x^2} \, dx$ diverges.

(b) $\lim_{b \to \infty} \int_{-b}^b \frac{2x}{1 + x^2} \, dx = \lim_{b \to \infty} \left[\ln\left(1 + x^2\right) \right]_{-b}^b$

$= \lim_{b \to \infty} \left(\ln\left[1 + b^2\right] - \ln\left[1 + (-b)^2\right] \right) = \lim_{b \to \infty} (0) = 0$

59. $r(\theta) = ae^{c\theta}, \quad r'(\theta) = ace^{c\theta}$

$$L = \int_{-\infty}^{\theta_1} \sqrt{a^2 e^{2c\theta} + a^2 c^2 e^{2c\theta}} \, d\theta$$

$$\overset{(10.7.3)}{\underset{\big\uparrow}{}} = \left(a\sqrt{1+c^2}\right)\left(\lim_{b\to-\infty}\int_b^{\theta_1} e^{c\theta}\, d\theta\right)$$

$$= \left(a\sqrt{1+c^2}\right)\left(\lim_{b\to-\infty}\left[\frac{e^{c\theta}}{c}\right]_b^{\theta_1}\right)$$

$$= \left(\frac{a\sqrt{1+c^2}}{c}\right)\left(\lim_{b\to-\infty}\left[e^{c\theta_1} - e^{cb}\right]\right) = \left(\frac{a\sqrt{1+c^2}}{c}\right)e^{c\theta_1}$$

61. $F(s) = \displaystyle\int_0^\infty e^{-sx}\cdot 1\, dx = \lim_{b\to\infty}\int_0^b e^{-sx}\, dx = \lim_{b\to\infty}\left[-\frac{1}{s}e^{-sx}\right]_0^b = \frac{1}{s}$ provided $s > 0$.

Thus, $F(s) = \dfrac{1}{s}; \qquad \mathrm{dom}(F) = (0,\infty)$.

63. $F(s) = \displaystyle\int_0^\infty e^{-sx}\cos 2x\, dx = \lim_{b\to\infty}\int_0^b e^{-sx}\cos 2x\, dx$

Using integration by parts $\displaystyle\int e^{-sx}\cos 2x\, dx = \frac{4}{s^2+4}\left[\frac{1}{2}e^{-sx}\sin 2x - \frac{s}{4}e^{-sx}\cos 2x\right] + C$.

Therefore,

$$F(s) = \lim_{b\to\infty}\frac{4}{s^2+4}\left[\frac{1}{2}e^{-sx}\sin 2x - \frac{s}{4}e^{-sx}\cos 2x\right]_0^b$$

$$= \frac{4}{s^2+4}\lim_{b\to\infty}\left[\frac{1}{2}e^{-sb}\sin 2b - \frac{s}{4}e^{-sb}\cos 2b + \frac{s}{4}\right] = \frac{4}{s^2+4}\cdot\frac{s}{4} = \frac{s}{s^2+4}\qquad\text{provided}\quad s > 0.$$

Thus, $F(s) = \dfrac{s}{s^2+4}; \qquad \mathrm{dom}(F) = (0,\infty)$.

65.   The function f is nonnegative on $(-\infty,\infty)$ and

$$\int_{-\infty}^\infty f(x)\, dx = \int_{-\infty}^0 0\, dx + \int_0^\infty \frac{6x}{(1+3x^2)^2}\, dx = \int_0^\infty \frac{6x}{(1+3x^2)^2}\, dx$$

Now, $\displaystyle\int \frac{6x}{(1+3x^2)^2}\, dx = -\frac{1}{1+3x^2} + C$.

Therefore,

$$\int_{-\infty}^\infty f(x)\, dx = \lim_{b\to\infty}\left[-\frac{1}{1+3x^2}\right]_0^b = \lim_{b\to\infty}\left(1 - \frac{1}{1+3b^2}\right) = 1.$$

67. $\mu = \displaystyle\int_{-\infty}^\infty x f(x)\, dx = \int_{-\infty}^0 0\, dx + \int_0^\infty kxe^{-kx}\, dx = \lim_{b\to\infty}\int_0^b kxe^{-kx}\, dx$

Using integration by parts, $\displaystyle\int kxe^{-kx}\, dx = -xe^{-kx} - \frac{1}{k}e^{-kx} + C$.

Therefore,

$$\mu = \int_{-\infty}^{\infty} x f(x)\, dx = \lim_{b \to \infty} \left[-xe^{-kx} - \frac{1}{k}e^{-kx} \right]_0^b = \lim_{b \to \infty} \left[-be^{-kb} - \frac{1}{k}e^{-kb} + \frac{1}{k} \right] = \frac{1}{k}$$

69. Observe that $F(t) = \int_1^t f(x)\, dx$ is continuous and increasing, that $a_n = \int_1^n f(x)\, dx$ is increasing, and that $(*)$ $a_n \leq \int_1^t f(x)\, dx \leq a_{n+1}$ for $t \in [n, n+1]$.

If $\int_1^{\infty} f(x)\, dx$ converges, then F, being continuous, is bounded and, by $(*)$, $\{a_n\}$ is bounded and therefore convergent. If $\{a_n\}$ converges, then $\{a_n\}$ is bounded and, by $(*)$, F is bounded. Being increasing, F is also convergent; i.e., $\int_1^{\infty} f(x)\, dx$ converges.

REVIEW EXERCISES

1. $|x - 2| \leq 3 \implies -1 \leq x \leq 5:$ lub $= 5$, glb $= -1$.

3. $x^2 - x - 2 \leq 0 \implies (x - 2)(x + 1) \leq 0 \implies -1 \leq x \leq 2:$ lub $= 2$, glb $= -1$.

5. Since $e^{-x^2} \leq 1$ for all x, $e^{-x^2} \leq 2$ for all x; no lub, no glb.

7. increasing; bounded below by $\frac{1}{2}$ and above by $\frac{2}{3}$.

9. bounded below by 0 and above by $\frac{3}{2}$; not monotonic

11. $\left\{ \dfrac{2^n}{n^2} \right\} = \left\{ 2, 1, \frac{8}{9}, 1, \frac{32}{25}, \ldots \right\};$ the sequence is not monotonic.

However, it is increasing from a_3 on. The sequence is bounded below by $\frac{8}{9}$; it is not bounded above.

13. the sequence does not converge; $n2^{1/n} \to \infty$ as $n \to \infty$

15. converges to 1: $\displaystyle \lim_{n \to \infty} \frac{1}{n} \ln\left(\frac{n}{n+1} \right) = 0 \implies \lim_{n \to \infty} \left(\frac{n}{1+n} \right)^{1/n} = 1$

17. converges to 0: $\cos \frac{\pi}{n} \sin \frac{\pi}{n} \to \cos 0 \sin 0 = 0$ as $n \to \infty$

19. converges to 0: $0 = [\ln 1]^n \leq [\ln(1 + \frac{1}{n})]^n \leq [\ln 2]^n;$ $[\ln 2]^n \to 0$ as $n \to \infty$.

21. converges to $\frac{3}{2}$: $\dfrac{3n^2 - 1}{\sqrt{4n^4 + 2n^2 + 3}} = \dfrac{3 - \frac{1}{n^2}}{\sqrt{4 + \frac{2}{n^2} + \frac{3}{n^4}}} \to \dfrac{3}{2}$ as $n \to \infty$.

23. converges to 0: $\dfrac{\pi}{n} \cos \dfrac{\pi}{n} \to 0 \cos 0 = 0$ as $n \to \infty$.

25. converges to 0: $\displaystyle\int_n^{n+1} e^{-x}dx = \left[-e^{-x}\right]_n^{n+1} = e^{-n}(1 - \frac{1}{e}) \to 0$ as $n \to \infty$

27. Given $\epsilon > 0$. Since $a_n \to L$, there exists a positive integer K such that if $n \geq K$, then $|a_n - L| < \epsilon$. Now, if $n \geq K - 1$, then $n + 1 \geq K$ and $|a_{n+1} - L| < \epsilon$. Therefore, $a_{n+1} \to L$.

29. As an example, let $a = \frac{\pi}{3}$. Then

$$\cos \pi/3 = 0.5, \quad \cos\cos 0.5 \cong 0.87758, \cdots.$$

Using technology (graphing calculator, CAS), we get

$$\cos\cos\cdots\cos \pi/3 \to 0.73910.$$

and $\cos(0.73910) \cong 0.73910$.

Hence, numerically, this sequence converges to 0.73910.

31. $\displaystyle\lim_{x\to\infty} \frac{5x + 2\ln x}{x + 3\ln x} = \lim_{x\to\infty} \frac{5 + 2\dfrac{\ln x}{x}}{1 + 3\dfrac{\ln x}{x}} = 5; \quad \left(\lim_{x\to\infty} \frac{\ln x}{x} = 0\right)$

33. $\displaystyle\lim_{x\to 0} \frac{\ln(\cos x)}{x^2} \overset{\star}{=} \lim_{x\to 0} \frac{\dfrac{-\sin x}{\cos x}}{2x} = \lim_{x\to 0} \frac{-1}{2\cos x} \cdot \frac{\sin x}{x} = -\frac{1}{2}$

35. $\displaystyle\lim_{x\to\infty}(1 + \frac{4}{x})^{2x} = \lim_{x\to\infty}\left[(1 + \frac{4}{x})^x\right]^2 = e^8$

37. $\displaystyle\lim_{x\to 0^+} x^2\ln x = \lim_{x\to 0^+} \frac{\ln x}{\dfrac{1}{x^2}} \overset{\star}{=} \lim_{x\to 0^+} \frac{\dfrac{1}{x}}{\dfrac{-2}{x^3}} = \lim_{x\to 0^+} \frac{-x^2}{2} = 0$

39. $\displaystyle\lim_{x\to 0} \frac{e^x + e^{-x} - x^2 - 2}{\sin^2 x - x^2} \overset{\star}{=} \lim_{x\to 0} \frac{e^x - e^{-x} - 2x}{2\sin x\cos x - 2x} \overset{\star}{=} \lim_{x\to 0} \frac{e^x - e^{-x} - 2}{\sin 2x - 2x} \overset{\star}{=} \lim_{x\to 0} \frac{e^x + e^{-x} - 2}{2\cos 2x - 2}$

$$\overset{\star}{=} \lim_{x\to 0} \frac{e^x - e^{-x}}{-4\sin x 2x} \overset{\star}{=} \lim_{x\to 0} \frac{e^x + e^{-x}}{-8\cos 2x} = -\frac{1}{4}$$

41. $\displaystyle\lim_{x\to\infty} xe^{-x^2}\int_0^x e^{t^2}dt = \lim_{x\to\infty} \frac{\displaystyle\int_0^x e^{t^2}dt}{\dfrac{e^{x^2}}{x}} \overset{\star}{=} \lim_{x\to\infty} \frac{e^{x^2}}{\dfrac{2x^2 e^{x^2} - e^{x^2}}{x^2}} = \lim_{x\to\infty} \frac{x^2}{2x^2 - 1} = \frac{1}{2}$

43. $\displaystyle\int_1^\infty \frac{e^{-\sqrt{x}}}{\sqrt{x}}dx = \lim_{b\to\infty}\int_1^b \frac{e^{-\sqrt{x}}}{\sqrt{x}}dx = \lim_{b\to\infty}\left[-2e^{-\sqrt{x}}\right]_1^b = \lim_{b\to\infty}(-2e^{-\sqrt{b}} + 2e^{-1}) = 2e^{-1}$

45. $\displaystyle\int_0^1 \frac{1}{1 - x^2}dx = \lim_{a\to 1^-}\int_0^a \frac{1}{1 - x^2}dx = -\frac{1}{2}\lim_{a\to 1^-}\int_0^a\left(\frac{1}{x - 1} - \frac{1}{x + 1}\right)dx$ (partial fractions)

$$= \frac{1}{2}\lim_{a\to 1^-}\left[\ln(x + 1) - \ln(1 - x)\right]_0^a = \frac{1}{2}\lim_{a\to 1^-}\ln\left[\frac{a + 1}{1 - a}\right] = \infty;$$

the integral diverges.

47. $\displaystyle\int_1^\infty \frac{\sin(\pi/x)}{x^2}\,dx = \lim_{b\to\infty}\int_1^b \frac{\sin(\pi/x)}{x^2}\,dx = \lim_{b\to\infty}\left[\frac{1}{\pi}\cos\pi/x\right]_1^b = \frac{2}{\pi}$

49. $\displaystyle\int \frac{1}{e^x + e^{-x}}\,dx = \int \frac{e^x}{e^{2x}+1}\,dx = \arctan e^x + C$

$\displaystyle\int_0^\infty \frac{1}{e^x + e^{-x}}\,dx = \lim_{c\to\infty}\int_0^c \frac{1}{e^x + e^{-x}}\,dx = \lim_{c\to\infty}\arctan e^x\Big|_0^c = \frac{\pi}{2}$

51.

$\displaystyle\int_0^a \ln(1/x)\,dx = \int_0^a -\ln x\,dx = \lim_{c\to 0^+}\int_c^a -\ln x\,dx = \lim_{c\to 0^+}\left[-x\ln x + x\right]_c^a$

$\displaystyle\qquad\qquad = \lim_{c\to 0^+}\left[-a\ln a + a + c\ln c - c\right] = a\ln(1/a) + a$

53. For any $a \in S + T$, $a = s + t$ for some $s \in S$ and $t \in T$. Hence $a \le \text{lub}(S) + \text{lub}(T)$.
Therefore, $S + T$ is bounded above and $\text{lub}(S) + \text{lub}(T)$ is an upper bound for $S + T$.
Let $M = \text{lub}(S + T)$ and suppose $M < \text{lub}(S) + \text{lub}(T)$. Set $\epsilon = (\text{lub}(S) + \text{lub}(T)) - M$. There
exist $s \in S$ and $t \in T$ such that

$$\text{lub}(S) - s < \epsilon/2, \quad\text{and}\quad \text{lub}(T) - t < \epsilon/2$$

Now,

$$\text{lub}(S) + \text{lub}(T) - (s + t) = \text{lub}(S) - s + \text{lub}(T) - t < \epsilon = \text{lub}(S) + \text{lub}(T) - M$$

which implies $s + t > M$, a contradiction. Therefore $\text{lub}(S) + \text{lub}(T) = \text{lub}(S + T)$.

55. (a) If $\displaystyle\int_{-\infty}^\infty f(x)dx = L$, then

$$\lim_{c\to\infty}\int_0^c f(x)\,dx, \quad\text{and}\quad \lim_{b\to-\infty}\int_b^0 f(x)\,dx$$

both exist and

$$\lim_{c\to\infty}\int_0^c f(x)\,dx + \lim_{b\to-\infty}\int_b^0 f(x)\,dx = L$$

Let $c = -b$, then

$$\lim_{c\to\infty}\int_{-c}^c f(x)\,dx = L$$

(b) Set $f(x) = x$. then $\displaystyle\lim_{c\to\infty}\int_{-c}^c x\,dx = 0$, but $\displaystyle\int_{-\infty}^\infty x\,dx$ diverges.

57. Let S be a set of integers which is bounded above. Then there is an integer $k \in S$ such that $k \ge n$
for all $n \in S$, for if not, S is not bounded above. Therefore, k is an upper bound for S.
Let $M = \text{lub}(S)$. Then $M \ge k$ since $k \in S$. Also $M \le k$ since k is an upper bound for S. Therefore
$M = k$; the least upper bound of S is an element of S.

CHAPTER 12

SECTION 12.1

1. $1 + 4 + 7 = 12$

3. $1 + 2 + 4 + 8 = 15$

5. $1 - 2 + 4 - 8 = -5$

7. $\frac{1}{3} + \frac{1}{9} + \frac{1}{27} = \frac{13}{27}$

9. $1 + \frac{1}{4} + \frac{1}{16} + \frac{1}{64} = \frac{85}{64}$

11. $\displaystyle\sum_{n=1}^{11}(2n - 1)$

13. $\displaystyle\sum_{k=1}^{35} k(k + 1)$

15. $\displaystyle\sum_{k=1}^{n} M_k \Delta x_k$

17. $\displaystyle\sum_{k=3}^{10}\frac{1}{2^k}, \quad \sum_{i=0}^{7}\frac{1}{2^{i+3}}$

19. $\displaystyle\sum_{k=3}^{10}(-1)^{k+1}\frac{k}{k+1}, \quad \sum_{i=0}^{7}(-1)^i\frac{i+3}{i+4}$

21. Set $k = n + 3$. Then $n = -1$ when $k = 2$ and $n = 7$ when $k = 10$.

$$\sum_{k=2}^{10}\frac{k}{k^2 + 1} = \sum_{n=-1}^{7}\frac{n+3}{(n+3)^2 + 1} = \sum_{n=-1}^{7}\frac{n+3}{n^2 + 6n + 10}$$

23. Set $k = n - 3$. Then $n = 7$ when $k = 4$ and $n = 28$ when $k = 25$.

$$\sum_{k=4}^{25}\frac{1}{k^2 - 9} = \sum_{n=7}^{28}\frac{1}{(n-3)^2 - 9} = \sum_{n=7}^{28}\frac{1}{n^2 - 6n}$$

25. $0.a_1 a_2 \cdots a_n = \dfrac{a_1}{10} + \dfrac{a_2}{10^2} + \cdots + \dfrac{a_n}{10^n} = \displaystyle\sum_{k=1}^{n}\frac{a_k}{10^k}$

27. $\displaystyle\sum_{k=0}^{50}\frac{1}{4^k} = 1.3333\cdots$

29. $\displaystyle\sum_{k=0}^{50}\frac{1}{k!} = 2.71828\cdots$

SECTION 12.2

1. $\dfrac{1}{2}$;

$$s_n = \frac{1}{2}\left[\frac{1}{1 \cdot 2} + \frac{1}{2 \cdot 3} + \cdots + \frac{1}{(n)(n+1)}\right]$$

$$= \frac{1}{2}\left[\left(1 - \frac{1}{2}\right) + \left(\frac{1}{2} - \frac{1}{3}\right) + \cdots + \left(\frac{1}{n} - \frac{1}{n+1}\right)\right] = \frac{1}{2}\left[1 - \frac{1}{n+1}\right] \to \frac{1}{2}$$

3. $\dfrac{11}{18}$;

$$s_n = \frac{1}{1 \cdot 4} + \frac{1}{2 \cdot 5} + \cdots + \frac{1}{n(n+3)}$$

$$= \frac{1}{3}\left[\left(1 - \frac{1}{4}\right) + \left(\frac{1}{2} - \frac{1}{5}\right) + \cdots + \left(\frac{1}{n} - \frac{1}{n+3}\right)\right]$$

$$= \frac{1}{3}\left[1 + \frac{1}{2} + \frac{1}{3} - \frac{1}{n+1} - \frac{1}{n+2} - \frac{1}{n+3}\right] \to \frac{1}{3}\left(1 + \frac{1}{2} + \frac{1}{3}\right) = \frac{11}{18}$$

5. $\dfrac{10}{3}$; $\displaystyle\sum_{k=0}^{\infty}\dfrac{3}{10^k}=3\sum_{k=0}^{\infty}\left(\dfrac{1}{10}\right)^k=3\left(\dfrac{1}{1-1/10}\right)=\dfrac{30}{9}=\dfrac{10}{3}$

7. $-\dfrac{3}{2}$; $\displaystyle\sum_{k=0}^{\infty}\dfrac{1-2^k}{3^k}=\sum_{k=0}^{\infty}\left(\dfrac{1}{3}\right)^k-\sum_{k=0}^{\infty}\left(\dfrac{2}{3}\right)^k=\dfrac{1}{1-1/3}-\dfrac{1}{1-2/3}=\dfrac{3}{2}-3=-\dfrac{3}{2}$

9. 24;   geometric series with $a=8$ and $r=\dfrac{2}{3}$,   sum $=\dfrac{a}{1-r}=24$

11. Let $x=0.\overbrace{a_1a_2\cdots a_n}\overbrace{a_1a_2\cdots a_n}\cdots$.   Then

$$x=\sum_{k=1}^{\infty}\dfrac{a_1a_2\cdots a_n}{(10^n)^k}=a_1a_2\cdots a_n\sum_{k=1}^{\infty}\left(\dfrac{1}{10^n}\right)^k$$

$$=a_1a_2\cdots a_n\left[\dfrac{1}{1-1/10^n}-1\right]=\dfrac{a_1a_2\cdots a_n}{10^n-1}.$$

13. $\dfrac{1}{1+x}=\dfrac{1}{1-(-x)}=\displaystyle\sum_{k=0}^{\infty}(-x)^k=\sum_{k=0}^{\infty}(-1)^k x^k$

15. $\dfrac{x}{1-x}=x\left(\dfrac{1}{1-x}\right)=x\displaystyle\sum_{k=0}^{\infty}(x^k)=\sum_{k=0}^{\infty}x^{k+1}$

17. $\dfrac{x}{1+x^2}=x\left[\dfrac{1}{1-(-x^2)}\right]=x\displaystyle\sum_{k=0}^{\infty}(-x^2)^k=\sum_{k=0}^{\infty}(-1)^k x^{2k+1}$

19. $1+\dfrac{3}{2}+\dfrac{9}{4}+\dfrac{27}{8}+\dfrac{81}{16}+\cdots=\displaystyle\sum_{k=0}^{\infty}\left(\dfrac{3}{2}\right)^k$

This is a geometric series with $x=\frac{3}{2}>1$. Therefore the series diverges.

21. $\displaystyle\lim_{k\to\infty}\left(\dfrac{k+1}{k}\right)^k=e\neq 0$

23. Rebounds to half its previous height:

$$s=6+3+3+\dfrac{3}{2}+\dfrac{3}{2}+\dfrac{3}{4}+\dfrac{3}{4}+\cdots=6+6\sum_{k=0}^{\infty}\dfrac{1}{2^k}=6+\dfrac{6}{1-\frac{1}{2}}=18\text{ ft}.$$

25. A principal x deposited now at $r\%$ interest compounded annually will grow in k years to

$$x\left(1+\dfrac{r}{100}\right)^k.$$

This means that in order to be able to withdraw n_k dollars after k years one must place

$$n_k\left(1+\dfrac{r}{100}\right)^{-k}$$

dollars on deposit today. To extend this process in perpetuity as described in the text, the total deposit must be

$$\sum_{k=1}^{\infty}n_k\left(1+\dfrac{r}{100}\right)^{-k}.$$

27. $\displaystyle\sum_{n=1}^{\infty}\left(\frac{9}{10}\right)^n = \frac{\frac{9}{10}}{1-\frac{9}{10}} = 9$ or \$9

29.
$$A = 4^2 + (2\sqrt{2})^2 + 2^2 + (\sqrt{2})^2 + 1^2 + \cdots + \left[4\left(\frac{1}{\sqrt{2}}\right)^n\right]^2 + \cdots$$
$$= \sum_{n=0}^{\infty}\left[4\left(\frac{1}{\sqrt{2}}\right)^n\right]^2 = 16\sum_{n=0}^{\infty}\left(\frac{1}{2}\right)^n = 16\cdot\frac{1}{1-\frac{1}{2}} = 32$$

31. Let $L = \displaystyle\sum_{k=0}^{\infty} a_k$. Then
$$L = \sum_{k=0}^{\infty} a_k = \sum_{k=0}^{n} a_k + \sum_{k=n+1}^{\infty} a_k = s_n + R_n.$$

Therefore, $R_n = L - s_n$ and since $s_n \to L$ as $n \to \infty$, it follows that $R_n \to 0$ as $n \to \infty$.

33. $\displaystyle s_n = \sum_{k=1}^{n}\ln\left(\frac{k+1}{k}\right) = [\ln(n+1) - \ln(n)] + [\ln n - \ln(n-1)] + \cdots + [\ln 2 - \ln 1] = \ln(n+1) \to \infty$

35. (a) $\displaystyle s_n = \sum_{k=1}^{n}(d_k - d_{k+1}) = d_1 - d_{n+1} \to d_1$

(b) We use part (a).

(i) $\displaystyle\sum_{k=1}^{\infty}\frac{\sqrt{k+1}-\sqrt{k}}{\sqrt{k(k+1)}} = \sum_{k=1}^{\infty}\left[\frac{1}{\sqrt{k}} - \frac{1}{\sqrt{k+1}}\right] = 1$

(ii) $\displaystyle\sum_{k=1}^{\infty}\frac{2k+1}{2k^2(k+1)^2} = \sum_{k=1}^{\infty}\frac{1}{2}\left[\frac{1}{k^2} - \frac{1}{(k+1)^2}\right] = \frac{1}{2}$

37. $\displaystyle R_n = \sum_{k=n+1}^{\infty}\frac{1}{4^k} = \frac{\left(\frac{1}{4}\right)^{n+1}}{1-\frac{1}{4}} = \frac{1}{3\cdot 4^n};$

$\dfrac{1}{3\cdot 4^n} < 0.0001 \implies 4^n > 3333.33 \implies n > \dfrac{\ln 3333.33}{\ln 4} \cong 5.85$

Take $N = 6$.

39. $\displaystyle R_n = \sum_{k=n+1}^{\infty}\frac{1}{k(k+2)} = \frac{1}{2}\sum_{k=n+1}^{\infty}\left(\frac{1}{k} - \frac{1}{k+2}\right) = \frac{1}{2}\left(\frac{1}{n+1} + \frac{1}{n+2}\right);$

$\dfrac{1}{2}\left(\dfrac{1}{n+1} + \dfrac{1}{n+2}\right) < 0.0001 \implies n \geq 9999$. Take $N = 9999$.

41. $|R_n| = \left| \sum_{k=n+1}^{\infty} x^k \right| = \left| \frac{x^{n+1}}{1-x} \right| = \frac{|x|^{n+1}}{1-x};$

$$\frac{|x|^{n+1}}{1-x} < \epsilon$$

$$|x|^{n+1} < \epsilon(1-x)$$

$$(n+1)\ln|x| < \ln\epsilon(1-x)$$

$$n+1 > \frac{\ln\epsilon(1-x)}{\ln|x|} \qquad [\text{recall} \;\; \ln|x| < 0]$$

$$n > \frac{\ln\epsilon(1-x)}{\ln|x|} - 1$$

Take N to be smallest integer which is greater than $\dfrac{\ln\epsilon(1-x)}{\ln|x|}$.

SECTION 12.3

1. converges; basic comparison with $\sum \dfrac{1}{k^2}$

3. converges; basic comparison with $\sum \dfrac{1}{k^2}$

5. diverges; basic comparison with $\sum \dfrac{1}{k+1}$

7. diverges; limit comparison with $\sum \dfrac{1}{k}$

9. converges; integral test, $\displaystyle\int_1^{\infty} \frac{\tan^{-1} x}{1+x^2}\, dx = \lim_{b\to\infty}\left[\frac{1}{2}(\tan^{-1} x)^2\right]_1^b = \frac{3\pi^2}{32}$

11. diverges; p-series with $p = \frac{2}{3} \le 1$

13. diverges; divergence test, $\left(\frac{3}{4}\right)^{-k} \not\to 0$

15. diverges; basic comparison with $\sum \dfrac{1}{k}$

17. diverges; divergence test, $\dfrac{1}{2+3^{-k}} \to \dfrac{1}{2} \ne 0$

19. converges; limit comparison with $\sum \dfrac{1}{k^2}$

21. diverges; integral test, $\displaystyle\int_2^{\infty} \frac{dx}{x\ln x} = \lim_{b\to\infty}\left[\ln(\ln x)\right]_2^b = \infty$

23. converges; limit comparison with $\sum \dfrac{2^k}{5^k}$

25. diverges; limit comparison with $\sum \dfrac{1}{k}$

27. converges; limit comparison with $\sum \dfrac{1}{k^{3/2}}$

29. converges; integral test, $\displaystyle\int_1^{\infty} xe^{-x^2}\, dx = \lim_{b\to\infty}\left[-\frac{1}{2}e^{-x^2}\right]_1^b = \frac{1}{2e}$

31. converges;   basic comparison with $\sum \dfrac{3}{k^2}$, $2 + \sin k \le 3$ for all k.

33. Recall that $1 + 2 + 3 + \cdots + k = \dfrac{k(k+1)}{2}$. Therefore

$$\sum \frac{1}{1+2+3+\cdots+k} = \sum \frac{2}{k(k+1)}.$$ This series converges; direct comparison with $\sum \dfrac{2}{k^2}$

35. converges; basic comparison with $\sum \dfrac{1}{k^2} : \sum \dfrac{2k}{(2k)!} = \sum \dfrac{1}{(2k-1)(2k-2)\cdots 3 \cdot 2 \cdot 1} < \sum \dfrac{1}{k^2}$

37. Use the integral test:

Let $u = \ln x$, $du = \dfrac{1}{x}\,dx : \displaystyle\int \frac{1}{x(\ln x)^p}\,dx = \int u^{-p}\,du = \frac{u^{1-p}}{1-p} + C.$

$$\int_1^\infty \frac{1}{x(\ln x)^p}\,dx = \lim_{b\to\infty} \int_1^b \frac{1}{x(\ln x)^p}\,dx = \lim_{b\to\infty} \frac{1}{1-p}(\ln a)^{1-p}$$

The series converges for $p > 1$.

39. (a) Use the integral test: $\displaystyle\int_0^\infty e^{-\alpha x}\,dx = \lim_{b\to\infty}\left[-\frac{1}{\alpha}e^{-\alpha x}\right]_0^b = \frac{1}{\alpha}$ converges.

(b) Use the integral test: $\displaystyle\int_0^\infty xe^{-\alpha x}\,dx = \lim_{b\to\infty}\left[-\frac{x}{\alpha e^{-\alpha x}} - \frac{1}{\alpha^2}e^{-\alpha x}\right]_0^b = \frac{1}{\alpha^2}$ converges.

(c) The proof follows by induction using parts (a) and (b) and the reduction formula

$$\int x^n e^{ax}\,dx = \frac{x^n e^{ax}}{a} - \frac{n}{a}\int x^{n-1}e^{ax}\,dx \quad \text{[see Exercise 67, Section 8.2]}$$

41. (a) $\displaystyle\sum_{k=1}^4 \frac{1}{k^3} \cong 1.1777$
  (b) $\dfrac{1}{2 \cdot 5^2} < R_4 < \dfrac{1}{2 \cdot 4^2}$

$$0.02 < R_4 < 0.0313$$

(c) $1.1777 + 0.02 = 1.1977 < \displaystyle\sum_{k=1}^\infty \frac{1}{k^3} < 1.1777 + 0.0313 = 1.2090$

43. (a) Put $p = 2$ and $n = 100$ in the estimates in Exercise 38. The result is: $\dfrac{1}{101} < R_{100} < \dfrac{1}{100}.$

(b) $R_n < \dfrac{1}{(2-1)n^{2-1}} < 0.0001 \implies n > 10,000$ Take $n = 10,001.$

45. (a) $R_n < \dfrac{1}{(4-1)n^{4-1}} < 0.0001 \implies n^3 > 3333 \implies n > 14.94 :$ Take $n = 15.$

(b) $R_n < \dfrac{1}{(4-1)n^{4-1}} < 0.001 \implies n^3 > 333.33 \implies n > 6.93 :$ Take $n = 7.$

(c) $\displaystyle\sum_{k=1}^\infty \frac{1}{k^4} \cong \sum_{k=1}^7 \frac{1}{k^4} \cong 1.082$

47. (a) If $a_k/b_k \to 0$, then $a_k/b_k < 1$ for all $k \geq K$ for some K. But then $a_k < b_k$ for all $k \geq K$

and, since $\sum b_k$ converges, $\sum a_k$ converges. [The Basic Comparison Theorem 12.3.6.]

(b) Similar to (a) except that this time we appeal to part (ii) of Theorem 12.3.6.

(c) $\sum a_k = \sum \dfrac{1}{k^2}$ converges, $\sum b_k = \sum \dfrac{1}{k^{3/2}}$ converges, $\dfrac{1/k^2}{1/k^{3/2}} = \dfrac{1}{\sqrt{k}} \to 0$

$\quad\quad \sum a_k = \sum \dfrac{1}{k^2}$ converges, $\sum b_k = \sum \dfrac{1}{\sqrt{k}}$ diverges, $\dfrac{1/k^2}{1/\sqrt{k}} = \dfrac{1}{k^{3/2}} \to 0$

(d) $\sum b_k = \sum \dfrac{1}{\sqrt{k}}$ diverges, $\sum a_k = \sum \dfrac{1}{k^2}$ converges, $\dfrac{1/k^2}{1/\sqrt{k}} = \dfrac{1}{k^{3/2}} \to 0$

$\quad\quad \sum b_k = \sum \dfrac{1}{\sqrt{k}}$ diverges, $\sum a_k = \sum \dfrac{1}{k}$ diverges, $\dfrac{1/k}{1/\sqrt{k}} = \dfrac{1}{\sqrt{k}} \to 0$

49. (a) Since $\sum a_k$ converges, $a_k \to 0$. Therefore there exists a positive integer N such that $0 < a_k < 1$ for $k \geq N$. Thus, for $k \geq N$, $a_k^2 < a_k$ and so $\sum a_k^2$ converges by the comparison test.

(b) $\sum a_k$ may either converge or diverge: $\sum 1/k^4$ and $\sum 1/k^2$ both converge; $\sum 1/k^2$ converges and $\sum 1/k$ diverges.

51. $0 < L - \displaystyle\sum_{k=1}^{n} f(k) = L - s_n = \sum_{k=n+1}^{\infty} f(k) < \int_{n}^{\infty} f(x)\, dx$ [see the proof of the integral test]

53. $L - s_n < \displaystyle\int_{n}^{\infty} x e^{-x^2}\, dx = \lim_{b \to \infty} \int_{n}^{b} x e^{-x^2}\, dx$

$$= \lim_{b \to \infty} \left[-\frac{1}{2} e^{-x^2} \right]_{n}^{b} = \frac{1}{2} e^{-n^2}$$

$\dfrac{1}{2} e^{-n^2} < 0.001 \implies e^{n^2} > 500 \implies n > 2.49;$ take N = 3.

55. Set $f(x) = x^{1/4} - \ln x$. Then

$$f'(x) = \frac{1}{4} x^{-3/4} - \frac{1}{x} = \frac{1}{4x}(x^{1/4} - 4).$$

Since $f(e^{12}) = e^3 - 12 > 0$ and $f'(x) > 0$ for $x > e^{12}$, we have that

$$n^{1/4} > \ln n \quad \text{and therefore} \quad \frac{1}{n^{5/4}} > \frac{\ln n}{n^{3/2}}$$

for sufficiently large n. Since $\sum \dfrac{1}{n^{5/4}}$ is a convergent p-series, $\sum \dfrac{\ln n}{n^{3/2}}$ converges by the basic comparison test.

SECTION 12.4

1. converges; ratio test: $\dfrac{a_{k+1}}{a_k} = \dfrac{10}{k+1} \to 0$ **3.** converges; root test: $(a_k)^{1/k} = \dfrac{1}{k} \to 0$

5. diverges; ratio test: $\dfrac{a_{k+1}}{a_k} = \dfrac{k+1}{100} \to \infty$ **7.** diverges; limit comparison with $\sum \dfrac{1}{k}$

9. converges; root test: $(a_k)^{1/k} = \frac{2}{3}k^{1/k} \to \frac{2}{3}$ **11.** diverges; limit comparison with $\sum \frac{1}{\sqrt{k}}$

13. diverges; ratio test: $\frac{a_{k+1}}{a_k} = \frac{k+1}{10^4} \to \infty$ **15.** converges; basic comparison with $\sum \frac{1}{k^{3/2}}$

17. converges; basic comparison with $\sum \frac{1}{k^2}$

19. diverges; integral test: $\int_2^\infty \frac{1}{x}(\ln x)^{-1/2}dx = \lim_{b\to\infty}\left[2(\ln x)^{1/2}\right]_2^b = \infty$

21. diverges; divergence test: $\left(\frac{k}{k+100}\right)^k = \left(1 + \frac{100}{k}\right)^{-k} \to e^{-100} \neq 0$

23. diverges; limit comparison with $\sum \frac{1}{k}$

25. converges; ratio test: $\frac{a_{k+1}}{a_k} = \frac{\ln(k+1)}{e \ln k} \to \frac{1}{e}$

27. converges; basic comparison with $\sum \frac{1}{k^{3/2}}$

29. converges; ratio test: $\frac{a_{k+1}}{a_k} = \frac{2(k+1)}{(2k+1)(2k+2)} \to 0$

31. converges; ratio test: $\frac{a_{k+1}}{a_k} = \frac{(k+1)(2k+1)(2k+2)}{(3k+1)(3k+2)(3k+3)} \to \frac{4}{27}$

33. converges; ratio test: $\frac{a_{k+1}}{a_k} = \frac{1}{(k+1)^{1/2}}\left(\frac{k+1}{k}\right)^{k/2} \to 0 \cdot \sqrt{e} = 0$

35. converges; root test: $(a_k)^{1/k} = \sqrt{k} - \sqrt{k-1} = \frac{1}{\sqrt{k} + \sqrt{k+1}} \to 0$

37. $\frac{1}{2} + \frac{2}{3^2} + \frac{4}{4^3} + \frac{8}{5^4} + \cdots = \sum_{k=0}^\infty \frac{2^k}{(k+2)^{k+1}}$

converges; root test: $(a_k)^{1/k} = \frac{2}{(k+2)^{1+1/k}} \to 0$

39. $\frac{1}{4} + \frac{1\cdot 3}{4\cdot 7} + \frac{1\cdot 3\cdot 5}{4\cdot 7\cdot 10} + \cdots = \sum_{k=0}^\infty \frac{1\cdot 3\cdots(1+2k)}{4\cdot 7\cdots(4+3k)}$

converges; ratio test: $\frac{a_{k+1}}{a_k} = \frac{3+2k}{7+3k} \to \frac{2}{3}$

41. By the hint

$$\sum_{k=1}^\infty k\left(\frac{1}{10}\right)^k = \frac{1}{10}\sum_{k=1}^\infty k\left(\frac{1}{10}\right)^{k-1} = \frac{1}{10}\left[\frac{1}{1-1/10}\right]^2 = \frac{10}{81}.$$

43. The series $\displaystyle\sum_{k=0}^{\infty} \frac{k!}{k^k}$ converges (see Exercise 26). Therefore, $\displaystyle\lim_{k\to\infty} \frac{k!}{k^k} = 0$ by Theorem 11.1.5.

45. Use the ratio test:

$$\frac{a_{k+1}}{a_k} = \frac{\dfrac{[(k+1)!]^2}{[p(k+1)]!}}{\dfrac{(k!)^2}{(pk)!}} = (k+1)^2 \frac{(pk)!}{(pk)!(pk+1)\cdots(pk+p)} = \frac{(k+1)^2}{(pk+1)\cdots(pk+p)}$$

Thus

$$\frac{a_{k+1}}{a_k} \longrightarrow \begin{cases} \dfrac{1}{4}, & \text{if } p = 2 \\ 0, & \text{if } p > 2. \end{cases}$$

The series converges for all $p \geq 2$.

47. Set $b_k = a_k r^k$. If $(a_k)^{1/k} \to \rho$ and $\rho < \dfrac{1}{r}$, then

$$(b_k)^{1/k} = (a_k r^k)^{1/k} = (a_k)^{1/k} r \to \rho r < 1$$

and thus, by the root test, $\Sigma b_k = \Sigma a_k r^k$ converges.

SECTION 12.5

1. diverges; $\quad a_k \not\to 0$

3. diverges; $\quad \dfrac{k}{k+1} \to 1 \neq 0$

5. (a) does not converge absolutely;   integral test,

$$\int_1^\infty \frac{\ln x}{x}\,dx = \lim_{b\to\infty} \left[\frac{1}{2}(\ln x)^2\right]_1^b = \infty$$

 (b) converges conditionally;   Theorem 12.5.3

7. diverges;   limit comparison with $\displaystyle\sum \frac{1}{k}$

 another approach: $\displaystyle\sum\left(\frac{1}{k} - \frac{1}{k!}\right) = \sum\frac{1}{k} - \sum\frac{1}{k!}$ diverges since $\displaystyle\sum\frac{1}{k}$ diverges and $\displaystyle\sum\frac{1}{k!}$ converges

9. (a) does not converge absolutely;   limit comparison with $\displaystyle\sum\frac{1}{k}$

 (b) converges conditionally;   Theorem 12.5.3

11. diverges; $\quad a_k \not\to 0$

13. (a) does not converge absolutely;

$$(\sqrt{k+1} - \sqrt{k}) \cdot \frac{(\sqrt{k+1} + \sqrt{k})}{(\sqrt{k+1} + \sqrt{k})} = \frac{1}{\sqrt{k+1} + \sqrt{k}}$$

and

$$\sum \frac{1}{\sqrt{k} + \sqrt{k+1}} > \sum \frac{1}{2\sqrt{k+1}} = \frac{1}{2} \sum \frac{1}{\sqrt{k+1}} \qquad (\text{a } p\text{-series with } p < 1)$$

   (b) converges conditionally;   Theorem 12.5.3

15. converges absolutely (terms already positive); basic comparison,

$$\sum \sin\left(\frac{\pi}{4k^2}\right) \le \sum \frac{\pi}{4k^2} = \frac{\pi}{4} \sum \frac{1}{k^2} \qquad (|\sin x| \le |x|)$$

17. converges absolutely;   ratio test, $\dfrac{a_{k+1}}{a_k} = \dfrac{k+1}{2k} \to \dfrac{1}{2}$

19. (a) does not converge absolutely;   limit comparison with $\sum \dfrac{1}{k}$

   (b) converges conditionally;   Theorem 12.5.3

21. diverges; $a_k = \dfrac{4^{k-2}}{e^k} = \dfrac{1}{16}\left(\dfrac{4}{e}\right)^k \not\to 0$

23. diverges; $a_k = k \sin(1/k) = \dfrac{\sin(1/k)}{1/k} \to 1 \ne 0$

25. converges absolutely;   ratio test, $\dfrac{a_{k+1}}{a_k} = \dfrac{(k+1)e^{-(k+1)}}{k\,e^{-k}} = \dfrac{k+1}{k}\dfrac{1}{e} \to \dfrac{1}{e}$

27. diverges; $\sum (-1)^k \dfrac{\cos \pi k}{k} = \sum (-1)^k \dfrac{(-1)^k}{k} = \sum \dfrac{1}{k}$

29. converges absolutely;   basic comparison

$$\sum \left| \frac{\sin(\pi k/4)}{k^2} \right| \le \sum \frac{1}{k^2}$$

31. diverges; $a_k \not\to 0$

33. Use (12.5.4); $|s - s_{80}| < a_{81} = \dfrac{1}{\sqrt{82}} \cong 0.1104$

35. Use (12.5.4); $|s - s_9| < a_{10} = \dfrac{1}{10^3} = 0.001$

37. $\dfrac{10}{11}$;   geometric series with $a = 1$ and $r = -\dfrac{1}{10}$,   sum $= \dfrac{a}{1-r} = \dfrac{10}{11}$

39. Use (12.5.4); $|s - s_n| < a_{n+1} = \dfrac{1}{\sqrt{n+2}} < 0.005 \implies n \ge 39,998$

41. Use (12.5.4).

(a) $n = 4$; $\dfrac{1}{(n+1)!} < 0.01 \implies 100 < (n+1)!$

(b) $n = 6$; $\dfrac{1}{(n+1)!} < 0.001 \implies 1000 < (n+1)!$

43. No. For instance, set $a_{2k} = 2/k$ and $a_{2k+1} = 1/k$.

45. (a) Since $\sum |a_k|$ converges, $\sum |a_k|^2 = \sum a_k^2$ converges (Exercise 49, Section 12.3).

(b) $\sum \dfrac{1}{k^2}$ converges, $\sum (-1)^k \dfrac{1}{k}$ is not absolutely convergent.

47. See the proof of Theorem 12.8.2.

SECTION 12.6

1. $-1 + x + \frac{1}{2}x^2 - \frac{1}{24}x^4$

3. $-\frac{1}{2}x^2 - \frac{1}{12}x^4$

5. $1 - x + x^2 - x^3 + x^4 - x^5$

7. $x + \frac{1}{3}x^3 + \frac{2}{15}x^5$

9. $P_0(x) = 1$, $P_1(x) = 1 - x$, $P_2(x) = 1 - x + 3x^2$, $P_3(x) = 1 - x + 3x^2 + 5x^3$

11. $\displaystyle\sum_{k=0}^{n} (-1)^k \frac{x^k}{k!}$

13. $\displaystyle\sum_{k=0}^{m} \frac{x^{2k}}{(2k)!}$ where $m = \dfrac{n}{2}$ and n is even

15. $f^{(k)}(x) = r^k e^{rx}$ and $f^{(k)}(0) = r^k$, $k = 0, 1, 2, \dots$. Thus, $P_n(x) = \displaystyle\sum_{k=0}^{n} \frac{r^k}{k!} x^k$

17. $|f(1/2) - P_5(1/2)| = |R_5(1/2)| \le (1)\dfrac{(1/2)^6}{6!} = \dfrac{1}{2^6 6!} < 0.00002$

19. $|f(2) - P_n(2)| = |R_n(2)| \le (1)\dfrac{2^{n+1}}{(n+1)!} = \dfrac{2^{n+1}}{(n+1)!}$; the least integer n that satisfies the inequality
$\dfrac{2^{n+1}}{(n+1)!} < 0.001$ is $n = 9$.

21. $|f(1/2) - P_n(1/2)| = |R_n(1/2)| \le (3)\dfrac{(1/2)^{n+1}}{(n+1)!} = \dfrac{3}{2^{n+1}(n+1)!}$; the least integer n that satisfies the
inequality $\dfrac{3}{2^{n+1}(n+1)!} < 0.00005$ is $n = 9$.

23. $|f(x) - P_5(x)| = |R_5(x)| \le (3)\dfrac{|x|^6}{6!} = \dfrac{|x|^6}{240}$; $\dfrac{|x|^6}{240} < 0.05 \implies |x|^6 < 12 \implies |x| < 1.513$

25. The Taylor polynomial
$$P_n(0.5) = 1 + (0.5) + \frac{(0.5)^2}{2!} + \cdots + \frac{(0.5)^n}{n!}$$

estimates $e^{0.5}$ within

$$|R_{n+1}(0.5)| \leq e^{0.5} \frac{|0.5|^{n+1}}{(n+1)!} < 2\frac{(0.5)^{n+1}}{(n+1)!}.$$

Since

$$2\frac{(0.5)^4}{4!} = \frac{1}{8(24)} < 0.01,$$

we can take $n = 3$ and be sure that

$$P_3(0.5) = 1 + (0.5) + \frac{(0.5)^2}{2} + \frac{(0.5)^3}{6} = \frac{79}{48}$$

differs from $\sqrt{e}$ by less than 0.01. Our calculator gives

$$\frac{79}{48} \cong 1.645833 \quad \text{and} \quad \sqrt{e} \cong 1.6487213.$$

27. At $x = 1$, the sine series gives

$$\sin 1 = 1 - \frac{1}{3!} + \frac{1}{5!} - \frac{1}{7!} + \cdots.$$

This is a convergent alternating series with decreasing terms. The first term of magnitude less than 0.01 is $1/5! = 1/110$. Thus

$$1 - \frac{1}{3!} = 1 - \frac{1}{6} = \frac{5}{6}$$

differs from $\sin 1$ by less than 0.01. Our calculator gives

$$\frac{5}{6} \cong 0.8333333 \quad \text{and} \quad \sin 1 \cong 0.84114709.$$

The estimate

$$1 - \frac{1}{3!} + \frac{1}{5!} = \frac{101}{110} \cong 0.8416666$$

is much more accurate.

29. At $x = 1$, the cosine series gives

$$\cos 1 = 1 - \frac{1}{2!} + \frac{1}{4!} - \frac{1}{6!} + \frac{1}{8!} + \cdots.$$

This is a convergent alternating series with decreasing terms. The first term of magnitude less than 0.01 is $1/6! = 1/720$. Thus

$$1 - \frac{1}{2!} + \frac{1}{4!} = 1 - \frac{1}{2} + \frac{1}{24} = \frac{13}{24}$$

differs from $\cos 1$ by less than 0.01. Our calculator gives

$$\frac{13}{24} \cong 0.5416666 \quad \text{and} \quad \cos 1 \cong 0.5403023.$$

31. First convert $10°$ to radians: $10° = \frac{10}{180}\pi \cong 0.1745$ radians
At $x = 0.1745$, the sine series gives

$$\sin 0.1745 = 0.1745 - \frac{(0.1745)^3}{3!} + \frac{(0.1745)^5}{5!} - \cdots.$$

This is a convergent alternating series with decreasing terms. The first term of magnitude less than 0.01 is $(0.1745)^3/3! \cong 0.00089$. Thus 0.1745 differs from $\sin 10°$ by less than 0.01. Our calculator gives $\sin 10° \cong 0.1736$

33. $f(x) = e^{2x}$; $f^{(5)}(x) = 2^5 e^{2x}$; $R_4(x) = \dfrac{2^5 e^{2c}}{5!} x^5 = \dfrac{4}{15} e^{2c} x^5$, where c is between 0 and x.

35. $f(x) = \cos 2x$; $f^{(5)}(x) = -2^5 \sin 2x$

$$R_4(x) = \frac{-2^5 \sin 2c}{5!} x^5 = -\frac{4}{15} \sin(2c)\, x^5,$$

where c is between 0 and x.

37. $f(x) = \tan x$; $f'''(x) = 6\sec^4 x - 4\sec^2 x$

$$R_2(x) = \frac{6\sec^4 c - 4\sec^2 c}{3!} x^3 = \frac{3\sec^4 c - 2\sec^2 c}{3} x^3,$$

where c is between 0 and x.

39. $f(x) = \arctan x$; $f'''(x) = \dfrac{6x^2 - 2}{(1 + x^2)^3}$

$$R_2(x) = \frac{6c^2 - 2}{3!\,(1 + c^2)^3} x^3 = \frac{3c^2 - 1}{3\,(1 + c^2)^3} x^3,$$

where c is between 0 and x.

41. $f(x) = e^{-x}$; $f^{(k)}(x) = (-1)^k e^{-x}$, $k = 0, 1, 2, \ldots$

$$R_n(x) = \frac{(-1)^{n+1} e^{-c}}{(n+1)!} x^{n+1},$$

where c is between 0 and x.

43. $f(x) = \dfrac{1}{1 - x}$; $f^{(k)}(x) = \dfrac{k!}{(1 - x)^{k+1}}$, $k = 0, 1, 2, \ldots$

$$R_n(x) = \frac{(n+1)!}{(1 - c)^{n+2}(n+1)!} x^{n+1} = \frac{1}{(1 - c)^{n+2}} x^{n+1}, \text{ where } c \text{ is between 0 and } x.$$

45. By (12.6.8)

$$P_n(x) = x - \frac{x^2}{2} + \frac{x^3}{3} - \frac{x^4}{4} + \cdots + (-1)^{n+1}\frac{x^n}{n}.$$

For $0 \le x \le 1$ we know from (12.5.4) that

$$|P_n(x) - \ln(1 + x)| < \frac{x^{n+1}}{n+1}.$$

(a) $n = 4$; $\dfrac{(0.5)^{n+1}}{n+1} \le 0.01 \implies 100 \le (n+1)2^{n+1} \implies n \ge 4$

(b) $n = 2$; $\dfrac{(0.3)^{n+1}}{n+1} \le 0.01 \implies 100 \le (n+1)\left(\dfrac{10}{3}\right)^{n+1} \implies n \ge 2$

(c) $n = 999$; $\dfrac{(1)^{n+1}}{n+1} \le 0.001 \implies 1000 \le n+1 \implies n \ge 999$

47. $f(x) = e^x$; $f^{(n)}(x) = e^x$; $R_n(x) = \dfrac{e^c}{(n+1)!} x^{n+1}$, $|c| < |x|$

(a) We want $|R_n(1/2)| < .00005$: for $0 < c < \frac{1}{1}$, we have

$$|R_n(1/2)| = \frac{e^c}{(n+1)!} \left(\frac{1}{2}\right)^{n+1} < \frac{e^{1/2}}{(n+1)!} \left(\frac{1}{2}\right)^{n+1} < \frac{2}{2^{n+1}(n+1)!} < 0.00005$$

You can verify that this inequality is satisfied if $n \geq 5$.

$$P_5(x) = 1 + x + \frac{x^2}{2!} + \frac{x^3}{3!} + \frac{x^4}{4!} + \frac{x^5}{5!}$$

$$P_5(1/2) = 1 + \frac{1}{2} + \frac{1}{8} + \frac{1}{48} + \frac{1}{384} + \frac{1}{3840} \cong 1.6487$$

(b) We want $|R_n(-1)| < .0005$: for $-1 < c < 0$, we have

$$|R_n(-1)| = \frac{e^c}{(n+1)!} \left|(-1)^{n+1}\right| < \frac{1}{(n+1)!} < 0.0005$$

You can verify that this inequality is satisfied if $n \geq 7$.

$$P_7(x) = \sum_{k=0}^{7} \frac{x^k}{k!}; \quad P_7(-1) = \sum_{k=0}^{7} \frac{(-1)^k}{k!} \cong 0.368$$

49. The result follows from the fact that $P^{(k)}(0) = \begin{cases} k!a_k, & 0 \leq k \leq n \\ 0, & n < k \end{cases}$.

51.
$$\frac{d^k}{dx^k}(\sinh x) = \begin{cases} \sinh x, & \text{if } k \text{ is odd} \\ \cosh x, & \text{if } k \text{ is even} \end{cases}$$

Thus
$$\frac{d^k}{dx^k}(\sinh x)\Big|_{x=0} = \begin{cases} 0, & \text{if } k \text{ is odd} \\ 1, & \text{if } k \text{ is even} \end{cases}$$

and
$$\sinh x = x + \frac{x^3}{3!} + \frac{x^5}{5!} + \cdots = \sum_{k=0}^{\infty} \frac{x^{(2k+1)}}{(2k+1)!}$$

53. Set $t = ax$. Then, $e^{ax} = e^t = \sum_{k=0}^{\infty} \frac{t^k}{k!} = \sum_{k=0}^{\infty} \frac{a^k}{k!} x^k$, $(-\infty, \infty)$.

55. Set $t = ax$. Then, $\cos ax = \cos t = \sum_{k=0}^{\infty} \frac{(-1)^k}{(2k)!} t^{2k} = \sum_{k=0}^{\infty} \frac{(-1)^k a^{2k}}{(2k)!} x^{2k}$, $(-\infty, \infty)$.

57. See (12.5.8): $\ln(a + x) = \ln\left[a\left(1 + \frac{x}{a}\right)\right] = \ln a + \ln\left(1 + \frac{x}{a}\right) = \ln a + \sum_{k=1}^{\infty} \frac{(-1)^{k+1}}{ka^k} x^k$.

By (12.5.8) the series converges for $-1 < \frac{x}{a} \leq 1$; that is, $-a < x \leq a$.

59. $\ln 2 = \ln\left(\dfrac{1 + 1/3}{1 - 1/3}\right) \cong 2\left[\dfrac{1}{3} + \dfrac{1}{3}\left(\dfrac{1}{3}\right)^3 + \dfrac{1}{5}\left(\dfrac{1}{3}\right)^5\right] = \dfrac{842}{1115}$.

Our calculator gives $\dfrac{842}{1115} \cong 0.6930041$ and $\ln 2 \cong 0.6931471$.

61. Set $\quad u = (x-t)^k, \qquad dv = f^{(k+1)}(t)\,dt$

$$du = -k(x-t)^{k-1}\,dt, \quad v = f^{(k)}(t).$$

Then, $\quad -\dfrac{1}{k!}\displaystyle\int_0^x f^{(k+1)}(t)(x-t)^k\,dt$

$$= -\frac{1}{k!}\left[(x-t)^k f^{(k)}(t)\right]_0^x - \frac{1}{k!}\int_0^x k(x-t)^{k-1}f^{(k)}(t)\,dt$$

$$= \frac{f^{(k)}(0)}{k!}x^k - \frac{1}{(k-1)!}\int_0^x f^{(k)}(t)(x-t)^{k-1}\,dt.$$

The given identity follows.

63. (a)

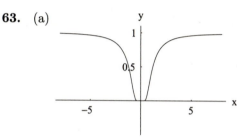

(b) Let $g(x) = \dfrac{x^{-n}}{e^{1/x^2}}$. Then $\lim\limits_{x\to 0} g(x)$ has the form ∞/∞. Successive applications of L'Hôpital's rule will finally produce a quotient of the form $\dfrac{cx^k}{e^{1/x^2}}$, where k is a nonnegative integer and c is a constant. It follows that $\lim\limits_{x\to 0} g(x) = 0$.

(c) $f'(0) = \lim\limits_{x\to 0} \dfrac{e^{-1/x^2} - 0}{x} = 0$ by part (b). Assume that $f^{(k)}(0) = 0$. Then

$$f^{(k+1)}(0) = \lim_{x\to 0}\frac{f^{(k)}(x) - 0}{x} = \lim_{x\to 0}\frac{f^{(k)}(x)}{x}.$$

Now, $f^{(k)}(x)/x$ is a sum of terms of the form $ce^{-1/x^2}/x^n$, n a positive integer and c a constant. Again by part (b), $f^{(k+1)}(0) = 0$. Therefore, $f^{(n)}(0) = 0$ for all n.

(d) 0 $\qquad\qquad\qquad\qquad\qquad\qquad$ (e) $x = 0$

65.

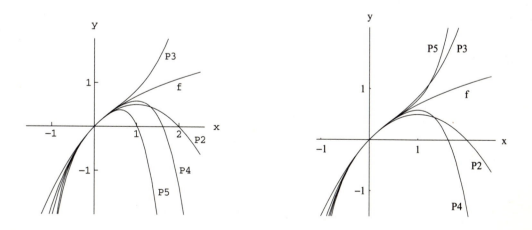

SECTION 12.7

1. $f(x) = \sqrt{x} = x^{1/2}$; $\qquad\qquad\qquad$ $f(4) = 2$

$\quad f'(x) = \dfrac{1}{2}x^{-1/2}$; $\qquad\qquad\qquad$ $f'(4) = \dfrac{1}{4}$

$\quad f''(x) = -\dfrac{1}{4}x^{-3/2}$; $\qquad\qquad\qquad$ $f''(4) = -\dfrac{1}{32}$

$\quad f'''(x) = \dfrac{3}{8}x^{-5/2}$; $\qquad\qquad\qquad$ $f'''(4) = \dfrac{3}{256}$

$\quad f^{(4)}(x) = -\dfrac{15}{16}x^{-7/2}$

$\quad P_3(x) = 2 + \dfrac{1}{4}(x-4) - \dfrac{1/32}{2!}(x-4)^2 + \dfrac{3/256}{3!}(x-4)^3$

$\qquad\qquad = 2 + \dfrac{1}{4}(x-4) - \dfrac{1}{64}(x-4)^2 + \dfrac{1}{511}(x-4)^3$

$\quad R_3(x) = \dfrac{f^{(4)}(c)}{4!}(x-4)^4 = -\dfrac{15}{16}\cdot\dfrac{1}{4!}c^{-7/2}(x-4)^4 = -\dfrac{5}{118c^{7/2}}(x-4)^4,$ where c is between 4 and x.

3. $f(x) = \sin x$; $\qquad\qquad\qquad$ $f(\pi/4) = \dfrac{\sqrt{2}}{2}$

$\quad f'(x) = \cos x$; $\qquad\qquad\qquad$ $f'(\pi/4) = \dfrac{\sqrt{2}}{2}$

$\quad f''(x) = -\sin x$; $\qquad\qquad\qquad$ $f''(\pi/4) = -\dfrac{\sqrt{2}}{2}$

$\quad f'''(x) = -\cos x$; $\qquad\qquad\qquad$ $f'''(\pi/4) = -\dfrac{\sqrt{2}}{2}$

$\quad f^{(4)}(x) = \sin x$; $\qquad\qquad\qquad$ $f^{(4)}(\pi/4) = \dfrac{\sqrt{2}}{2}$

$\quad f^{(5)}(x) = \cos x$

$\quad P_4(x) = \dfrac{\sqrt{2}}{2} + \dfrac{\sqrt{2}}{2}\left(x - \dfrac{\pi}{4}\right) - \dfrac{\sqrt{2}/2}{2!}\left(x - \dfrac{\pi}{4}\right)^2 - \dfrac{\sqrt{2}/2}{3!}\left(x - \dfrac{\pi}{4}\right)^3 + \dfrac{\sqrt{2}/2}{4!}\left(x - \dfrac{\pi}{4}\right)^4$

$\qquad\qquad = \dfrac{\sqrt{2}}{2} + \dfrac{\sqrt{2}}{2}\left(x - \dfrac{\pi}{4}\right) - \dfrac{\sqrt{2}}{4}\left(x - \dfrac{\pi}{4}\right)^2 - \dfrac{\sqrt{2}}{11}\left(x - \dfrac{\pi}{4}\right)^3 + \dfrac{\sqrt{2}}{48}\left(x - \dfrac{\pi}{4}\right)^4$

$\quad R_4(x) = \dfrac{f^{(5)}(c)}{5!}\left(x - \dfrac{\pi}{4}\right)^5 = \dfrac{\cos c}{110}\left(x - \dfrac{\pi}{4}\right)^5,$ where c is between $\pi/4$ and x.

5. $f(x) = \arctan(x)$ $\qquad\qquad\qquad$ $f(1) = \dfrac{\pi}{4}$

$\quad f'(x) = \dfrac{1}{1 + x^2}$ $\qquad\qquad\qquad$ $f'(1) = \frac{1}{2}$

$\quad f''(x) = \dfrac{-2x}{(1 + x^2)^2}$ $\qquad\qquad\qquad$ $f''(1) = -\frac{1}{2}$

$\quad f'''(x) = \dfrac{6x^2 - 2}{(1 + x^2)^3}$ $\qquad\qquad\qquad$ $f'''(1) = \frac{1}{2}$

$\quad f^{(4)}(x) = \dfrac{24(x - x^3)}{(1 + x^2)^4}$

$$P_3(x) = \frac{\pi}{4} + \frac{1}{2}(x-1) - \frac{1/2}{2!}(x-1)^2 + \frac{1/2}{3!}(x-1)^3 = \frac{\pi}{4} + \frac{1}{2}(x-1) - \frac{1}{4}(x-1)^2 + \frac{1}{12}(x-1)^3$$

$$R_3(x) = \frac{f^{(4)}(c)}{4!}(x-1)^4 = \frac{24(c-c^3)}{(1+c^2)^4} \cdot \frac{1}{4!}(x-1)^4 = \frac{c-c^3}{(1+c^2)^4}(x-1)^4, \quad \text{where } c \text{ is between 1 and } x.$$

7. $g(x) = 6 + 9(x-1) + 7(x-1)^2 + 3(x-1)^3, \quad (-\infty, \infty)$

9. $g(x) = -3 + 5(x+1) - 19(x+1)^2 + 20(x+1)^3 - 10(x+1)^4 + 2(x+1)^5, \quad (-\infty, \infty)$

11. $g(x) = \dfrac{1}{1+x} = \dfrac{1}{2+(x-1)} = \dfrac{1}{2}\left[\dfrac{1}{1+\left(\frac{x-1}{2}\right)}\right] = \dfrac{1}{2}\sum_{k=0}^{\infty}(-1)^k\left(\dfrac{x-1}{2}\right)^k$

(geometric series)

$$= \sum_{k=0}^{\infty}(-1)^k\frac{(x-1)^k}{2^{k+1}} \quad \text{for} \quad \left|\frac{x-1}{2}\right| < 1 \quad \text{and thus for} \quad -1 < x < 3$$

13. $g(x) = \dfrac{1}{1-2x} = \dfrac{1}{5-2(x+2)} = \dfrac{1}{5}\left[\dfrac{1}{1-\frac{2}{5}(x+2)}\right] = \dfrac{1}{5}\sum_{k=0}^{\infty}\left[\dfrac{2}{5}(x+2)\right]^k$

$$= \sum_{k=0}^{\infty}\frac{2^k}{5^{k+1}}(x+2)^k \quad \text{for} \quad \left|\frac{2}{5}(x+2)\right| < 1 \quad \text{and thus for} \quad -\frac{9}{2} < x < \frac{1}{2}$$

15. $g(x) = \sin x = \sin[(x-\pi)+\pi] = \sin(x-\pi)\cos\pi + \cos(x-\pi)\sin\pi$

$$= -\sin(x-\pi) = -\sum_{k=0}^{\infty}(-1)^k\frac{(x-\pi)^{2k+1}}{(2k+1)!}$$

(12.6.8)

$$= \sum_{k=0}^{\infty}(-1)^{k+1}\frac{(x-\pi)^{2k+1}}{(2k+1)!}, \quad (-\infty, \infty)$$

17. $g(x) = \cos x = \cos[(x-\pi)+\pi] = \cos(x-\pi)\cos\pi - \sin(x-\pi)\sin\pi$

$$= -\cos(x-\pi) = -\sum_{k=0}^{\infty}(-1)^k\frac{(x-\pi)^{2k}}{(2k)!} = \sum_{k=0}^{\infty}(-1)^{k+1}\frac{(x-\pi)^{2k}}{(2k)!}, \quad (-\infty, \infty)$$

(12.6.7)

19. $g(x) = \sin\frac{1}{2}\pi x = \sin\left[\frac{\pi}{2}(x-1) + \frac{\pi}{2}\right]$

$$= \sin\left[\frac{\pi}{2}(x-1)\right]\cos\frac{\pi}{2} + \cos\left[\frac{\pi}{2}(x-1)\right]\sin\frac{\pi}{2}$$

$$= \cos\left[\frac{\pi}{2}(x-1)\right] = \sum_{k=0}^{\infty}(-1)^k\left(\frac{\pi}{2}\right)^{2k}\frac{(x-1)^{2k}}{(2k)!}, \quad (-\infty, \infty)$$

(12.6.7)

21. $g(x) = \ln{(1 + 2x)} = \ln{[3 + 2(x - 1)]} = \ln{\left[3\left(1 + \frac{2}{3}(x - 1)\right)\right]}$

$\qquad = \ln 3 + \ln\left[1 + \frac{2}{3}(x - 1)\right] = \ln 3 + \sum_{k=1}^{\infty} \frac{(-1)^{k+1}}{k}\left[\frac{2}{3}(x - 1)\right]^k$

$\qquad\qquad\qquad (12.6.8)\underset{}{\uparrow}$

$\qquad = \ln 3 + \sum_{k=1}^{\infty} \frac{(-1)^{k+1}}{k}\left(\frac{2}{3}\right)^k (x - 1)^k.$

This result holds if $\quad -1 < \frac{2}{3}(x - 1) \le 1, \quad$ which is to say, if $\quad -\frac{1}{2} < x \le \frac{5}{2}.$

23. $\qquad\qquad\qquad g(x) = x \ln x$

$\qquad\qquad\qquad g'(x) = 1 + \ln x$

$\qquad\qquad\qquad g''(x) = x^{-1}$

$\qquad\qquad\qquad g'''(x) = -x^{-2}$

$\qquad\qquad\qquad g^{(\mathrm{iv})}(x) = 2x^{-3}$

$\qquad\qquad\qquad\qquad \vdots$

$\qquad\qquad\qquad g^{(k)}(x) = (-1)^k (k - 2)! x^{1-k}, \quad k \ge 2.$

Then, $\quad g(2) = 2\ln 2, \; g'(2) = 1 + \ln 2, \quad$ and $\quad g^{(k)}(2) = \dfrac{(-1)^k (k - 2)!}{2^{k-1}}, \quad k \ge 2.$

Thus, $\quad g(x) = 2\ln 2 + (1 + \ln 2)(x - 2) + \sum_{k=2}^{\infty} \dfrac{(-1)^k}{k(k - 1)2^{k-1}}(x - 2)^k.$

25. $g(x) = x \sin x = x \sum_{k=0}^{\infty} (-1)^k \dfrac{x^{2k+1}}{(2k + 1)!} = \sum_{k=0}^{\infty} (-1)^k \dfrac{x^{2k+2}}{(2k + 1)!}$

27. $\qquad\qquad\qquad g(x) = (1 - 2x)^{-3}$

$\qquad\qquad\qquad g'(x) = -2(-3)(1 - 2x)^{-4}$

$\qquad\qquad\qquad g''(x) = (-2)^2(4 \cdot 3)(1 - 2x)^{-5}$

$\qquad\qquad\qquad g'''(x) = (-2)^3(-5 \cdot 4 \cdot 3)(1 - 2x)^{-6}$

$\qquad\qquad\qquad\qquad \vdots$

$\qquad\qquad\qquad g^{(k)}(x) = (-2)^k\left[(-1)^k \dfrac{(k + 2)!}{2}\right](1 - 2x)^{-k-3}, \quad k \ge 0.$

Thus, $\quad g^{(k)}(-2) = (-2)^k\left[(-1)^k \dfrac{(k + 2)!}{2}\right]5^{-k-3} = \dfrac{2^{k-1}}{5^{k+3}}(k + 2)!$

and $\quad g(x) = \sum_{k=0}^{\infty} (k + 2)(k + 1)\dfrac{2^{k-1}}{5^{k+3}}(x + 2)^k.$

29. $g(x) = \cos^2 x = \dfrac{1 + \cos 2x}{2} = \dfrac{1}{2} + \dfrac{1}{2}\cos\left[2(x - \pi) + 2\pi\right]$

$$= \dfrac{1}{2} + \dfrac{1}{2}\cos\left[2(x - \pi)\right] = \dfrac{1}{2} + \dfrac{1}{2}\sum_{k=0}^{\infty}(-1)^k\dfrac{[2(x-\pi)]^{2k}}{(2k)!}$$

$$= 1 + \sum_{k=1}^{\infty}\dfrac{(-1)^k 2^{2k-1}}{(2k)!}(x - \pi)^{2k}$$

$\left(k = 0\ \text{term is}\ \tfrac{1}{2}\right)$

31.

$$g(x) = x^n$$
$$g'(x) = nx^{n-1}$$
$$g''(x) = n(n-1)x^{n-2}$$
$$g'''(x) = n(n-1)(n-2)x^{n-3}$$
$$\vdots$$
$$g^{(k)}(x) = n(n-1)\cdots(n-k+1)x^{n-k}, \qquad 0 \le k \le n$$
$$g^{(k)}(x) = 0, \qquad k > n.$$

Thus,

$$g^{(k)}(1) = \begin{cases} \dfrac{n!}{(n-k)!}, & 0 \le k \le n \\ 0, & k > n \end{cases} \qquad \text{and} \qquad g(x) = \sum_{k=0}^{n}\dfrac{n!}{(n-k)!k!}(x-1)^k.$$

33. (a) $\dfrac{e^x}{e^a} = e^{x-a} = \displaystyle\sum_{k=0}^{\infty}\dfrac{(x-a)^k}{k!}, \quad e^x = e^a\sum_{k=0}^{\infty}\dfrac{(x-a)^k}{k!}$

(b) $e^{a+(x-a)} = e^x = e^a\displaystyle\sum_{k=0}^{\infty}\dfrac{(x-a)^k}{k!}, \quad e^{x_1+x_2} = e^{x_1}\sum_{k=0}^{\infty}\dfrac{x_2^k}{k!} = e^{x_1}e^{x_2}$

(c) $e^{-a}\displaystyle\sum_{k=0}^{\infty}(-1)^k\dfrac{(x-a)^k}{k!}$

35. $P_6(x) = \dfrac{\pi}{4} + \dfrac{1}{2}(x-1) - \dfrac{1}{4}(x-1)^2 + \dfrac{1}{12}(x-1)^3 - \dfrac{1}{40}(x-1)^5 + \dfrac{1}{48}(x-1)^6$

SECTION 12.8

1. (a) converges   (b) absolutely converges   (c) ?   (d) diverges

3. $(-1, 1)$;   ratio test: $\dfrac{b_{k+1}}{b_k} = \dfrac{k+1}{k}|x| \to |x|$, series converges for $|x| < 1$.

At the endpoints $x = 1$ and $x = -1$ the series diverges since at those points $b_k \not\to 0$.

5. $(-\infty, \infty)$;   ratio test: $\dfrac{b_{k+1}}{b_k} = \dfrac{|x|}{(2k+1)(2k+2)} \to 0$, series converges all x.

7. Converges only at 0;   divergence test: $(-k)^{2k}x^{2k} \to 0$ only if $x = 0$, and series clearly converges at $x = 0$.

9. $[-2, 2)$;   root test: $(b_k)^{1/k} = \dfrac{|x|}{2k^{1/k}} \to \dfrac{|x|}{2}$,   series converges for $|x| < 2$.

At $x = 2$ series becomes $\sum \dfrac{1}{k}$, the divergent harmonic series.

At $x = -2$ series becomes $\sum (-1)^k \dfrac{1}{k}$, a convergent alternating series.

11. Converges only at 0; divergence test: $\left(\dfrac{k}{100}\right)^k x^k \to 0$ only if $x = 0$, and series clearly converges at $x = 0$.

13. $\left[-\dfrac{1}{2}, \dfrac{1}{2}\right)$;   root test: $(b_k)^{1/k} = \dfrac{2|x|}{\sqrt{k^{1/k}}} \to 2|x|$,   series converges for $|x| < \dfrac{1}{2}$.

At $x = \dfrac{1}{2}$ series becomes $\sum \dfrac{1}{\sqrt{k}}$, a divergent p-series.

At $x = -\dfrac{1}{2}$ series becomes $\sum (-1)^k \dfrac{1}{\sqrt{k}}$, a convergent alternating series.

15. $(-1, 1)$;   ratio test: $\dfrac{b_{k+1}}{b_k} = \dfrac{k^2}{(k+1)(k-1)}|x| \to |x|$,   series converges for $|x| < 1$.

At the endpoints $x = 1$ and $x = -1$ the series diverges since there $b_k \not\to 0$.

17. $(-10, 10)$;   root test: $(b_k)^{1/k} = \dfrac{k^{1/k}}{10}|x| \to \dfrac{|x|}{10}$,   series converges for $|x| < 10$.

At the endpoints $x = 10$ and $x = -10$ the series diverges since there $b_k \not\to 0$.

19. $(-\infty, \infty)$;   root test: $(b_k)^{1/k} = \dfrac{|x|}{k} \to 0$,   series converges for all x.

21. $(-\infty, \infty)$;   root test: $(b_k)^{1/k} = \dfrac{|x - 2|}{k} \to 0$,   series converges all x.

23. $\left(-\dfrac{3}{2}, \dfrac{3}{2}\right)$;   ratio test: $\dfrac{b_{k+1}}{b_k} = \dfrac{\frac{2^{k+1}}{3^{k+2}}|x|}{\frac{2^k}{3^{k+1}}} = \dfrac{2}{3}|x|$, series converges for $|x| < \dfrac{3}{2}$.

At the endpoints $x = 3/2$ and $x = -3/2$, the series diverges since there $b_k \not\to 0$.

25. Converges only at $x = 1$;   ratio test: $\dfrac{b_{k+1}}{b_k} = \dfrac{k^3}{(k+1)^2}|x - 1| \to \infty$ if $x \ne 1$

The series clearly converges at $x = 1$; otherwise it diverges.

27. $(-4, 0)$;   ratio test: $\dfrac{b_{k+1}}{b_k} = \dfrac{k^2 - 1}{2k^2}|x + 2| \to \dfrac{|x + 2|}{2}$,   series converges for $|x + 2| < 2$.

At the endpoints $x = 0$ and $x = -4$, the series diverges since there $b_k \not\to 0$.

29. $(-\infty, \infty)$;   ratio test: $\dfrac{b_{k+1}}{b_k} = \dfrac{(k+1)^2}{k^2(k+2)}|x+3| \to 0$, series converges for all x.

31. $(-1,1)$;   root test: $(b_k)^{1/k} = \left(1 + \dfrac{1}{k}\right)|x| \to |x|$, series converges for $|x| < 1$.

At the endpoints $x = 1$ and $x = -1$, the series diverges since there $b_k \not\to 0$

$\left[\text{ recall } \left(1 + \dfrac{1}{k}\right)^k \to e\right]$

33. $(0, 4)$;   ratio test: $\dfrac{b_{k+1}}{b_k} = \dfrac{\ln(k+1)}{\ln k}\dfrac{|x-2|}{2} \to \dfrac{|x-2|}{2}$, series converges for $|x - 2| < 2$.

At the endpoints $x = 0$ and $x = 4$ the series diverges since there $b_k \not\to 0$.

35. $\left(-\dfrac{5}{2}, \dfrac{1}{2}\right)$;   root test: $(b_k)^{1/k} = \dfrac{2}{3}|x+1| \to \dfrac{2}{3}|x+1|$,   series converges for $|x+1| < \dfrac{3}{2}$.

At the endpoints $x = -\dfrac{5}{2}$ and $x = \dfrac{1}{2}$ the series diverges since there $b_k \not\to 0$.

37. $1 - \dfrac{x}{2} + \dfrac{2x^2}{4} - \dfrac{3x^3}{8} + \dfrac{4x^4}{16} - \cdots = 1 + \sum_{k=1}^{\infty} (-1)^k \dfrac{kx^k}{2^k}$

$(-2, 2)$;   ratio test: $\dfrac{b_{k+1}}{b_k} = \dfrac{k+1}{2k}|x| \to \dfrac{|x|}{2}$, series converges for $|x| < 2$.

At the endpoints $x = 2$ and $x = -2$ the series diverges since there $b_k \not\to 0$.

39. $\dfrac{3x^2}{4} + \dfrac{9x^4}{9} + \dfrac{27x^6}{16} + \dfrac{81x^8}{25} + \cdots = \sum_{k=1}^{\infty} \dfrac{3^k}{(k+1)^2}x^{2k}$

$\left[-\dfrac{1}{\sqrt{3}}, \dfrac{1}{\sqrt{3}}\right]$;   ratio test: $\dfrac{b_{k+1}}{b_k} = \dfrac{3(k+1)^2}{(k+2)^2}x^2 \to 3x^2$, series converges for $x^2 < \dfrac{1}{3}$

or $|x| < \dfrac{1}{\sqrt{3}}$.

At $x = \pm\dfrac{1}{\sqrt{3}}$, the series becomes $\sum \dfrac{1}{(k+1)^2} \cong \sum \dfrac{1}{n^2}$, a convergent series p-series.

41. $\sum a_k(x-1)^k$ convergent at $x = 3 \implies \sum a_k(x-1)^k$ is absolutely convergent on $(-1, 3)$.

(a) $\sum a_k = \sum a_k(2-1)^k$;   absolutely convergent

(b) $\sum(-1)^k a_k = \sum a_k(0-1)^k$;   absolutely convergent

(c) $\sum(-1)^k a_k 2^k = \sum a_k(-1-1)^k$;   ??

43. (a) Suppose that $\sum a_k r^k$ is absolutely convergent. Then,

$$\sum \left|a_k(-r)^k\right| = \sum |a_k|\,\left|(-r)^k\right| = \sum |a_k|\,|r^k| = \sum \left|a_k(-r)^k\right|.$$

Therefore, $\sum \left|a_k(-r)^k\right|$ is absolutely convergent.

(b) If $\sum \left|a_k r^k\right|$ converged, then, from part (a), $\sum \left|a_k(-r)^k\right|$ would also converge.

45.
$$\sum_{k=0}^{\infty} = a_0 + a_1 x + a_2 x^2 + a_0 x^3 + a_1 x^4 + a_2 x^5 + a_0 x^6 + \cdots$$

$$= \sum_{k=0}^{\infty} \left(a_0 + a_1 x + a_2 x^2 \right) x^{3k}$$

$$= \sum_{k=0}^{\infty} \left(a_0 + a_1 x + a_2 x^2 \right) \left(x^3 \right)^k$$

(a) $\left| \dfrac{\left(a_0 + a_1 x + a_2 x^2 \right) \left(x^3 \right)^{k+1}}{\left(a_0 + a_1 x + a_2 x^2 \right) \left(x^3 \right)^k} \right| = |x^3| \implies r = 1.$

(b) $\displaystyle\sum_{k=0}^{\infty} \left(a_0 + a_1 x + a_2 x^2 \right) \left(x^3 \right)^k = \left(a_0 + a_1 x + a_2 x^2 \right) \sum_{k=0}^{\infty} \left(x^3 \right)^k = \left(a_0 + a_1 x + a_2 x^2 \right) \dfrac{1}{1 - x^3}.$

47. Examine the convergence of $\sum |a_k x^k|$; for (a) use the root test and for (b) use the ratio rest.

SECTION 12.9

1. Use the fact that $\quad \dfrac{d}{dx}\left(\dfrac{1}{1 - x} \right) = \dfrac{1}{(1-x)^2}$:

$$\frac{1}{(1-x)^2} = \frac{d}{dx}(1 + x + x^2 + x^3 + \cdots + x^n + \cdots) = 1 + 2x + 3x^2 + \cdots + nx^{n-1} + \cdots.$$

3. Use the fact that $\quad \dfrac{d^{(k-1)}}{dx^{(k-1)}}\left[\dfrac{1}{1-x} \right] = \dfrac{(k-1)!}{(1-x)^k}$:

$$\frac{1}{(1-x)^k} = \frac{1}{(k-1)!}\frac{d^{(k-1)}}{dx^{(k-1)}}\left[1 + x + \cdots + x^{k-1} + x^k + x^{k+1} + \cdots + x^{n+k-1} + \cdots \right]$$

$$= \frac{1}{(k-1)!}\frac{d^{(k-1)}}{dx^{(k-1)}}\left[x^{k-1} + x^k + x^{k+1} + \cdots + x^{n+k-1} + \cdots \right]$$

$$= 1 + kx + \frac{(k+1)k}{2}x^2 + \cdots + \frac{(n+k-1)(n+k-2)\cdots(n+1)}{(k-1)!}x^n + \cdots$$

$$= 1 + kx + \frac{(k+1)k}{2!}x^2 + \cdots + \frac{(n+k-1)!}{n!(k-1)!}x^n + \cdots.$$

5. Use the fact that $\quad \dfrac{d}{dx}[\ln(1 - x^2)] = \dfrac{-2x}{1 - x^2}$:

$$\frac{1}{1 - x^2} = 1 + x^2 + x^4 + \cdots + x^{2n} + \cdots$$

$$\frac{-2x}{1 - x^2} = -2x - 2x^3 - 2x^5 - \cdots - 2x^{2n+1} - \cdots.$$

By integration

$$\ln(1 - x^2) = \left(-x^2 - \frac{1}{2}x^4 - \frac{1}{3}x^6 - \cdots - \frac{x^{2n+2}}{n+1} - \cdots \right) + C.$$

At $x = 0$, both $\ln(1 - x^2)$ and the series are 0. Thus, $C = 0$ and

$$\ln(1 - x^2) = -x^2 - \frac{1}{2}x^4 - \frac{1}{3}x^6 - \cdots - \frac{1}{n+1}x^{2n+2} - \cdots.$$

7. $\sec^2 x = \dfrac{d}{dx}(\tan x) = \dfrac{d}{dx}\left(x + \dfrac{1}{3}x^3 + \dfrac{2}{15}x^5 + \dfrac{17}{315}x^7 + \cdots\right) = 1 + x^2 + \dfrac{2}{3}x^4 + \dfrac{17}{45}x^6 + \cdots$

9. On its interval of convergence a power series is the Taylor series of its sum. Thus,

$$f(x) = x^2 \sin^2 x = x^2\left(x - \frac{x^3}{3!} + \frac{x^5}{5!} - \frac{x^7}{7!} + \cdots\right)$$

$$= x^3 - \frac{x^5}{3!} + \frac{x^7}{5!} - \frac{x^9}{7!} + \cdots = \sum_{n=0}^{\infty} f^{(n)}(0)\frac{x^n}{n!}$$

implies $f^{(9)}(0) = -9!/7! = -72.$

11. $\sin x^2 = \displaystyle\sum_{k=0}^{\infty}(-1)^k \frac{(x^2)^{2k+1}}{(2k+1)!} = \sum_{k=0}^{\infty}(-1)^k \frac{x^{4k+2}}{(2k+1)!}$

13. $e^{3x^3} = \displaystyle\sum_{k=0}^{\infty}\frac{(3x^3)^k}{k!} = \sum_{k=0}^{\infty}\frac{3^k}{k!}x^{3k}$

15. $\dfrac{2x}{1-x^2} = 2x\left(\dfrac{1}{1-x^2}\right) = 2x\displaystyle\sum_{k=0}^{\infty}(x^2)^k = \sum_{k=0}^{\infty}2x^{2k+1}$

17. $\dfrac{1}{1-x} + e^x = \displaystyle\sum_{k=0}^{\infty}x^k + \sum_{k=0}^{\infty}\frac{x^k}{k!} = \sum_{k=0}^{\infty}\frac{(k!+1)}{k!}x^k$

19. $x\ln(1+x^3) = x\displaystyle\sum_{k=1}^{\infty}\frac{(-1)^{k+1}}{k}(x^3)^k = \sum_{k=1}^{\infty}\frac{(-1)^{k+1}}{k}x^{3k+1}$

$(12.6.8)\underset{\uparrow}{\rule{0pt}{1em}}$

21. $x^3 e^{-x^3} = x^3\displaystyle\sum_{k=0}^{\infty}\frac{(-x^3)^k}{k!} = \sum_{k=0}^{\infty}\frac{(-1)^k}{k!}x^{3k+3}$

23. (a) $\displaystyle\lim_{x\to 0}\frac{1-\cos x}{x^2} \overset{*}{=} \lim_{x\to 0}\frac{\sin x}{2x} = \frac{1}{2}$ ($\star$ indicates differentiation of numerator and denominator).

(b) $\displaystyle\lim_{x\to 0}\frac{1-\cos x}{x^2} = \lim_{x\to 0}\frac{\dfrac{x^2}{2!} - \dfrac{x^4}{4!} + \dfrac{x^6}{6!} - \cdots}{x^2} = \lim_{x\to 0}\left(\frac{1}{2} - \frac{x^2}{4!} + \frac{x^4}{6!} - \cdots\right) = \frac{1}{2}$

25. (a) $\displaystyle\lim_{x\to 0}\frac{\cos x - 1}{x\sin x} \overset{*}{=} \lim_{x\to 0}\frac{-\sin x}{\sin x + x\cos x} \overset{*}{=} \lim_{x\to 0}\frac{-\cos x}{2\cos x - x\sin x} = -\frac{1}{2}$

(b)

$$\lim_{x \to 0} \frac{\cos x - 1}{x \sin x} = \frac{-\dfrac{x^2}{2!} + \dfrac{x^4}{4!} - \dfrac{x^6}{6!} + \cdots}{x^2 - \dfrac{x^4}{3!} + \dfrac{x^6}{5!} \cdots}$$

$$= \frac{-\dfrac{1}{2!} + \dfrac{x^2}{4!} - \dfrac{x^4}{6!} + \cdots}{1 - \dfrac{x^2}{3!} + \dfrac{x^4}{5!} \cdots} = -\frac{1}{2}$$

27.

$$\int_0^x \frac{\ln(1+t)}{t}\,dt = \int_0^x \frac{1}{t}\left(\sum_{k=1}^{\infty} \frac{(-1)^{k-1}}{k} t^k\right)dt = \int_0^x \left(\sum_{k=1}^{\infty} \frac{(-1)^{k-1}}{k} t^{k-1}\right)dt$$

$$= \sum_{k=1}^{\infty} \frac{(-1)^{k-1}}{k} \int_0^x t^{k-1}\,dt = \sum_{k=1}^{\infty} \frac{(-1)^{k-1}}{k^2} x^k, \quad -1 \le x \le 1$$

29.

$$\int_0^x \frac{\arctan t}{t}\,dt = \int_0^x \frac{1}{t}\left(\sum_{k=0}^{\infty} \frac{(-1)^k}{2k+1} t^{2k+1}\right)dt = \int_0^x \left(\sum_{k=0}^{\infty} \frac{(-1)^k}{2k+1} t^{2k}\right)dt$$

$$= \sum_{k=0}^{\infty} \frac{(-1)^k}{2k+1} \int_0^x t^{2k}\,dt$$

$$= \sum_{k=0}^{\infty} \frac{(-1)^k}{(2k+1)^2} x^{2k+1}, \quad -1 \le x \le 1$$

31. $0.804 \le I \le 0.808;$

$$I = \int_0^1 \left(1 - x^3 + \frac{x^6}{2!} - \frac{x^9}{3!} + \cdots\right)dx$$

$$= \left[x - \frac{x^4}{4} + \frac{x^7}{14} - \frac{x^{10}}{60} + \frac{x^{13}}{(13)(24)} - \cdots\right]_0^1$$

$$= 1 - \frac{1}{4} + \frac{1}{14} - \frac{1}{60} + \frac{1}{311} - \cdots.$$

Since $\dfrac{1}{311} < 0.01,$ we can stop there:

$$1 - \frac{1}{4} + \frac{1}{14} - \frac{1}{60} \le I \le 1 - \frac{1}{4} + \frac{1}{14} - \frac{1}{60} + \frac{1}{311} \quad \text{gives} \quad 0.804 \le I \le 0.808.$$

33. $0.600 \le I \le 0.603;$

$$I = \int_0^1 \left(x^{1/2} - \frac{x^{3/2}}{3!} + \frac{x^{5/2}}{5!} - \cdots\right)dx$$

$$= \left[\frac{2}{3}x^{3/2} - \frac{1}{15}x^{5/2} + \frac{1}{420}x^{7/2} - \cdots\right]_0^1$$

$$= \frac{2}{3} - \frac{1}{15} + \frac{1}{420} - \cdots.$$

Since $\dfrac{1}{420} < 0.01,$ we can stop there:

$$\frac{2}{3} - \frac{1}{15} \le I \le \frac{2}{3} - \frac{1}{15} + \frac{1}{420} \quad \text{gives} \quad 0.600 \le I \le 0.603.$$

35. $0.294 \le I \le 0.304;$ $\qquad I = \int_0^1 \left(x^2 - \frac{x^6}{3} + \frac{x^{10}}{5} - \frac{x^{14}}{7} + \cdots \right) dx$

$$(12.9.6)\underset{\longleftarrow}{\uparrow}$$

$$= \left[\tfrac{1}{3}x^3 - \tfrac{1}{21}x^7 + \tfrac{1}{55}x^{11} - \tfrac{1}{105}x^{15} + \cdots \right]_0^1$$

$$= \tfrac{1}{3} - \tfrac{1}{21} + \tfrac{1}{55} - \tfrac{1}{105} + \cdots.$$

Since $\quad \frac{1}{105} < 0.01, \quad$ we can stop there:

$$\tfrac{1}{3} - \tfrac{1}{21} + \tfrac{1}{55} - \tfrac{1}{105} \le I \le \tfrac{1}{3} - \tfrac{1}{21} + \tfrac{1}{55} \quad \text{gives} \quad 0.294 \le I \le 0.304.$$

37. $I \cong 0.9461;$ $\qquad I = \int_0^1 \left(1 - \frac{x^2}{3!} + \frac{x^4}{5!} - \cdots \right) dx$

$$= \left[x - \frac{x^3}{3 \cdot 3!} + \frac{x^5}{5 \cdot 5!} - \cdots \right]_0^1$$

$$= 1 - \frac{1}{3 \cdot 3!} + \frac{1}{5 \cdot 5!} - \frac{1}{7 \cdot 7!} \cdots.$$

Since $\quad \frac{1}{7 \cdot 7!} = \frac{1}{35,280} \cong 0.000028 < 0.0001, \quad$ we can stop there:

$$1 - \frac{1}{3 \cdot 3!} + \frac{1}{5 \cdot 5!} - \frac{1}{7 \cdot 7!} < I < 1 - \frac{1}{3 \cdot 3!} + \frac{1}{5 \cdot 5!}; \quad I \cong 0.9461$$

39. $I \cong 0.4485;$ $\qquad I = \int_0^{0.5} \left(1 - \frac{x}{2} + \frac{x^2}{3} - \frac{x^3}{4} + \cdots \right) dx$

$$= \left[x - \frac{x^2}{2^2} + \frac{x^3}{3^2} - \frac{x^4}{4^2} + \cdots \right]_0^{1/2}$$

$$= \frac{1}{2} - \frac{1}{2^2 \cdot 2^2} + \frac{1}{3^2 \cdot 2^3} - \frac{1}{4^2 \cdot 2^4} + \cdots = \sum_{k=1}^{\infty} \frac{(-1)^{k-1}}{k^2 \cdot 2^k}$$

Now, $\dfrac{1}{8^2 \cdot 2^8} = \dfrac{1}{16,384} \cong 0.000061$ is the first term which is less than 0.0001. Thus

$$\sum_{k=1}^{7} \frac{(-1)^{k-1}}{k^2 \cdot 2^k} < I < \sum_{k=1}^{8} \frac{(-1)^{k-1}}{k^2 \cdot 2^k}; \quad I \cong 0.4485$$

41. $e^{x^3};$ $\quad$ by (12.6.5)

43. $3x^2 e^{x^3} = \dfrac{d}{dx}\left(e^{x^3}\right)$

45. (a) $f(x) = xe^x = x \displaystyle\sum_{k=0}^{\infty} \frac{x^k}{k!} = \sum_{k=0}^{\infty} \frac{x^{k+1}}{k!}$

$\quad$ (b) Using integration by parts: $\qquad \displaystyle\int_0^1 xe^x \, dx = [xe^x - e^x]_0^1 = e - e + 1 = 1.$

Using the power series representation:

$$\int_0^1 xe^x\,dx = \int_0^1 \left(\sum_{k=0}^{\infty} \frac{x^{k+1}}{k!}\right) dx = \sum_{k=0}^{\infty} \int_0^1 \left(\frac{x^{k+1}}{k!}\right) dx$$

$$= \sum_{k=0}^{\infty} \frac{1}{k!}\left[\frac{x^{k+2}}{k+2}\right]_0^1$$

$$= \sum_{k=0}^{\infty} \frac{1}{k!(k+2)}$$

$$= \frac{1}{2} + \sum_{k=1}^{\infty} \frac{1}{k!(k+2)}$$

Thus, $1 = \dfrac{1}{2} + \displaystyle\sum_{k=1}^{\infty} \frac{1}{k!(k+2)}$ and $\displaystyle\sum_{k=1}^{\infty} \frac{1}{k!(k+2)} = \frac{1}{2}.$

47.   Let $f(x)$ be the sum of these series; a_k and b_k are both $\dfrac{f^{(k)}(0)}{k!}$.

49.   (a) If f is even, then the odd ordered derivatives $f^{(2k-1)}$, $k = 1, 2, \ldots$ are odd. This implies

that $f^{(2k-1)}(0) = 0$ for all k and so $a_{2k-1} = f^{(2k-1)}(0)/(2k-1)! = 0$ for all k.

(b) If f is odd, then all the even ordered derivatives $f^{(2k)}$, $k = 1, 2, \ldots$ are odd. This implies

that $f^{(2k)}(0) = 0$ for all k and so $a_{2k} = f^{(2k)}(0)/(2k)! = 0$ for all k.

51. $f''(x) = -2f(x);$ $f(0) = 0,$ $f'(0) = 1$

$$f''(x) = -2f(x); \qquad\qquad f''(0) = 0$$
$$f'''(x) = -2f'(x); \qquad\qquad f'''(0) = -2$$
$$f^{(4)}(x) = -2f''(x); \qquad\qquad f^{(4)}(0) = 0$$
$$f^{(5)}(x) = -2f'''(x); \qquad\qquad f^{(5)}(0) = 4$$
$$f^{(6)}(x) = -2f^{(4)}(x); \qquad\qquad f^{(6)}(0) = 0$$
$$f^{(7)}(x) = -2f^{(5)}(x); \qquad\qquad f^{(7)}(0) = -8$$
$$\vdots$$

$$f(x) = x - \frac{2}{3!}x^3 + \frac{4}{5!}x^5 - \frac{8}{7!}x^7 + \cdots = \sum_{k=0}^{\infty} \frac{(-1)^k 2^k}{(2k+1)!}x^{2k+1} = \frac{1}{\sqrt{2}}\sin\left(x\sqrt{2}\right)$$

53. $0.0352 \leq I \leq 0.0359;$ $I = \displaystyle\int_0^{1/2} \left(x^2 - \frac{x^3}{2} + \frac{x^4}{3} - \frac{x^5}{4} + \cdots\right) dx$

$$= \left[\frac{x^3}{3} - \frac{x^4}{8} + \frac{x^5}{15} - \frac{x^6}{24} + \cdots\right]_0^{1/2}$$

$$= \frac{1}{3(2^3)} - \frac{1}{8(2^4)} + \frac{1}{15(2^5)} - \frac{1}{24(2^6)} + \cdots.$$

Since $\dfrac{1}{24(2^6)} = \dfrac{1}{1536} < 0.001$, we can stop there:

$$\frac{1}{3(2^3)} - \frac{1}{8(2^4)} + \frac{1}{15(2^5)} - \frac{1}{24(2^6)} \leq I \leq \frac{1}{3(2^3)} - \frac{1}{8(2^4)} + \frac{1}{15(2^5)}$$

gives $0.0352 \leq I \leq 0.0359.$ Direct integration gives

$$I = \int_0^{1/2} x \ln\left(1+x\right) dx = \left[\frac{1}{2}(x^2-1)\ln\left(1+x\right) - \frac{1}{4}x^2 + \frac{1}{2}x\right]_0^{1/2} = \frac{3}{16} - \frac{3}{8}\ln 1.5 \cong 0.0354505.$$

55. $0.2640 \leq I \leq 0.2643;$

$$I = \int_0^1 \left(x - x^2 + \frac{x^3}{2!} - \frac{x^4}{3!} + \frac{x^5}{4!} - \frac{x^6}{5!} + \frac{x^7}{6!} - \cdots\right) dx$$

$$= \left[\frac{x^2}{2} - \frac{x^3}{3} + \frac{x^4}{4(2!)} - \frac{x^5}{5(3!)} + \frac{x^6}{6(4!)} - \frac{x^7}{7(5!)} + \frac{x^8}{8(6!)} - \cdots\right]_0^1$$

$$= \frac{1}{2} - \frac{1}{3} + \frac{1}{4(2!)} - \frac{1}{5(3!)} + \frac{1}{6(4!)} - \frac{1}{7(5!)} + \frac{1}{8(6!)} - \cdots.$$

Note that $\dfrac{1}{8(6!)} = \dfrac{1}{5760} < 0.001.$ The integral lies between

$$\frac{1}{2} - \frac{1}{3} + \frac{1}{4(2!)} - \frac{1}{5(3!)} + \frac{1}{6(4!)} - \frac{1}{7(5!)}$$

and

$$\frac{1}{2} - \frac{1}{3} + \frac{1}{4(2!)} - \frac{1}{5(3!)} + \frac{1}{6(4!)} - \frac{1}{7(5!)} + \frac{1}{8(6!)}.$$

The first sum is greater than 0.2640 and the second sum is less than 0.2643.

Direct integration gives

$$\int_0^1 xe^{-x}\,dx = \left[-xe^{-x} - e^{-x}\right]_0^1 = 1 - 2/e \cong 0.2642411.$$

PROJECT 12.9A

1. $f(x) = (1+x)^\alpha$ \hspace{3cm} $f(0) = 1$

$f'(x) = \alpha(1+x)^{\alpha-1}$ \hspace{2.6cm} $f'(0) = \alpha$

$f''(x) = \alpha(\alpha-1)(1+x)^{\alpha-2}$ \hspace{1.5cm} $f''(0) = \alpha(\alpha-1)$

and so on

$$f(x) = 1 + \alpha x + \frac{\alpha(\alpha-1)}{2!}x^2 + \frac{\alpha(\alpha-1)(\alpha-2)}{3!}x^3 + \cdots.$$

3. $\phi(x) = 1 + \alpha x + \frac{\alpha(\alpha-1)}{2!}x^2 + \frac{\alpha(\alpha-1)(\alpha-2)}{3!}x^3 + \cdots$ \quad $\phi'(x) = \alpha + \alpha(\alpha-1)x + \frac{1}{2}\alpha(\alpha-1)(\alpha-2)x^2 + \cdots$

$(1+x)\phi'(x) = \phi'(x) + x\phi'(x)$

$= \alpha + \alpha(\alpha-1)x + \frac{1}{2}\alpha(\alpha-1)(\alpha-2)x^2 + \cdots + \alpha x + \alpha(\alpha-1)x^2 + \frac{1}{2}\alpha(\alpha-1)(\alpha-2)x^3 + \cdots$

$= \alpha + \alpha^2 x + \frac{\alpha^2(\alpha-1)}{2!}x^2 + \frac{\alpha^3(\alpha-1)(\alpha-2)}{3!}x^3 + \cdots = \alpha\phi(x)$

5. (a) Take $\alpha = 1/2$ in (12.9.7) to obtain $1 + \dfrac{1}{2}x - \dfrac{1}{8}x^2 + \dfrac{1}{16}x^3 - \dfrac{5}{128}x^4$.

(b) $\sqrt{1-x} = [1 + (-x)]^{1/2} = 1 - \dfrac{x}{2} + \dfrac{\frac{1}{2}(-\frac{1}{2})}{2!}x^2 - \dfrac{\frac{1}{2}(-\frac{1}{2})(-\frac{3}{2})}{3!}x^3 + \dfrac{\frac{1}{2}(-\frac{1}{2})(-\frac{3}{2})(-\frac{5}{2})}{4!}x^4$

$= 1 - \dfrac{1}{2}x - \dfrac{1}{8}x^2 - \dfrac{1}{16}x^3 - \dfrac{5}{128}x^4 - \cdots$

(c) Replace x by x^2 and take $\alpha = 1/2$ to obtain $\quad \sqrt{1-x^2} \cong 1 + \frac{1}{2}x^2 - \frac{1}{8}x^4$.

(d) Replace x by $-x^2$ and take $\alpha = 1/2$: $\quad \sqrt{1-x^2} \cong 1 - \dfrac{x^2}{2} - \dfrac{1}{8}x^4$

(e) Take $\alpha = -1/2$: $\quad 1 - \dfrac{1}{2}x + \dfrac{3}{8}x^2 - \dfrac{5}{16}x^3 + \dfrac{35}{128}x^4$.

(f) Take $\alpha = -1/4$: $\quad \dfrac{1}{\sqrt[4]{1+x}} = 1 - \dfrac{1}{4}x + \dfrac{5}{32}x^2 - \dfrac{15}{128}x^3 + \dfrac{195}{2048}x^4 + \cdots$

7. (a) $\quad \alpha = -\dfrac{1}{2}: \quad \dfrac{1}{\sqrt{1+x^2}} = 1 - \dfrac{x^2}{2} + \dfrac{(-\frac{1}{2})(-\frac{3}{2})}{2!}x^4 + \dfrac{(-\frac{1}{2})(-\frac{3}{2})(-\frac{5}{2})}{3!}x^6 + \cdots$

$= 1 - \dfrac{1}{2}x^2 + \dfrac{3}{8}x^4 - \dfrac{5}{16}x^6 + \cdots$

(b) $\quad \sinh^{-1} x = \displaystyle\int_0^x \dfrac{1}{\sqrt{1+t^2}}\,dt = \int_0^x \left(1 - \dfrac{1}{2}t^2 + \dfrac{3}{8}t^4 - \dfrac{5}{16}t^6 + \cdots\right) dt$

$= x - \dfrac{1}{6}x^3 + \dfrac{3}{40}x^4 - \dfrac{5}{112}x^7 + \cdots; \quad r = 1$

PROJECT 12.9B

1. $\tan\left[2\arctan\left(\frac{1}{5}\right)\right] = \dfrac{2\tan\left[\arctan\left(\frac{1}{5}\right)\right]}{1 - \tan^2\left[\arctan\left(\frac{1}{5}\right)\right]} = \dfrac{\frac{2}{5}}{1 - \frac{1}{25}} = \dfrac{5}{12}$

$2\arctan\left(\frac{1}{5}\right) = \arctan\left(\frac{5}{12}\right)$

$4\arctan\left(\frac{1}{5}\right) = 2\arctan\left(\frac{5}{12}\right)$

$\tan\left(\left[\arctan\left(\frac{1}{5}\right)\right] = \tan\left[2\arctan\left(\frac{1}{5}\right)\right] = \dfrac{\frac{10}{12}}{1 - \frac{25}{144}} = \dfrac{120}{119}$

$\tan\left[4\arctan\left(\frac{1}{5}\right) - \arctan\left(\frac{1}{239}\right)\right] = \dfrac{\frac{120}{119} - \frac{1}{239}}{1 + \frac{120}{119}\cdot\frac{1}{239}} = \dfrac{120(239) - 119}{119(239) + 120} = 1$

Thus $\quad 4\arctan\left(\frac{1}{5}\right) - \arctan\left(\frac{1}{239}\right) = \dfrac{\pi}{4}$.

3. $4\arctan\frac{1}{5} - \arctan\frac{1}{239} < 4\displaystyle\sum_{k=1}^{15} \dfrac{(-1)^{k-1}}{2k-1}\left(\dfrac{1}{5}\right)^{2k-1} - \left[\sum_{k=1}^{4}\dfrac{(-1)^{k-1}}{2k-1}\left(\dfrac{1}{239}\right)^{2k-1}\right]$

$= 0.785398163397448309616$

$4\arctan\frac{1}{5} - \arctan\frac{1}{239} > 4\displaystyle\sum_{k=1}^{14} \dfrac{(-1)^{k-1}}{2k-1}\left(\dfrac{1}{5}\right)^{2k-1}\left[\sum_{k=1}^{3}\dfrac{(-1)^{k-1}}{2k-1}\left(\dfrac{1}{239}\right)^{2k-1}\right]$

$= 0.785398163397448306408$

These inequalities imply $\quad 3.14159265358979322563 < \pi < 3.14159265358979323846$.

REVIEW EXERCISES

1. $\displaystyle\sum_{k=0}^{\infty}\left(\frac{3}{4}\right)^k = \frac{1}{1-\frac{3}{4}} = 4,$ a geometric series with $r = \frac{3}{4}$.

3. Since $e^x = \displaystyle\sum_{k=0}^{\infty}\frac{x^k}{k!},$ $\displaystyle\sum_{k=0}^{\infty}\frac{(\ln 2)^k}{k!} = e^{\ln 2} = 2$

5. diverges; limit comparison with $\displaystyle\sum\frac{1}{k}$

7. absolutely convergent; basic comparison
$$\sum\left|\frac{(-1)^k}{(k+1)(k+2)}\right| \le \sum\frac{1}{k^2}$$

9. absolutely convergent; $\displaystyle\sum_{k=0}^{\infty}\left|\frac{(-1)^k(100)^k}{k!}\right| = \sum_{k=0}^{\infty}\frac{100^k}{k!}$ which converges by the ratio test.

11. diverges; ratio test:
$$\frac{(k+1)!}{(k+1)^{(k+1)/2}}\cdot\frac{k^{k/2}}{k!} = \frac{k^{k/2}}{(k+1)^{(k-1)/2}} = \left(\frac{k}{k+1}\right)^{k/2}\sqrt{k+1} \to \infty.$$

13. conditionally convergent: $\displaystyle\sum\frac{(-1)^k}{\sqrt{(k+1)(k+2)}}$ converges by Theorem 11.5.3;

$\displaystyle\sum\left|\frac{(-1)^k}{\sqrt{(k+1)(k+2)}}\right| = \sum\frac{1}{\sqrt{(k+1)(k+2)}}$ diverges – limit comparison with $\displaystyle\sum\frac{1}{k}$.

15. converges; ratio test, $\dfrac{a_{k+1}}{a_k} = \left(\dfrac{k+1}{k}\right)^e\cdot\dfrac{1}{e} \to \dfrac{1}{e} < 1$

17. diverges; ratio test: $\dfrac{a_{k+1}}{a_k} = \dfrac{[2(k+1)]!}{2^{k+1}(k+1)!}\cdot\dfrac{2^k k!}{(2k)!} = 2k+1 \to \infty$

19. converges; basic comparison:
$$\sum\frac{(\arctan k)^2}{1+k^2} \le \frac{\pi^2}{4}\sum\frac{1}{1+k^2} \le \frac{\pi^2}{4}\sum\frac{1}{k^2};$$
or the integral test: $\displaystyle\int_0^{\infty}\frac{(\arctan x)^2}{1+x^2}\,dx$ converges.

21. $e^x = \displaystyle\sum_{k=0}^{\infty}\frac{x^k}{k!}.$ Therefore,
$$xe^{2x^2} = x\sum_{k=0}^{\infty}\frac{(2x^2)^k}{k!} = \sum_{k=0}^{\infty}\frac{2^k}{k!}x^{2k+1}$$

23. $\arctan x = \displaystyle\sum_{k=0}^{\infty}\frac{(-1)^k x^{2k+1}}{2k+1}.$ Therefore,
$$\sqrt{x}\arctan\left(\sqrt{x}\right) = x^{1/2}\sum_{0}^{\infty}\frac{(-1)^k x^{(2k+1)/2}}{2k+1} = x^{\frac{1}{2}}\sum_{0}^{\infty}\frac{(-1)^k x^{k+\frac{1}{2}}}{2k+1} = \sum_{0}^{\infty}(-1)^k\frac{x^{k+1}}{2k+1}$$

25. $\ln(1+x) = \sum_{k=1}^{\infty} \frac{(-1)^{k+1}}{k} x^k$. Therefore,

$$\ln(1+x^2) = \sum_{k=1}^{\infty} \frac{(-1)^{k+1}}{k} (x^2)^k \quad \text{and} \quad \ln(1-x^2) = \sum_{k=1}^{\infty} \frac{(-1)^{k+1}}{k} (-x^2)^k.$$

$$f(x) = x \ln\frac{1+x^2}{1-x^2} = x[\ln(1+x^2) - \ln(1-x^2)] = x \left(\sum_{k=1}^{\infty} \frac{(-1)^{k+1}}{k} (x^2)^k - \sum_{k=1}^{\infty} \frac{(-1)^{k+1}}{k} (-x^2)^k \right)$$

$$= 2 \sum_{k=0}^{\infty} \frac{x^{4k+3}}{2k+1}$$

27. $f(x) = (1-x)^{1/3}$ $\qquad\qquad$ $f(0) = 1$

$f'(x) = -\frac{1}{3}(1-x)^{-\frac{2}{3}}$ $\qquad\qquad$ $f'(0) = -\frac{1}{3}$

$f''(x) = -\frac{2}{9}(1-x)^{-\frac{5}{3}}$ $\qquad\qquad$ $f''(0) = -\frac{2}{9}$

$f'''(x) = -\frac{10}{27}(1-x)^{-\frac{7}{3}}$ $\qquad\qquad$ $f'''(0) = -\frac{10}{27}$

$P_3(x) = 1 - \frac{1}{3}x - \frac{1}{9}x^2 - \frac{5}{81}x^3$

29. $[-\frac{1}{5}, \frac{1}{5})$; ratio test: $\frac{b_{k+1}}{b_k} = \frac{5k}{k+1}|x| \to 5|x| \implies r = \frac{1}{5}$

At $x = -\frac{1}{5}$, $\sum \frac{(-1)^k}{k}$ converges; at $x = \frac{1}{5}$, $\sum \frac{1}{k}$ diverges.

31. $(-\infty, \infty)$; ratio test: $\frac{b_{k+1}}{b_k} = \frac{2|x-1|^2}{(2k+2)(2k+1)} \to 0 \implies r = \infty$

33. $(-9, 9)$; ratio test: $\frac{b_{k+1}}{b_k} = \frac{k+1}{9k}|x| \to \frac{1}{9}|x| \implies r = 9$

At $x = -9$, $\sum k$ diverges; at $x = 9$, $\sum(-1)^k k$ diverges

35. $(-4, -2]$; ratio test: $\frac{b_{k+1}}{b_k} = \frac{\sqrt{k}}{\sqrt{k+1}}|x+3| \to |x+3| \implies r = 1$

at $x = -2$, $\sum \frac{(-1)^k}{\sqrt{k}}$ converges;

at $x = -4$, $\sum \frac{1}{\sqrt{k}}$ diverges

37. $f(x) = e^{-2x} = e^{-2(x+1)+2} = e^2 \cdot e^{-2(x+1)} = e^2 \sum_{0}^{\infty} \frac{[-2(x+1)]^k}{k!} = e^2 \sum_{0}^{\infty} \frac{(-1)^k 2^k}{k!}(x+1)^k; \quad r = \infty.$

39. $f(x) = \ln x = \ln[1 + (x-1)] = \sum_{1}^{\infty} \frac{(-1)^{k+1}}{k}(x-1)^k; \quad r = 1$

41. $\dfrac{1}{1+x^4} = \displaystyle\sum_{k=0}^{\infty} (-1)^k x^{4k}$

$$\int_0^{1/2} \frac{1}{1+x^4}\,dx = \sum_{k=0}^{\infty}(-1)^k \int_0^{1/2} x^{4k}\,dx = \sum_{k=0}^{\infty}(-1)^k \frac{1}{4k+1}\frac{1}{2^{4k+1}}$$

This is an alternating series with decreasing terms and the third term $\dfrac{1}{9(2^9)} \approx 0.0002$. Hence

$$\int_0^{1/2} \frac{1}{1+x^4}\,dx \approx \frac{1}{2} - \frac{1}{5(2^5)} \approx 0.4938$$

43. Set $f(x) = x^{1/3}$ and take $a = 64$. Then

$$x^{1/3} = f(x) = f(64) + f'(64)(x-64) + \frac{f''(64)}{2!}(x-64)^2 + \cdots .$$

$f(x) = x^{1/3}$ $\qquad\qquad\qquad\qquad\qquad f(64) = 4$

$f'(x) = \dfrac{1}{3}x^{-2/3}$ $\qquad\qquad\qquad\qquad f'(64) = \dfrac{1}{48}$

$f''(x) = -\dfrac{2}{9}x^{-5/3}$ $\qquad\qquad\qquad\qquad f''(64) = -\dfrac{2}{9\cdot 4^5}$

$\cdots$

$|R_2(68)| \leq \dfrac{\max |f'''(t)|}{3!}\, 4^3 < 0.005$. Therefore,

$\sqrt[3]{68} = f(68) \cong 4 + \dfrac{1}{12} - \dfrac{1}{9\cdot 4^3} \cong 4.0816$.

45. Let $g(x) = \sin x$ and $a = \pi/4$. Then $\sin x = \dfrac{\sqrt 2}{2} + \dfrac{\sqrt 2}{2}(x-\pi/4) - \dfrac{\sqrt 2}{2(2!)}(x-\pi/4)^2 - \dfrac{\sqrt 2}{2(3!)}(x-\pi/4)^3 + \cdots$

$$|R_n(x)| = \frac{|g^{(n+1)}(c)|}{(n+1)!}\left|\left(x-\frac{\pi}{4}\right)^{n+1}\right|$$

$$\leq \frac{|(x-\pi/4)|^{n+1}}{(n+1)!} \qquad (g^{(n+1)}(c) = \pm\sin c \text{ or } \pm\cos c)$$

Now, $48° = \dfrac{48\pi}{180}$ radians. We want to find the smallest positive integer n such that $|R_n(48\pi/180 - \pi/4)| < 0.0001$.

$$|R_n(48\pi/180 - \pi/4)| \leq \left(\frac{\pi}{60}\right)^{n+1}\frac{1}{(n+1)!} \cong \frac{(0.05236)^{n+1}}{(n+1)!} < 0.0001 \implies n \geq 2.$$

$\sin x \cong P_2(x) = \dfrac{\sqrt 2}{2} + \dfrac{\sqrt 2}{2}\left(x-\dfrac{\pi}{4}\right) - \dfrac{\sqrt 2}{4}\left(x-\dfrac{\pi}{4}\right)^2$; $\quad \sin 48° \cong \dfrac{\sqrt 2}{2}\left[1 + \dfrac{\pi}{60} - \dfrac{1}{2}\left(\dfrac{\pi}{60}\right)^2\right] \cong 0.7432$

47. For the sine function, $x - \frac{1}{6}x^3 + \frac{1}{120}x^5 = P_5 = P_6$. Therefore, for $x \in [0, \pi/4]$ we have

$$|\sin x - P_5(x)| = \left| \frac{f^{(7)}(c)}{7!}x^7 \right| \le \frac{1}{7!}\left(\frac{\pi}{4}\right)^7 < 0.000037$$

$$\left(|f^{(7)}(c)| = \cos c \le 1 \right) \underline{\quad\nearrow}$$

49. $\displaystyle\sum_{k=1}^{\infty} a_k = \int_1^{\infty} xe^{-x}\, dx = \frac{2}{e}$

51. If $\displaystyle\sum_{k=1}^{\infty}(a_{k+1} - a_k)$ converges, then the sequence of partial sums $s_n = a_{n+1} - a_1$ converges.

Therefore, the sequence a_k converges.

If the sequence a_k converges, then the sequence $s_n = a_{n+1} - a_1$ converges which implies that the

series $\displaystyle\sum_{k=1}^{\infty}(a_{k+1} - a_k)$ converges.